Movement Disorders

Movement Disorders

Edited by

NANDKUMAR S. SHAH, Ph.D. †

Director, Ensor Foundation Research Laboratory
William S. Hall Psychiatric Institute
Research Professor, Department of Neuropsychiatry and Behavioral Science
University of South Carolina School of Medicine
Columbia, South Carolina

and

ALEXANDER G. DONALD, M.D.

Director, William S. Hall Psychiatric Institute
Professor and Chairman, Department of Neuropsychiatry and Behavioral Science
University of South Carolina School of Medicine
Columbia, South Carolina

PLENUM MEDICAL BOOK COMPANY
New York and London

Library of Congress Cataloging in Publication Data

Movement disorders.

Includes bibliographies and index.
1. Movement disorders. 2. Neuropharmacology. I. Shah, Nandkumar S. II. Donald, Alexander G., 1928- . [DNLM: 1. Movement Disorders. WL 390 M9352]
RC385.M69 1986 616.7′4 86-4884
ISBN-13: 978-1-4684-5040-8 e-ISBN-13: 978-1-4684-5038-5
DOI: 10.1007/978-1-4684-5038-5

Softcover reprint of the hardcover 1st edition 1986
233 Spring Street, New York, N.Y. 10013

Plenum Medical Book Company is an imprint of Plenum Publishing Corporation

In Memoriam

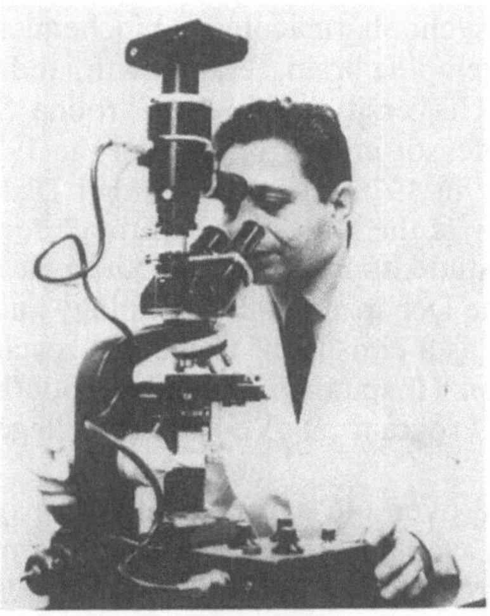

Nandkumar S. Shah
(1928–1983)

Nandkumar S. Shah died on May 23, 1983, at the age of 55. Dr. Shah was Chief of Research Services and Director of the Ensor Research Laboratory at the William S. Hall Psychiatric Institute and Research Professor, Department of Neuropsychiatry and Behavioral Science, at the University of South Carolina School of Medicine. He had completed the majority of the work involved in the publication of this volume at the time of his death.

Dr. Shah, a son of the late Shankarbhai and Parvati Shah, was born in Nandurbar, India. He received his B.S. and M.S. in biochemistry from Poona University, India. He completed his Ph.D. in pharmacology at the University of Florida, Gainesville, Florida, in 1965.

Dr. Shah was the epitome of a scholar and an excellent teacher. In addition, he was a superb researcher. He made significant contributions to the medical community of Columbia, South Carolina, through his research and teaching.

Through his many publications and presentations at national and international symposia, his work became known worldwide.

Dr. Shah's professional career began in 1955 at Poona University and at the Gandhi Memorial Hospital, where he conducted studies on lathyrism. He was later selected by the Atomic Energy Commission to continue his research in clinical biochemistry. He came to the United States in 1961 and studied at the College of Medicine at the University of Florida. On completion of his studies, he joined the staff at the Thudichum Psychiatric Research Laboratory in Galesburg, Illinois, working with the late Dr. Harold Himwich. He subsequently became Chief of the Psychopharmacology Radioisotope Laboratory, conducting studies on biogenic amines and hallucinogens: uptake, distribution, metabolism, and excretions under various conditions.

From 1970 to the time of his death, Dr. Shah directed research programs including projects in psychopharmacology, biochemical mechanisms, neurochemical aspects of developing brain, fetal growth, and drug metabolism at the Hall Institute. At the University of South Carolina School of Medicine he became a Research Professor in Neuropsychiatry in 1975 and an Adjunct Professor in Pharmacology in 1978. He was Adjunct Distinguished Professor in the College of Pharmacy of the University. Through formal lectures, he taught residents and medical students in psychiatry and pharmacology. He directed graduate programs in the Department of Psychology and in the School of Pharmacy. Dr. Shah served as a consultant to the Neuroscience Laboratory at the Veterans Administration Hospital in Columbia, South Carolina, and to the Biomedical Research Program at Voorhees College in Denmark, South Carolina.

Dr. Shah published over 140 papers related to psychopharmacology and the biological basis of psychiatry in national and international journals. In addition, he co-edited two major books entitled *Endorphins and Opiate Antagonists in Psychiatric Research: Clinical Implications* and *GABA Neurotransmission: Current Developments in Physiology and Neurochemistry*.

For the last 13 years, Dr. Shah conducted an annual research symposium of international significance on topics of pharmacological interest, under the auspices of the William S. Hall Psychiatric Institute. These symposia, over the years, have brought scientists of international stature to Columbia, providing the local scientific, medical, and psychiatric communities with a unique opportunity to personally exchange ideas with these outstanding men and women.

Dr. Shah earned several awards from national and international scientific and pharmaceutical associations: (1) from the American Society for Pharmacology and Experimental Therapeutics (1975–1976); (2) from the National Institute for General Medical Sciences (1980); (3) from the Deutscher Akademischer Austauschdienst (German Academic Exchange Service) to work at Max Planck Institute for Psychiatry, Munich, West Germany, in psychoneuroendocrinology (1983); and (4) the Director's Award, from the William S. Hall Psychiatric Institute. He was listed in The Marquis *Who's Who, Who's Who in the Biomedical Sciences,* and the *International Directory of Investigators in*

Psychopharmacology published by the World Health Organization and the National Institute of Mental Health. He was also a member of the Ad Hoc Committee of the NASA Biochemical Research Program in space and an editor/advisor for several professional journals.

Dr. Shah was strongly attached to his family. He had affection for his brothers, nephews, and nieces, helping to guide and shape their careers and financing their education when needed. Neeta, his wife of 22 years, is a remarkable person who shared his experiences, successes, and disappointments. His daughter shares her father's love of science and is pursuing a career in medicine.

Dr. Shah was an excellent teacher, an internationally known researcher, a humane colleague, and a gentleman who is missed by all. In June, 1983, when he was selected as the first recipient (posthumously) of the Director's Award of the William S. Hall Psychiatric Institute for the most significant contribution to its mission and purpose during the preceding year, Dr. Shah was described as "always interested in and willing to help students . . . the epitome of the scholar and the teacher—an extremely diligent and hard-working faculty member who set a high standard of excellence for students in all disciplines."

This volume is respectfully dedicated to the memory of Dr. Nandkumar S. Shah in appreciation of his accomplishments and his compassion for his fellow man.

Alexander G. Donald

Contributors

LARRY D. ALPHS • Maryland Psychiatric Research Center, Department of Psychiatry, University of Maryland, Baltimore, Maryland

GEORGE M. ANDERSON • Child Study Center and Department of Laboratory Medicine, Yale University, New Haven, Connecticut

ROSS J. BALDESSARINI • Department of Psychiatry and Neuroscience Program, Harvard Medical School and Massachusetts General Hospital, and Mailman Research Center, McLean Hospital, Belmont, Massachusetts

THOMAS R. E. BARNES • Department of Psychiatry, Charing Cross and Westminster Medical School, London, England

HEMENDRA N. BHARGAVA • Department of Pharmacodynamics, College of Pharmacy, The University of Illinois at Chicago, Health Sciences Center, Chicago, Illinois

J. R. BIANCHINE • Department of Pharmacology, College of Medicine, The Ohio State University, Columbus, Ohio. *Present address*: Medical Affairs, American Critical Care, McGaw Park, Illinois

RICHARD L. BORISON • Department of Psychiatry, Downtown Veterans Administration Medical Center, and Medical College of Georgia, Augusta, Georgia

DONALD B. CALNE • Experimental Therapeutics Branch, National Institute of Neurological and Communicative Disorders and Stroke, National Institutes of Health, Bethesda, Maryland. *Present address*: Health Sciences Center Hospital, Department of Medicine, University of British Columbia, Vancouver, British Columbia, Canada

DANIEL E. CASEY • Departments of Psychiatry, Research, and Neurology, Veterans Administration Medical Center and Oregon Health Sciences University, Portland, Oregon

C. C. CLOUGH • Department of Neurology, Queen Elizabeth Hospital, Birmingham, England

DONALD J. COHEN • Child Study Center, Children's Clinical Research Center, and Departments of Psychiatry, Pediatrics, and Psychology, Yale University, New Haven, Connecticut

JONATHAN O. COLE • Psychopharmacology Services, McLean Hospital, Belmont, Massachusetts

JOHN M. DAVIS • Illinois State Psychiatric Institute, Chicago, Illinois

BRUCE I. DIAMOND • Department of Psychiatry, Downtown Veterans Administration Medical Center, and Medical College of Georgia, Augusta, Georgia

RICHARD DORSEY • Department of Psychiatry, Doctors Hospital West, Columbus, Ohio

GEORGE GARDOS • West-Ros-Park Mental Health Center, Boston, Massachusetts

JES GERLACH • Department H, Saint Hans Hospital, Roskilde, Denmark

CHRISTOPHER G. GOETZ • Department of Neurological Sciences, Rush–Presbyterian–St. Lukes Medical Center, Chicago, Illinois

THOMAS J. GOLDSCHMIDT • Department of Neuropsychiatry and Behavioral Science, University of South Carolina School of Medicine, Columbia, South Carolina. *Present address:* Department of Neurology and Psychiatry, Tulane University School of Medicine, New Orleans, Louisiana

MENEK GOLDSTEIN • Department of Neurochemistry, New York University School of Medicine, New York, New York

ANA HITRI • Department of Psychiatry, Downtown Veterans Administration Medical Center, and Medical College of Georgia, Augusta, Georgia

DILIP V. JESTE • Neuropsychiatry Branch, National Institute of Mental Health, Saint Elizabeths Hospital, Washington, D.C.

CHARLES A. KAUFMANN • Neuropsychiatry Branch, National Institute of Mental Health, Saint Elizabeths Hospital, Washington, D.C.

HAROLD L. KLAWANS • Department of Neurological Sciences, Rush–Presbyterian–St. Lukes Medical Center, Chicago, Illinois

JOSEPH KNOLL • Department of Pharmacology, Semmelweis University of Medicine, Budapest, Hungary

SØREN KORSGAARD • Department H, Saint Hans Hospital, Roskilde, Denmark

JAMES F. LECKMAN • Child Study Center, Children's Clinical Research Center, and Departments of Psychiatry and Pediatrics, Yale University, New Haven, Connecticut

ABRAHAM N. LIEBERMAN • Department of Neurology, New York University School of Medicine, New York, New York

RUUD B. MINDERAA • Eramus Universiteit Rotterdam, Rotterdam, The Netherlands

RICHARD P. NEWMAN • Experimental Therapeutics Branch, National Insitute of Neurological and Communicative Disorders and Stroke, National Institutes of Health, Bethesda, Maryland. *Present address*: 1004 Beverly Drive, Rockledge, Florida

DAVID L. PAULS • Child Study Center and Department of Human Genetics, Yale University, New Haven, Connecticut

J. M. S. PEARCE • Department of Neurology, Hull Royal Infirmary, Hull, England

EDMOND H. PI • Department of Psychiatry, University of Southern California School of Medicine, Los Angeles, California

R. ARLEN PRICE • Department of Psychiatry, University of Pennsylvania, Philadelphia, Pennsylvania

C. W. RICHARD III • Department of Pharmacology, College of Medicine, The Ohio State University, Columbus, Ohio

MARK A. RIDDLE • Child Study Center, Children's Clinical Research Center, and Departments of Psychiatry and Pediatrics, Yale University, New Haven, Connecticut

ROBERT C. SCHNACKENBERG • Department of Neuropsychiatry and Behavioral Science, University of South Carolina School of Medicine, Columbia, South Carolina

R. D. SCHWARTZ • Department of Pharmacology, College of Medicine, The Ohio State University, Columbus, Ohio

ARTHUR K. SHAPIRO • Tourette and Tic Laboratory and Clinic, Mount Sinai School of Medicine, New York, New York

ELAINE SHAPIRO • Tourette and Tic Laboratory and Clinic, Mount Sinai School of Medicine, New York, New York

GEORGE M. SIMPSON • Department of Psychiatry, The Medical College of Pennsylvania, Philadelphia, Pennsylvania

STEPHEN M. STAHL • VA–Stanford Mental Health Clinical Research Center, Department of Psychiatry and Behavioral Sciences, Stanford University School of Medicine, Stanford, California, and Laboratory of Neuropsychopharmacology, Schizophrenia Biologic Research Center, Veterans Administration Medical Center, Palo Alto, California. *Present address:* Neuroscience Research Centre, Merck Sharp and Dohme Research Laboratories, Terlings Park-Eastwick Road, Harlow, Essex, England

CHARLES N. STILL • Department of Neuropsychiatry and Behavioral Science, University of South Carolina School of Medicine, Columbia, South Carolina

CAROLINE M. TANNER • Department of Neurological Sciences, Rush–Presbyterian–St. Lukes Medical Center, Chicago, Illinois

DANIEL TARSY • Department of Neurology, Boston University and Harvard Medical School, and Neurology Section, New England Deaconess Hospital, Boston, Massachusetts

MALCOLM P. I. WELLER • Academic Department of Psychiatry, Royal Free Hospital School of Medicine, The Royal Free Hospital, London, England

RICHARD JED WYATT • Neuropsychiatry Branch, National Institute of Mental Health, Saint Elizabeths Hospital, Washington, D.C.

Preface

The human nervous system—that most complex organization of energy and matter—has yielded a few glimmers of understanding of its operational mechanics during the last two decades. These have mostly been at the biochemical level of structure and function. Throughout history, as one of the mysteries of nature begins to yield some insights into its function, it has been beneficial to look at it from different points of view.

We have developed a volume on movement disorders that is primarily directed toward the biochemical understanding of these disorders and their treatment. Each disorder is presented from several points of view. Although this approach leads to some repetition, it is our aim that the final outcome be a more complete understanding.

Much has been written about movement: the beauty of the prima ballerina, the strength of the olympic athlete, and the agility of the surgeon. Seldom do we stop to look beneath the surface—the coordination of muscle groups, the finely tuned balance allowing rapid response in either direction, the individual muscle fibers coordinated to maximize strength and agility, and the nerve fibers connecting muscle with nerve centers. Some of these communicate sensory input of position to the centers while others communicate directions of movement to muscles.

We encourage our readers to be constantly alert to the possibility of increasing their understanding of other nervous system functions, including thought disorder, through an understanding of movement, either in general principle or by specific chemical interaction.

Could the grotesquely bizarre movements of acute dystonia in the form of oculogyric crisis bear a relationship to the finely coordinated facial expression of emotions such as joy and sorrow—somewhat similar to the relationship that a psychotic thought process bears to intellectual problem solving? Although the comparison is inexact, an understanding of the pathological mechanism of one may shed light on that of the other. Nowhere is the exquisite balance of nature more apparent than in the coordination of movement in the human being. The equilibrium which responds with relaxation and tension in exact proportions of specific muscle fibers and groups of fibers resulting in the desired movement exemplifies the finely tuned orchestration of nature.

One is impressed repeatedly with the infinitesimal amounts of some substances that can cause severe malfunction of the nervous system. Caution should be used with any drug that alters neurotransmission within the brain. Any need for repeated prescriptions should be reviewed regularly by a physician knowledgeable in the nuances of neuropharmacology.

Man's foremost challenge today is the development of understanding of the central nervous system—how it works, what can go wrong with it, and what can be done about it. This is truly the frontier of medicine.

ACKNOWLEDGMENTS. I wish to express appreciation to our wives, Neeta N. Shah and Emma Louise Donald, for their patient understanding during the editing of this book; to Annette G. Bonneau for her diligent work in corresponding, compiling, and proofreading; to Gregory C. Smith for preparation of the index; and to Janice Stern, Senior Medical Editor at Plenum Publishing Corporation, for her editorial assistance.

Alexander G. Donald

Contents

II. TARDIVE DYSKINESIA

III. TOURETTE SYNDROME AND TIC

IV. HUNTINGTON'S DISEASE

V. NEUROLEPTIC DRUGS IN THE PRODUCTION OF MOVEMENT DISORDERS

INTRODUCTION

Neuropharmacology of Movement Disorders

Comparison of Spontaneous and Drug-Induced Movement Disorders

STEPHEN M. STAHL

1. INTRODUCTION

1.1. Neuropharmacological Approach to Movement Disorders

Disorders of voluntary movement are common neurological problems that have unusual importance to both clinician and scientist. Movement disorders are important to the clinician because they are generally among the most treatable neuropsychiatric disorders, yet are also among the most frequent iatrogenic complications of drug therapies. Thus, the clinician is in a powerful position both to create and to alleviate movement disorders. Proper understanding of these conditions will help the clinician distinguish spontaneously occurring diseases from the ones he creates with drugs. Proper understanding of neuropharmacology will help the clinician to use drug treatments to alleviate movement disorders while avoiding or reducing complications from improper drug usage. For the neuroscientist, movement disorders represent simultaneously the greatest triumphs as well as the greatest failures of basic brain research. The triumphs include the first breakthroughs in applications of neurotransmitter mechanisms to neuronal functioning. Knowledge of basal ganglia anatomy, biochemistry, and pharmacology has led, for example, to the development of treatments for Parkinson's disease. On the other hand, the failures include the

STEPHEN M. STAHL • VA–Stanford Mental Health Clinical Research Center, Department of Psychiatry and Behavioral Sciences, Stanford University School of Medicine, Stanford, California 94305, and Laboratory of Neuropsychopharmacology, Schizophrenia Biologic Research Center, Veterans Administration Medical Center, Palo Alto, California 94304. *Present address:* Neuroscience Research Centre, Merck Sharp and Dohme Research Laboratories, Terlings Park-Eastwick Road, Harlow, Essex CM 20 2QR, England.

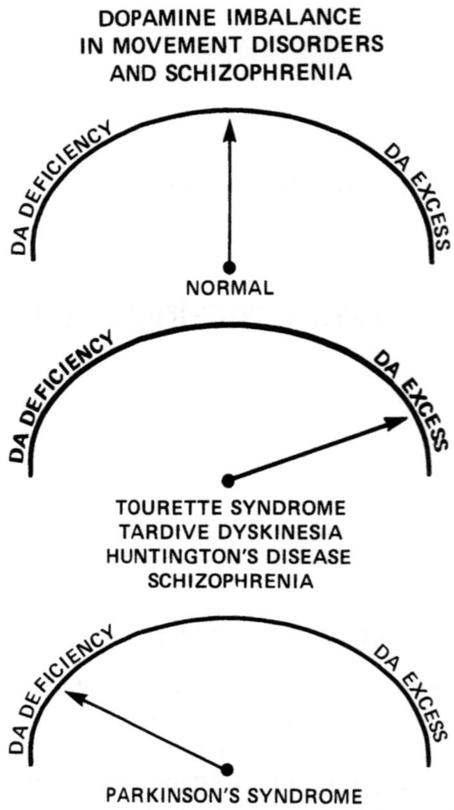

FIGURE 1. Central nervous system disorders characterized by relative dopamine excess or by relative dopamine deficiency.

production of unwanted movement disorders by numerous drugs, including the irreversible and disabling disease tardive dyskinesia. Understanding these lessons, both good and bad, from the field of movement disorders puts the basic scientist in a strong position to develop future treatments for other disorders of the human central nervous system.

Abnormal involuntary movement disorders are classified in several ways. Movement disorders can be usefully distinguished by categorizing them: (1) on the basis of clinical phenomenology to establish an accurate diagnosis; (2) by distinguishing drug-induced disorders from spontaneous (idiopathic) disorders to specify etiology; and (3) by pharmacological characteristics to aid in understanding pathophysiology, prescribing treatment, and generating hypotheses for future research.

In the second section of this chapter, movement disorders are described phenomenologically. The third and fourth sections of this chapter separate drug-induced movement disorders (Section 3) from spontaneous (idiopathic) disorders (Section 4). Within these last two sections, we classify movement disorders according to current pharmacological concepts. Since most classifications of movement disorders emphasize their differential diagnosis and ep-

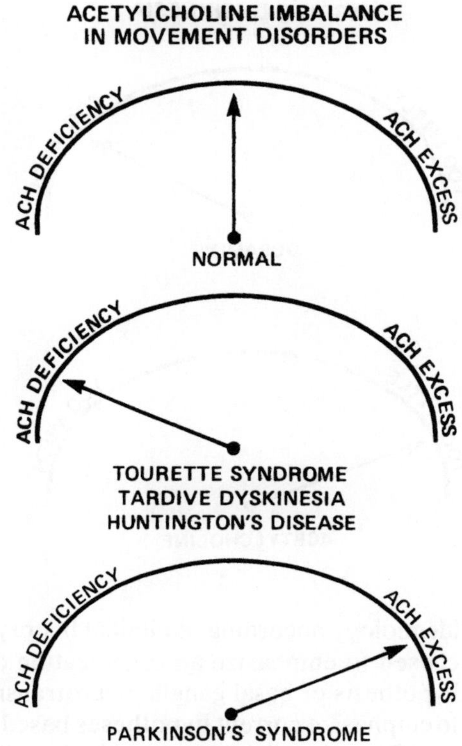

FIGURE 2. Central nervous system disorders characterized by relative acetylcholine deficiency or by relative acetylcholine excess.

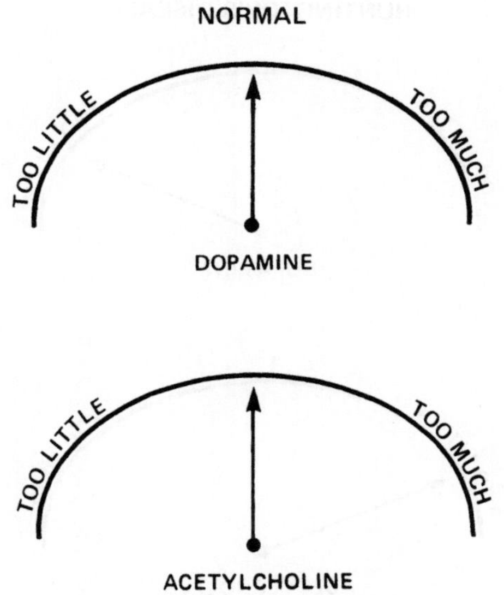

FIGURE 3. Normal balance between dopamine and acetylcholine.

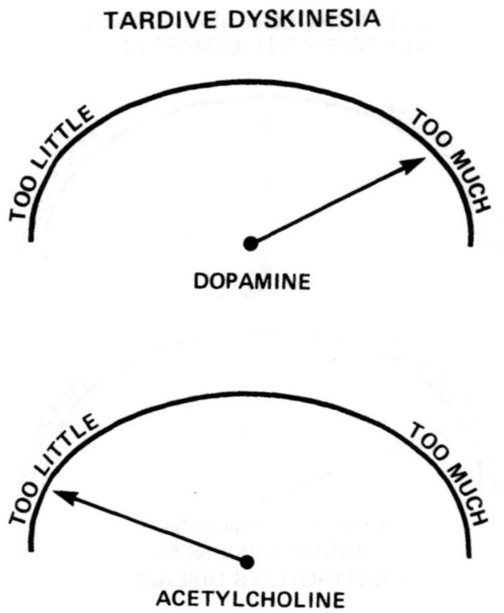

FIGURE 4. Neurotransmitter imbalance in tardive dyskinesia is characterized by relative dopamine excess and by relative acetylcholine deficiency.

idemiology according to clinical history and neurological examination, we have chosen to emphasize an organization of movement disorders according to the hypothesis of basal ganglia neurotransmitter imbalance (Figs. 1 and 2) in order to emphasize current hypotheses based on the pharmacology of these disorders.

Over the past 10 years, the hypothesis of an imbalance between the neurotransmitters dopamine (DA) and acetylcholine (ACh) in the basal ganglia has been a unifying (if oversimplifying) concept in attempting to understand many

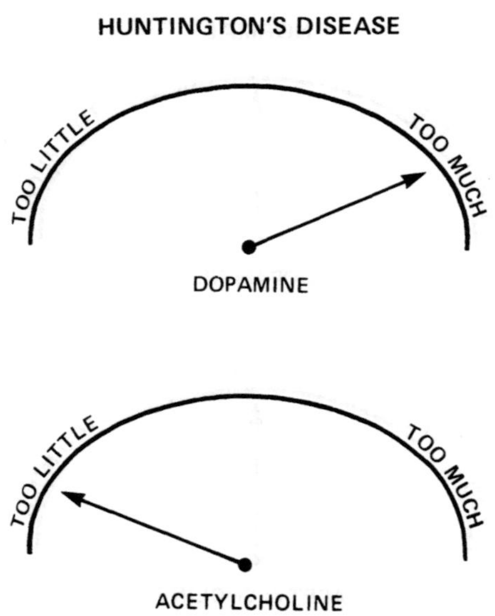

FIGURE 5. Neurotransmitter imbalance in Huntington's chorea is characterized by relative dopamine excess and by relative acetylcholine deficiency.

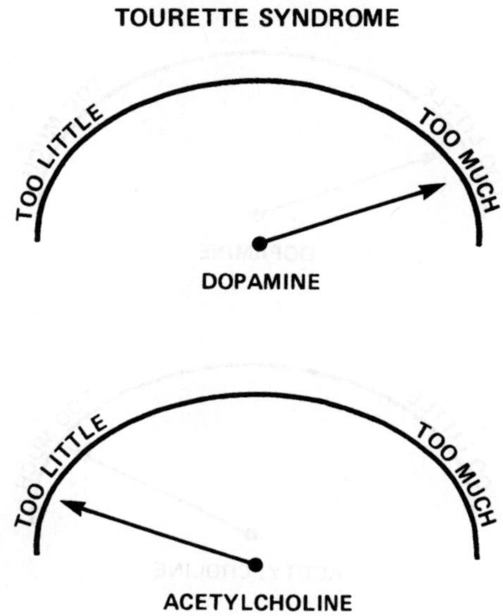

FIGURE 6. Neurotransmitter imbalance in Tourette syndrome is characterized by relative dopamine excess and by relative acetylcholine deficiency.

movement disorders (Figs. 3–9)(Klawans, 1973; Tarsy, 1976; Stahl *et al.*, 1982a; Marsden, 1982). More recently, this concept has been expanded to include γ-amino-butyric acid (GABA) as a component of the neurotransmitter imbalance in movement disorders (Roberts *et al.*, 1976; Stahl *et al.*, 1985d). While several other neurotransmitters such as the neuropeptides may ultimately prove to have important roles in the pathophysiology of movement disorders,

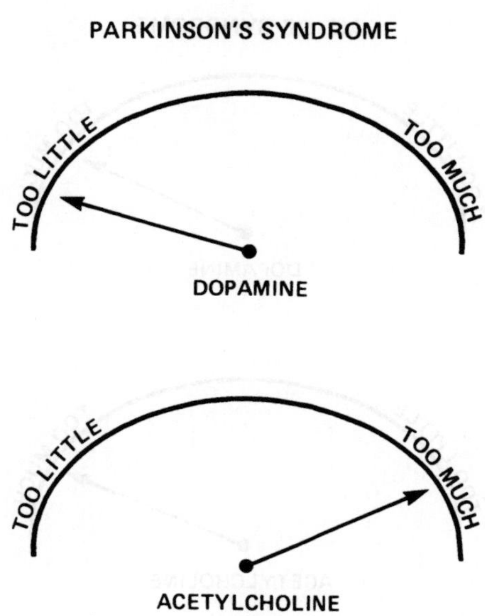

FIGURE 7. Neurotransmitter imbalance in Parkinson's disease is characterized by relative dopamine deficiency and by relative acetylcholine excess.

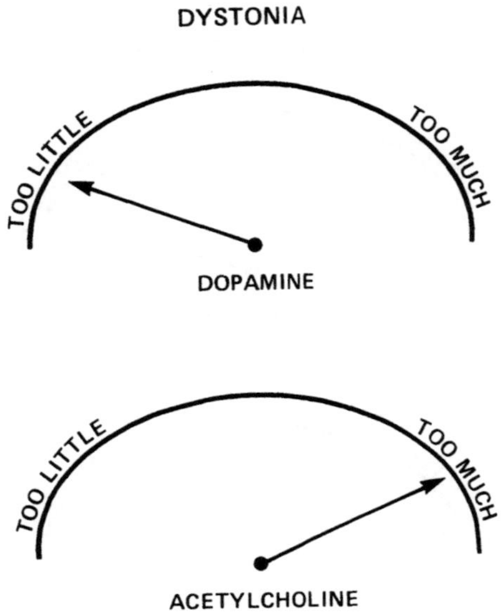

FIGURE 8. Neurotransmitter imbalance in dystonia may be characterized in part by relative dopamine deficiency and by relative acetylcholine excess.

DA, ACh, and GABA are currently the most widely investigated neurotransmitters in movement disorders (Figs. 1 and 2).

Specifically, DA excess and ACh deficiency may characterize part of the pathology of several hyperkinetic movement disorders (both spontaneous and drug-induced). On the other hand, DA deficiency and ACh excess may characterize several rigid–dystonic movement disorders (both spontaneous and drug-induced). This review not only presents these DA-ACh neurotransmitter

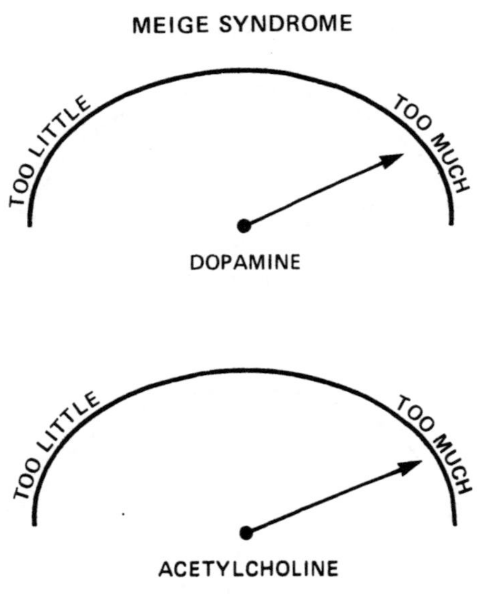

FIGURE 9. Neurotransmitter imbalance in idiopathic orofacial dyskinesia (Meige syndrome) may be characterized by relative excesses in both dopamine and acetylcholine.

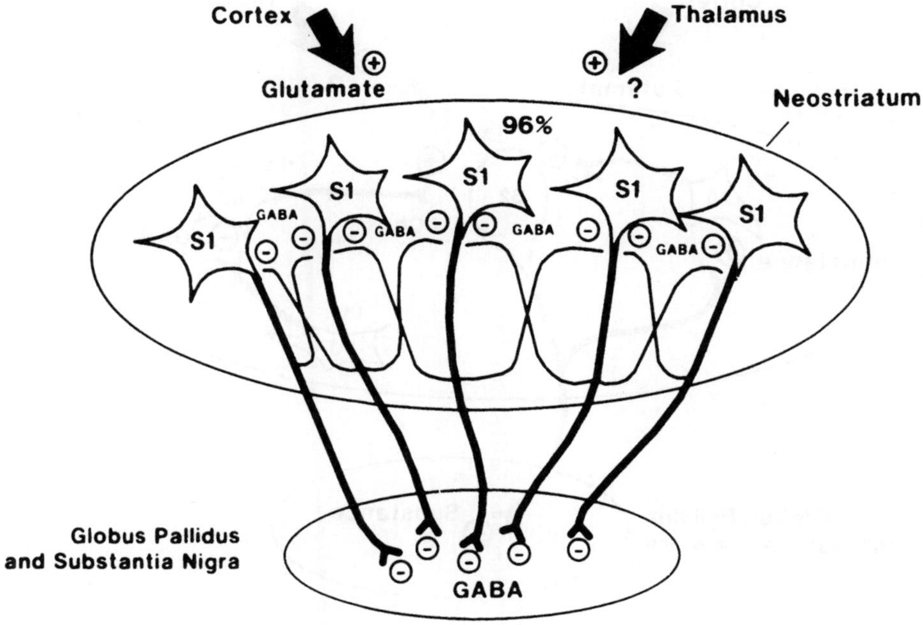

FIGURE 10. GABA efferents from the striatal spiney 1 (S1) neurons comprise 96% of striatal neurons and project an inhibitory output to neighboring S1 neurons as well as to globus pallidus and substantia nigra.

imbalance paradigms in movement disorders, but also includes paradigms involving serotonin and norepinephrine. Although of heuristic value, the neurotransmitter imbalance hypothesis does not provide a complete and accurate pharmacological characterization of movement disorders in all patients, since it fails to explain many data. Nevertheless, it does provide a starting point for understanding the pharmacology of movement disorders.

1.2. The Chemical Anatomy of the Basal Ganglia

The basal ganglia is widely considered to be the neuroanatomical site of dysfunction in many of the movement disorders. A great deal of new knowledge now extends our perspective of the basal ganglia beyond the roles of DA, ACh, and GABA. Not only are the *locations* and *interconnections* of the neuronal pathways that flow into and out of the basal ganglia being characterized, but so are the specific *neurotransmitters, receptors,* and *morphological* cell types of each pathway. This "chemical anatomy" of the basal ganglia now provides a rich and colorful variety of cell types, neurotransmitters, and receptors to enhance the neuropharmacological understanding of diseases and drugs that act in the basal ganglia.

The original handful of known neurotransmitters of a few years ago has

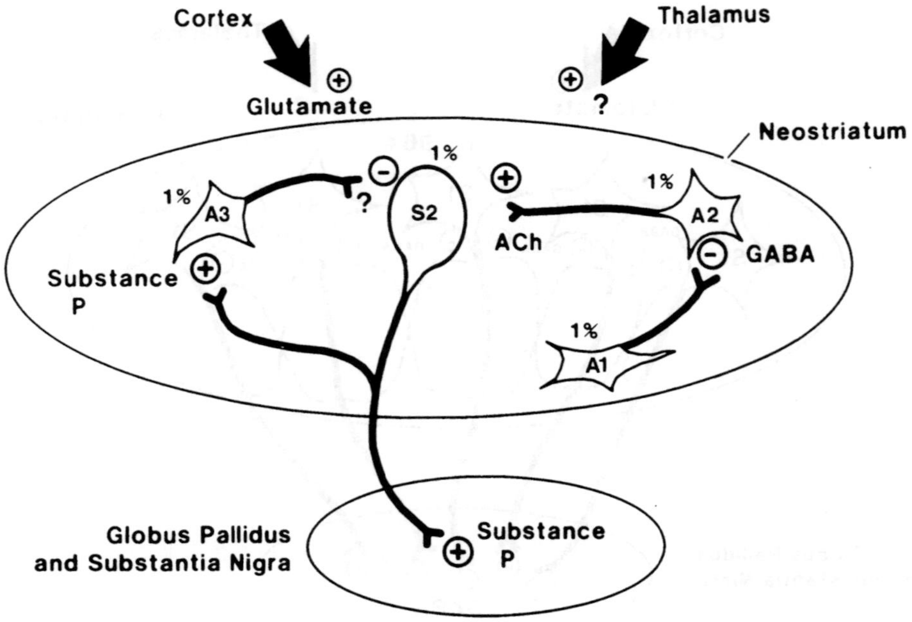

FIGURE 11. Substance P efferents from the striatal spiney 2 (S2) neurons comprise 1% of striatal neurons and project an excitatory output to globus pallidus and neostriatum. S2 neurons are also linked by a recurrent collateral to a negative feedback Aspiney 3 (A3) neuron within the striatum. Aspiney 1, 2, and 3 neurons all comprise 1% of striatal neurons. These interneurons use different neurotransmitters (see text).

now expanded to include more than three dozen, including more than 30 that are peptides (Iversen, 1983). Also, it appears that many neuronal pathways contain not just one neurotransmitter, but two (Lundberg and Hokfelt, 1983). That is, a small-molecular-weight amine or amino acid (such as DA) is often located together in the same neuron with a larger-molecular-weight peptide (such as cholecystokinin) (Lundberg and Hokfelt, 1983). The role of such peptide neurotransmitters in the function or dysfunction of the basal ganglia is virtually unknown, and there are essentially no drugs that are known to act on such peptides. In addition to this plethora of neurotransmitters, now perhaps two to a neuron, it appears that each of the neurotransmitters may have multiple receptors. Thus, the 40 neurotransmitters each act on several different receptors, resulting in the number of possible interactions of drugs, diseases, and physiological neurotransmitters at receptor sites being in the hundreds. Where this expansion will end no one knows, but the possibilities for understanding diseases and for discovering new drugs now appears more promising than ever.

One attempt to integrate current concepts of basal ganglia "chemical anatomy" is shown in Figs. 10–12. Undoubtedly, this attempt will be revised many times in the future. Even though we have very little knowledge today as to how these newer concepts relate to drugs or diseases of the basal ganglia, they

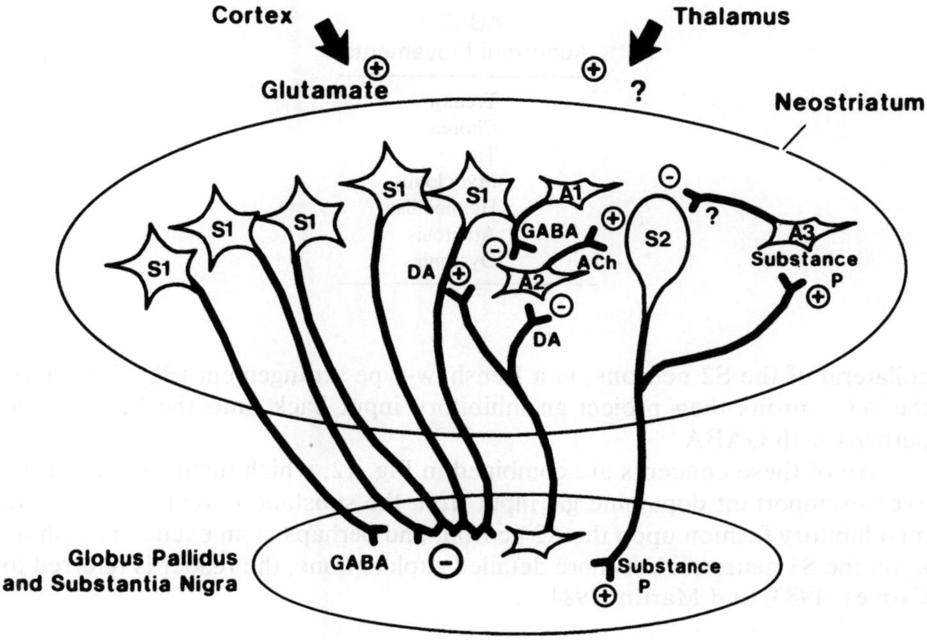

FIGURE 12. Summary of the chemical anatomy of the basal ganglia. Figures 10 and 11 are combined with a dopaminergic input from substantia nigra to show an overview of contemporary hypotheses of the basal ganglia (see Figures 10, 11, and text).

serve as a template upon which future clinical applications will be superimposed.

Thus, Groves (1983) and Martin (1984), among others, have integrated several complex concepts of basal ganglia neurochemistry and neuroanatomy, which are represented in Figs. 10–12. Briefly, the neostriatum receives several major inputs, including one (excitatory) from the cortex using glutamate as neurotransmitter and one (excitatory) from the thalamus (unknown neurotransmitter). The neostriatum has two major outputs. One of these comes from spiney 1 (S1) neurons, which project to globus pallidus and substantia nigra, which are inhibitory, and which utilize GABA and perhaps enkephalin as messengers (see Fig. 10). This S1 output comprises 96% of striatal neurons. The second major output of the neostriatum comes from 1% of its neurons, called the spiney 2 (S2) neurons. They project to globus pallidus and substantia nigra, are excitatory, and utilize substance P as neurotransmitter (Fig. 11).

Three other cells, each 1% of the total populations, are interneurons within the striatum: the aspiney 1, 2, and 3 (A1, A2, A3) neurons (Fig. 11). The A1 neurons project onto the A2 neurons. The A1 neurons are inhibitory and have perhaps GABA and/or somatostatin as neurotransmitters. The A2 neurons project onto the S2 neurons, are excitatory, and utilize ACh as neurotransmitter. Finally, the A3 neurons receive an excitatory substance P input from a recurrent

TABLE 1
Abnormal Movements

Tremor
Chorea
Tic
Myoclonus
Hemiballism
Athetosis
Dystonia

collateral of the S2 neurons, in a Renshaw-type arrangement (Fig. 11). Thus, the A3 neurons then project an inhibitory input back onto the S2 neurons, perhaps with GABA.

All of these concepts are combined in Fig. 12, which includes as well the ever-so-important dopaminergic input from the substantia nigra, projecting in an inhibitory fashion upon the A2 neuron, and perhaps in an excitatory fashion upon the S1 neurons. For more detailed explanations, the reader is referred to Groves (1983) and Martin (1984).

2. DESCRIPTION OF ABNORMAL MOVEMENTS

A great number of confusing terms have been used to describe abnormal movements, and no two clinicians seem to use the same terms in exactly the same manner. General terms include *dyskinesia, hyperkinesia,* and *bradykinesia*. Almost any abnormal movement can be called a *dyskinesia* (literally, "abnormal movement"). Movements that are too fast are *hyperkinesias,* and the slowing of movements is called *bradykinesia* (sometimes also called akinesia). This section lists several specific abnormal movements (see also Table 1). Clusters of these abnormal movements make up a *syndrome*.

2.1. Tremor

All tremors are rhythmical, alternating contractions of opposing muscles. There are three major types of tremor. Parkinsonian tremor is a resting tremor of moderate to high amplitude (about four to six oscillations in 1 sec). Parkinsonian tremor is reduced in amplitude by voluntary movements. It is one of the manifestations of the DA deficiency–ACh excess syndrome (see Section 3.3). The second major type of tremor is postural tremor, which is also called action tremor because it is minimally present at rest, but exaggerated by actions and by sustaining a posture. It is very fine, rapid (about 10–12 oscillations in 1 sec), and associated with β-adrenergic excess (see Section 3.4). The third type of tremor is intention tremor, which is absent at rest and during the first parts of a voluntary movement, but as the movement continues and fine adjustments are made (e.g., in touching the finger to the nose), a jerky, rhythmical, coarse, high-amplitude oscillation appears. The neuropharmacology of intention tremor is unknown.

2.2. Chorea

The rapid, darting, unpredictable movements of chorea look like fidgety, purposeless activity. Chorea (literally, "dance") is quick, is purposeless, can affect any part of the body (especially the extremities), and is a caricature of normal movements. Chorea is often associated with DA excess (see Section 3.1).

2.3. Tic

A tic is a rapid, spasmodic, stereotyped movement that can have many of the features of a choreiform movement. However, tics can be vocal as well as motor, are much more stereotyped than chorea, and are usually accompanied by an irresistible urge to perform the tic. Motor tics can be a simple contraction of a muscle group, usually in the face (e.g., blinking), but are often complex stereotyped movements, such as touching something or protruding the chin. Many motor tics are accompanied by forced vocalizations, such as sniffing, grunting, belching, or howling. Some vocal tics are rapid expulsion of formed words, usually expletives (called coprolalia). Tics are often associated with DA excess (see Section 3.1).

2.4. Myoclonus

Myoclonus is the term for movements that are abrupt, shocklike contractions of muscles, irregular in rhythm and irregular in amplitude. Myoclonus resembles chorea in that it is brief and arrhythmical, but is much faster. Myoclonus is a confusing concept that includes a large number of abnormal movements ranging from pathological startle responses (startle myoclonus), to certain seizure disorders (myoclonic epilepsy), to certain cerebellar disorders, to many others. The neuropharmacology of one type of myoclonus, postanoxic intention myoclonus, is related to serotonin deficiency (see Section 4.3).

2.5. Hemiballism

Hemiballism is a severe form of chorea characterized by wild flinging movements of the extremities on one side of the body. The proximal limb muscles are affected by incessant, exhaustive, flinging, and throwing motions. Hemiballism, like its related lesser movements of chorea, is related to DA excess.

2.6. Athetosis

Athetosis is a moving spasm, a slow, writhing, wormlike, relatively continuous postural change in the distal limbs. Hands and fingers are most frequently affected. When associated with chorea (choreoathetosis), it is part of a DA excess syndrome. Athetosis can also be associated with dystonia.

2.7. Dystonia

Dystonia is a widely used term for many different things. There exists a confusing array of dystonic movements, dystonic postures, specific dystonic syndromes, and specific dystonic diseases. In general, dystonia refers to an abnormal, slow, twisting movement of any muscle (dystonic movement) that causes the body to assume a contorted, twisted, contracted posture (dystonic posture). Focal dystonias are named for the specific muscle group affected (e.g., torticollis, blepharospasm, writer's cramp, oromandibular dystonia). Generalized dystonia is the involvement of many muscle groups throughout the body. The generalized dystonia known as dystonia musculorum deformans may be related in part to DA deficiency–Ach excess pharmacology (see Section 3.3.2).

2.8. Akathisia

Akathisia is another confusing term. It literally means "to be unable to sit" and refers to various forms of motor and subjective restlessness. Some authors emphasize the *movement disorder* component of akathisia, which includes "muscular impatience" and is seen to an observer as jittery, restless movements, often of the feet or legs. Other authors, however, emphasize the *mental or subjective* component of akathisia, which is the observation that patients with overt motor restlessness also experience strong, subjective, distinctly unpleasurable feelings of the need to move, without any relief of this need after making the movement.

3. DRUG-INDUCED MOVEMENT DISORDERS

This section discusses movement disorders caused by specific drugs (Table 2). Drug-induced movement disorders are caused by the ability of a drug to disrupt basal ganglia neurotransmitter balance in a specific manner. This can induce the equivalent imbalance seen in a number of spontaneous movement disorders and, not surprisingly, an equivalent clinical syndrome. By first understanding movement disorders caused by drugs whose mechanisms of action are known (Section 3), one can then better appreciate the pharmacology of the spontaneously occurring movement disorders (Section 4).

TABLE 2
Drugs That Induce Movement Disorders

Dopamine agonists (e.g., levodopa, amphetamine, methylphenidate, pemoline, bromocriptine, lisuride, pergolide)
Dopamine antagonists (e.g., haloperidol, fluphenazine, chlorpromazine, thioridazine)
Beta adrenergic enhancing drugs (e.g., caffeine, lithium, tricyclic antidepressants, theophyllin, albuterol)
Hormones (e.g., thyroid, estrogen)
MPTP (1-methyl-4-phenyl-1,2,3,6-tetrahydropyridine)

3.1. DA Excess Syndromes and DA Agonists

Chorea, Tics, and Stereotypies

Receptor stimulating agents can precipitate various hyperkinetic movement disorders *de novo* (Tables 2 and 3) (Klawans, 1973; Stahl *et al.*, 1982a). Thus, choreatic movements, orobuccal facial dyskinesias, and tics may be precipitated by levodopa, amphetamine, methylphenidate, pemoline, bromocriptine, and other DA agonists (Fann *et al.*, 1980; Chase *et al.*, 1979; Friedhoff and Chase, 1982). These toxic dyskinesias have several names including levodopa dyskinesia and toxic Tourette syndrome (discussed in detail in Section 4.1.3) (Stahl, 1980; Klawans *et al.*, 1978; Fog and Pakkenberg, 1980; DeVeaugh-Geiss, 1980). Another manifestation of an excess of DA agonists is the production of stereotypies. These are repetitive, semipurposeful or senseless, ritualistic movements, often quite complex (such as assembling and disassembling an object). Stereotypies are seen in humans when intoxicated with amphetamine or cocaine (Rubovits and Klawans, 1972; Meltzer and Stahl,

TABLE 3
Dopamine Excess–Acetylcholine Deficiency Syndromes

Drug-induced
Chorea
Tics
Stereotypy
Tardive dyskinesia
Idiopathic or spontaneous
Hemiballism
Huntington's chorea
Sydenham's chorea
Hyperthyroid chorea
Chorea gravidarum
Senile chorea
Tourette syndrome

1976). These syndromes are all probably related to excessive stimulation of DA receptors and subside when DA agonists are discontinued or reduced in dosage. Interestingly, toxic DA hyperkinesias are also reduced when physostigmine is administered (Klawans, 1973; Tarsy *et al.*, 1974). Since physostigmine is a centrally acting cholinergic drug (blocking acetylcholinesterase and allowing ACh to increase), the DA excess syndromes may all be relative ACh defiency syndromes as well. Enhancing central ACh tends to oppose the overactivity of DA. This observation fits with the hypothesis that DA and ACh are counterbalancing in the basal ganglia, and that an excess of DA is accompanied by a relative deficiency of ACh in hyperkinetic movement disorders (see Fig. 3–9).

3.2. DA Excess Syndromes and Chronic DA Antagonists

Tardive Dyskinesia (Fig. 4)

Tardive dyskinesia (TD) is a collection of movement disorders produced by chronic treatment with DA receptor blocking neuroleptic medications (Berger and Rexroth, 1980; Jeste *et al.*, 1979; DeVeaugh-Geiss, 1982; Stahl, 1985). For the purposes of this discussion, the term tardive dyskinesia includes both persistent and remitting forms, as well as neuroleptic withdrawal dyskinesias. A unique form of TD, the tardive Tourette syndrome, is also included. The *tardive* Tourette syndrome is a rare clinical manifestation of TD in which prolonged neuroleptic treatment produces motor and vocal tics indistinguishable from those observed in *spontaneous* Tourette syndrome (see Section 4) rather than producing the classical orobuccal facial movements usually seen in TD (Stahl, 1980; Klawans *et al.*, 1978; Fog and Pakkenberg, 1980; DeVeaugh-Geiss, 1980).

The evidence for DA–ACh–GABA imbalance in TD comes from two sources: animal models of biochemistry and behavior after prolonged treatment with neuroleptic agents, and pharmacological manipulation of symptoms in patients by selective drug administration. Direct biochemical measurements in tissues from patients have not provided supporting evidence for this concept. These biochemical studies are complicated by the fact that almost all patients have concomitant psychiatric disorders. Thus, a biochemical finding may be related to the movement disorder, to the psychiatric disorder, or to unrelated effects of chronic drug treatment. Prolonged treatment of animals with neuroleptics causes specific animal behaviors that are mediated by DA to become supersensitive to DA agonists (Fann *et al.*, 1980). Prolonged neuroleptic treatment also enhances the capacity of apomorphine to reduce DA turnover and increases [^{3}H]-apomorphine and [^{3}H]-haloperidol binding in the striatum (Fann *et al.*, 1980; DeVeaugh-Geiss, 1982). These animal models suggest that prolonged administration of neuroleptics could induce "supersensitivity" of striatal DA receptors.

The DA–ACh imbalance hypothesis of hyperkinetic movement disorders proposes that normal striatal neurotransmitter balance (Fig. 3) is disrupted when striatal DA receptors become supersensitive in man after prolonged neuroleptic treatment and that the clinical manifestation is TD (Fig. 4) (Fann *et al.*, 1980; Stahl *et al.*, 1982a). When striatal DA receptors become supersensitive in man as striatal neurons degenerate, the clinical manifestation may be Huntington's chorea (Fig. 5) (see Section 4) (Chase *et al.*, 1979; Stahl *et al.*, 1982a). If striatal DA receptors are supersensitive in childhood because of a genetic abnormality, the clinical manifestation may be Tourette syndrome (Fig. 6; see also Section 4) (Friedhoff and Chase, 1982; Stahl *et al.*, 1982a; Stahl and Berger, 1982b; Uhr *et al.*, 1984, 1985). Finally, if striatal DA receptors are supersensitive or overstimulated by chronic DA agonists, the clinical result could be a toxic dyskinesia, such as levodopa dyskinesia (discussed in Section 3.1) (Klawans, 1973; Stahl *et al.*, 1982a). These hypotheses are based on analogies with the animal models discussed previously. However, direct evidence for supersensitive striatal DA receptors in TD and other movement disorders is lacking.

Clinical responses to pharmacological agents also suggest DA excess and ACh and GABA deficiency in TD. This inference is based on observations of symptom improvement in TD symptoms following the administration of agents that *decrease* DA, *increase* ACh, or *increase* GABA (Berger and Rexroth, 1980; Fann *et al.*, 1980; Stahl *et al.*, 1982a; DeVeaugh-Geiss, 1982). The exacerbation of the symptoms of TD by agents that *increase* DA or *decrease* ACh is a less consistent finding, but also supports the hypothesis of DA excess and ACh deficiency.

Most (Gerlach *et al.*, 1974; Fann *et al.*, 1974; Klawans and Rubovits, 1974; Davis and Berger, 1978; Tamminga *et al.*, 1980; Stahl *et al.*, 1982a) but not all (Tarsy *et al.*, 1974) investigators have found that TD patients improve following physostigmine administration. Other investigators report that physostigmine improves TD movements in some patients, but worsens TD movements in others (Moore and Bowers, 1980). Most investigators report improvement of symptoms in TD patients during treatment with the ACh precursors choline chloride or lecithin (phosphatidyl choline) (Fann *et al.*, 1980; Davis and Berger, 1978, 1979; Davis *et al.*, 1975, 1976; Gelenberg *et al.*, 1979; Barbeau *et al.*, 1979; Stahl *et al.*, 1982a). Although these results with cholinergic agonists are promising, there are several problems with this approach. The variable course of movement frequency in TD is an important methodological obstacle in trying to determine whether any treatment is effective. In addition, the available form of choline chloride is a bitter-tasting liquid that is metabolized in the gastrointestinal tract by bacteria to trimethylamine, giving the patient the foul odor of "dead fish." Lecithin, even in the more pure forms now available, adds considerable caloric intake. Thus, a long-acting, safe, and palatable cholinergic agonist that is potent and centrally active is needed to further test the role of cholinergic agonist treatments in TD.

Drugs that enhance the action of GABA may improve TD in some patients. Thus, the benzodiazepines as weak GABA agonists suppress TD movements

TABLE 4
Dopamine Deficiency–Acetylcholine Excess Syndromes

Drug-induced
Neuroleptic-induced parkinsonism
Neuroleptic-induced dystonia
Neuroleptic-induced akathisia
Spontaneous or idiopathic
Parkinson's disease
Dystonic musculorum deformans

in some patients (O'Flanagan, 1975; Sedman, 1976). The GABA-mimetic muscimol suppresses TD but at the expense of exacerbating psychosis since it is also a psychotomimetic (Tamminga *et al.,* 1979). Two drugs that block the degradative enzyme GABA-transaminase cause GABA levels to rise in the central nervous system and may improve TD in some patients. These drugs are γ-acetylenic-GABA and γ-vinyl-GABA (Casey *et al.,* 1980; Stahl *et al.,* 1985a, 1985d; Tamminga *et al.,* 1983; Tell *et al.,* 1981). Other possible GABAergic drugs such as baclofen and THIP [4,5,6,7-tetrahydro-isoxazolo-(5,4-c)-pyridin-3-ol] may reduce hyperkinetic movements in some TD patients (Berger and Rexroth, 1980; Baldessarini and Tarsy, 1978). Unfortunately, the results with GABA-enhancing agents to date are not dramatically useful for clinical treatment purposes. However, these experiments do help to implicate a contribution of relative GABA deficiency to the pathophysiology of TD and suggest that the development of better GABAergic agents may be more clinically useful in the future.

3.3. DA Deficiency Syndromes and Acute Neuroleptic Administration

3.3.1. Drug-Induced Parkinsonism (Table 4)

Pharmacological agents that deplete DA or block DA postsynaptic receptors have long been known to produce a pseudoparkinsonism with tremor, bradykinesia, and rigidity (Tables 2 and 4) (Klawans, 1973; Stahl *et al.,* 1982a). A less well-known form of drug-induced parkinsonism is the "rabbit syndrome," manifested by parkinsonian tremor of the orbicularis oris muscles (Villaneuve *et al.,* 1973; Sovner and Dimascio, 1977). The casual observer may mistake this tremor for the orobuccal facial movements of TD. Application of the DA–ACh imbalance paradigms described here can easily establish the diagnostic distinction between TD and the rabbit syndrome on the basis of differential pharmacology. The rabbit syndrome, with DA deficiency–ACh excess, is improved by anticholinergic agents, whereas TD, with DA excess–ACh deficiency, is worsened by anticholinergic agents. A physostigmine infusion may also be of diagnostic utility in difficult cases, since physostigmine may worsen the rabbit syndrome yet improve TD (see Section 3.2) (Weiss *et al.*, 1980; Stahl *et al.*, 1982a). Another form of drug-induced parkinsonism is that seen after abrupt discontinuation of chronic, high-dose anticholinergic drugs. Presumably, cholinergic muscarinic receptors are rendered supersensitive by chronic administration of anticholinergic agents, creating parkinsonism by relative ACh excess when anticholinergic agents are abruptly withdrawn (Gardos *et al.*, 1978; Luchins *et al.*, 1980).

These forms of drug-induced parkinsonism are by and large *reversible* upon discontinuation of the specific drug and the passage of time. Recently, however, a type of *irreversible* parkinsonism has been described in humans using an illicit drug intravenously (Langston *et al.*, 1983). This drug is actually a toxin called MPTP (i.e., 1-methyl-4-phenyl-1,2,5,6-tetrahydropyridine), which is a by-product of the synthesis of a "new heroin" for narcotic addicts by illegal laboratories. MPTP is structurally related to meperidine (Demerol), a narcotic used legally to treat severe pain and used illegally by narcotic addicts. When one illegal laboratory in California began producing MPTP as a "new heroin" for narcotic addicts, several addicts who injected it intravenously promptly developed relatively severe parkinsonism that is clinically identical to spontaneous (idiopathic) Parkinson's disease, and which, unfortunately, has proven to be irreversible. MPTP-induced parkinsonism is, however, treatable with levodopa and with DA agonists, just like idiopathic Parkinson's disease (see Section 4.2.1.) (Langston *et al.*, 1983).

It has now been shown that MPTP is a toxin that specifically destroys the same small group of brain cells—those of the substantia nigra—that degenerate in naturally occurring Parkinson's disease (Langston *et al.*, 1983; Lewin, 1984a, 1984b). Furthermore, these same changes can be produced experimentally in animals such as primates and mice, thus producing a very powerful animal model for Parkinson's disease (Lewin 1984a, 1984b; Heikkila *et al.*, 1984; Langston *et al.*, 1984).

Now that this model is available, it should greatly enhance the development of new drugs not only for MPTP-induced parkinsonism, but for Parkinson's disease as well. The model has already proved valuable in demonstrating the therapeutic efficacy of a new DA agonist +-PHNO(+-4-propyl-9-hydroxynaphthoxazine) (Martin *et al.*, 1984) in MPTP-induced parkinsonism in cynomolgus primates (A. Crossman, G. Woodruff, and S. Stahl, unpublished data). Clinical trials with this drug in human Parkinson's disease are currently in progress (Stahl, unpublished data, 1985).

3.3.2. Drug-Induced Dystonias (Table 4)

Neuroleptic drugs can induce focal or generalized muscular dystonias, especially of the face and neck, which closely resemble the spontaneously occurring dystonias of dystonia musculorum deformans (see Section 4) (Klawans, 1973; Korczyn, 1978; Stahl *et al.*, 1982a). These drug-induced dystonias obey classical pharmacological DA deficiency and ACh excess syndromes, as they are produced by DA receptor blocking neuroleptic drugs and are rapidly relieved by anticholinergic agents (Klawans, 1973; Korczyn, 1978; Stahl *et al.*, 1982a). The drug-induced dystonias can serve as clinical and pharmacological models for the spontaneously occurring dystonias, discussed in Section 4.2.

3.3.3. Akathisia (Table 4)

Akathisia is a confusing concept and a much neglected subject in the psychiatric literature (Stahl, 1985). Even today, there appears to be no consensus as to the definition of this term. At the turn of the century, Haskovec (1904) observed restless movements in his patients, which he thought were of *hysterical* origin, and he coined the term akathisia (which literally means to be unable to sit) to describe such movements. In the 1920s, Bing observed a curious restlessness, a "muscular impatience," and an inability to sit still in patients with *Parkinson's disease,* and adopted Haskovec's term akathisia to describe this phenomenon in patients with disorders of the basal ganglia (Bing, 1923, 1939; Adams and Victor, 1981). Ekbom (1944, 1960) wrote that restless movements of the legs could occur "spontaneously" in patients who did not have hysteria or basal ganglia disease. Fortunately, he called this phenomenon *restless legs syndrome* (or Ekbom's syndrome) rather than akathisia.

In the 1950s, clinicians began to notice jittery, restless movements, often of the feet, in patients taking neuroleptics (Steck, 1954). Since these movements resembled those described by Haskovec (1904) and by Bing (1923, 1939), they were called *neuroleptic-induced akathisia* and were understood originally as a *movement disorder* and one of the extrapyramidal side effects of the neuroleptics that fit in the constellation of dystonia–parkinsonism–akathisia (Ayd, 1961). Indeed, the improvement of neuroleptic-induced akathisia by anticholinergic drugs (Freedman and DeJong, 1961a, 1961b) in many patients solidified the association of akathisia with other extrapyramidal side effects since all of these have been conceptualized as a result of DA–ACh imbalance produced by neuroleptic blockade of DA receptors and relieved by anticholinergic blockade of muscarinic receptors (see Fig. 7) (Marsden *et al.*, 1975; Stahl *et al.*, 1982a).

Soon, however, sensitive and critical clinicians listening to their patients realized that neuroleptic-treated patients with restless movements also experienced strong subjective, distinctly unpleasurable feelings of the need to move, without any relief of this need after making the movement (Van Putten, 1975; Kendler, 1976). Thus, akathisia was described as a *mental disorder,* distinct from the psychotic patient's familiar state of anxiety and discomfort, distinct from dysphoric, dissociative, or hyperactive reactions to neuroleptics, and indeed distinct from observations of overt restless movements seen by an examiner. Thus, a dilemma: is neuroleptic-induced akathisia a movement disorder or a mental disorder? Some authors emphasize one, or the other, leading to confusion in the literature and lack of consensus.

Today, it seems reasonable to suggest that neuroleptic-induced akathisia

TABLE 5
β-Adrenergic Excess Syndromes

Drug-induced postural tremor (e.g., lithium, caffeine, tricyclic antidepressants, theophyllin, beta agonists)
Alcohol and opiate withdrawal
Essential tremor
Tourette syndrome
Stiff-man syndrome

is *both* a movement disorder *and* and a mental disorder (Stahl, 1985). To diagnose it, a clinician must not only observe a patient's movements, but also elicit the patient's subjective experiences. Having both objective and subjective data in hand, the clinician must then make some diagnostic judgments without much guidance from the literature. Barnes and Braude (Braude *et al.*, 1983; Barnes and Braude, 1985) suggest operational definitions for both objective and subjective components of neuroleptic-induced akathisia. Van Putten *et al.* (1985) have expanded the criteria for the subjective component. No matter whose criteria the clinician chooses, the clinician must nevertheless make two judgments: (1) Are the movements I am observing normal movements, restless movements, dyskinesias, or other abnormalities? (2) Is the patient having subjective distress, and if so, is it restlessness, anxiety, depression, or some other mental disorder?

Having decided that the patient has objective and/or subjective restlessness, the current perplexity is to decide whether *both* components must be present to make a diagnosis of akathisia, or whether these components are dissociable into clinically useful subtypes.

3.4. β-Adrenergic Excess Syndromes

3.4.1. Postural Tremor

A number of drugs can induce a postural tremor characterized by rapid, low-amplitude oscillations made worse with action or holding a posture (Tables 2 and 5). This tremor is similar to "physiological tremor" seen with nervousness or anxiety. It can be caused by caffeine, lithium, exogenous thyroid, and tricyclic antidepressants, and by any number of drugs used to treat asthma, including theophylline and β-adrenergic agonists (Hollister, 1978; Goodman and Gillman, 1982; Barchas *et al.*, 1977; Halaris, 1983). These drugs appear to share in common the ability to drive central β-adrenergic receptors and thereby produce this tremor. β-Adrenergic antagonists such as propranolol reduce or stop the postural tremor induced by these drugs (Kirk *et al.*, 1972).

3.4.2. Alcohol and Opiate Withdrawal

Chronic abuse of alcohol or opiates can lead to physiological addiction defined as the development of withdrawal symptoms upon abrupt drug discontinuation. The withdrawal state in both instances is characterized by a hyperadrenergic state, including postural tremor, sweating, tachycardia, fever, irritability, and even seizures (Adams and Victor, 1981). Classically, withdrawal symptoms are treated by readministering the drug of abuse and then slowly tapering it. The β-adrenergic withdrawal symptoms of alcohol abuse can be suppressed in part by propranolol (Victor, 1966; Zilm *et al.*, 1975). The hyperadrenergic symptoms of opiate abuse can be suppressed by α-2 adrenergic agonists such as clonidine (Gold *et al.*, 1978, 1982; Washton and Resnick, 1982) and lofexidine (Washton *et al.*, 1981; Gold *et al.*, 1981).

4. SPONTANEOUS MOVEMENT DISORDERS

4.1. DA Excess and ACh Deficiency in Hyperkinetic Movement Disorders

A number of disorders manifesting hyperkinetic movements, whether spontaneous or drug-induced, can be understood as pathological syndromes of relative DA excess and relative ACh deficiency (Figs. 4–6). GABA deficiency may also accompany these syndromes of DA excess and ACh deficiency. Under drug-induced disorders (Section 3), we have discussed DA agonist choreas, tics, and stereotypies, as well as neuroleptic-induced tardive dyskinesia syndromes (Fig. 4). Here we will review spontaneous choreas (including Huntington's disease) and Tourette syndrome (Figs. 5 and 6).

4.1.1. Huntington's Disease (Fig. 5)

The evidence for DA excess and ACh and GABA deficiency in the movement disorder of Huntington's disease (HD) comes both from direct biochemical assays of postmortem brains from HD patients and from pharmacological trials in HD patients. Direct assay of DA and metabolites in HD yields inconclusive results. DA in basal ganglia of HD patients is reportedly increased, decreased, or normal, and the DA metabolite homovanillic acid (HVA) may be normal or decreased in cerebrospinal fluid (CSF) of HD patients (Chase *et al.*, 1979). We have found that CSF HVA is low in HD, but that this decrease in HVA does not correlate with the severity of the movement disorder (Stahl *et al.*, 1985b). Rather, it correlates with the severity of the dementia that accompanies the movement disorder (Stahl *et al.*, 1985c). One must always bear in mind that spontaneously occurring disorders are rarely confined to a single neuronal system. Thus, DA–ACh imbalance in the basal ganglia may be applicable to the movement disorder component of a disease, but neurotransmitter imbalance in other brain regions may need to be considered to explain the mental status changes of that disease.

The cholinergic and GABA systems have also been studied in HD brains. The ACh-synthesizing enzyme choline acetyltransferase (CAT) has decreased activity, and cholinergic muscarinic receptors are decreased in numbers in HD brains (Chase *et al.*, 1979). Acetylcholinesterase activity is normal (Chase *et al.*, 1979). The concentrations of GABA, the activity of its synthetic enzyme glutamic acid decarboxylase (GAD), and the number of receptors for GABA are all decreased in HD brains; GABA concentrations may also be decreased in CSF of HD patients (Chase *et al.*, 1979).

For more than 10 years, the concept of "functional DA hyperactivity" has dominated pharmacological approaches to HD (Barbeau *et al.*, 1973). This hypothesis has been extensively reviewed, and is based primarily on observations of symptom improvement in HD patients given a variety of drugs that deplete DA or block DA receptors (Barbeau *et al.*, 1973; Chase *et al.*, 1979). DA agonists also exacerbate HD symptoms in some cases. The modification of the concept of DA hyperactivity in HD to the hypothesis of an imbalance among DA, ACh, and GABA is based in part on findings of decreased CAT, GABA, GAD, cholinergic muscarinic, and GABA receptors in HD brains discussed previously. The imbalance hypothesis also receives support from preliminary evidence that ACh and GABA replacement therapies may improve movements in some HD patients (Chase *et al.*, 1979).

Many investigators have given cholinomimetics such as physostigmine, choline, and lecithin to patients with HD, and have reported variable results, with some patients improving and others not improving (Chase *et al.*, 1979; Davis and Berger, 1978, 1979; Tarsy *et al.*, 1974; Davis *et al.*, 1975, 1976; Barbeau *et al.*, 1979; Klawans and Rubovits, 1972; Growdon and Wurtman, 1979; Stahl *et al.*, 1982a). In addition to the problems with ACh replacement treatments discussed in reference in TD patients, there is another potential limitation for ACh replacement treatments in HD patients. HD is characterized by a progressive degeneration of neurons upon which the receptors responsive to cholinergic agents are located. Thus, HD patients may respond to cholinomimetics at one stage of their illness and not respond as the disease progresses.

The use of GABA-enhancing agents has been similarly disappointing (Bird, 1980). Only inconsistent benefits have been derived from treatments with agents such as THIP, gamma-vinyl-GABA, and isoniazid (Shoulson *et al.*, 1975, 1976; Perry *et al.*, 1979; Paulson *et al.*, 1979; Foster *et al.*, 1983).

Future therapies may be able to capitalize on recent discoveries regarding additional neurotransmitters. That is, in addition to GABA (and GAD) and ACh (and CAT) being *decreased* in HD, apparently substance P, enkephalins, and cholecystokinin are also all decreased (Martin, 1984). In addition to the DA being possibly *increased* in some studies of HD, so are thyrotropin-releasing hormone, somatostatin, and neurotensin increased (Martin, 1984). The neurotransmitters serotonin, vasoactive intestinal polypeptide, and norepinephrine may all be *unchanged* in HD (Martin, 1984). Perhaps new drugs aimed at these additional neurotransmitter imbalances may become effective therapies in HD.

4.1.2. Chorea (Table 3)

Several other neurological syndromes are associated with chorea in addition to Huntington's chorea, discussed in Section 4, and DA agonist chorea, discussed in Section 3. *Sydenham's chorea* (Weissberg and Friedrich, 1978; Adams and Victor, 1981) is the movement disorder that follows infection with group B *Streptcoccus,* presumably secondary to immunological changes in the basal ganglia that render the receptors there hyperactive to DA. Sydenham's chorea can be suppressed by DA receptor blockers. *Chorea gravidarum* (Lewis and Parsons, 1966) is a movement disorder produced in pregnant women presumably by the high circulating estrogens of pregnancy. It can also be suppressed by neuroleptics and abates spontaneously after delivery. A drug-induced analogue to this is *birth control pill chorea* (Adams and Victor, 1981) observed in some women taking oral contraceptives. *Hyperthyroid chorea* (Adams and Victor, 1981) is a rare complication of high circulating thyroxine levels, is reduced by neuroleptics, and resolves in the euthyroid state. *Senile chorea* (Wells and Duncan, 1980; Adams and Victor, 1981) is an idiopathic benign chorea of the elderly and is supprressed by neuroleptics. *Benign paroxysmal choreoathetosis* (Kertesy, 1967; Wells and Duncan, 1980) is another idiopathic chorea, but this one is probably not related to DA hyperactivity. This chorea may be worsened by movements (*kinesogenic chorea*) and may be related to a seizure disorder, since it is suppressed by the anticonvulsant phenytoin (Kertesy, 1967; Wells and Duncan, 1980).

4.1.3. Tourette Syndrome (Fig. 6)

Tourette syndrome is a disorder of motor and vocal tics with onset during childhood (Friedhoff and Chase, 1982). We have recently reviewed the evidence for DA excess and ACh deficiency in this disorder (Stahl and Berger, 1982a, 1982b). This evidence is based on animal models and on our own data (Stahl and Berger, 1981a) as well as that of others showing clinical responses to cholinergic agents such as physostigmine, choline choride, and lecithin. In addition, the DA–ACh hypothesis of Tourette syndrome (Fig. 6) is supported by clinical responses to DA agonists and antagonists and by correlations between the ability of neuroleptics to block Tourette syndrome symptoms and the ability of neuroleptics to block DA and ACh receptors (Stahl and Berger, 1982b). An interesting link between idiopathic Tourette syndrome and tardive dyskinesia is the identification of tardive Tourette syndrome cases (see Section 3.2.3). This unique form of tardive dyskinesia and Tourette syndrome suggests that the pathophysiology of idiopathic Tourette syndrome may be similar to that of tardive dyskinesia.

We have also hypothesized that symptoms of Tourette syndrome may represent hyperactivity selectively of the D2 subtype of DA receptor (Stahl and Berger, 1982b; Uhr *et al.*, 1984). That is, D1 receptors are linked to the enzyme adenylate cyclase, and D2 receptors are not (Iversen, 1975; Kebabian and Calne, 1979). Tourette syndrome is improved proportionally to the ability of neuroleptics to block D2 receptors, but *not* proportionately to their ability to block D1 receptors (Stahl and Berger, 1982b). We therefore proposed that a selective D2 antagonist would not only be effective in Tourette syndrome, but might simultaneously have fewer side effects.

As mentioned previously, pharmacological evidence links DA excess to Tourette syndrome and is based on exacerbation of symptoms with DA agonists such as amphetamine, methylphenidate, and levodopa and improvement of symptoms with DA antagonists such as neuroleptics (Friedhoff and Chase, 1982). Haloperidol is the current treatment of choice for Tourette syndrome (Friedhoff and Chase, 1982). However, haloperidol is a nonspecific DA blocker (D1 and D2 receptors) that is associated with extrapyramidal side effects presumably from nigrostriatal DA blockade (Friedhoff and Chase, 1982). Hence, its efficacy in ameliorating the motor and vocal tics of Tourette syndrome is limited by adverse effects that can either prevent adequate dosage for symptom control or lead to noncompliance.

Piquindone (R022-1319), a pyrroloisoquinoline derivative [rac.-3-ethyl-2, 6-dimethyl-4, 4a, 5, 6, 7, 8, 8a, 9-octahydro-4a, 8a-trans-1H-pyrrolo(2,3-g)isoquinolin-4-one,hydrochloride, dihydrate] designed via a three-dimensional computer model of receptors (Olsen *et al.*, 1981), is a novel D2 receptor antagonist (Davidson *et al.*, 1983). In animal experiments, piquindone blocks discrete avoidance response at doses corresponding to a relatively potent antipsychotic agent (approximately half that of haloperidol) (Davidson *et al.*, 1983). However, it is less potent (approximately one-sixth to one-tenth that of haloperidol) in producing the acute stereotypy and catalepsy that are behavioral models used to predict extrapyramidal symptoms (Davidson *et al.*, 1983). Piquindone may therefore preferentially block DA receptors in the mesolimbic system rather than in the nigrostriatal system (Davidson *et al.*, 1983) in addition to being a more selective blocker of D2 receptors (Nakajima *et al.*, 1983).

In both open trials and in double-blind placebo-controlled trials, we have demonstrated not only that piquindone is effective in ameliorating the symptoms of TS, but also that this efficacy was attained without the production of unwanted extrapyramidal side effects (Uhr *et al.*, 1984). These observations further support the notion that Tourette syndrome is asociated with D2 hyperactivity, and that selective D2 receptor antagonists may be more selective treatments for Tourette syndrome.

4.2. DA Deficiency and ACh Excess in Rigid–Dystonic Movement Disorders

A number of disorders manifesting rigid, akinetic, dystonic movements or postures, whether spontaneous or drug-induced, can be understood as pathological syndromes of relative DA deficiency and relative ACh excess. This is the pathological mirror image of hyperkinetic movement disorders discussed in Section 4.1. The DA deficiency–ACh excess syndromes are listed in Table

4 and include Parkinson's disease, drug-induced parkinsonism (including the rabbit syndrome), drug-induced dystonias, and some cases of spontaneous dystonias of focal or generalized type.

4.2.1. Parkinson's Disease (Fig. 7)

Parkinson's disease (PD) is the movement disorder first to be understood as an imbalance between neurotransmitters (Table 4 and Fig. 7). The DA deficiency from nigrostriatal neuronal degeneration in PD, with resultant relative ACh excess, has been the subject of numerous reviews over the years (Klawans, 1973; Tarsy, 1976; Yahr, 1973; Calne, 1973; McDowell and Barbeau, 1974; Calne *et al.*, 1975; Marsden, 1982).

PD serves as a model for hypotheses of neurotransmitter imbalance in numerous other movement disorders. PD itself refers classically to the condition produced by loss of the substantia nigra. Specifically, the dopaminergic cell bodies in the zona compact of the substantia nigra degenerate, lose their black neuromelanin pigment, and develop eosinophilic inclusions known as Lewy bodies. It is probably necessary to lose more than 80% of the nigral neurons (and therefore 80% of the striatal DA) before the symptoms of akinesia, rigidity, and tremor are produced. Recently, other biochemical changes have been described in PD, including loss of DA from other neurons in the brain, as well as loss of norepinephrine, serotonin, GABA, and ACh from certain brain areas (Marsden, 1982). The clinical correlations to these additional neurotransmitter imbalances are not known, but may well explain the parkinsonism-plus syndromes described in Section 4.2.2.

Evidence for DA–ACh imbalance in the movement disorder of PD is firmly established (Marsden, 1982; Spehlmann and Stahl, 1976; Stahl *et al.*, 1982a). DA activity can be enhanced either by replacing deficient DA by the precursor levodopa, or by administering synthetic drugs that directly stimulate DA receptors. These DA agonists include pergolide, bromocriptine, and lisuride (Marsden, 1982). Fortunately, it does *not* appear that the DA receptors on striatal neurons are lost when the substantia nigra neurons (which deliver DA to the striatum) degenerate. Thus, the receptors remain in place to be stimulated with drugs. This is not the case in Huntington's disease (discussed in Section 4.1.1), since both receptors and neurons degenerate. Long-term replacement therapy of PD does have its problems. Not only do the drugs appear to lose some of their effectiveness over time, but they also seem to develop shorter durations of action ("end-of-dose deterioration" or "wearing-off effect"). With time, such fluctuations become increasingly rapid, from mobility to immobility, and unpredictable ("on–off effect"). Managing these fluctuations in responsiveness to DA agents is now the most troublesome issue in the treatment of PD.

One of the hopes for managing these disabling fluctuations in symptoms is to administer the drug treatments by novel drug delivery mechanisms. For example, the wide swings in symptoms of PD patients undergoing treatment with oral levodopa can be eliminated by giving the drug by a constant intravenous infusion (Nutt *et al.*, 1984; Quinn *et al.*, 1984).

However, even this approach is limited, since the effects of intravenous levodopa can be nullified when the patient eats a meal containing neutral amino acids that compete (and therefore block) levodopa's entering the brain (Nutt *et al.*, 1984). Thus, a meal can produce an "off" signal to the brain, since levodopa is no longer delivered there.

New hope is emerging for the use of DA agonists in novel drug delivery systems (Stahl, 1984). Thus, the DA agonists could be given in oral osmotic pumps that are swallowed and then release the drug slowly and steadily within the gut (Urquhart *et al.*, 1984). A very exciting possibility is the hope that a dopamine agonist might be administered *transdermally*. That is, an adhesive patch containing a DA agonist applied to the skin could deliver the drug in a manner very much analogous to an intravenous infusion, but without a needle (Stahl, 1984). The constant delivery of drug would prevent fluctuations in the level of drug in the blood. Furthermore, since foods and amino acids do not compete with the agonists for uptake into the brain, the patient would be able to eat without worrying about sending an "off" signal to his brain. The drug +-PHNO mentioned previously (Martin *et al.*, 1984) is the most promising candidate for these novel drug delivery systems, since it is extremely potent and lipid soluble (thus allowing penetration of the skin). Clinical trials utilizing these concepts of novel drug delivery of the depression agonist +PHNO are in progress (Stahl and Marsden, unpublished data, 1985) and offer hope for the PD patient with the "wearing-off" and "on-off" complications of longterm treatment and severe disease.

Anticholinergic agents also reduce parkinsonian symptoms. These agents are less effective in general than the dopaminergic agents, but they tend to have an additive effect and are thus used conjointly. Thus, DA deficiency in PD can be ameliorated in part by inducing ACh deficiency with anticholinergic drugs (Fig. 7).

4.2.2. Parkinsonism Plus (Table 6)

A plethora of neurological disorders contain parkinsonism as a part of a more global neurological condition that includes other neurological symptoms due to degeneration of other neuronal systems. Some of these syndromes are listed in Table 6. The parkinsonism of these disorders can often be treated in the same manner as PD itself (Fig. 7). The ancillary symptoms are probably not related to DA–ACh imbalance in the basal ganglia. As the neurotransmitter imbalances for the ancillary symptoms come to be understood in the future, specific therapies should be possible for these additional disabling clinical features.

4.2.3. Dystonia Musculorum Deformans (Fig. 8)

The spontaneous dystonias are a group of movement disorders of unknown cause with numerous subtypes. Dystonia can be inherited in an autosomal

TABLE 6
Parkinsonism-Plus Syndromes

Shy-Drager syndrome
Strionigral degeneration
Olivopontocerebellar degeneration
Progressive supranuclear palsy
Wilson's disease
Hallervorden–Spatz disease
Pallidopyramidal degenerations
Parkinsonism–amyotrophic lateral sclerosis–dementia complex
Symptomatic parkinsonism with multiinfarct dementia
Symptomatic parkinsonism with Alzheimer's dementia

recessive pattern, or can be spontaneous and idiopathic. Dystonias can develop in childhood or adult life and can be focal (such as blepharospasm or torticollis), or generalized and stable, or progressive in clinical course (Eldridge and Fahn, 1976).

The role of DA–ACh imbalance in dystonia has been investigated by administering agonists and antagonists of both DA and ACh (Fig. 8 and Table 4). We have used this strategy in five patients with generalized dystonia and one patient with torticollis (Stahl and Berger, 1981b, 1982c). Physostigmine worsened symptoms in all five patients, while anticholinergic agents improved symptoms in the five patients receiving them (benzotropine or trihexphenidyl). The DA receptor blocking neuroleptic haloperidol exacerbated dystonia, while the DA receptor stimulating agent bromocriptine improved dystonia. Others have also seen observed improvement in dystonia after agents that enhance DA activity including levodopa, bromocriptine, lisuride, and tetrahydrobiopterin (Eldridge and Fahn, 1976; Newman *et al.,* 1982; LeWitt *et al.,* 1983; Stahl *et al.,* 1984). These pharmacological results are consistent with a DA deficiency and ACh excess pathology parallel to that of Parkinson's disease (Fig. 8).

Other investigators have also reported clinical improvement in some dystonia patients after anticholinergic agents, and that the anticholinergic-induced improvement can be reversed by physostigmine (Fahn, 1979; Tanner *et al.,* 1979, 1981). These observations are all consistent with the hypothesis of relative ACh excess in dystonia patients that can be symptomatically worsened by physostigmine and improved by anticholinergics.

4.3. Serotonin Deficiency in Myoclonus

One form of myoclonus can be produced when the brain sustains injury from anoxia. This type of movement disorder, called postanoxic intention myoclonus, responds, often dramatically, to the serotonin precursor 5-hydroxytryptophan, or to the combination of tryptophan plus a monoamine oxidase inhibitor (Van Woert *et al.*, 1977; Magnussen *et al.*, 1978; Chadwick *et al.*, 1977). Although no absolute deficiency of serotonin has been demonstrated in brains of patients with postanoxic intention myoclonus (Marsden, 1982), the fact that serotoninergic agents so effectively relieve symptoms has led to the speculation that there may be a relative deficiency of serotonin in these patients.

4.4. Norepinephrine Excess (Table 6)

4.4.1. Postural Tremor

A spontaneously occurring postural or action tremor is known by several names, including essential tremor, familial tremor, and senile tremor (Critchley, 1949, 1972; Adams and Victor, 1981; Marsden, 1982). It resembles the tremor caused by drugs that enhance β-adrenergic activity (discussed in Section 2) and is relieved by β-adrenergic blockers such as propranolol (Sweet *et al.*, 1974; Wells and Duncan, 1980; Adams and Victor, 1981). Interestingly, the essential tremor is often relieved by alcohol (Wells and Duncan, 1980; Adams and Victor, 1981).

4.4.2. Tourette Syndrome

Tourette syndrome was discussed previously as a DA excess–ACh deficiency syndrome. In addition, it appears that the symptoms of certain Tourette patients, particularly their obsessive–compulsive behavioral symptoms, may be relieved by the α-2 adrenergic agonist clonidine (Cohen *et al.*, 1979, 1980; Friedhoff and Chase, 1982). This has led to speculation that Tourette syndrome may be related in part to excessive noradrenergic activity that is reduced by administration of the presynaptic agonist clonidine, which acts to turn off the noradrenergic neuron.

4.4.3. Stiff-Man Syndrome (Tables 5 and 7)

The stiff-man syndrome is a rare disorder characterized by progressive but fluctuating rigidity and spasm (Moersch and Woltman, 1956). Its cause is unknown, but it clinically resembles strychnine and tetanus poisonings (Olafson *et al.*, 1964). No direct demonstration of neurotransmitter dysfunction has been made, but pharmacological studies suggest a possible imbalance between GABA and norepinephrine (Schmidt *et al.*, 1975). The weak GABAergic agents diazepam and baclofen improve the stiff-man syndrome, and levodopa slightly worsens it (Howard, 1963; Ricker and Mertens, 1968; Schmidt *et al.*, 1975). Central nervous system norepinephrine metabolism is increased in the stiff-man syndrome, and this increase is reduced when symptoms are improved by diazepam (Schmidt *et al.*, 1975). Thus, preliminary evidence suggests a GABA deficiency–norepinephrine excess as one hypothesis to explain this disorder.

TABLE 7
Movement Disorders Not Well Classified Pharmacologically

Myoclonus (? serotonin deficiency)
Dystonias (? DA–ACh imbalance)
Stiff-man syndrome (? GABA–NE imbalance)
Cataplexy
Spasticity and paralysis
Ataxia and intention tremor

4.5. Movement Disorders Not Well Classified Pharmacologically (Table 7)

4.5.1. Focal and Generalized Dystonias

Although some patients with generalized dystonia obey DA deficiency–ACh excess pharmacology, many do not. Despite the often pervasively disabling features of this disorder, no structural brain abnormality is known to exist in dystonia, and no known chemical abnormality has yet been demonstrated. If ever a disease was a good candidate for pharmacological treatment, it would be dystonia because of its normal brain gross anatomy and thus the potential for the reversibility of this condition. However, we remain largely ignorant of its neurotransmitter imbalance and are thus powerless in suggesting appropriate therapy. Perhaps the answer will be found in one of the more recently discovered neurotransmitter peptides (Jaquet and Abrams, 1982).

The search for the explanation of the neurochemical abnormality in dystonia could potentially benefit many different kinds of disorders. *Benign essential blepharospasm* is forceful dystonic contractions of the muscles around the eyes. *Torticollis* is one of the most common forms of dystonia and is characterized by abnormal involuntary twisting of the neck. *Writer's cramp* is an action dystonia of the hand when the subject attempts to write. *Spasmodic dysphonia* is an action of dystonia of the pharyngeal muscles when the patient attempts to speak, sing, or swallow. Other focal dystonias can involve the trunk or the leg. Virtually all known pharmacological agents have been given to patients with these various forms of focal dystonia and all with virtually no consistently good effects.

A syndrome of spontaneously occurring, symmetrical dystonic face and neck spasms of unknown cause is recognized by several names, including idiopathic orofacial dyskinesia, blepharospasm–oromandibular dystonia, Brueghel's syndrome, and Meige dystonia (Eldridge and Fahn, 1976; Marsden, 1976; Paulson, 1980; Casey, 1980; Tolosa and Lai, 1979). The movements of this disorder can closely resemble the facial dyskinesias induced by levodopa or by chronic neuroleptic treatment, and differential diagnosis among these conditions can be difficult. We have administered physostigmine to two patients with Meige dystonia, and found deterioration of symptoms similar to that observed in patients with generalized dystonia (Stahl *et al.*, 1982a, 1982b). One of these patients subsequently improved significantly on anticholinergic medication. Neither of these patients experienced any symptom changes with bromocriptine. The one Meige dystonia patient given haloperidol improved greatly, and this improvement was enhanced by the addition of the anticholinergic agent benztropine to the haloperidol treatment.

These preliminary findings are consistent with those of most (Marsden, 1976; Paulson, 1973; Tolosa and Lai, 1979) but not all (Casey, 1980) investigators who have also found improvement after anticholinergic agents, as well as worsening after physostigmine in Meige dystonia. Other investigators also consistently report reduced facial dystonia during haloperidol administration (Marsden, 1976; Casey, 1980; Tolosa and Lai, 1979). Levodopa may worsen facial movements in these patients with Meige Dystonia (Casey, 1980).

Our pharmacological studies and those of others suggest relative DA overactivity and relative ACh overactivity in some patients with Meige Dystonia (Fig. 9). This hypothesized pathology is unique among movement disorders studied to date. The unique pharmacological profile suggests that this hypothesis may help to differentiate Meige dystonia from TD when the diagnosis is in doubt. According to this hypothesis, a physostigmine infusion should worsen symptoms in patients with Meige dystonia and improve symptoms in TD.

4.5.2. Cataplexy

Cataplexy is part of the larger syndrome known as narcolepsy. Narcolepsy is a clinical tetrad comprised of (1) daytime attacks of rapid-eye-movement sleep; (2) sleep paralysis; (3) hallucinations while falling asleep or waking up; and (4) cataplexy (Adams and Victor, 1981). The cataplexy part of the narcolepsy tetrad is the sudden loss of muscular tone causing complete paralysis (except for the eyes). This can be caused by sudden emotions such as surprise or laughter, and usually lasts a few moments. Although the neurotransmitter imbalance of cataplexy is unknown, it may be related to disruption of noradrenergic signals from the locus coeruleus to the serotoninergic raphe nucleus, since brain tumors in this area of the brain can produce cataplexy (Bonduelle and Degos, 1976; Stahl *et al.*, 1980).

4.5.3. Spasticity and Paralysis

Probably the most disabling neurological condition is the loss of voluntary control of motor movements by damage to pyramidal neurons from stroke, trauma, and multiple sclerosis (to name just a few) (Young and Delwaide, 1981; Adams and Victor, 1981). We have virtually no clue as to the neurotransmitter messages that control this neuronal system. Gaining some command over the function of the intact lower motor neurons after the upper motor neurons are destroyed by disease is one of the supreme challenges of neuropharmacological therapy for the future.

4.5.4. Ataxia and Intention Tremor

Ataxia (incoordination of movement) and intention tremor are motor disorders that are thought to arise from abnormalities in cerebellar signals (Adams and Victor, 1981). The neurotransmitter imbalances in these disorders are also poorly understood. Multiple sclerosis plaques commonly disrupt these signals (Adams and Victor, 1981). Alcohol is a cerebellar poison that can produce both reversible and irreversible ataxia (incoordination of movement) (Adams and Victor, 1981). These are the major clues in our understanding of neurotransmitters in ataxia and intention tremor, and the cerebellar movement disorders are another clinical area in great need of investigation in order to provide rational pharmacological therapies in the future.

5. SUMMARY

This chapter describes and defines both the neurology and the pharmacology of abnormal involuntary movement disorders. Drug-induced movement disorders are compared and contrasted with spontaneous movement disorders. The pharmacology of all movement disorders is discussed along the lines of the neurotransmitter imbalance hypothesis, emphasizing particularly the role of DA and ACh. Further therapeutic strategies will depend on our ability to discover and integrate the potential roles of other neurotransmitters (such as GABA, substance P, and other neuropeptides) into the pathophysiology of movement disorders. New neurotransmitters could be the targets for new drugs to treat the disorders of movement. Lessons learned from the notion of DA–ACh imbalance in movement disorders may be applicable in a general sense as our understanding of new neurotransmitters and new drugs advances.

6. REFERENCES

Adams, R. D., and Victor, M., 1981, *Principles of Neurology,* McGraw-Hill, New York.

Ayd, F. J., 1961, A survey of drug induced extrapyramidal reactions, *J.A.M.A.* **175:**1054–1060.

Baldessarini, R. J., and Tarsy, D., 1978, Tardive dyskinesia, in: *Psychopharmacology: A Generation of Progress* (M. A. Lipton, A. DiMascio, and K. R. Killam, eds.), pp. 993–1004, Raven Press, New York.

Barbeau, A., Chase, T. N., and Paulson, G. W. (eds.), 1973, *Huntington's Chorea, 1872–1972, Advances in Neurology Series,* Vol. 1, Raven Press, New York.

Barbeau, A., Growdon, J. H., and Wurtman, R. J. (eds.), 1979, *Nutrition and the Brain,* Volume 5, *Choline and Lecithin in Brain Disorders,* Raven Press, New York.

Barchas, J. D., Berger, P. A., Ciaranello, R. D., and Elliott, G. R. (eds.), 1977, *Psychopharmacology: From Theory to Practice,* Oxford University Press, New York.

Barnes, T. R. E., and Braude, W. M., 1985, Akathisia variants and tardive dyskinesia, *Arch. Gen. Psychiatry,* **42**:874–878.

Berger, P. A., and Rexroth, K., 1980, Tardive dyskinesia: Clinical, biological and pharmacological perspectives, *Schizophr. Bull.* **6**:102–116.

Bing, R., 1923, Ueber einige bemerkenswerte begleitersche inungen der "extrapyramidalen rigiditat (akathisie-mikrographie-kinesia paradoxa), *Schweiz. Med. Wochenschr.* **4**:167–171.

Bing, R., 1939, *Textbook of Nervous Diseases,* 5th ed., Haymaker, W. (translation and enlarged), pp. 169, 758–759, Henry Limpton Press, London.

Bird, E. D., 1980, Chemical pathology of Huntington's Disease, *Ann. Rev. Pharmacol. Toxicol.* **20**:533–551.

Bonduellle, M., and Degos, C., 1976, Symptomatic narcolepsies: A critical study, in: *Narcolepsy* (C. Guilleminault, W. C. Dement, P. Passauant, eds.), pp. 313–332, Spectrum Publications, Inc., New York.

Braude, W. M., Barnes, T. R. E., and Gore, S. M., 1983, Clinical characteristics of akathisia: A systematic investigation of acute psychiatric inpatient admissions. *Br. J. Psychiatry* **143**:139–150.

Calne, D. B., Chase, T. N., and Barbeau, A. (eds), 1975, *Dopaminergic Mechanisms, Advances in Neurology Series,* Vol. 9, Raven Press, New York.

Calne, D. M., (ed.), 1973, *Progress in the Treatment of Parkinsonism, Advances in Neurology Series,* Vol. 3. Raven Press, New York.

Casey, D. E., 1980, Pharmacology of blepharospasm-oromandibular dystonia syndrome, *Neurology* **30**:690–695.

Casey, D. E., Gerlach, J., Magelund, G., and Christensen, T. R., 1980, Gamma-acetylenic GABA in tardive dyskinesia. *Arch. Gen. Psychiatry* **37**:1376–1379.

Chadwick, D., Hallett, M., Harris, R., Jenner, P., Reynolds, E. H., and Marsden, C. D., 1977, Clinical, biochemical and physiological features distinguishing myoclonus responsive to 5-hydroxytryptophan, tryptophan with a monoamine oxidase inhibitor, and clonazepam, *Brain* **100**:455–487.

Chase, T. N., Wexler, N. S., Barbeau, A. (eds.), 1979, *Huntington's Disease, Advances in Neurology Series,* Vol. 23, Raven Press, New York.

Cohen, D. J., Young, J. G., Nathanson, J. A., and Shaywitz, B. A., 1979, Clonidine in Tourette's syndrome, *Lancet* **2**:551–553.

Cohen, D. J., Detlor, J., Young, J. G., and Shaywitz, B. A., 1980, Clonidine ameliorates Gilles de la Tourette syndrome, *Arch. Gen. Psychiatry* **37**:1350–1357.

Critchley, E., 1949, Observations on essential (heredofamilial) tremor, *Brain* **72**:113–139.

Critchley, E., 1972, Clinical manifestations of essential tremor, *J. Neurol. Neurosurg. Psychiatry* **35**:365–372.

Davidson, A. B., Bofe, E., MacNeil, D. A., and Cock, L., 1983, Pharmacologic effects of R022-1319: A new antipsychotic agent, *Psychopharmacology* **79**:32–39.

Davis, K. L., and Berger, P. A., 1978, Pharmacological investigations of the cholinergic imbalance hypotheses of movement disorders and psychosis, *Biol. Psychiatry* **13**:23–49.

Davis, K.L., and Berger, P.A. (eds.), 1979, *Brain Acetylcholine and Neuropsychiatric Disease,* Plenum Press, New York.

Davis, K. L., Hollister, L. E., Berger, P. A., and Barchas, J. D., 1975, Cholinergic imbalance hypotheses of psychoses and movement disorders: Strategies for evaluation, *Psychopharmacol. Commun.* **1**:533–543.

Davis, K. L., Hollister, L. E., Barchas, J. D., and Berger, P. A., 1976, Choline in tardive dyskinesia and Huntington's disease, *Life Sci.* **19**:1507–1516.

DeVeaugh-Geiss, J., 1980, Tardive Tourette syndrome, *Neurology* **30**:562–563.

DeVeaugh-Geiss, J. (ed.), 1982, *Tardive Dyskinesia and Related Involuntary Movement Disorders,* John Wright, Boston.

Ekbom, K. W., 1944, Asthensia Crurum parathestica (irritable legs), *Acta Med. Scand.* **118**:197.

Ekbom, K. A., 1960, Restless legs syndrome, *Neurology* **10**:868–873.

Eldridge, R., and Fahn, S. (eds.), 1976, *Dystonia, Advances in Neurology Series,* Vol. 14, Raven Press, New York.
Fahn, S., 1979, Treatment of dystonia with high-dosage anticholinergic medication, *Neurology* **29:**605.
Fann, W. E., Lake, C. R., Gerber, C. J., and McKenzie, G. M., 1974, Cholinergic suppression of tardive dyskinesia, *Psychopharmacology* **37:**101–107.
Fann, W. E., Smith, R. C., Davis, J. M., and Domino, E. F. (eds.), 1980, *Tardive Dyskinesia,* SP Medical and Scientific Books, New York.
Fog, R., and Pakkenberg, H., 1980, Theoretical and clinical aspects of the Tourette Syndrome (chronic multiple tics), *J. Neural. Transm.* **16**(Suppl):211–215.
Foster, N. L., Chase, T. N., Denaro, A., Hare, T. A., and Tamminga, C. A., 1983, THIP treatment of Huntington disease, *Neurology* **33:**637–639.
Freedman, D. X., and DeJong, J., 1961a, Thresholds for drug-induced akathisia, *Am. J. Psychiatry* **117:**930–931.
Freedman, D. X. and DeJong, J., 1961b, Factors that determine drug-induced akathisia, *Dis. Nerv. Sys.* **22:**2–8.
Friedhoff, A., and Chase, T. N. (eds.), 1982, *Tourette Syndrome, Advances in Neurology Series,* Raven Press, New York.
Gardos, G., Cole, J. O., and Tarsy, D., 1978, Withdrawal syndromes associated with antipsychotic drugs *Am. J. Psychiatry* **135:**1321–1324.
Gelenberg, A. J., Doller-Wojcik, J. C., and Growdon, J. H., 1979, Choline and lecithin in the treatment of tardive dyskinesia: preliminary results from a pilot study. *Am. J. Psychiatry* **36:**772–776.
Gerlach, J., Reisby, N., and Randrup, A., 1974, Dopaminergic hypersensitivity and cholinergic hypofunction in the pathophysiology of tardive dyskinesia, *Psychopharmacology* **34:**21–25.
Gold, M. S., Redmond, D. E., and Kleber, H. D., 1978, Clonidine blocks acute opiate-withdrawal symptoms, *Lancet* **2:**599–602.
Gold, M. S., Pottash, A. C., Annitto, W. J., Extein, I., and Kleber, H. D., 1981, Lofexidine, a clonidine analogue effective in opiate withdrawal, *Lancet* **1:**992–993.
Gold, M. S., Pottash, A. C., and Extein, I., 1982, Clonidine: inpatient studies from 1978 to 1981, *J. Clin. Psychiatry* **43**(6)(sec. 2):35–38.
Goodman, L. S., and Gilman, A. (eds.), 1982, *The Pharmacological Basis of Therapeutics,* Macmillan, New York.
Groves, P. M., 1983, A theory of the functional organization of the neostriatum and the neostriatal control of voluntary movement, *Brain Res. Rev.* **5:**198–132.
Growdon, J. H., and Wurtman, R. J., 1979, Oral choline administration to patients with Huntington's Disease, in: *Huntington's Disease. Advances in Neurology Series,* Vol. 23 (T. N. Chase, N. S. Wexler, and A. Barbeau, eds.), pp. 765–776, Raven Press, New York.
Halaris, A. E., 1983, The use of lithium in psychiatric practice, *Psychiatry Ann.* **12:**53–64.
Haskovec, L., 1904, Weitere bemerkungen ueber die akathisie, *Wien. Med. Wochenschr.* **54:**526–530.
Heikkila, R. E., Hess, A., and Duvoisin, R. C., 1984, Dopaminergic neurotoxicity of 1-methyl-4-phenyl-1,2,5,6-tetrahydropyridine in mice, *Science* **224:**1451–1453.
Hollister, L. E., 1978, Tricyclic antidepressants. *N. Engl. J. Med.* **299:**1168–1172.
Howard, F. M., 1963, A new and effective drug in the treatment of the stiff-man syndrome: preliminary report, *Mayo Clin. Proc.* **38:**203–212.
Iversen, L. L., 1975, Dopamine receptors in the brain, *Science* **188:**1084–1089.
Iversen, L. L., 1983, Neuropeptides—What next? *Trends Neurosci.* **6:**293–294.
Jacquet, Y. F., and Abrams, G. M., 1982, Postural asymmetry and movement disorder after unilateral microinjection of adrenocorticotropin 1-24 in rat brainstem, *Science* **218:**175–177.
Jeste, D. V., Potkin, S. G., Sinha, S., Feder, S., and Wyatt, R. J., 1979, Tardive dyskinesia—reversible and persistent, *Arch. Gen. Psychiatry* **36:**585–590.
Kebabian, J. W., and Calne, D. B., 1979, Multiple receptors for dopamine, *Nature* **277:**93–96.
Kendler, K. S., 1976, A medical student's experience with akathisia, *Am. J. Psychiatry,* **133:**4.

Keretesy, A., 1967, Paroxysmal kinesigenic choreoathetosis, *Neurology* **17**:680–684.
Kirk, L., Baastrup, P. C., and Schou, M., 1972, Propranolol and lithium-induced tremor, *Lancet* **1**:839.
Klawans, H. L., 1973, *The Pharmacology of Extrapyramidal Movement Disorders*, Karger, Basel.
Klawans, H. L., and Rubovits, R., 1972, Central cholinergic-anticholinergic antagonism in Huntington's chorea, *Neurology* **22**:107–112.
Klawans, H. L., and Rubovits, R., 1974, Effect of cholinergic and anticholinergic agents on tardive dyskinesia, *J. Neurol. Neurosurg. Psychiatry* **27**:941–947.
Klawans, H. L., Falk, D. K., Nausieda, P. A., and Wiener, W. J., 1978, Gilles de la Tourette syndrome after long-term chlorpromazine therapy, *Neurology* **28**:1064–1066.
Korczyn, A. D., 1978, The pathophysiology of dystonia, *J. Neural. Transm.* **42**:245–250.
Langston, J. W., Ballard, P., Tetrud, J. W., and Irwin, I., 1983, Chronic parkinsonism in humans due to a product of meperidine-analog synthesis, *Science* **219**:979–980.
Langston, J. W., Irwin, I., Langston, E. B., and Forno, L. S., 1984, Pargyline prevents MPTP-induced Parkinsonism in primates, *Science* **255**:1480–1482.
Lewin, R., 1984a, Trail of ironies to Parkinson's disease: Sloppy chemical synthesis by an illicit drug producer has led to important insights into the basic cause of Parkinson's disease, *Science* **224**:1083–1085.
Lewin, R., 1984b, Brain enzyme is the target of drug toxin: A chemical known as MPTP causes a Parkinson-like state in humans and monkeys; biochemical and autoradiographical studies are closing in on the mechanism, *Science* **225**:1460–1462.
Lewis, B. V., and Parsons, M., 1966, Chorea gravidarum, *Lancet* **1**:284.
LeWitt, P. A., Newman, R. P., Miller, L. P., Lovenberg, W., and Eldridge, R., 1983, Treatment of dystonia with tetrahydrobiopterin, *N. Engl. J. Med.* **308**:157–158.
Luchins, D. J., Freed, W. J., and Wyatt, R. J., 1980, The role of cholinergic supersensitivity in the medical symptoms associated with withdrawal of antipsychotic drugs, *Am. J. Psychiatry* **137**:1395–1398.
Lundberg, J. M., and Hokfelt, T., 1983, Coexistence of peptides and classical neurotransmitters, *Trends Neurosci.* **6**:325–333.
Magnussen, I., Dupont, E., Engback, F., and de Fine Olivarius, B., 1978, Posthypoxic intention myoclonus treated with 5-hydroxytryptophan and an extracerebral decarboxylase inhibitor, *Acta Neurolog. Scand.* **57**:289–294.
Marsden, C. D., 1976, Blepharospasm-oromandibular dystonia syndrome (Brueghel's syndrome), *J. Neurol. Neurosurg. Psychiatry* **39**:1204–1209.
Marsden, C. D., 1982, Basal ganglia disease, *Lancet* **2**:21–27.
Marsden, C. D., Tarsy, D., and Baldessarini, R. J., 1975, Spontaneous and drug-induced movement disorders in psychotic patients, in: *Psychiatric Aspects of Neurological Disease* (D. F. Benson and D. Blumer, eds.), pp. 219–265, Grune and Stratton, New York.
Martin, G. E., Williams, M., Pettibone, D. J., Yarbrough, G. G., Clineschmidt, B. V., and Jones, J. H., 1984, Pharmacologic profile of a novel potent direct-acting dopamine agonist, +-4-propyl-9-hydroxynaphthoxazine (+-PHNO), *J. Pharmacol. Exp. Ther.* **229**:569–576.
Martin, J. B., 1984, Huntington's disease: New approaches to an old problem, *Neurology* **34**:1059–1072.
McDowell, F., and Barbeau, A. (eds.), 1974, *Second Canadian-American Conference on Parkinson's Disease, Advances in Neurology Series,* Vol. 5, Raven Press, New York.
Meltzer, H. Y., and Stahl, S. M., 1976, The dopamine hypothesis of schizophrenia: A review, *Schizophr. Bull.* **2**:19–76.
Moersch, F. P., and Woltman, H. W., 1956, Progressive fluctuating rigidity and spasm ("stiff-man" syndrome): Report of a case and some observations in 13 other cases, *Mayo Clin. Proc.* **31**:421–427.
Moore, D. C., and Bowers, M. B., 1980, Identification of a subgroup of tardive dyskinesia patients by pharmacologic probes, *Am. J. Psychiatry* **137**:1202–1205.
Nakajima, T., Kurlima, I., Nakamura, K., and Davidson, A. B., 1983, Interaction of pyrrolois-

oquinoline antipsychotics with a D-2 dopamine receptor subtype evaluation by ^{3}H-R022-1319 binding, *Soc. Neurosci.* **9**:431.

Newman, R. P., LeWitt, P. A., Raphaelson, M. I., Foster, N. L., Calne, D. B., and Chase, T. N., 1982, Dystonia: Therapy with dopaminergic ergot derivatives, *Ann. Neurol.* **12**:82.

Nutt, J. G., Woodward, W. R., Hammerstad, J. P., Carter, J. H., and Anderson, J. L., 1984, the "on-off" phenomenon in Parkinson's disease: Relation to levodopa absorption and transport, *N. Engl. J. Med.* **310**:483–488.

O'Flanagan, P. M., 1975, Clonazepam in the treatment of drug-induced dyskinesia, *Br. Med. J.* **1**:269–270.

Olafson, R. A., Mulden, D. W., and Howard, F. M., 1964, Stiff-man syndrome: A review of the literature, report of three additional cases and discussion of pathophysiology and treatment, *Mayo Clin. Proc.* **39**:131–143.

Olsen, G. L., Cheung, H. C., Morgan, M. D., Blount, J. F., Todaro, L., Berger, L., Davidson, A. B., and Boff, E., 1981, A dopamine model and its application in the design of a new class of rigid pyrrolo[2,3.g] isoquinoline antipsychotics, *J. Med. Chem.* **24**:1026–1034.

Paulson, G. W., 1973, Meige's Syndrome, *Geriatrics* **27**:69–73.

Paulson, G. W., Shaw, G., and Malarkey, W. B., 1979, Huntington's Disease, INH and prolactin levels, *Adv. Neurol.* **23**:797–801.

Perry, T. L., Wright, J. M., Hansen, S., and McLeod, P. M., 1979, Isoniazid therapy of Huntington's disease, *Adv. Neurol.* **23**:785–796.

Quinn, N., Parkes, J. D., and Marsden, C. D., 1984, Control of on/off phenomenon by continuous intravenous infusion of levodopa, *Neurology* **34**:1131–1136.

Ricker, R., and Mertens, H. G., 1968, EMG phenomena in the stiff-man syndrome, *Electroencephalogr. Clin. Neurosphysiol.* **25**:413.

Roberts, E., Chase, T. N., Tower, D. B. (eds.), 1976, *GABA in Nervous System Function,* Raven Press, New York.

Rubovits, R., and Klawans, H. L., 1972, Implications of amphetamine-induced stereotyped behavior as a model for tardive dyskinesia, *Arch. Gen. Psychiatry* **27**:502–507.

Schmidt, R. T., Stahl, S. M., and Spehlmann, R., 1975, A pharmacologic study of the stiff-man syndrome: Correlation of clinical symptoms with urinary 3-methoxy-4-hydroxy-phenyl glycol (MHPG) excretion, *Neurology* **35**:622–626.

Sedman, G., 1976, Clonazepam in the treatment of tardive oral dyskinesia, *Br. Med. J.* **2**:583.

Shoulson, I., Chase, T. N., Roberts, E., and van Balgooy, J. N., 1975, Huntington's disease: Treatment with imidazole-4-acetic acid, *N. Engl. J. Med.* **293**:504–505.

Shoulson, I., Kartzinel, R., and Chase, T. N., 1976, Huntington's disease: Treatment with dipropylacetic acid and gamma-aminobutyric acid, *Neurology* **26**:61–63.

Sovner, R., and Dimascio, A., 1977, The effect of benztropine mesylate in the rabbit syndrome and tardive dyskinesia, *Am. J. Psychiatry* **134**:1301–1302.

Spehlmann, R., and Stahl, S. M., 1976, Dopamine-acetylcholine imbalance in Parkinson's disease: Possible regenerative overgrowth of cholinergic axon terminals, *Lancet* **1**:724–726.

Stahl, S. M., 1980, Tardive Tourette Syndrome in an autistic patient after long-term neuroleptic administration, *Am. J. Psychiatry* **137**:1267–1269.

Stahl, S. M., 1984, Constant drug delivery of dopamine agonists and antagonists: Possible prevention or treatment of tardive dyskinesia and levodopa dyskinesia. Abstracts of the American College of Neuropsychopharmacology Meeting, American College of Neuropsychopharmacy Press, Vanderbilt, Tennessee.

Stahl, S. M., 1985, Akathisia and tardive dyskinesia: Changing concepts, *Arch. Gen. Psychiatry* **42**:915–917.

Stahl, S. M., and Berger, P. A., 1981a, Physostigmine in Tourette Syndrome: evidence for cholinergic underactivity, *Am. J. Psychiatry* **138**:240–242.

Stahl, S. M., and Berger, P. A., 1981b, Bromocriptine in dystonia, *Lancet* **2**:745.

Stahl, S. M., and Berger, P.A., 1982a, Cholinergic and dopaminergic mechanisms in Tourette syndrome, in: *Tourette Syndrome. Advances in Neurology Series,* Vol. 35 (T. N. Chase and A. J. Friedhoff, eds.), pp. 141–150, Raven Press, New York.

Stahl, S. M., and Berger, P. A., 1982b, Neuroleptic effects in Tourette syndrome predict dopamine excess and acetylcholine deficiency, *Biol. Psychiatry* **17**:1047–1053.

Stahl, S. M., and Berger, P. A., 1982c, Bromocriptine, physostigmine and neurotransmitter mechanisms in the dystonias, *Neurology* **32**:889–892.

Stahl, S. M., Layzer, R. B., Aminoff, M. J., Townsend, J. J., and Feldon, S., 1980, Continuous cataplexy in a patient with a midbrain tumor: The limp man syndrome, *Neurology* **30**:1115–1118.

Stahl, S. M., Davis, K. L., and Berger, P. A., 1982a, Neuropharmacology of tardive dyskinesia, spontaneous dyskinesia and other dystonias, *J. Clin. Psychopharmacol.* **2**:321328.

Stahl, S. M., Yesavage, J. A., and Berger, P. A., 1982b, Pharmacologic characteristics of idiopathic orofacial dyskinesia (Meige dystonia): Differentiation from tardive dyskinesia, *J. Clin. Psychiatry* **43**:445–446.

Stahl, S. M., Berger, P. A., Newman, R. P., and LeWitt, P., 1984, Bromocriptine and lisuride in dystonias, *Neurology* **34**:135–136.

Stahl, S. M., Thornton, J. E., Simpson, M. L., Berger, P. A., and Napoliello, M., 1985a, Gamma-vinyl-GABA in tardive dyskinesia and other movement disorders, *Biol. Psychiatry* **20**:888–893.

Stahl, S. M., Faull, K. F., Barchas, J. D., and Berger, P. A., 1985b, Cerebrospinal fluid monoamine metabolites in movement disorders and normal aging, *Arch. Neurol.* **42**:166–169.

Stahl, S. M., Thiemann, S., Faull, K. F., Barchas, J. D., and Berger, P. A., 1985c, Neurochemistry of dopamine in Huntington Dementia and normal aging, *Arch. Gen. Psychiatry*, in press.

Stahl, S. M., Thornton, J. E., and Berger, P. A., 1985d, Gamma-aminobutyric acid and movement disorders: Role of gamma-vinyl-GABA, in: *Clinical and Pharmacological Studies in Psychiatric Disorders* (G. D. Burrows and T. R. Norman, eds.), John Libby, London, in press.

Steck, H., 1954, Le syndrome extrapyramidal et diencephalique au cours des traitments au largactil et au serpasil, *Ann. Med-Psych.* **112**:737–743.

Sweet, R. D., Blumberg, J., Lee, J. M., and McDowell, F. H., 1974, Propranolol treatment of essential tremor, *Neurology* **24**:64–67.

Tamminga, C. A., Crayton, J. W., and Chase, T. N., 1979, Improvement in tardive dyskinesia after muscimol therapy, *Arch. Gen. Psychiatry* **36**:595–598.

Tamminga, C. A., Smith, R. C., and Davis, J. M., 1980, The effects of cholinergic drugs on the involuntary movements of tardive dyskinesia, in: *Tardive Dyskinesia* (W. E. Fann, R. C. Smith, J. M. Davis and E. E. Domino, eds.), pp. 411–418, SP Medical and Scientific Books, New York.

Tamminga, C. A., Thaker, G. K., Ferraro, T. N., and Hare, T. A., 1983, GABA agonist treatment improves tardive dyskinesia, *Lancet* **II**:97–98.

Tanner, C. M., Goetz, C. G., Weiner, W. J., Nausieda, P. A., Wilson, R., and Klawans, H. L., 1979, The role of cholinergic mechanisms in spasmodic torticollis, *Neurology* **29**:604–605.

Tanner, C. M., Goetz, C. G., Glantz, R. H., Nausieda, P. A., Weiner, W. J., and Klawans, H. L., 1981, Cholinergic mechanisms in idiopathic partial and segmented dystonias, *Ann. Neurol.* **10**:94.

Tarsy, D., 1976, Dopamine-acetylcholine interaction in the basal ganglia, in: *Basic and Clinical Aspects of Neurotransmitter Function* (W. S. Fields, ed.), pp. 1–15, S. Karger, Basel.

Tarsy, D., Leopold, N., and Sax, D. S., 1974, Physostigmine in choreiform movement disorders, *Neurology* **24**:28–33.

Tell, G. P., Schecter, P. J., Kock-Weser, J., Cantiniaux, P., Chabannes, J-P., and Lambert, P. A., 1981, Effects of gamma-vinyl GABA, *N. Engl. J. Med.* **305**:581–582.

Tolosa, E. S., and Lai, C., 1979, Meige disease: Striatal dopaminergic preponderance, *Neurology* **29**:1126–1130.

Uhr, S. B., Berger, P. A., Pruitt, B., and Stahl, S. M., 1984, R022-1319, a D-2 receptor antagonist for the treatment of Tourette syndrome, *N. Engl. J. Med.* **311**:989.

Uhr, S. B., Pruitt, B., Berger, P. A., and Stahl, S. M., 1985, Case report of four patients with Tourette syndrome treated with piquindone, a D-2 receptor antagonist, *J. Clin. Psychophamacol.* in press.

Urquhart, J., Fara, J. W., and Willis, K. L., 1984, Rate-controlled delivery systems in drug and hormone research, *Ann. Rev. Pharmacol. Toxicol.* **24**:199–236.

Van Putten, T., 1975, The many faces of akathisia, *Compr. Psychiatry* **16**:43–47.

Van Putten, T., May, P. R. A., and Marder, S. R., 1985, Akathisia with haloperidol and thiothixene, *Arch. Gen. Psychiatry*, in press.

Van Woert, M. H., Rosenbaum, D., Howieson, J., and Bowers, M. B., 1977, Long-term therapy of myoclonus and other neurological disorders with L-5-hydroxytryptophan and carbidopa, *N. Engl. J. Med.* **296**:70–75.

Victor, M., 1966, Treatment of alcoholic intoxication and the withdrawal syndrome. A critical analysis of the use of drugs and other forms of therapy, *Psychosom. Med.* **28**(4, part 2):636–650.

Villaneuve, A., Jus, K., and Jus, A., 1973, Polygraphic studies of tardive dyskinesia and of the rabbit syndrome during different stages of sleep, *Biol. Psychiatry* **6**:259–274.

Washton, A. M., and Resnick, R. B., 1982, Outpatient opiate detoxification with clonidine, *J. Clin. Psychiatry* **436**(sec. 2):39–41.

Washton, A. M., Resnick, R. B., Perzel, J. F., and Garwood, J., 1981, Lofexidine, a clonidine analogue effective in opiate withdrawal, *Lancet* **1**:991–992.

Weiss, K. J., Ciranlo, D. A., and Shader, R. I., 1980, Physostigmine test in the rabbit syndrome and tardive dyskinesia, *Am. J. Psychiatry* **137**:627–628.

Weissberg, M. P., and Friedrich, E. V., 1978, Sydenham's chorea: Case report of a diagnostic dilemma, *Am. J. Psychiatry* **135**:607–609.

Wells, C. E., and Duncan, G. W., 1980, *Neurology for Psychiatrists*, F. A. Davis, Philadelphia.

Yahr, M. D. (ed.), 1973, *The Treatment of Parkinsonism—The Role of DOPA Decarboxylase Inhibitors, Advances in Neurology Series,* Vol. 2, Raven Press, New York.

Young, R. R., and Delwaide, P. J., 1981, Drug therapy of spasticity, *N. Engl. J. Med.* **304**:28–33.

Zilm, D. H., Sellers, E. M., and MacLeod, S. M., 1975, Propranolol effect on tremor in alcohol withdrawal, *Ann. Intern. Med.* **83**:234–236.

PART I

Parkinson's Disease

CHAPTER 1

Parkinsonism

Physiology and Pharmacology

RICHARD P. NEWMAN and DONALD B. CALNE

1. INTRODUCTION

Major advances have been made in the last two decades in understanding the biochemical pharmacology and pathophysiology of Parkinson's disease. The value of levodopa as the mainstay for chronic therapy of parkinsonism is well established. Aspects of the cardinal features of parkinsonism, rigidity, tremor, and bradykinesia, have become better elucidated by physiological analysis. Recent insights into receptor mechanisms provide new approaches to pharmacotherapy.

In considering extrapyramidal function, one profits from an understanding of nigrostrial interconnections (Teravainen and Calne, 1979). Nigral neurons synapse with striatal nuclei via a dopaminergic pathway; the former appear to exert a tonic modulatory influence on the striatum in relation to movement. There is a feedback GABAergic pathway from the striatum which appears to synapse on interneurons in the pars reticulata of the substantia nigra; these transmit inhibitory signals to the dopaminergic cells of the pars compacta. There is also evidence for an additional, independent striatonigral pathway for which substance P is the neurotransmitter or neuromodulator; the nature of its role is not well understood. The major output of the striatum is to the globus pallidus. Motor cortex receives input from the ventral lateral nucleus of the thalamus as the link of diseased extrapyramidal function to pyramidal output.

2. PATHOPHYSIOLOGY

A striking feature of the motor disorder of parkinsonism is bradykinesia, or slowness of movement (Hallett and Khoshbin, 1980). Its components include

RICHARD P. NEWMAN and DONALD B. CALNE • Experimental Therapeutics Branch, National Institute of Neurological and Communicative Disorders and Stroke, National Institutes of Health, Bethesda, Maryland 20205. *Present address of R. P. N.:* 1004 Beverly Drive, Rockledge, Florida 32955. *Present address of D. B. C.:* Health Sciences Center Hospital, Department of Medicine, University of British Columbia, Vancouver, British Columbia V6T1W5, Canada.

prolongation of reaction time in initiating movements, arresting false movements, and changing motor patterns, as well as excessive fatiguing. Normal individuals can move their limbs with a continuous spectrum of velocities; the fastest movements are often called ballistic. It is hypothesized that bradykinesia involves an abnormality of ballistic movements. To generate a correct ballistic movement, subjects must be able to make an accurate perceptual prediction of what size, force, and duration of movement are required, and send a sufficiently precise command signal to the motor system to produce it (Flowers, 1976). The motor system must also respond accurately and consistently enough to the command signal for the movement to be realized as intended. It follows that ballistic movement control becomes largely predictive, in that subjects act with a repertoire of predetermined movement or movement sequences which produce a predictable effect. This implies in part a deficit in perceptual judgment in controlling movement or unreliability in motor programming. It seems that parkinsonian patients must monitor their actions more closely than normals; there is the tendency to perform all actions at a slow and steady pace so that a reasonable degree of control may be maintained. On clinical examination, one notes masked facies, decreased frequency of blinking, and speech that is monotonous, low intensity, and slurred. There is reduction of automatic and synergistic movements (such as arm swing and postural reactions). There is impairment in initiating involuntary movements, as occurs in freezing episodes. In addition, disorders of posture may be found with dominance of the progravity or flexor muscles producing a forward tilt of the head and back from flexion of the hips and knees and adduction of the arms and thighs. Postural stability may be disturbed, the patients being unable to maintain their center of gravity over their feet.

Rigidity, defined as an increase in the resistance to passive muscle stretch, is derived from increased activity of lower motor neurons innervating extrafusal muscle fibers. Rigidity must be differentiated from spasticity, in which hypertonia is dependent on limb displacement and is velocity dependent (Burke and Lance, 1973). In rigidity, there is normal α–γ coactivation but the tone is increased, even in the resting state; it appears that some long-loop reflex pathways are not operative (Andrews *et al.*, 1973). In parkinsonism, rigidity may be dissociated from bradykinesia; it is the latter which is usually responsible for most of the gait disturbance and loss of dexterity.

Tremor is characterized by "involuntary rhythmic oscillations about a position of equilibrium which involve either the whole body or part of it" (Dejerine, 1914). In parkinsonism, there is tremor at rest which occurs in muscle in a relaxed state and usually stops as soon as muscle voluntarily contracts, whether or not contraction brings about movement of a body part (Rondot *et al.*, 1978). In the resting state in parkinsonism, unlike the normal situation, the motor units discharge in a relatively synchronized manner at regular intervals. Further disturbance of the motor unit is evidenced by abnormal patterns during active contraction; synchronization of motor unit potentials is lost, and the firing rate increases beyond normal discharge rates. Parkinsonian tremor manifests itself mainly in the distal extremities, but also occasionally in muscles of mastication and the tongue (Struppler *et al.*, 1978). Electromyography of parkinsonian tremor shows a reciprocal alternating innervation if antagonists are being recorded. If the tremor can be readily suppressed by voluntary activity (such as finger-to-nose testing using the affected limb), then it is a true parkinsonian tremor. If voluntary innervation does not lead to suppression of the tremor, then conditions such as benign essential tremor must be considered in the differential diagnosis. Nevertheless, postural tremor may occur in Parkinson's disease. As tremor is a major feature of parkinsonism, one can suppose that a lesion of the dopaminergic nigrostriatal pathway is responsible. In addition, administration of dopamine via levodopa, its precursor, may ameliorate and even stop the tremor.

It appears that the neurophysiological basis for tremor originates in the striatum (Rondot and Bathien, 1978). This oscillatory activity is transmitted to the globus pallidus and subsequently to the ventral lateral nucleus of the thalamus and from there to the cortex where corticospinal pathways project to the spinal motor neurons. It is known that destruction of pallidal efferent fibers or the ventral lateral nucleus of the thalamus halts parkinsonian tremor. The parkinsonian tremor is characterized by a frequency between 3 and 7 Hz, most often from 4 to 5 Hz. With neuroleptic treatment, the appearance of a parkinsonian syndrome with akinesia and rigidity is often observed. Tremor is also seen and has the characteristics of parkinsonian tremor with a similar frequency of 4–6 Hz (Rondot *et al.*, 1978). This contrasts with normal or accentuated physiological tremor, which is not present at rest and has a higher frequency, 8–12 Hz.

An advance in the quantitative assessment of tremor involves a technique known as triaxial accelerometry (Jankovic and Frost, 1981). The accelerometer is a device which is attached to the anatomical site being studied; it consists of three identical piezoelectric units whose responsive axes are mutually perpendicular such that acceleration in any direction is detected by at least one of the elements. The outputs from the three elements are combined vectorially by computer and displayed as absolute magnitudes of acceleration and accelerational change in three dimensions. It can be easily applied to measure response to medical therapy.

Positron emission tomography (PET) is a new technique which has given us new understanding of some of the metabolic changes in the brain in parkinsonism. Asymmetrical metabolic rates of glucose utilization in the inferior basal ganglia have been demonstrated in hemiparkinsonians; the affected side (i.e., contralateral to the affected limbs) showed increased metabolism. This is consistent with the concept of increased disinhibition resulting from progressive loss of neurons in the substantia nigra (Martin *et al.*, 1984). Other PET studies with radiolabeled fluoro-L-dopa have revealed changes in hemiparkinsonians; the accumulation of precursor was reduced in the striatum contralateral to symptoms. There were also irregularities ipsilaterally; this is consistent with the idea that the physical signs of Parkinson's disease develop relatively late in the natural history of the disease and represent a state of striatal decompensation (Garnett *et al.*, 1984).

3. PHARMACOLOGY

A direct agonist (also referred to as simply an agonist) is a substance that acts directly on postsynaptic receptors, producing effects similar to those of the physiological transmitter. The term "indirect agonist" has been employed in the broad sense as an agent that stimulates neurotransmitter release (e.g., amphetamines), blocks reuptake (e.g., benztropine), or behaves as a precursor (e.g., levodopa). Here the term agonist will be restricted to agents which mimic the normal transmitter at the receptor site. Antagonists are substances which act on synaptic receptor sites to block normal physiological transmitters. Certain substances behave as agonists at low tissue concentrations but as antagonists at higher concentrations. These substances are referred to as "incomplete" or "partial agonists." Many normal transmitters and most artificial agonists have been shown to possess such biphasic properties so here the term agonist is employed for those substances which activate receptor mechanisms in tissue concentrations normally obtained in therapeutic situations. There is evidence that autoreceptors are located at both cell body and presynaptic axonal endings of cells producing dopamine as a specific transmitter; they probably regulate the synthesis, storage, or release of transmitter.

There is general agreement that, as with other neurotransmitters, multiple receptors exist for dopamine and these may be separated by biochemical and/or pharmacological study (Kebabian and Calne, 1979; Creese *et al.*, 1981). Such distinctions may be of pathophysiological and therapeutic significance. For dopamine the major criterion for classification is the presence or absence of linkage to adenylate cyclase. Prototype adenylate cyclase-linked (D1) dopamine receptors are found in rat caudate nucleus (Kebabian and Calne, 1979), bovine parathyroid gland (Brown *et al.*, 1977), and carp retina (Watling *et al.*, 1979). In the striatum in Parkinson's disease, the activity of this cyclase is known to be decreased (Shibuya, 1979). A prototype D2 receptor, which is not linked to adenylate cyclase, is found in the anterior pituitary mammotroph (prolactin-secreting cell) (Kebabian and Calne, 1979), in the intermediate lobe of the pituitary (Cote *et al.*, 1981), and in the striatum (Kebabian and Calne, 1979; Schachter *et al.*, 1980). Apomorphine, dopamine, and the dopaminergic ergots act as D2 receptor agonists in nanomolar concentrations (Kebabian and Calne, 1979; Kebabian and Cote, 1981). In contrast, at the D1 receptor, dopamine is an agonist micromolar concentrations. A benzazepine (SK&F 38393) has been reported to have relatively selective agonist activity at the D1 receptor (Setler *et al.*, 1978). It does not induce stereotypy, emesis, or inhibition of prolactin release. It should prove extremely useful in analyzing dopaminergic mechanisms in both animals and man.

Further understanding of receptor mechanisms has been gained by studying the effects of dopamine antagonists. Metoclopramide, molindone, and sulpiride are relatively selective D2 receptor antagonists, while other phenothiazines and butyrophenones are less selective D1 and D2 antagonists (Kebabian and Calne, 1979; Sulser and Robinson, 1978). At present there are no selective D1 receptor antagonists. *cis*-Flupenthixol, a thioxanthine, preferentially (75–90%) (Cross and Owen, 1980), but not exclusively, binds to D1 receptors.

It is of great interest that dopaminergic mechanisms are involved in both motor function and psychosis. It is clear that parkinsonism and schizophrenia have different anatomical substrates; the former is a manifestation of basal ganglia dysfunction and the latter probably of cortical dysfunction, limbic or otherwise. It is also clear that dopaminergic neurons are found not only in the striatum but in the cortex as well (Moore and Kelly, 1978). When the pharmacological properties and clinical effects of these drugs are evaluated, there is evidence that the prominent deficit in Parkinson's disease, as well as in hyperprolactinemic states, appears to be reduced activation of D2 receptors (Schachter *et al.*, 1980). The importance of the different receptors in schizophrenia is not clear. Attempts have been made to subclassify D2 receptors. Techniques include radioligand binding and inhibition by various agonists and antagonists, effects of guanosine triphosphate, effects of kainic acid and 6-hydroxydopamine lesions of the substantia nigra, and effects of thermal exposure (Sokoloff *et al.*, 1980).

Apomorphine has been in clinical use for almost a century, mainly to induce vomiting (Corsini *et al.*, 1981). Parkinsonian patients have increased tone as a cardinal feature of their illness. Apomorphine was given to such patients before it was known to be a dopamine agonist; it was known to reduce decerebrate rigidity in animals (Burns and Calne, 1981; Schwab *et al.*, 1951). Apomorphine has been shown to improve many of the features of parkinsonism; it reduces akinesia, rigidity, and tremor (Teravainen and Calne, 1979). However, it has been abondoned as an antiparkinson agent because of nephrotoxicity.

N-Propylnorapomorphine is a structural analogue of apomorphine (Menon *et al.*, 1978). It has been shown to be an active antiparkinson agent, effective on oral administration, with significantly less nephrotoxicity (Schwab *et al.*, 1951; Tamminga *et al.*, 1981). However, this drug has not achieved a place in routine therapy (Tolosa *et al.*, 1977).

The significance of ergot, a fungus that infests the rye plant, has undergone a complete transformation over the past several hundred years (Hauth, 1979). What was a notorious poison now provides a host of valuable pharmacological derivatives. The major goups include the ergolines, simple derivatives of the tetracyclic lysergic acid, which are known to have dopaminergic properties (Pieri *et al.*, 1974; Von Hunger *et al.*, 1974); the clavines, ergolines with a methyl or hydroxymethyl group at the 8 position; and the ergopeptines, which have a linkage of the lysergic acid ring to a cyclic peptide (Teravainen and Calne, 1979; Hauth, 1979). Ergot compounds are structurally related not only to dopamine but also the the neurotransmitters serotonin and norepinephrine; substitutents on the ergoline skeleton provide a diversity of biological activities.

Bromocriptine is a semisynthetic ergopeptine whose dopaminergic properties are well established (Corrodi *et al.*, 1973). *In vitro* studies of dopamine-sensitive adenylate cyclase activity show that bromocriptine is completely devoid of stimulating effects; conversely, it antagonizes the stimulating action of dopamine (Markstein, 1981). Bromocriptine inhibits prolactin secretion *in vivo*, displaces [^{3}H]-haloperidol and [^{3}H]-spiroperidol from specific striatal binding sites, and inhibits stimulation-evoked acetycholine release; these findings imply that this agent blocks D1 receptors and acts as an agonist for D2 receptors (Euvrard *et al.*, 1979; Stoof *et al.*, 1979). Bromocriptine has been shown to be effective in the treatment of parkinsonism (Teravainen and Calne, 1979; Lewitt and Calne, 1981). The dosage and administration of bromocriptine are controversial. In one study, patients with mild Parkinson's disease responded to "lower doses" (20–40 mg/day) while "higher doses" (40–90 mg/day) were required in patients with severe deficits (Larsen *et al.*, 1984). Another study concluded that the response rate with low dose and high dose were as good for both *de novo* and levodopa-treated patients (Grimes, 1984). Low dose and slow escalation of bromocriptine dose appears to increase the number of patients tolerating the drug (Teychenne, 1983). Compared with levodopa (with or without carbidopa or benzserazide), bromocriptine causes less dyskinesia and ameliorates early-morning dystonia. There is reduced end-of-dose akinesia and a decrease in the severity and frequency of "wearing-off" reactions seen with long-term levodopa use. When levodopa is used concomitantly with bromocriptine, the dose should be decreased to about half the previous intake. Psychiatric reactions are more common with bromocriptine, and these may persist for up to several weeks after the drug is stopped. There are several types of adverse reactions peculiar to bromocriptine; these include an erythromelalgia-like vasculopathy (Eisler *et al.*, 1981), which is a mononuclear infiltration of dermal vessels, and rarely the following have been suggested to result from use of the drug: pleuropulmonary fibrosis (LeWitt and Calne, 1981), urinary incontinence (Gopinathan and Calne, 1980), distal vasospasm (Duvoisin, 1976), and bleeding from peptic ulcer (LeWitt and Calne, 1981). Concomitant administration of the dopamine antagonist domperidone to prevent nausea and vomiting usually allows more rapid increase in the dosage of bromocriptine and a higher total daily dosage in some patients (Quinn *et al.*, 1981).

Lisuride is an ergoline with dopaminergic activity similar to bromocriptine; it inhibits dopamine-sensitive adenylate cyclase and prolactin release (Spano *et al.*, 1979; Thorner *et al.*, 1981). Thus, it acts as a D1 antagonist and a D2 agonist. Differences in activity, however, occur at other types of receptors including central adrenergic receptors (Cote *et al.*, 1979). Although there is less clinical experience with lisuride, it has been shown to be an effective antiparkinson agent, both alone and in combination with levodopa (Gopinathan *et al.*, 1981). Clinical uses are similar to those of bromocriptine. Lisuride appears to give greatest benefit in ameliorating bradykinesia and rigidity. Central adverse reactions, including delusions and hallucinations, are similar to those seen with levodopa and bromocriptine.

Pergolide is an ergoline with similar D2 agonist properties (Lemberger and Crabtree, 1979; Koller *et al.*, 1980); it inhibits prolactin release, displaces [^{3}H]-spiroperidol from striatal binding sites, and induces stereotypic behavior in animal models of parkinsonism. In contrast to bromocriptine and lisuride, however, pergolide stimulates dopamine-sensitive adenylate cyclase, indicating D1 agonism (Goldstein *et al.*, 1980). Pergolide appears to have a more prolonged effect than other ergot derivatives on prolactin suppression. Clinical experience is still limited, but pergolide has been found to be useful in patients with advanced Parkinson's disease and in wearing-off or on–off phenomena (Lieberman *et al.*, 1981). The psychiatric side effects are similar to those of bromocriptine and lisuride. A similar spectrum of clinical effects was found with pergolide, lisuride, and bromocriptine, despite neurochemical differences (LeWitt *et al.*, 1983). The clinical significance of the D1 agonist properties is not known.

Certain agents exist which have D2 receptor agonist properties but with neither agonist nor antagonist properties at D1 receptors. The drug LY-141865 is a tricyclic derivative of ergot which is of considerable interest because its structure represents a major departure from other dopaminergic ergot derivatives; in particular, it has a three-ring configuration instead of four (Tsuruta *et al.*, 1981). It has not been tested clinically, but it may be valuable as a tool for studying D2 receptors without contaminating the experimental system with D1 effects.

ADTN is a rigid analogue of dopamine and is a potent and selective dopamine receptor agonist (Cannon *et al.*, 1978). It stimulates dopamine-sensitive adenylate cyclase and decreases striatal HVA (probably by autoreceptor stimulation); however, although it suppresses prolactin release, it does not induce stereotypy. Thus, there is clearly D1 agonism; however, there are somewhat conflicting effects in terms of D2 receptor function. Further study may help elucidate the concept of multiple D2 receptor subclasses.

The drug-induced extrapyramidal syndromes are mainly associated with neuroleptic drug therapy. Acute dystonic reactions occur in some patients, often after the first dose or within the first 4 days of treatment (Sovner and DiMascio, 1978). The clinical features of acute drug-induced dystonia are numerous and variable, depending on the muscle groups involved. When the head and face are involved, one may see oculogyric crises, blepharospasm, grimacing, tongue twisting, or tics. Involvement of the neck may produce retrocollis or torticollis. Dysphagia or respiratory distress may result from pharyngeal involvement. Truncal and limb dystonias are also common. Most acute dystonic reactions are self-limited; they are sufficiently dramatic to be confused with meningism, tetanus, or hysteria. There is evidence that such reactions may be due to enhanced dopamine release onto supersensitive postsynaptic receptors; tolerance is observed on repeated neuroleptic administration (Kolbe *et al.*, 1981). It is of interest that increased dopaminergic receptor activity has been reported in the brains of schizophrenics never treated with neuroleptics

(Creese *et al.*, 1981). It has been suggested that dystonia results from impairment of dopaminergic–noradrenergic balance, in which the latter predominates (Korczyn, 1978). In idiopathic torsion dystonia there are elevated levels of dopamine-β-hydroxylase (which converts dopamine to norepinephrine) (Korczyn, 1978). This relative imbalance of activity may be induced by neuroleptics.

Another adverse reaction that may be encountered in neuroleptic treatment is drug-induced parkinsonism, manifested most frequently by tremor, akinesia, and rigidity, alone or in combination (Sovner and DiMascio, 1978). Seborrhea, sialorrhea, dysarthria, and dysphagia may occur but less commonly. The time of onset of drug-induced parkinsonism may vary; although it usually occurs after several weeks of therapy, signs and symptoms may appear within days or even hours if acute loading doses are used. Akinesia may be severe, occasionally taking the form of akinetic mutism or a catatoniclike state.

In addition to the neuroleptics, reserpine and tetrabenazine may produce parkinsonism by depleting dopamine stores. Lithium may produce similar symptoms, possibly by induction of receptor subsensitivity (Creese *et al.*, 1981).

A most disturbing extrapyramidal reaction seen with neuroleptic therapy is tardive dyskinesia. This is a complex syndrome of hyperkinetic involuntary movement, primarily of the tongue, mouth, and face, although other parts of the body can be affected (Baldessarini and Tarsy, 1978; Crane, 1971; Berger and Rexroth, 1980). There may be varying combinations of chorea, athetosis, myoclonus, dystonia, tics, and orofacial dyskinesias, such as blepharospasm, grimacing, chewing, and tongue rolling. The movements are usually coordinated, stereotyped, and arrhythmical, and they disappear during sleep. Axial dystonia and gait disturbance are more common in younger patients and limb movements more conspicuous after the age of 50. In contrast to acute drug-induced dystonia, this persistent form of dyskinesia evolves gradually without producing subjective distress. Tardive dyskinesia has been reported following exposure to virtually all classes of neuroleptic agents, and only rarely with less than 6 months of neuroleptic treatment. The etiology is not clear, but it appears that chronic neuroleptic administration produces a state of denervation supersensitivity of nigrostriatal dopamine receptors (Jeste and Wyatt, 1979). Based on animal experiments and human pharmacological observations, it has been suggested that there may be cholinergic hypofunction in the face of dopaminergic receptor supersensitivity (Creese *et al.*, 1981; Jeste and Wyatt, 1979).

The major challenge it presents is the potential for irreversibility of the syndrome, despite discontinuation of antipsychotic drugs. Following drug withdrawal, the majority of patients remain symptomatic even after 6–24 months (Crane, 1971). It is reported that on continued neuroleptic treatment, however, the movements may actually disappear. Fluctuations in plasma neuroleptic levels, possibly produced by alterations of metabolism during chronic use, may be responsible for the appearance of the syndrome (Nasrallah, 1980).

A wide range of drugs has been used to treat tardive dyskinesia with varying degrees of success. There does not appear to be a universally effective treatment (Berger and Rexroth, 1980; Jeste and Wyatt, 1979). Other hyperkinetic movement disorders must be considered in the differential diagnosis of tardive dyskinesia. Meige syndrome is a spontaneously occurring disorder in which there is bilaterally symmetrical oromandibulofacial dystonia, sometimes associated with tremor and mild torticollis. Patients with Huntington's disease have generalized chorea, usually in association with a dementing disorder. The "rabbit syndrome" is a late-onset neuroleptic-induced syndrome, characterized by arrhythmical involuntary masticatory movements, resembling the chewing movements of a rabbit (Berger and Rexroth, 1980). Unlike tardive dyskinesia, it persists during sleep, responds to anticholinergics, and readily reverses on discontinuation of neuroleptics. Patients with ill-fitting dentures may have abnormal movements of the mouth. There are studies which suggest that a small percentage of patients with tardive dyskinesia actually have spontaneous dyskinesia (Baldessarini and Tarsy, 1978).

Of the neuroleptics, extrapyramidal reactions are more likely to occur with haloperidol and fluphenazine and less likely with thioridazine (Sovner and DiMascio, 1978). The difference may relate, in part, to the anticholinergic properties of these drugs. Haloperidol also has relatively greater potency in antagonizing D2 receptors (those felt to be responsible for motor dysfunction) compared to D1 receptors (Creese *et al.*, 1981). Molindone, another nonphenothiazine, nonbutyrophenone neuroleptic with relatively selective D2 antagonist properties, is also known to produce extrapyramidal reactions (Kebabian and Calne, 1979; Clark *et al.*, 1970).

A major development in the understanding of the pathogenesis of Parkinson's disease occurred when it was discovered that several individuals developed persistent symptoms of parkinsonism (e.g., hypokinesia, resting tremor, postural disturbance) following intravenous self-administration of an illicit preparation containing an analogue of meperidine and a synthetic side product (Langston *et al.*, 1983; Burns *et al.*, 1984). 1-Methyl-4-phenyl-1,2,3,6-tetrahydropteridine (MPTP) was shown to be the responsible neurotoxic agent by reproducing the syndrome in the rhesus monkey (Burns *et al.*, 1983). MPTP selectively and irreversibly damages dopaminergic neurons in the substantia nigra of the rhesus monkey with loss of striatal dopamine content but without damaging other dopaminergic neurons. The occurrence of MPTP-induced parkinsonism leads one to consider a toxic cause of Parkinson's disease; it is unlikely that MPTP is the sole chemical that can produce such a toxic effect (Burns *et al.*, 1984).

The discovery of the opiate peptides, which may function as neurotransmitters in the nervous system, has led to speculation about their roles in brain function. Opiate peptides and receptors are normally found in high concentrations in the basal ganglia (Simantov *et al.*, 1976; Pert *et al.*, 1976). Several lines of evidence suggest that these pathways may be important in parkinsonism and other neurological diseases, particularly those involving motor dysfunction. First, these opioids as well as their receptors are not homogeneously distributed throughout the brain, but are concentrated in striatum and midbrain. This suggests that the opioids may have specific roles as the neurotransmitters or modulators in areas of brain concerned with motor activity and that alterations of their activity may have an effect on motor function. Second, a cataleptic syndrome can be produced in rats by the major tranquilizers, reserpine, and cholinergic agents, all of which produce or exacerbate parkinsonism in man (Segal *et al.,* 1977). Catalepsy in the rat may therefore be an animal model of some features of parkinsonism. The opiate peptide β-endorphin and morphine both may produce catalepsy in rats, but characterized by muscular rigidity rather than hypotonia seen with the major tranquilizers (Izumi *et al.,* 1977; Elliot *et al.,* 1981; Agnoli *et al.,* 1980). The morphine-induced catalepsy is enhanced by striatal lesions and reversed by levodopa, apomorphine, and naloxone.

Although studies with lower doses of naloxone (0.8–2.0 mg) (Price *et al.,* 1979) and higher doses of naltrexone (100 mg) (Nutt *et al.,* 1978) have failed to show any benefit in Parkinson's disease, it has been reported that higher doses of naloxone (8.0 mg) significantly reduce tremor, bradykinesia, and rigidity (Agnoli *et al.,* 1980). In addition, a decrease of metenkephalin (Taquet *et al.,* 1981) and of striatal opiate receptors has been found in parkinsonian brains (Reisine *et al.,* 1979; Larsen, personal communication, 1981).

4. SUMMARY

Parkinsonism is a manifestation of dopaminergic dysfunction in the basal ganglia. There are multiple types of dopamine receptors in the brain, whose activities are likely to be responsible for different aspects of motor and endocrine function. Continued detailed study using specific dopamine receptor agonists and antagonists should help elucidate some of these mechanisms.

Dopaminergic ergots represent a new class of drugs which are of value in the treatment of Parkinson's disease, with fairly specific indications. Drug-induced extrapyramidal reactions resemble spontaneous Parkinson's disease and choreatic and dystonic disorders. The endogenous opiatelike peptides and receptors represent a new area for study. Their significance in parkinsonism remains to be established.

5. REFERENCES

Agnoli, A., Ruggieri, S., Falaschi, P., Mearelli, S., DelRoscio, S., D'Urso, R., and Frajes, G., 1980, Are the enkephalins involved in Parkinson's disease? Clinical and neuroendocrine responses to naloxone administration, in: *Neural Peptides and Neuronal Communication* (E. Costa and M. Trabucchi, eds.), pp. 511–521, Raven Press, New York.

Andrews, C. J., Neilson, P. D., and Lance, J. W., 1973, Comparison of stretch reflexes and shortening reactions in activated normal subjects with those in Parkinson's disease, *J. Neurol. Neurosurg. Psychiatry* **36:**329–333.

Baldessarini, R. J., and Tarsy, D., 1978, Tardive dyskinesia, in: *Psychopharmacology: A Generation of Progress* (M. A. Lipton, A. DiMascio, and K. F. Killam, eds.), pp. 993–1004, Raven Press, New York.

Berger, P. A., and Rexroth, K., 1980, Tardive dyskinesia: Clinical, biological, and pharmacological perspectives, *Schizophrenia Bull.* **6**:102–116.

Brown, E. M., Carroll, R. J., and Aurbach, G. D., 1977, Dopaminergic stimulation of cyclic AMP accumulation and parathyroid hormone release from dispersed bovine parathyroid cells, *Proc. Natl. Acad. Sci. USA* **74**:4210–4213.

Burke, D., and Lance, J. W., 1973, Studies of the reflex effects of primary and secondary spindle endings in spasticity, in: *New Developments in Electromyography and Clinical Neurophysiology,* Vol. 3 (J. E. Desmedt, ed.), pp. 479–495, Karger, Basel.

Burns, R. S., and Calne, D. B., 1981, Treatment of parkinsonism with artificial dopaminomimetics: Pharmacokinetic considerations, in: *Apomorphine and Other Dopaminomimetics, Vol. 2, Clinical Pharmacology* (G. U. Corsini and G. L. Gessa, eds.), pp. 93–106, Raven Press, New York.

Burns, R. S. Chiueh, C. C., Markey, S. P., Ebert, M. H., Jacobowitz, D. M., and Kopin, I. J., 1983, A primate model of parkinsonism: Selective destruction of dopaminergic neurons in the pars compacta of the substantia nigra by *N*-methyl-4-phenyl-1,2,3,6-tetrahydropteridine, *Proc. Natl. Acad. Sci. USA* **80**:4546–4550.

Burns, R. S., Markey, S. P., Phillips, J. M., and Chiueh, C. C., 1984, The neurotoxicity of 1-methyl-4-phenyl-1,2,3,6-tetrahydropteridine in the monkey and man, *Can. J. Neurol. Sci.* **11**:166-168.

Cannon, J. G., Costall, B., Laduron, P. M., Leysen, J. E., and Naylor, R. J., 1978, Effects of some derivatives of 2-aminotetralin on dopamine-sensitive adenylate cyclase and the binding of (3-H)-haloperidol to neuroleptic receptors in the rat striatum, *Biochem. Pharmacol.* **27**:1417–1420.

Clark, M. L., Huber, W. K., Sakata, K., Fowles, D. C., and Serafetinides, E. A., 1970, Molindone in chronic schizophrenia, *Clin. Pharmacol. Ther.* **11**:680–688.

Corrodi, H., Fuxe, K., Hokfelt, T., Lidbrink, P., and Ungerstedt, U., 1973, Effects of ergot drugs on central catecholamine neurons: Evidence for a stimulation of central dopamine neurons. *J. Pharm. Pharmacol.* **25**:409–412.

Corsini, G. U., Piccardi, M. P., Bocchetta, A., Bernardi, F., and Del Zompo, M., 1981, Behavioral effects of apomorphine in man: Dopamine receptor implication, in: *Apomorphine and Other dopaminomimetics,* Vol. 2, *Clinical Pharmacology* (G. U. Corsini and G. L. Gessa, eds.), pp. 13–24, Raven Press, New York.

Cote, T., Munemura, M., and Kebabisn, J., 1979, Lisuride hydrogen maleate: An ergoline with beta-adrenergic antagonist activity, *Eur. J. Pharmacol.* **59**:303–306.

Cote, T. E., Grewe, C. W., and Kebabian, J. W., 1981, Stimulation of a D-2 dopamine receptor in the intermediate lobe of the rat pituitary gland decreases the responsiveness of the beta-adrenoceptor: Biochemical mechanism, *Endocrinology* **108**:420–426.

Crane, G. E., 1971, Persistence of neurological symptoms due to neuroleptic drugs, *Am. J. Psychiat.* **127**:143–146.

Creese, I., Sibley, D. R., Leff, S., and Hamblin, M., 1981, Dopamine receptors: Subtypes, localization, and regulation, *Fed. Proc.* **40**:147–152.

Cross, A. J., and Owen, F., 1980, Characteristics of 3-H-*cis*-flupenthixol binding to calf brain membranes, *Eur. J. Pharmacol.* **65**:341–347.

Dejerine, J., 1914, *Semeiologie des Affections du Systeme Nerveux,* p. 464, Masson, Paris.

Duvoisin, R. C., 1976, Digital vasospasm with bromocriptine, *Lancet,* **2**:204.

Eisler, T., Hall, R. P., Kalavar, K. A. R., and Calne, D. B., 1981, Erythromelalgia-like eruption in parkinsonian patients treated with bromocriptine, *Neurology* **31**:1368–1370.

Elliot, P. N. C., Jenner, P., and Marsden, C. D., 1981, Akinesia, physiologic rest mechanisms, opiates, and the basal ganglia, in: *Research Progress in Parkinson's Disease* (F. C. Rose and R. Capildes, eds.), pp. 98–104, Pittman Medical Ltd., Kent, Great Britain.

Euvrard, C., Premont, J., Oberlander, C., Boissier, J. R., and Bockaert, J., 1979, Is dopamine-sensitive adenylate cyclase involved in regulating the activity of striatal cholinergic neurons? *Naunyn. Schmiedeberg. Arch. Pharmacol.* **309**:241–245.

Flowers, K. A., 1976, Visual 'closed loop' and 'open-loop' characteristics of voluntary movement in patients with parkinsonism and intention tremor, *Brain* **99**:269–310.

Garnett, E. S., Nahmias, C., and Firnau, G., 1984, Central dopaminergic pathways in hemiparkinsonism examined by positron emmision tomography, *Can. J. Neurol. Sci.* **11**:174–179.

Goldstein, M., Lieberman, A., Lew, J. Y., Asano, T., Rosenfeld, M. R., and Makman, M. H., 1980, Interaction of pergolide with central dopaminergic receptors, *Proc. Natl. Acad. Sci. USA* **77**:3725–3728.

Gopinathan, G., and Calne, D. B., 1980, Incontinence of urine with long-term bromocriptine therapy, *Ann. Neurol.* **8**:204.

Gopinathan, G., Teravainen, H., Dambrosia, J. M., Ward, C. D., Sanes, J. N., Stuart, W. K., Evarts, E. V., and Calne, D. B., 1981, Lisuride in parkinsonism, *Neurology* **31**:371–376.

Grimes, J. D., 1984, Bromocriptine in Parkinson's disease: Results obtained with high and low dose therapy, *Can. J. Neurol. Sci.* **11**:225–228.

Hallett, M., and Khoshbin, S., 1980, A physiological mechanism of bradykinesia, *Brain* **103**:301–314.

Hauth, H., 1979, Chemical aspects of ergot derivative with central dopaminergic activity, in: *Dopaminergic Ergots and Motor Function* (K. Fuxe and D. B. Calne, eds.), pp. 23–32, Pergamon Press, Oxford.

Izumi, K., Motomatsu, T., Chretien, M., Butterworth, R. F., Lis, M., Seidah, N., and Barbeau, A., 1977, Beta-endorphin induced akinesia in rats: Effect of apomorphine and alpha-methyl-p-tyrosine and related modifications of dopamine turnover in the basal ganglia, *Life Sci.* **20**:1149–1156.

Jankovic, J., and Frost, J. D., 1981, Quantitative assessment of parkinsonian and essential tremor: Clinical application of triaxial accelerometry, *Neurology* **31**:1235–1240.

Jeste, D. V. and Wyatt, R. J., 1979, In search of treatment for tardive dyskinesia: Review of the literature. *Schizophrenia Bull.* **5**:251–293.

Kebabian, J. W.. and Calne, D. B., 1979, Multiple receptors for dopamine, *Nature* (*London*) **277**:93–96.

Kebabian, J. W. and Cote, T. E., 1981, Dopamine receptors and cyclic AMP: A decade of progress, *Trends Pharmacol. Sci.* **2**:69–71.

Kolbe, H., Clow, A., Jenner, P., and Marsden, C. D., 1981, Neuroleptic-induced acute dystonic reactions may be due to enhanced dopamine release on to supersensitive postsynaptic receptors, *Neurology* **31**:434–439.

Koller, W. L., Weiner, W. J., Diamond, B. I., Nausieda, P. A., and Klawans, H. L., 1980, The pharmacologic evaluation of pergolide mesylate as a potential anti-parkinsonian agent, *Neuropharmacology* **19**:831–837.

Korczyn, A. D., 1978, The pathophysiology of dystonia, *J. Neural Transm.* **42**:245–250.

Langston, J. W., Ballard, P., Tetrud, J. W., and Irwin, I., 1983, Chronic parkinsonism in humans due to a product of meperidine-analog synthesis, *Science* **219**:979–980.

Larsen, T. A., Newman, R., LeWitt, P., and Calne, D. B., 1984, Severity of Parkinson's disease and the dosage of bromocriptine, *Neurology* **34**:795–797.

Lemberger, L. and Crabtree, R. E., 1979, Pharmacologic effects in man of a potent, long acting dopamine receptor agonist, *Science* **205**:1151–1152.

LeWitt, P. A. and Calne, D. B., 1981, Recent advances in the treatment of Parkinson's disease: The role of bromocriptine, *J. Neural Transm.* **51**:175–184.

LeWitt, P. A., Ward, C. D., Larsen, T. A., Raphaelson, M. I., Newman, R. P., Foster, N. L., Dambrosia, J., and Calne, D. B., 1983, Comparison of pergolide and bromocriptine therapy in parkinsonism, *Neurology,* **33**:1009–1014.

Lieberman, A., Goldstein, M., Liebowitz, M., Neophytides, A., Kupersmith, M., Pact, V., and Kleinberg, D., 1981, Treatment of advanced Parkinson disease with pergolide, *Neurology* **31**:675–682.

Markstein, R., 1981, Neurochemical effects of some ergot derivatives: A basis for their antiparkinsonian actions, *J. Neural Transm.* **51**:39–59.

Martin, W. R. W., Beckman, J. H., Calne, D. B., Adam, M. J., Harrop, R., Rogers, J. G., Ruth,

T. J., Sayre, C. I., and Pate, B. D., 1984, Cerebral glucose metabolism in Parkinson's disease, *Can. J. Neurol. Sci.* **11**:169–173.
Menon, M. K., Clark, W. G., and Neurmeyer, J. L., 1978, Comparison of the dopaminergic effects of apomorphine and (-)-*N-n*-propylnorapomorphine, *Eur. J. Pharmacol.* **52**:1–9.
Moore, K. E., and Kelly, P. H., 1978, Biochemical pharmacology of mesolimbic and mesocortical dopaminergic neurons, in: *Psychopharmacology: A Generation of Progress* (M. A. Lipton, A. DiMascio, and K. F. Killam, eds.), pp. 221–234. Raven Press, New York.
Nasrallah, H. A., 1980, Neuroleptic plasma levels and tardive dyskinesia: A possible link? *Schizophrenia Bull.* **6**:4–7.
Nutt, J. G., Rosin, A. J., Eisler, T., Calne, D. B., and Chase, T. N., 1978, Effect of an opiate antagonist on movement disorders, *Arch. Neurol.* **35**:810–811.
Pert, C. B., Kuhar, M. J., and Snyder, S. H., 1976, Opiate receptors: Autoradiographic localization in rat brain, *Proc. Natl. Acad. Sci. USA* **73**:3729–3733.
Pieri, L., Pieri, M., and Haefely, W., 1974, LSD as an agonist of dopamine receptors in the striatum, *Nature (London)* **252**:586–588.
Price, P., Baxter, R. C. H., Parkes, J. D., and Marsden, C. D., 1979, Opiate antagonists and Parkinson's disease, *Arch. Neurol.* **36**:661.
Quinn, N., Illas, A., L'Hermitte, F., and Agid, Y., 1981, Bromocriptine and domperidone in the treatment of Parkinson's disease, *Neurology* **31**:662–667.
Reisine, T. D., Rossor, M., Spokes, E., Iversen, L. L., and Yamamura, H. I., 1979, Alterations in brain opiate receptors in Parkinson's disease, *Brain Res.* **173**:378–382.
Rondot, P., and Bathien, N., 1978, Pathophysiology of parkinsonian tremor, in: *Progress in Clinical Neurophysiology,* Vol. 5, *Physiological Tremor, Pathological Tremors, and Clonus* (J. E. Desmedt, ed.), pp. 138–149, Karger, Basel.
Rondot, P., Jednyak, C. P., and Ferry, G., 1978, Pathological tremors: Nosological correlates, in: *Progress in Clinical Neurophysiology,* Vol. 5, *Physiological Tremor, Pathological Tremors and Clonus* (J. E. Desmedt, ed.), pp. 95–113, Karger, Basel.
Schachter, M., Bedard, P., Debono, A. G., Jenner, P., Marsden, C. D., Price, P., Parkes, J. D., Keenan, J., Smith, B., Rosenthaler, J., Horowski, R., and Dorow, R., 1980, The role of D-1 and D-2 receptors, *Nature (London)* **286**:137–138.
Schwab, R. S., Amados, L. V., and Lettvian, J. U., 1951, Apomorphine in Parkinson's disease, *Trans. Am. Neurol. Assoc.* **76**:251–253.
Segal, D., Browne, R. G., Bloom, F., Ling, N., and Guillemin, R., 1977, Beta-endorphin: Endogenous opiate or neuroleptic? *Science* **198**:411–413.
Setler, P. E., Sarau, H. M., Zirkle, C. L., and Saunders, H. L., 1978, The central effects of a novel dopamine agonist, *Eur. J. Phamacol.* **50**:419–430.
Shibuya, M., 1979, Dopamine-sensitive adenylate cyclase activity in the striatum in Parkinson's disease, *J. Neural Transm.* **44**:287–295.
Simantov, R., Kuhar, M. J., Pasternak, G. W., and Snyder, S. H., 1976, The regional distribution of a morphine-like factor enkephalin in monkey brain, *Brain Res.* **106**:189–197.
Sokoloff, P., Martres, M. P., and Schwartz, J. C., 1980, Three classes of dopamine receptor (D-2, D-3, D-4) identified by binding studies with 3-H-apomorphine and 3-H-domperidone, *Naunyn-Schmiedeberg Arch. Pharmacol.* **315**:89–102.
Sovner, R. and DiMascio, A., 1978, Extrapyramidal syndromes and other neurological side effects of psychotropic drugs, in: *Psychopharmacology: A Generation of Progress* (M. A. Lipton, A. DiMascio, and K. F. Killam, eds.), pp. 1021–1032, Raven Press, New York.
Spano, P. F., Frattola, L., Govoni, S., Tonon, G. C., and Trabucchi, M., 1979, Dopaminergic ergot derivatives: Selective agonists of a new class of dopamine receptors, in: *Dopaminergic Ergot Derivatives and Motor Functions* (K. Fuxe and D. B. Calne, eds.), pp. 159–171, Pergamon Press, Oxford.
Stoof, J. C., Thieme, R. E., Vrijmoed de Vries, M. C., and Mulder, A. H., 1979, In vitro acetylcholine release from rat caudate nucleus as a new model for testing drugs with dopamine receptor activity, *Naunyn-Schmiedeberg Arch. Pharmacol.* **309**:119–124.
Struppler, A., Erbel, F., and Velho, F., 1978, An overview of the pathophysiology of parkinsonian

and other pathological tremors, in: *Progress in Clinical Neurophysiology, Vol. 5, Physiologic Tremor, Pathologic Tremors, and Clonus* (J. E. Desmedt, ed.), pp. 114–128, Karger, Basel.

Sulser, F., and Robinson, S. E., 1978, Clinical implications of pharmacologic differences among antipsychotic drugs with particular emphasis on biochemical central synaptic mechanisms, in: *Psychopharmacology: A Generation of Progress* (M. A. Lipton, A. DiMascio and K. F. Killam, eds.), pp. 943–954, Raven Press, New York.

Tamminga, C. A., DeFraites, E. G., Gotts, M. D., and Chase, T. N., 1981, Apomorphine and N-n-propylnorapomorphine in treatment of schizophrenia, in: *Apomorphine and Other Dopaminomimetics,* Vol. 2, *Clinical Pharmacology* (G. U. Corsini and G. L. Gessa, eds.), pp. 49–55, Raven Press, New York.

Taquet, H., Javoy-Agid, F., Cesselin, F., and Agid, Y., 1981, Methionine–enkephalin deficiency in brains of patients with Parkinson's disease, *Lancet* **1**:1367–1368.

Teravainen, H., and Calne, D. B., 1979, Developments in understanding the physiology and pharmacology of parkinsonism, *Acta Neurol. Scand.* **60**:1–11.

Teychenne, P. F., 1983, Parkinson's disease: The current status of drug treatment: The role of peptides, in: *Neurotransmitters and Neuropeptides, Asia Pacific Congress Series No. 24,* Exerpta Medica.

Thorner, M. O., Perryman, R. L., Cronin, M., Logan, I. S., Chun, J. K., and MacLeod, R. M., 1981, Dopaminergic drugs in the treatment of prolactin disorders, in: *Apomorphine and Other Dopaminomimetics,* Vol. 2, *Clinical Pharmacology* (G. U. Corsini and R. L. Gessa, eds.), pp. 203–213, Raven Press, New York.

Tolosa, E. S., Cotzias, G. C., Burckhardt, P. G., Tang, L. C., and Dahl, K. E., 1977, The dopaminergic and antidopaminergic effects of some aporphines, *Exp. Neurol.* **55**:56–66.

Tsuruta, K., Frey, E. A., Grewe, C. W., Cote, T. E., Eskay, R. L., and Kebabian, J. W., 1981, Evidence that LY-141865 specifically stimulates the D-2 receptor, *Nature* (*London*) **292**:463–469.

Von Hunger, K., Roberts, S., and Hill, D. F., 1974, LSD as an agonist and antagonist at central dopamine receptors, *Nature* (*London*) **252**:588–589.

Watling, K. J., Dowling, J. E., and Iversen, L. L., 1979, Dopamine receptors in the retina may all be linked to adenylate cyclase, *Nature* (*London*) **281**:578–580.

CHAPTER 2

Role of B-Type Monoamine Oxidase Inhibition in the Treatment of Parkinson's Disease

An Update

JOSEPH KNOLL

1. INTRODUCTION

Parkinson's disease (PD), first described in 1817 as paralysis agitans by James Parkinson in his "Essay on the Shaking Pulsy," is characterized by tremor, bradykinesia, rigidity, and a postural defect. The usual distinction between postencephalitic and idiopathic PD is of little practical importance as about 85–90% of the patients suffer from the idiopathic form of the illness.

The onset of PD is extremely rare in the first decades of life, and only about 10% of the patients begin having the disease before the age of 50 years. According to regional prevalence studies (Pollock and Hornabrook, 1966; Martilla, 1974; Dupont 1977) about 0.1% of the population over 40 years of age develop PD. The prevalence increases sharply with age. In a Finnish study (Martilla, 1974), the prevalence rate was found to be 22.7 per 100,000 in the age group 40–44 years, whereas 796.9 per 100,000 had the disease between 70 and 74 years.

Population-based epidemiological studies have shown an even distribution of the disease in the two sexes, and heredity does not seem to play a significant role in the manifestation of the illness.

Several hypotheses (for review see Martilla, 1980), including premature aging, viral infection, immunological abnormality, enzyme defects and toxins, were proposed as putative causes leading to PD, but the fundamental cause of the illness is still unknown. However, it is probably the only prevalent serious

---

JOSEPH KNOLL • Department of Pharmacology, Semmelweis University of Medicine, Budapest, Hungary.

neurological disease the biochemical pathology of which seems to be firmly established by now.

2. NIGROSTRIATAL DOPAMINE DEFICIENCY: THE ESSENTIAL BIOCHEMICAL LESION IN PARKINSON'S DISEASE

Tretiakoff demonstrated in 1919 the loss of pigmentation in the substantia nigra of patients suffering from either idiopathic or postencephalitic PD. This was the first proof of the involvement of this nucleus in the disease, and his findings were corroborated by Hassler in 1938, who demonstrated nerve cell loss in the substantia nigra in fully developed parkinsonian syndromes, independent of their etiology.

The observation that reserpine, which depletes dopamine from the corpus striatum, produced in man a syndrome very similar to parkinsonism was in good agreement with the pioneering assumption of Carlsson, based on the presence of large amounts of dopamine in the striatum (Bertler and Rosengren, 1959), that this amine is involved in the control of motor functions (Carlsson, 1959). The hypothesis was substantially supported by the discovery of Ehringer and Hornykiewicz in 1960 that dopamine concentration was strikingly reduced in the striatum of parkinsonians, which led to the first trials of the effect of levodopa in PD (Birkmayer and Hornykiewicz, 1961) and paved the way for the hitherto most effective medicinal treatment of this illness (for review see Birkmayer and Riederer, 1983; Rinne, 1983).

There is no longer any doubt regarding the rate-limiting role of the nigrostriatal dopaminergic neuron in the control of motor functions. Dopamine cell bodies are dominant in the A9 and A10 areas and the zona compacta of the substantia nigra and produce such large amounts of dopamine that the striatum, in which the neurons terminate, contains the highest amount of dopamine in the brain. The physiological role of dopamine released in the striatum is the continuous inhibition of the release of acetylcholine (ACh) from the cholinergic interneurons of the caudate nucleus.

3. SPECIFIC NEUROTOXIN, 1-METHYL-4-PHENYL-1,2,5,6-TETRAHYDROPYRIDINE-INDUCED PARKINSONIANLIKE CONDITION: AN EXPERIMENTAL MODEL FOR STUDYING THE BASIC CAUSE OF PARKINSON'S DISEASE

During the synthesis with a method described by Ziering *et al.* (1947) of a meperidine analogue, 1-methyl-4-phenyl-4-propionoxy-piperidine (MPPP), a by-product, 1-methyl-4-phenyl-1,2,5,6-tetrahydropyridine (MPTP), admixes with varying amounts of the end material.

A graduate student who produced and injected himself intravenously with MPPP which was contaminated with MPTP developed parkinsonian symptoms.

He died 2 years after the onset of his parkinsonism. Nerve cell degeneration in the substantia nigra, comparable in severity to that usually seen in Parkinson's disease, was detected (Davis *et al.*, 1979).

A few years later MPPP, again contaminated with MPTP, was synthetized in a clandestine laboratory in northern California and sold as "synthetic heroin." Langston *et al.* (1983) detected four patients (one female and three males) who injected amounts of the material corresponding to 160–640 mg/kg MPTP per day intravenously and developed severe parkinsonian symptoms within a week. All patients responded to L-dopa plus carbidopa therapy, and none of them have shown signs of recurrence.

Langston and Ballard (1983) also described the interesting case of a chemist who, starting in 1964, synthesized a total of 3 kg MPTP for a major pharmaceutical company and was hospitalized in 1971, when he was 38 years old, for Parkinson's disease.

Further studies in monkeys by Burns *et al.* (1983) corroborated that MPTP is a specific neurotoxin which causes cell death in the substantia nigra and, as a consequence, loss of dopamine in the striatum.

Some animal species, rat, cat, and guinea pig, were described to be almost insensitive to the neurotoxin (Chieuh *et al.* 1983; Kolata, 1983; Burns *et al.*, 1984), whereas mice, (especially some strains) were found to be responsive to the substance (Heikkila *et al.*, 1984; Hallman *et al.*, 1984). Mytilineou and Cohen (1984) found that in explants of rat embryo mesencephalon MPTP exerted a selective destructive action in dopamine neurons *in vitro*, similar to the action observed in humans and monkeys *in vivo*. Thus, MPTP is a neurotoxin having highly selective affinity for the substantia nigra. The recently observed explosion in the number of papers dealing with MPTP clearly indicates the unusual interest in this new line of research.

High-affinity binding sites for [^{3}H]MPTP which correspond to monoamine oxidase (MAO), of which MPTP is a nonselective inhibitor of modest potency, were described (Parsons and Rainbow, 1984). MPTP is about 350 times less potent in inhibiting MAO-B than (–)deprenyl and about 240 times less potent than clorgyline in inhibiting MAO-A.

It is, however, the peculiar metabolism of MPTP which deserves special attention.

Chiba *et al.* (1984) proposed the hypothesis that the metabolic activation of MPTP by B-type MAO forms electrophilic intermediates which react with nucleophilic functionalities of neuronal macromolecules. MPTP is oxidized by B-type MAO leading to the final stable product, 1-methyl-4-phenylpyridinium (MPP+), which is a highly polar substance. As MPP+ cannot cross the blood–brain barrier, it is ineffective when given parenterally. During the oxidation process, however, a short-lived intermediate, dihydropyridinium ion, is formed, a part of which might not be further oxidized to MPP+, thus forming a powerful oxidation center. This oxidation center is thought to convert dopamine to toxic oxidation products, which kill the cells of the substantia nigra.

The metabolism of dopamine is extremely complicated. Biotransformation

TABLE 1
Effect of Dopamine on the Release of Acetylcholine (ACh) from Isolated Striatal Slices Dissected from Untreated and 6-Hydroxydopamine (6-OHDA) Pretreated Rats[a]

Treatment		*In vitro* concentration of dopamine (mole/liter)	Ouabain (2×10^{-5} mole/liter) induced release of ACh (pmole/g per min)	Significance
In vivo	*In vitro*			
Saline	None	—	366.7 ± 57.3 (5)	—
Saline	Dopamine	2.6×10^{-4}	591.0 ± 52.5 (5)	1:2 <0.02
6-OHDA	None	—	706.0 ± 60.0 (5)	1:3 <0.001
6-OHDA	Dopamine	2.6×10^{-4}	372.9 ± 93.8 (4)	3:4 <0.05

[a] Mean ± SEM, two-tailed *t* statistic, number of experiments in parentheses. The experiments were carried out on the fifth day following intraventricular injection of either saline or 6-OHDA (250 μg in 20 μl).

Levodopa, though it is a symptomatic replacement therapy, revolutionized of dopamine proceeds rapidly to yield the principal excretion products, 3,4-dihydroxyphenyl-acetic acid (DOPAC) and homovanillic acid (HVA), but dozens of complicated oxidation products are also formed in very small amounts. The specific dark pigment, neuromelanin, in the substantia nigra is also a dopamine oxidation product, which, if involved in the formation of MTPT-induced toxic oxidation products, may explain the selectivity of MPTP to the substantia nigra.

It now seems to be a fruitful working hypothesis that environmental factors, of which MPTP is an example, as well as unidentified metabolic errors leading to the formation or accumulation of toxic endogenous dopamine metabolites might play a part in the etiology of Parkinson's disease. On the other hand, it is also conceivable that endogenous toxic dopamine metabolites may be involved in the age-related decline of striatal dopaminergic activity.

4. THE ISOLATED STRIATUM DISSECTED FROM A 6-HYDROXY-DOPAMINE-TREATED RAT: AN EXPERIMENTAL MODEL FOR SCREENING COMPOUNDS WITH POTENTIAL THERAPEUTIC BENEFIT IN PARKINSON'S DISEASE

The physiological consequences of dopaminergic regulation in the caudate nucleus are detectable by measuring the basal and ouabain-stimulated release of ACh of the rat as shown by us in a number of papers (Vizi *et al.*, 1977; Knoll, 1978c; Harsing *et al.*, 1979).

Table 1 shows the substantial increase in the rate of ACh released from isolated striatal slices dissected from 6-hydroxydopamine (6-OHDA) pretreated rats. The ouabain-stimulated release of ACh from isolated striatal slices amounted to 366.7 pmole/g per min, and 706 pmole/g per min ACh was released from slices dissected from 6-OHDA-treated rats. This is clear-cut evidence that the nigrostriatal dopaminergic neurons play a rate-limiting role in controlling the release of ACh in the caudate nucleus.

Dopamine (2.6×10^{-4} M) significantly enhanced the release of ACh from the isolated striatum of untreated rats. This effect proves the importance of the presynaptic dopamine "autoreceptor" situated on dopamine neurons (Carlsson, 1975) in the local feedback control of dopamine release. Via these autoreceptors exogenous dopamine inhibits transmitter release from the terminals of the nigrostriatal dopaminergic neurons; consequently, the release of ACh is increased.

Dopamine, however, strongly inhibits the release of ACh from striatum slices dissected from a rat pretreated with 6-OHDA. Table 1 shows that the highly increased release of ACh from the striatum slices dissected from 6-OHDA-treated rats (706 pmole/g per min) could be suppressed to the normal level by the addition of 2.6×10^{-4} M dopamine. The amount of ACh, 372.9 pmole/g per min, released from the striatum taken from 6-OHDA-treated rats was, in the presence of dopamine, the same as from striatum slices dissected from untreated rats (366.7 pmole/g per min). The strong inhibitory effect of dopamine on ACh release in the striatum of 6-OHDA-treated animals is convincing evidence for the presence of two kinds of dopamine receptor in the caudate nucleus: a presynaptic autoreceptor on the dopaminergic nerve terminals and a postsynaptic dopamine receptor on the cholinergic interneurons. After the chemical lesioning of the nigrostriatal dopaminergic neurons dopamine autoreceptors are absent and exogenous dopamine acts on the postsynaptic dopamine receptors located on the cholinergic caudate interneurons, stimulation of which inhibits the release of ACh.

The situation is similar in PD. Loss of the nigrostriatal dopaminergic neurons leads to an uninhibited release of ACh in the striatum (as in our model), and exogenous dopamine (administered in the form of the precursor levodopa) inhibits the release of ACh.

5. LEVODOPA, THE BEST AVAILABLE TREATMENT OF PD

PD, the cause of which is unknown, is at present an incurable disease, and none of the agents used in treating this illness stops its progression. The different families of drugs used as antiparkinson agents represent different kinds of symptomatic therapy which improve more or less and for shorter or longer duration the quality of life of the patient. This is also true for levodopa, which is now considered to be the best available therapy of PD.

the medication of PD, and alone or in combination with a decarboxylase inhibitor it is now the standard drug throughout the world.

The discovery by Ehringer and Hornykiewicz in 1960 that dopamine deficit in the striatum is the biochemical lesion in PD inspired Birkmayer and Hornykiewicz in 1961 to inject L-dopa, the precursor of dopamine, into parkinsonian patients. In agreement with their expectations, the intravenous injection of 50 mg of levodopa increased enormously the energy of motor function in aged parkinsonians.

Simultaneously Barbeau *et al.* found in 1961 that the excretion of dopamine is decreased in PD, and they were also able to demonstrate the dramatic antiakinetic effect of levodopa in parkinsonian patients.

Within a few years the levodopa therapy was elaborated, and many groups proved, in large-scale studies, the therapeutic benefit of this treatment in akinetic PD (Barbeau, 1969; Birkmayer, 1969; Cotzias *et al.*, 1969; Yahr *et al.*, 1969; Rinne *et al.*, 1970).

The essential problem with levodopa therapy lies in the large number of adverse effects of mainly central origin associated with a progressive deterioration of parkinsonian symptoms during the long-term administration of the drug. Unfortunately, only the initial responses to levodopa in a new patient are excellent, but in the long run the improvement in function following levodopa decreases.

For example, Barbeau (1980) followed 80 severely akinetic patients from November 1967 to August 1979. Initially 19% of these patients were functionally independent in their activity of daily living. After 2 months of levodopa treatment 41% of the patients were independent, which was an important improvement in the quality of life of the patients. After 6 years, however, 24% and after 11 years only 15% of the patients maintained functional independence.

In all studies with long-term administration of levodopa, progressive reduction in the duration of action of a single dose, gradual impairment of mental function, and, in many patients, a final unresponsiveness to levodopa was observed. Probably not only the progression of the disease, but also some long-term effects of levodopa play a role in the described phenomena. Portin and Rinne (1980), who analyzed the nature and intensity of cognitive and emotional disturbances in 79 parkinsonians in an 8 to 10 year study, concluded that during the course of treatment progression of the disease led to a general decline in mental ability. The cognitive functions improved only during the first 2–3 months of levodopa treatment, but this change disappeared after 2–3 years of administration. After 8–10 years of levodopa treatment a deterioration in motor, visuospatial, verbal, and memory processes, serious emotional disturbances, and an overall breakdown in the integrative aspects of brain mechanisms was detected.

The periodic reemergence of various forms of akinesia (wearing-off-effect, on–off phenomenon, parosystic hypotonic freezing) after 3 or more years of continuous levodopa administration is one of the most serious long-term side effects of levodopa treatment. Besides the widely observed oscillations in performance, postural instability, manifested by the high frequency of falls, is another serious adverse effect of long-lasting levodopa therapy.

The long-term side effects that accompany the continuous administration of levodopa initiated the search for adjuvants with levodopa-sparing effect.

6. THE LEVODOPA-SPARING EFFECT OF EXTRACEREBRAL DECARBOXYLASE INHIBITORS

Birkmayer and Mentasti demonstrated in 1967 that N^1-DL-seryl-N^2-2,3,4-trihydroxybenzylhydrazid HCl (Ro-4-4602, later named Benserazide), which it was originally hoped would be an antihypertensive agent, unexpectedly potentiated and increased the duration of the effect of levodopa in parkinsonians. The explanation was soon given by Bartholini *et al.* (1967) as they were able to show that benserazide is a peripheral decarboxylase inhibitor which, when given concurrently with levodopa, increases the catecholamine levels in the brain. Another peripheral decarboxylase inhibitor, carbidopa, acts similarly.

Comparative clinical trials between levodopa alone and in combination with either benserazide (Madopar) or carbidopa (Sinemet) clearly demonstrated the advantages of combined therapy with a decarboxylase inhibitor and levodopa. Inhibition of the decarboxylation of levodopa in peripheral tissues allows a greater proportion of levodopa to reach the brain, and as a consequence, the effective dose of levodopa can be reduced by about 75%. The peripheral side effects of levodopa are largely eliminated. Abnormal involuntary movements, however, tend to develop earlier and in more severe form when levodopa is administered concurrently with a peripheral decarboxylase inhibitor, and mental side effects occur with the same frequency as in patients treated with levodopa alone.

According to Rinne (1980a), who analyzed the concentrations of dopamine and homovanillic acid in postmortem brain samples taken from patients treated with levodopa alone or in combination with extracerebral decarboxylase inhibitors, the combination increased dopamine turnover more selectively in the nigrostriatal system than levodopa alone.

7. THE LEVODOPA-SPARING EFFECT OF (−)DEPRENYL, SELECTIVE INHIBITOR OF B-TYPE MONOAMINE OXIDASE

It is common knowledge that the physiological functions of mitochondrial MAO are

1. Intraneuronal inactivation of transmitter in noradrenergic, dopaminergic, and serotonergic neurons.
2. Gastrointestinal inactivation of pressor amines present in foodstuffs.

3. Inactivation of a large variety of endogenously formed and exogenous amines in liver and blood vessels.

Accordingly, the consequences of MAO inhibition are

1. Increase of transmitters to new steady-state levels in catecholaminergic and serotonergic neurons.
2. Facilitated access of pressor amines into circulation ("cheese effect").
3. Increased levels of endogenous "trace amines" (e.g., phenylethylamine, octopamine, tryptamine). Endogenously formed pressor amines as well as those absorbed from the intestines (mainly tyramine) circulate and reach the organs without control.
4. The pharmacological effects of the precursors of transmitters which are substrates to MAO are strongly potentiated and the duration of their effect is significantly prolonged (for review see Knoll, 1980, 1983).

The pharmacological effects of levodopa, the precursor of dopamine, which is also oxidized by mitochondrial MAO, are known to be strongly potentiated and prolonged in duration in animals pretreated with a potent MAO inhibitor.

It was reasonable to expect a potentiation and prolongation of the effect of levodopa in parkinsonians with the concurrent administration of MAO inhibitors. The attempt was made by Birkmayer and Hornykiewicz in 1962, who tried a number of irreversible MAO inhibitors in combination with levodopa and were able to demonstrate the potentiation of the antiakinetic effect of levodopa by the enzyme inhibitors, but distressing side effects, including strongly increased involuntary movements, hypertensive reactions, and toxic delirium, terminated their further work in this line. A later careful clinical study of levodopa in combination with MAO inhibitors also demonstrated the serious adverse effects of this regimen (Hunter *et al.*, 1970). There was agreement that to give MAO inhibitors concurrently with levodopa is contraindicated.

The conclusion, however, was challenged by the development of (−)deprenyl (Jumex, Eldepryl), a highly potent irreversible inhibitor of MAO, which could be safely administered with levodopa without distressing side effects.

The unique behavior of (−)deprenyl lies in its peculiar spectrum of pharmacological activity.

7.1. (−)Deprenyl, the MAO Inhibitor without the Cheese Effect

(−)Deprenyl, a highly potent, irreversible inhibitor of MAO, was developed in 1964 as a new-spectrum psychic energizer by Knoll *et al.* (1964, 1965).

In contrast to the MAO inhibitors in medicinal use, such as tranylcypromine, phenelzine, pargyline and isocarboxazide, which strongly potentiate the effect of tyramine, (−)deprenyl was found to inhibit the uptake of tyramine and was predicted to be an MAO inhibitor without the "cheese effect" in man (Knoll *et al.*, 1967, 1968). Clinical studies confirmed this claim, and (−)deprenyl is now considered to be a safe MAO inhibitor for human use (for review see Knoll, 1978c, 1979a, 1980, 1983; Birkmayer and Riederer, 1983; Reynolds *et al.*, 1978).

It is well known that the cheese effect is the most serious side reaction of the MAO inhibitors in clinical use. Soon after the introduction of MAO inhibitors as antidepressants, serious hypertensive crises, similar to a paroxysm induced by pheochromocytoma, leading in certain instances to fatal intracranial bleeding, were observed (for review see Ayd and Blackwell, 1970). Potentiation of the pressor effect of tyramine is likely to be the main cause of dangerous hypertensive reactions which supervene after the intake of certain food materials containing high amounts of free amine (e.g., cheeses, yeast products, beans, Chianti wines, pickled herring, and chicken liver) in patients treated with MAO inhibitors. This cheese reaction seriously discredited the MAO inhibitors and restricted their therapeutic use, which requires careful medical control. It is now the general view that "the potential toxic effects of the MAO inhibitors are greater and more serious than those of any other group of psychotherapeutic agents" (Goodman and Gilman, 1980, p. 429). The cheese effect, first described by Blackwell (1963), is thought to be primarily a consequence of inhibition of the intestinal enzyme (Blackwell *et al.*, 1967).

(−)Deprenyl is devoid of this adverse effect for two reasons: (1) it inhibits the uptake of tyramine (Knoll *et al.*, 1967, 1968; Knoll, 1978a), noradrenaline (Knoll and Magyar, 1972), and dopamine (Knoll, 1978c; Harsing *et al.*, 1979), and (2) as a selective inhibitor of MAO-B (Knoll and Magyar, 1972) it leaves (within the therapeutic dose range) the MAO activity of the intestine practically unchanged (for review see Knoll, 1980, 1983).

7.2. The High Selectivity of (−)Deprenyl for Inhibiting B-Type MAO *in Vitro* and *in Vivo*

The discovery that two main forms of mitochondrial MAO exist and the development of our present knowledge concerning the dual nature of MAO are inseparable from the introduction of two substrate-selective, highly potent, irreversible inhibitors, deprenyl (Knoll *et al.*, 1965) and clorgyline (Johnston, 1968).

Clorgyline (2,4-dichlorophenoxypropyl-*N*-methylpropargylamine), a compound with structural similarities to deprenyl, was found by Johnston to be a selective inhibitor of that type of MAO which deaminates 5-HT. To distinguish the two forms of MAO, one highly sensitive to clorgyline and one relatively insensitive to it, he introduced the terms "type A" and "type B" MAO. This nomenclature has become widely accepted; MAO-A is selectively inhibited by clorgyline, and MAO-B by (−)deprenyl (Knoll and Magyar, 1972).

MAO-A is thought to be specialized for binding and metabolizing the ethylamine side chain of the substrate if it is attached to a 5-hydroxyindole ring (serotonin oxidase), and MAO-B is specialized for recognizing and metabolizing phenylethylamine, PEA (phenylethylamine oxidase). There are many other amines that, because of structural similarities, are substrates of either MAO-A or MAO-B or are common substrates of both enzymes (for review see Knoll, 1976, 1978b, 1980; Knoll and Magyar, 1972).

The high selectivity of (−)deprenyl to MAO-B *in vivo* was demonstrated in the cat by a special method developed for continuous monitoring of liver MAO-B activity (Knoll, 1976). The metabolism of intravenously injected PEA was continuously monitored in anesthetized cats using the nictitating membrane as detector. MAO-B in the liver was found to control the serum concentration of the injected amine. Neither the pretreatment of a cat with 10 mg/kg clorgyline nor the intravenous injection of high doses (5–10 mg/kg) of this MAO inhibitor changed the metabolism of intravenously injected PEA in the cat, as did 0.1 mg/kg of (−)deprenyl. Another selective inhibitor of MAO-A (Lilly 51641) was also found to be ineffective in the nictitating membrane test (Knoll, 1979b).

As the intestinal tissue contains, in many species, including man, mainly MAO-A (Squires, 1972), the finding that clorgyline was 60 times more potent than (−)deprenyl in blocking the oxidation of tyramine in intestines of the rat *in vitro* (Knoll, 1978a) and has a similar difference in man (Squires, 1972) is of practical importance. It means that the main barrier controlling the access of tyramine to the circulation by effecting its degradation is practically unaffected in (−)deprenyl-treated subjects but is blocked in clorgyline-treated ones. Thus, the selective MAO-A inhibitors have no advantage over nonselective MAO inhibitors (such as tranylcypromine, phenelzine, and isocarboxazide), but the selective inhibition of MAO-B is more promising with respect to the hazards of combination with a variety of foods and drugs.

(−)Deprenyl has a higher affinity to the B form of MAO than to the A form (Knoll and Magyar, 1972). In high concentration, or when the enzyme is exposed to the inhibitor for a long time, the A form is also inhibited. For this reason the question must be considered whether (−)deprenyl loses its selectivity to the B form when patients are treated daily with the drug. We found that the selectivity of 0.25 mg/kg (−)deprenyl when rats were treated with daily doses for 21 days was the same as that in single-dose experiments (Ekstedt *et al.*, 1979). The uptake of deprenyl in brain is a fast procedure reaching the peak concentration within seconds after an intravenous dose, and almost all of the inhibitor is eliminated from the tissue after 20–30 min (Magyar *et al.*, 1971). Thus, it seems likely that the enzyme is not exposed to the inhibitor in brain for a long enough time and at a high enough concentration to establish irreversible binding to the A form if the dose of (−)deprenyl is not higher than 0.25 mg/kg. At higher doses, however, (−)deprenyl reaches a concentration in the brain high enough to inhibit a portion of the A form too. The inhibition of both the A and B forms is higher in the brain than in the liver, which may also be explained by the fast penetration of (−)deprenyl in the brain, rendering a higher concentration in this tissue in the initial phase of its distribution.

If (−) deprenyl was given in one single dose instead of in daily doses, but with the same total amount of the inhibitor, a higher degree of inhibition occurred, which means that a large proportion of both the A and B forms has been resynthesized during the period of treatment. No difference in the degree of inhibition with 0.25 mg/kg (−)deprenyl occurred when rats were treated for 14 to 21 days compared to the treatment for 7 days, which means that the daily synthesis of the enzyme was as high as the inhibition after each dose. The major portion of the inhibited activity was restored 1 week after treatment with three weekly injections of (−)deprenyl (Ekstedt *et al.*, 1979), which shows a great fluctuation of the enzyme activity when weekly doses are given instead of daily doses.

As (−)deprenyl is usually administered in daily oral doses of 5–10 mg (for review see Birkmayer and Riederer, 1983), the dose range in man seems to be in accordance with the selective dose range of the drug in animal studies. It is worth mentioning that (−)deprenyl inhibits MAO-B activity with a very good safety margin. Only 0.17–0.26% of the LD_{50} was needed in different species (mouse, rat, cat, dog) to block completely MAO-B activity in the brain (see Knoll, 1978c, Table IV.).

The action of low doses of (−)deprenyl on the selective inhibition of MAO-B in human brain was demonstrated by Riederer *et al.* (1978). They compared the sensitivity of human brain mitochondrial MAO to (−)deprenyl and clorgyline in 15 brain areas *in vitro*. (−)Deprenyl was found to be many thousand times more powerful an inhibitor than clorgyline. For (−)deprenyl the ID_{50} varied between 0.05 μM (hypothalamus) and 0.95 μM (caudate nucleus); values for clorgyline varied between 500 μM (pineal gland) and 3200 μM (raphe plus reticular formation). They also demonstrated that the deamination of dopamine was almost completely inhibited in the brain of seven patients with Parkinson's disease treated with 10 mg of (−)deprenyl daily, 6.0 ± 1.8 days before death, while 34% of MAO activity against 5-HT still remained present.

7.3. Inhibition of the Uptake of Monoamines *in Vitro* and *in Vivo* by (−)Deprenyl

(−)Deprenyl inhibits the uptake of monoamines into the nerve endings of catecholaminergic neurons, and the effect is independent from the MAO-inhibiting property of the compound.

That deprenyl is a potent inhibitor of the noradrenaline-releasing effect of tyramine in smooth muscle *in vivo* and *in vitro* was first demonstrated by Knoll *et al.* in 1967 and studied in detail in a number of papers (Knoll *et al.*, 1968; Knoll, 1976, 1978a, 1978b, 1978c; Knoll and Magyar, 1972). (−)Deprenyl proved to be highly efficient in blocking the uptake of tyramine into the noradrenaline nerve terminals in different isolated smooth muscle tests (e.g., nictitating membrane of the cat, central ear artery and main pulmonal artery strip of the rabbit, rat vas deferens).

Studies with structural relatives of deprenyl in these tests revealed that inhibition of MAO and the effect on the uptake of monoamines are completely independent from each other. For example, TZ-650, which differs from deprenyl only in the lack of a methyl group at the α carbon and is as potent as

TABLE 2
Inhibitory Effect of (−)Deprenyl on the Release of Acetylcholine (ACh) from Isolated Striatal Slices Dissected from Untreated or 6-Hydroxydopamine (6-OHDA) Pretreated Rats[a]

Treatment *in vivo*	*In vitro* concentration of (−)deprenyl (mole/liter)	Ouabain-stimulated release of ACh (pmole/g per min) ± SEM	Significance
None	None	481.05 ± 83.25 (4)	—
None	9.0×10^{-5}	243.27 ± 54.66 (4)	$P < 0.05$
None	2.2×10^{-4}	226.10 ± 26.84 (6)	$P < 0.01$
None	4.5×10^{-4}	122.02 ± 20.34 (4)	$P < 0.01$
None	9.0×10^{-4}	36.46 ± 7.37 (4)	$P < 0.01$
6-OHDA	None	777.20 ± 91.04 (4)	—
6-OHDA	2.2×10^{-4}	280.68 ± 38.20 (4)	$P < 0.01$

[a] Mean ± SEM, two-tailed *t* statistic, number of experiments in parentheses. The experiments were carried out on the fifth day following intraventricular injection of either saline or 6-OHDA (250 μg in 20 μl).

deprenyl in inhibiting MAO-B, was found to be, in contrast to deprenyl, a potent releaser of noradrenaline in vascular smooth muscle and this effect could be blocked by deprenyl (Knoll, 1978a). Thus, (−)deprenyl, in contrast to the non-selective and A-selective MAO inhibitors as well as to many selective MAO-B inhibitors, is unique in its ability to inhibit tyramine uptake in different tests.

Deprenyl was found to inhibit the uptake of noradrenaline in neural tissue. This was first demonstrated by Knoll and Magyar in 1972, using mouse cortical slices, and corroborated in studies performed with rat cortex (Braestrup *et al.*, 1975). Direct evidence was provided by Simpson (1978) that (−)deprenyl inhibits the uptake of noradrenaline also in heart tissue.

(−)Deprenyl was found to be an inhibitor of the uptake of dopamine in isolated striatum slices of the rat (for details see Knoll, 1978c; Harsing *et al.*, 1979). Table 2 shows the dose-dependent inhibitory effect of (−)deprenyl on the ouabain-stimulated release of ACh from isolated striatal slices dissected from untreated or 6-OHDA-pretreated rats. Table 3 demonstrates that a single subcutaneous dose of 0.25 mg/kg of (−)deprenyl, which inhibits B-type MAO in the rat brain irreversibly, leaves the uptake of dopamine in the striatum unaltered 24 hr after the injection of the drug, whereas the subcutaneous administration of the same dose of (−)deprenyl for 14 days inhibits the uptake of dopamine in the striatum significantly, and this condition can be demonstrated in the tissue taken 24 hr after the last injection of (−)deprenyl.

7.4. Effect of (−)Deprenyl on the Brain Content and Turnover Rate of Catecholamines

To better understand the mechanism of the effect of (−)deprenyl on the catecholaminergic system, we have measured the content and turnover rate of

TABLE 3
The Uptake of Dopamine in Striatal Preparations Taken from Rats Pretreated with Saline or (−)Deprenyl[a]

	$[^3H]$dopamine taken up (pmole/mg protein)
Saline	430 ± 66
(−)Deprenyl, 0.25 mg/kg, s.c.	469 ± 45
Saline	462 ± 32
(−)Deprenyl, 0.25 mg/kg, s.c. daily for 14 days	359 ± 31[b]

[a] Striatum taken 24 hr after the last injection. For methodological details see Zsilla *et al.* (1986).
[b] $p < 0.05$.

dopamine in the striatum and noradrenaline in the brain stem of the rat under the influence of the subcutaneous injection of 0.25 mg/kg (−)deprenyl. We found that 24 hr after the injection of a single dose of 0.25 mg/kg (−)deprenyl, which is known to inhibit MAO-B activity selectively in the brain, the contents and turnover rates of the catecholamines remained unchanged. We were able, however, to detect important changes in the metabolism of monoamines after daily subcutaneous injections of 0.25 mg/kg (−)deprenyl for 14 days (Zsilla and Knoll, 1982).

Table 4 shows that the repeated administration of small daily doses of (−)deprenyl significantly increased the turnover rate of dopamine in the rat striatum, whereas no significant change in the teldiencephalon striatum was to be detected, as measured 24 hr after the last injection. The increase in the turnover rate of dopamine in the striatum was due to the enhancement of the fractional rate constant of dopamine efflux and the slight, but significant, increase in the dopamine content (60.3 versus 52.7 nmoles/g). With regard to noradrenaline, a significant decrease in the turnover rate and unchanged level of this amine in the brain stem was found. No change in the turnover rate of noradrenaline was detected in the teldiencephalon striatum (Table 5).

This finding substantially supports our view (Knoll, 1981) that (−)deprenyl facilitates dopaminergic modulation in the brain.

An increase in the turnover rate of dopamine due to a change in the efflux rate of this amine from its storage is in keeping with the possibility that (−)deprenyl, which increases the content of dopamine by inhibiting MAO-B, also increases the rate of the utilization of this important modulator in the striatum. Presumably, an increase in the rate of utilization is due to an increase in the firing rate of dopaminergic neurons. As, in contrast to the rat, MAO-B seems to be the main form of the enzyme in the nigrostriatal dopaminergic neurons

TABLE 4
The Effect of (−)Deprenyl on Dopamine (DA) Content, Turnover Rate of Dopamine (TR_{DA}), and Fractional Rate Constant (k_b) of Dopamine Efflux[a]

Striatum	DA (nmoles/g)	TR_{DA} (nmoles/g per hr)	k_b (per hr)
Control (saline)	52.7 ± 1.6	13.7 ± 1.3	0.26
Deprenyl, 14 × 0.25 mg/kg	60.3 ± 2.2[b]	20.4 ± 0.98*	0.34
Teldiencephalon striatum			
Control (saline)	2.8 ± 0.1	1.3 ± 0.05	0.45
Deprenyl 14 × 0.25 mg/kg	33.5 ± 0.2[b]	1.6 ± 0.09	0.46

[a] Animals were injected subcutaneously with 0.25 mg/kg (−)deprenyl daily for 14 days. Controls were treated with saline. Animals were killed 24 hr after the last injection of (−)deprenyl and saline, respectively. Rats were injected with 250 mg/kg i.p., α-methyl-*p*-tyrosine methyl ester HCl 1 and 2 hr before decapitation. Each time point had six values. TR = (steady state) × k_b. Brains were dissected according to Glowinski and Iversen (1966). Catecholamine contents were determined fluorimetrically according to Carlsson and Waldeck (1958). Turnover rates were measured according to Tozer *et al.* (1966).
[b] $p < 0.05$.

in man (Glover *et al.*, 1977), the conditions for (−)deprenyl to facilitate dopaminergic modulation in human brain are particularly favorable.

The peculiar effect of (−)deprenyl on the turnover rate of dopamine in the rat striatum is due to its specific spectrum of pharmacological activity. This is clearly shown by comparing the effect of (−)deprenyl to that of clorgyline, the highly selective inhibitor of A-type MAO. Table 6 shows that in contrast to (−)deprenyl, the daily subcutaneous administration of 0.25 mg/kg clorgyline significantly decreased the turnover rate of dopamine. As shown in Table 7,

TABLE 5
The Effect of (−)Deprenyl on Noradrenaline (NA) Content, Turnover Rate of Noradrenaline (TR_{NA}), and Fractional Rate Constant (k_b) of Noradrenaline Efflux[a]

Brainstem	NA (nmoles/g)	TR_{NA} (nmoles/g per hr)	k_b (per hr)
Control (saline)	3.1 ± 0.2	0.81	0.26
Deprenyl, 14 × 0.25 mg/kg	3.2 ± 0.1	0.39	0.12
Teldiencephalon striatum			
Control (saline)	1.2 ± 0.1	0.30	0.26
Deprenyl, 14 × 0.25 mg/kg	1.5 ± 0.1[b]	0.36	0.24

[a] For details see Table 4.
[b] $p < 0.05$.

TABLE 6
The Effect of Clorgyline on Dopamine (DA) Content, Turnover Rate of Dopamine (TR_{DA}), and Fractional Rate Constant (k_b) of Dopamine Efflux[a]

Striatum	DA (nmoles/g)	TR_{DA} (nmoles/g per hr)	k_b (per hr)
Control (saline)	63.9 ± 2.9	19.3 ± 0.9	0.30
Clorgyline, 14 × 0.25 mg/kg	81.6 ± 4.1[b]	6.4 ± 0.3[b]	0.08
Teldiencephalon striatum			
Control (saline)	3.7 ± 0.5	1.1 ± 0.1	0.30
Clorgyline, 14 × 0.25 mg/kg	4.6 ± 0.4	0.5 ± 0.04[b]	0.11

[a] For details see Table 4.
[b] $p < 0.05$.

clorgyline leaves the activity of B-type MAO in the brain unchanged, whereas A-type MAO activity is almost completely blocked.

We have to consider here that in the rat A-type MAO is the main form of the enzyme in the nigrostriatal dopaminergic neuron, whereas the glial enzyme is mainly B-type (for review see Knoll, 1980). Earlier we suggested that (−)deprenyl might influence the fate of dopamine in the striatum by inhibiting glial MAO (Knoll, 1980). The nigrostriatal dopaminergic neuron ends in the striatum with free terminals; thus, the dopamine molecules continuously emitted from these terminals have to diffuse and travel long distances to reach their target, the cholinergic interneurons, in the caudate nucleus. Dopamine here plays the role of the main physiological inhibitor of acetylcholine release. We may thus assume that the blockade of glial B-type MAO by (−)deprenyl acts as an endogenous dopamine-sparing mechanism, which in combination with the inhibition of the reuptake of dopamine creates appropriate conditions for the facilitation of dopaminergic tone in the striatum.

TABLE 7
The Inhibition in Percent of MAO-A and MAO-B Activity in Brain Mitochondria Prepared from Rats Treated with Daily Subcutaneous Injections of 0.25 mg/kg (−)Deprenyl or Clorgyline, respectively, for 14 Days[a]

	MAO-A substrate: 5HT	MAO-B substrate: PEA
(−)Deprenyl	21.5 ± 1.0	75 ± 4.0
Clorgyline	90 ± 0.8	5.3 ± 0.03

[a] For methodological details see Zsilla *et al.* (1986).

7.5. The Effect of (−)Deprenyl on the Release of Catecholamines from Nerve Terminals and the Potential Releasing Effect of the Metabolites of (−)Deprenyl *in Vivo*

We studied first the effect of (−)deprenyl on the release of noradrenaline in the microsomal fraction of rat heart homogenate (Knoll and Magyar, 1972). We were able to demonstrate that in contrast to MAO inhibitors in clinical use (nialamide, tranylcypromine, pargyline), which facilitated the outflow of the transmitter from the nerve terminals, deprenyl strongly inhibited the efflux of labeled noradrenaline.

We later found that in the rat heart preparation clorgyline, the selective inhibitor of MAO-A, enhanced the release of noradrenaline (Knoll, 1976). Clorgyline enhanced the release of dopamine, too, from the synaptosomes of the striatum and acted in this test like tyramine, whereas (−)deprenyl left the efflux of labeled dopamine unchanged (Knoll, 1978c).

We concluded that (−)deprenyl *per se* is devoid of catecholamine-releasing effect; it may even, as in the heart tissue, inhibit the efflux of noradrenaline from the nerve terminals. The *in vivo* effect of (−)deprenyl on the release of catecholamines should be noted, as the propargyl group seems to be split off in the liver, and the main metabolites of (−)deprenyl are amphetamine and methamphetamine (Reynolds *et al.*, 1978), although with high probability these are the pharmacologically less active levorotatory forms of these compounds.

As we usually measure the effects of (−)deprenyl 24 hr after the injection of the drug, only traces of either the unchanged drug molecule or the metabolites can be present in the organism, as the elimination of (−)deprenyl and its metabolites is practically complete within a day (Magyar *et al.*, 1971). The possibility, however, of the retention of trace amounts of the metabolites of (−)deprenyl in brain tissue, which might still exert a small, continuous release of dopamine and noradrenaline, respectively, from the nerve terminals, cannot be ruled out completely. The assumption of a certain share of traces of amphetaminelike metabolites of (−)deprenyl in the complex pharmacological effects of the drug (Karoum *et al.*, 1982) is not necessarily inconsistent with the congruent experimental and clinical observations of the complete lack of amphetamine like symptoms during long-term administration of the usual daily doses of (−)deprenyl. It must be borne in mind, however, that even if we do not rule out the possibility of the accumulation of traces of amphetaminelike metabolites of (−)deprenyl and also take their releasing effect into account, the well-known accumulation of PEA, an endogenous trace amine with higher catecholamine-releasing potency than (−)amphetamine in the (−)deprenyl-treated animal or man, will evidently surpass the releasing effect of the respectable traces of (−)deprenyl metabolites.

As the facilitation of the dopaminergic modulation in the brain by (−)deprenyl seems to be of high therapeutic importance, we checked the possible share of the accumulation of amphetaminelike metabolites in this effect.

As was demonstrated in Table 4, the daily administration of 0.25 mg/kg

(−)deprenyl for 14 days significantly increased the turnover rate of dopamine in the rat striatum. Therefore, we checked also the effect of (±)amphetamine on the striatal dopaminergic system, by injecting 0.25 mg/kg of this drug daily for 14 days (Zsilla *et al.*, 1983). In contrast to (−)deprenyl, amphetamine significantly decreased the content of dopamine in the striatum, and even the turnover rate showed a decreasing tendency. (±)Amphetamine influenced the turnover rate of noradrenaline in the brain stem like (−)deprenyl. Thus, we may conclude that the metabolites of (−)deprenyl cannot play a substantial role in facilitation of the dopaminergic modulation in the brain induced by this drug.

Our main conclusion that (−)deprenyl facilitates the activity of the dopaminergic neuron to physiological and pharmacological stimuli was further proved by direct measurement of dopamine and DOPAC from the striatum (Kerecsen *et al.*, 1985). Reverse-phase high-performance liquid chromatography with electrochemical detection is a specific and sensitive analytical method for measuring a few hundred pmoles of catecholamines in samples of biological origin (for review see Kissinger *et al.*, 1981). Using this method, we were able to find further proof of the peculiar enhancement of striatal dopaminergic activity in rats treated with (−)deprenyl.

Striata, removed according to Glowinski and Iversen (1966), were either halved into two parts and soaked in Krebs solution or cut into 0.3 to 0.5 mm thin slices and superfused. DA and DOPAC efflux from the resting tissue was first measured, and then the efflux following 20 mmole/liter KCl stimulation was estimated.

The thin slice released 44.5 ± 7.6 pmole/g per min DA and 56.3 ± 13.1 pmole/g per min DOPAC as resting value, which changed to 115.0 ± 9.1 DA and 55.4 DOPAC, respectively, following KCl stimulation. The lack of increase in DOPAC level after stimulation speaks in favor of the rapid escape of DA from the thin slice. In agreement with this review, the bulky Glowinsky–Iversen preparation released 91.1 ± 7.2 pmole/g per min DA and 258.3 ± 18.5 pmole/g per min DOPAC under resting conditions, and these values changed to 200 ± 25.8 and 291.5 ± 29.5, respectively, following KCl stimulation.

When (−)deprenyl (1–100 μg/ml) was added to thin-slice preparation *in vitro*, the release of both DA and DOPAC remained unchanged. In the bulky preparation (−)deprenyl *in vitro* decreased the release of DOPAC dose-dependently, but left DA outflow unchanged, within a reasonable dose range (0.25–1 μg/ml).

However, the amount of DA emitted under resting or KCl-stimulated conditions significantly increased if the striatum was taken from rats pretreated with daily subcutaneous injection of 0.25 mg/kg (−)deprenyl for 3 weeks and the striatum was removed 24 hr after the last injection. Table 8 shows the changes in DA and DOPAC efflux due to a 3-week treatment with (−)deprenyl, which are consistent with the experience with (−)deprenyl in Parkinson's disease.

TABLE 8
The Effect of a 3-Week Treatment with (−)Deprenyl on DA and DOPAC Release from Rat Striata[a]

		Resting		20 nmole/liter KCl		
		DA	DOPAC	DA	DOPAC	*n*
Thin-slice preparation	Control	44.5 ± 7.6	56.3 ± 13.1	115.0 ± 9.1	55.4 ± 36.3	10
	(−)Deprenyl	106.9 ± 9.1[b]	58.7 ± 17.9	263.0 ± 57.3[b]	33.3 ± 6.2	6
Glowinsky–Iversen preparation	Control	91.1 ± 7.2	258.3 ± 18.5	200.0 ± 25.8	291.5 ± 29.5	10
	(−)Deprenyl	503.1 ± 20.6[b]	71.8 ± 10.1[b]	1451.2 ± 183.1[b]	120.8 ± 36.3[b]	6

[a] Values are pmole/g per min; 0.25 mg/kg (−)deprenyl was administered subcutaneously once a day for 3 weeks. Striata were removed 24 hr after the last injection. Control animals were injected similarly once a day with saline. For methodological details see Kerecsen *et al.* (1985).
[b] Marks significance ($p < 0.05$).

7.6. The Effect of (−)Deprenyl on Monoamine Receptors

The effects of clorgyline and deprenyl, the selective inhibitors of MAO-A and MAO-B, respectively, on postsynaptic noradrenaline and serotonin receptors were studied on isolated organs (Knoll, 1976). (−)Deprenyl is devoid of significant inhibitory effect on postsynaptic noradrenaline or serotonin receptors unless extremely high concentrations, without practical meaning, are used. In striking contrast to (−)deprenyl, clorgyline proved to be a very potent, selective, noncompetitive inhibitor of the 5-HT receptor, leaving the acetylcholine and histamine receptors unaltered. In some preparations clorgyline itself stimulated the 5-HT receptors before blocking it. In the guinea pig vas deferens (−)deprenyl usually stimulated and clorgyline inhibited motor transmission. In this test, too, clorgyline was found to be a potent, noncompetitive, selective inhibitor of the 5-HT receptor.

On the other hand, very high doses of (−)deprenyl proved to interact with dopamine receptors and inhibit the effect of apomorphine in the rat (Knoll, 1978c). To exert such an effect on postsynaptic dopamine receptors *in vivo*, doses 40–100 times higher than those needed for the selective inhibition of MAO-B were needed. Thus, it may be concluded that (−)deprenyl leaves the function of the postsynaptic monoamine receptors in the therapeutic dose range unchanged. However, the fact that (−)deprenyl has an affinity to the dopamine receptors and its binding results in a blockade of the receptors led to the proposition that the drug may, even in the doses used in therapy, inhibit the presynaptic dopamine receptors which are known to be more sensitive to inhibitors. As the presynaptic dopamine receptors serve as regulators of the release of dopamine, inhibition of these receptors may represent an additional factor in the facilitation of dopaminergic modulation by (−)deprenyl (Knoll, 1978c).

The effect of long-term treatment with low doses of (−)deprenyl on postsynaptic catecholamine receptors, however, needs further careful analyses in the future. Especially, receptor binding studies following (−)deprenyl treatment with different doses and different durations are needed to get final information regarding the effects of (−)deprenyl on the metabolism and function of the catecholamine receptors.

Daily injections of (−)deprenyl in the MAO-B selective dose range (1 μmole/kg, s.c.) attenuated the noradrenaline-dependent stimulation of cortical adenylate cyclase and reduced the number of brain recognition sites for β-adrenergic receptor ligands. Such an effect is typically elicited by antidepressants and is not at all specific to (−)deprenyl, and the semiselective MAO-B inhibitor pargyline (2.5 μmole/kg per day for 3 weeks) acted similarly (Zsilla *et al.*, 1983).

(−)Deprenyl, however, exerted a quite unique effect which was not shared with pargyline; it increased the number of [^{3}H]imipramine recognition sites (Zsilla *et al.*, 1983), the physiological relevance of which remains obscure.

7.7. The Prevention of MPTP-Induced Parkinsonian Condition by Pretreatment with (−)Deprenyl

MPTP, which induces parkinsonian condition in humans and monkeys by selectively killing the nigrostriatal dopaminergic neuron, exerts its neurotoxic effect through hitherto unidentified, probably short-lived metabolite(s), the formation of which depends on the presence of B-type MAO.

In a series of experiments Chiba *et al.* (1984) demonstrated that rat brain mitochondrial fraction metabolized 18.5 nmoles/mg protein per 30 min.

Interestingly, A-type MAO, which is the main form of the enzyme in the rat nigrostriatal dopaminergic neuron, had no influence on the biotransformation of MPTP. In the presence of 10^{-7} M clorgyline the metabolism of MPTP remained unchanged (19.5 ± 2.4 nmoles/mg protein per 30 min).

Preincubation with the same concentration of (−)deprenyl, however, inhibited the oxidation of MPTP almost completely (2.9 ± 2.3 nmoles/mg protein per 30 min).

Cohen *et al.* (1985) demonstrated that (−)deprenyl also prevents the neurotoxic effect of MPTP on the nigrostriatal dopaminergic neuron in a strain of monkeys, *Macaca fascicularis*. In these monkeys 0.35 mg free base/kg MPTP was injected intravenously on 4 successive days, and at 11–12 days after the last dose the dopamine and homovanillic acid content in the head of the caudate, as well as the [^{3}H]dopamine uptake into synaptosomes of the tissue, was measured. Whereas the control caudate contained 11.4 ± 1.0 μg/g dopamine and 9.1 ± 0.7 μg/g HVA and 0.53 ± 0.14 pmole/mg [^{3}H]dopamine was taken into the synaptosomes, in MPTP-treated monkeys the dopamine and HVA content was practically nil and there was no sign of DA uptake into synaptosomes. Those monkeys, however, which were injected i.m. as priming doses of 10 mg free base/kg (−)deprenyl for 4 days, followed by daily maintenance doses of 2 mg/kg to the end of the experiment, and in which MPTP was injected at 2 hr after (−)deprenyl starting with the fourth priming dose, were protected against the neurotoxic effect of MPTP.

Both the DA content of the caudate (11.4 ± 1.4 μg/g) and the DA uptake into the synaptosomes prepared from the tissue (0.53 ± 0.13 pmole/mg tissue) remained completely unchanged.

Based on our pharmacological data, we proposed earlier that for the best possible protection of the nigrostriatal dopaminergic neuron, treating parkinsonian patients in the early stage of the disease with (−)deprenyl and giving it concurrently with any other drug, until death, must be seriously considered (Knoll, 1983). Now the new findings with MPTP seem to support this view. Cohen *et al.* (1985) end their paper with the statement ". . . treatment of patients with deprenyl could have the beneficial effect of preserving dopamine neuron." Langston *et al.* (1984) end their paper with the conclusion ". . . it seems that testing the early use of deprenyl . . . to alter the course of the disease may be warranted."

8. THE PLACE OF (−)DEPRENYL IN THE CURRENT THERAPY OF PARKINSON'S DISEASE

In excellent agreement with the peculiar pharmacological spectrum observed in animal experiments, Birkmayer *et al.* showed in 1975 that (−)deprenyl potentiated the antiakinetic effect of levodopa in parkinsonians, and they also proved that a single oral dose of (−)deprenyl alone could reduce functional disability within 60 min and this benefit might last up to 3 days.

In 1977 Birkmayer and his co-workers published the first long-term study with (−)deprenyl and demonstrated that the therapeutic benefit with this drug could be maintained for more than 2 years when combined with Madopar. The important observation of Birkmayer and his group (1977) was first corroborated by Stern and his co-workers (Lees *et al.*, 1977) and confirmed in a number of clinical papers (Csanda *et al.*, 1978, 1980; Stern *et al.*, 1978, 1982; Yahr, 1978, 1980; Wajsbort and Youdim, 1980; Rinne *et al.*, 1978; Rinne, 1980b, 1983; Yahr *et al.*, 1983; Birkmayer, 1983; Csanda and Tarczy, 1983; Gerstenbrand *et al.*, 1983; Presthus and Hajba, 1983; Portin and Rinne, 1983).

(−)Deprenyl proved to be a safe drug in man. This was again in excellent agreement with the experimental data on animals. In different species (mouse, rat, cat, dog) less than 0.5% of the LD_{50} sufficed to block completely B-type MAO in the brain, and the inhibition of dopamine uptake as well as the increase in the turnover rate of dopamine in the striatum could be maintained with a daily administration of such low doses of (−)deprenyl.

(−)Deprenyl was originally described as a new-spectrum MAO inhibitor, essentially differing from the previously known MAO inhibitors in its ability to counter in vascular smooth muscle the effect of tyramine (Knoll *et al.*, 1967, 1968), the substance responsible for the cheese effect (Blackwell, 1963).

Our claim that (−)deprenyl is an MAO inhibitor without the cheese effect was substantially supported by the clinicians who, even during long-term administration of the drug, never observed hypertensive reactions and there was no need for any dietary restriction. In some volunteers treated with deprenyl,

Varga, who performed the first clinical trial with deprenyl proving its antidepressant effect (Varga and Tringer, 1967), tried to provoke the cheese reaction in 1966, by administering huge amounts of tyramine-rich cheeses, but no significant change in the blood pressure could be detected (E. Varga, unpublished results). Elsworth *et al.* (1978) and Sandler *et al.* (1978) followed the suggestion of Knoll (1976) and investigated the relationship between tyramine and (−)deprenyl in humans. They found that subjects were able to consume up to 200 mg of tyramine during (−)deprenyl treatment before a rise of blood pressure and slowing pulse occurred. This amount of tyramine is unlikely to be consumed in a normal diet.

Our claim that (−)deprenyl is an MAO inhibitor without the cheese effect was also confirmed by Stern *et al.* (1978), who demonstrated that long-term (−)deprenyl treatment with 10 mg daily up to 18 months did not change tyramine response in the patients. According to Stern *et al.* (1978), "the safety factor remained well beyond the dose of oral tyramine that might be contained in an average diet."

The safeness of (−)deprenyl was further demonstrated by Pare *et al.* (1978) and Mendis *et al.* (1981) in patients. The authors used the tyramine pressor test (Ghose *et al.*, 1975) and found that even high doses of (−)deprenyl failed to increase the sensitivity to intravenous tyramine in the patients, whereas moderate doses of standard MAO inhibitors increased it considerably.

Regarding the place of (−)deprenyl in the pharmacotherapy of PD, the following must be considered.

Barbeau (1980), who performed one of the first large-scale studies with levodopa alone and in combination with decarboxylase inhibitor, analyzed his experiences based on a total of 1052 parkinsonian patients over a period of 12 years and concluded that "long-term side effects are numerous. . . . For all these reasons although we recognize that levodopa is still the best available therapy, we prefer to delay its onset until absolutely necessary." Considering this opinion and the peculiar pharmacological spectrum of (−)deprenyl, Yahr's opinion with regard to (−)deprenyl—"It may have additional usefulness as a primary drug in the early phases of parkinsonism" (Yahr, 1980)—deserves careful clinical scrutiny.

The value of (−)deprenyl as an adjuvant to levodopa alone or in combination with a decarboxylase inhibitor was firmly established by Birkmayer and his group. Birkmayer *et al.* (1983) reported results of a highly interesting study comparing 323 outpatients with PD who received Madopar or Sinemet daily, with a mean duration of 8.2 years, without (−)deprenyl and 285 outpatients on combined levodopa therapy for 8.7 years plus (−)deprenyl 10 mg/day at breakfast and lunch. The duration of (−)deprenyl medication lasted from 2 to 8 years. Another comparison was made between 141 deceased patients on conventional therapeutic regimen, with a mean duration of 8 years of combined levodopa treatment without (−)deprenyl, and a group of 96 patients with a mean duration of 9 years of combined levodopa treatment plus (−)deprenyl for 2–6 years.

In both the living and the deceased groups of outpatients a significant improvement of disability in the (−)deprenyl-treated patients compared to those with conventional therapy only was found. These data are in agreement with earlier ones showing that (−)deprenyl is a valuable adjuvant to usual antiparkinson therapy, especially in patients with on–off symptomatology (for review see Birkmayer and Riederer, 1983). The more important and highly remarkable observation of this recent study of Birkmayer *et al.* (1983), however, lies in comparison of the duration of the illness in the groups with and without (−)deprenyl. *The additional supplement of (−)deprenyl in PD led to a significant prolongation of the duration of the illness, which means that the life-span of the patients was prolonged.*

This seems to be an extremely important observation which might be the consequence of the described effect of (−)deprenyl in increasing the dopaminergic tone in the brain.

In this context it is worth mentioning that based on animal experiments, we proposed to check the possibility of a drug strategy to improve the quality of life in senescence by the continuous administration of (−)deprenyl. The hypothesis of this drug strategy was first proposed in 1980 (Knoll, 1981) and extended in 1981 (Knoll, 1982). The essence of this hypothesis is as follows: In the aging brain, there is a loss of neurons, compensated for by a proliferation of glial cells. We might thus predict that dopaminergic and "trace aminergic" modulation in the brain declines in senescence because of the loss of neurons and because of the increased MAO-B activity present in the glia. The significant increase of the incidence of depression in the elderly, the age-dependent decline in male sexual vigor, and the frequent appearance of parkinsonian symptoms in the latter decades of life might be attributed to a decrease of dopamine and "trace amines" in the brain. Thus, it may be possible to counteract these biochemical lesions of aging by chronic administration of (−)deprenyl, a selective inhibitor of MAO-B, which facilitates, by a complex pharmacological spectrum of activity, dopaminergic and "trace-aminergic" activity in the brain and is a safe drug in man.

The restitution and long-term maintenance of full-scale sexual activity in aged male rats continuously treated with (−)deprenyl was demonstrated as an experimental model in support of the view that the long-term administration of small doses of (−)deprenyl may improve the quality of life in senescence by counteracting the age-related decline of a dopamine-dependent brain function (Knoll, 1981, 1982; Knoll *et al.*, 1983).

The recent observation of Birkmayer *et al.* in parkinsonians that long-term administration of (−)deprenyl prolongs the duration of the incurable disease shows that in parkinsonians, known to have a very low dopaminergic tone in the brain, the facilitation of dopaminergic activity in the central nervous system by (−)deprenyl is highly favorable. Up to the present only aged parkinsonians got (−)deprenyl daily for years, and, as we see it, their life-span grew longer. The possibility that the same mechanism works in natural aging of the brain, without PD, as proposed earlier (Knoll, 1981, 1982), remains open for clinical research.

Birkmayer *et al.* concluded that ". . . the prolongation of the Parkinson's disease process via long-term (−)deprenyl treatment shows for the first time that the degeneration of the dopaminergic nigro-striatal fibres can be depressed to some extent. This has not been observed in other antiparkinson drugs so far. Combined L-dopa treatment suppresses the so called early death rate of patients with Parkinson's disease, but has, however, not been shown to influence the duration of the disease." It seems to be obvious that "prolongation of the PD" means the increase of the life-span of the patients, owing, with high probability, to the described facilitation of the central dopaminergic tone by chronic administration of (−)deprenyl.

The value of (−)deprenyl as an adjuvant in PD, as well as a special tool for countering the consequences of decreased central dopaminergic tone, an unavoidable biochemical lesion of aging in man, deserves extensive clinical investigation in the future.

With regard to PD, the prediction of Stern (1980) might prove to be true: "of the numerous agents and adjuvants that have been tried in the past decade . . . very few seem likely . . . to find an enduring place in current therapeutics. However, . . . deprenyl may prove to be an exception." Stern ended with the statement, "Among the many unresolved problems concerning the value of deprenyl as an effective adjuvant in the treatment of Parkinson's disease remains the question whether patients who are treated in this manner throughout the course of their illness will run a more benign course with fewer late complications." In Birkmayer's patients treated with (−)deprenyl for years, the duration of the incurable disease, i.e., the life-span, was significantly prolonged. Thus, the illness runs an undeniably more benign course. Should these findings be corroborated in further clinical studies, the significance of such an effect of (−)deprenyl is obviously farther reaching than the unresolved problems of treatment of PD. It may catalyze a new drug strategy of much broader importance. The aging of the nigrostriatal dopaminergic neuron is peculiar. The decline of the striatal dopaminergic activity seems to be an unavoidable biochemical lesion of aging. As the dopamine content of the human caudate nucleus decreases by 13% per decade after the age 45, we may really look at Parkinson's disease as a kind of selective, highly accelerated "premature aging" of the nigrostriatal dopaminergic neuron, when the dopamine content of this neuron shrinks within a very short time to less than 10% of the normal level in the premorbid state (Knoll, 1983).

The recent findings that an exogenous toxin, MPTP, destroys the nigrostriatal dopaminergic neuron in a few days and the enzyme that plays the main role in catalyzing this change is B-type MAO and the previous findings suggesting that the selectively increased B-type MAO activity in the aging brain plays the key role in the age-related decline of nigrostriatal dopaminergic activity (Knoll, 1982) all led to the search for a basically common mechanism in these processes.

MPTP is oxidized by B-type MAO to that short-lived intermediate, probably dihydropyridinium ion, which destroys the nigrostriatal dopaminergic neuron. We may assume that B-type MAO catalyzes the production of a number of short-lived dopamine oxidation products with different complexity of which compounds with specific neurotoxic effect on the nigrostriatal dopaminergic neuron may exist. The unavoidable age-related increase in B-type MAO activity would then mean the formation of increased amounts of these endogenous toxic agents which slowly destroy a progressively increasing number of nigrostriatal dopaminergic neurons. The concept that natural Parkinson's disease, the MPTP-induced disease, and the age-related decline of dopaminergic tone in the striatum have a common meeting point—B-type MAO—deserves attention.

To prove this working hypothesis requires extremely complicated analytical work, because it is highly probable that the toxic agents formed during the oxidation processes catalyzed by B-type MAO are short-lived. We may, however, approach the problem from a practical angle. If B-type MAO plays the main role in formation of the putative endogenous toxic metabolites, the safe inhibition of this enzyme is of crucial practical importance.

(−)Deprenyl, the only potent selective inhibitor of B-type MAO, which is safe enough to be administered for years without any harm, seems to me to be the appropriate experimental tool to check the validity of this assumption.

9. REFERENCES

Ayd, F. J. and Blackwell, B. (eds.), 1970, *Discoveries in Biological Psychiatry*, J. B. Lippincott, Philadelphia.

Barbeau, A., 1969, L-Dopa therapy in Parkinson's disease: A critical review of nine years experience, *Can. Med. Assoc. J.* **101**:791.

Barbeau, A., 1980, High level therapy in severely akinetic parkinsonian patients: twelve years later, in: *Parkinson's Disease. Current Progress, Problems and Management* (U.K. Rinne, M. Klinger, and G. Stamm, eds.) pp. 229–239, Elsevier/North Holland Biomedical Press, Amsterdam, New York.

Barbeau, A., Murphy, C. F., and Sourkes, T. L., 1961, Excretion of dopamine in diseases of basal glanglia, *Science* **133**:1706.

Bartholini, G., Burkhard, W. P., Pletscher, A., and Bates, V. M., 1967, Increase of cerebral catecholamines by L-DOPA after inhibition of peripheral decarboxylase, *Nature* (*London.*) **215**:852.

Bertler, A., and Rosengren, E., 1959, On the distribution in brain monoamines and of enzymes responsible for their formation, *Experientia* **15**:10.

Birkmayer, W., 1969, Experimentelle Ergebnisse Uber die Kombinationsbehandlung des Parkinson-Syndromes mit L-Dopa und einem Decarboxylasehemmer (Ro 4-4602), *Wien. klin. Wschr.* **81**:677.

Birkmayer, W., 1983, Deprenyl (selegiline) in the treatment of Parkinson's disease, *Acta Neurol. Scand.* **95**(Suppl.):103.

Birkmayer, W., and Hornykiewicz, O., 1961, Der L-Dioxyphenylalanin (L-DOPA)-effect bei der Parkinson-Akinesia, *Wien. klin. Wschr.* **73**:787.

Birkmayer, W., and Hornykiewicz, O., 1962, Der-L-dioxyphenylalanin-Effect beim Parkinson-Syndrom des Menschen, *Arch. Psychiatr. Nervenkr.* **2P3**:560.

Birkmayer, W., and Mentasti, M., 1967, Weitere experimentelle Unterschungen uber des Katecholaminstoffwechsel bei extrapyramidalen Erkrankugen (Parkinson-und Choreasyndrom). *Arch. Psych. Z. ges. Neurol.* **210**:29.

Birkmayer, W., and Riederer, P., 1983, *Parkinson's Disease Biochemistry, Clinical Pathology and Treatment*, Springer-Verlag, Wien, New York.

Birkmayer, W., Riederer, P., Youdim, M. B. H., and Linauer, W., 1975, The potentiation of the anti-akinetic effect after L-DOPA treatment by an inhibitor of MAO-B, Deprenyl, *J. Neural. Transm.*, **36**:303.

Birkmayer, W., Riederer, P., Ambrozi, L., and Youdim, M. B. H., 1977, Implications of combined treatment with "Madopar" and L-Deprenyl in Parkinson's disease, *Lancet* **(I) 8009**:439.

Birkmayer, W., Knoll, J., Riederer, P. and Youdim, M. B. H., 1983, (−)Deprenyl leads to prolongation of L-dopa efficacy in Parkinson's disease, *Mod. Probl. Pharmacopsychiatr.* **19**:170.

Blackwell, B., 1963, Hypertensive crisis due to monoamine oxidase inhibitors, *Lancet* **I**:849.

Blackwell, B., Marley, E., Price, J. and Taylor, D., 1967, Hypertensive interactions between monoamine oxidase inhibitors and foodstuffs, *Br. J. Psychiatry.* **113**:349.

Braestrup, C., Andersen, H., and Randrup, A., 1975, The monoamine oxidase B inhibitor deprenyl potentiates phenylethylamine behavior in rats without inhibition of catecholamine metabolite formation, *Eur. J. Pharmacol.* **34**:181.

Burns, R. S., Chiueh, C. C., Markey, S. P., Ebert, M. H., Jacobowitz, D. M., and Kopin, I. J., 1983, A primate model of parkinsonism: Selective destructions in the pars compacta of the substantia nigra by *N*-methyl-4-phenyl-1,2,3,6-tetrahydropyridine, *Proc. Natl. Acad. Sci. USA* **80**:4546.

Burns, R. S., Markey, S. P., Phillips, J. M., and Chieuh, C. C., 1984, The neurotoxicity of 1-methyl-4-phenyl-1,2,3,6-tetrahydropyridine in the monkey and man, *Can. J. Neurol. Sci.* **11**:166.

Carlsson, A., 1959, The occurrence distribution and physiological role of catecholamines in the nervous system, *Pharmacol. Rev.* **11**:490.

Carlsson, A., 1975, Receptor mediated control of dopamine metabolism, in: *Pre- and Postsynaptic Receptors* (E. Usdin and W. E. Bunney, eds.), p. 49, Marcel Dekker, New York.

Carlsson, A., and Waldeck, B., 1958, A fluorimetric method for the determination of DA (3-hydroxytyramine), *Acta Physiol. Scand.* **44**:293.

Chiba, K., Trevor, A., and Castagnoli, N., Jr., 1984, Metabolism of the neurotoxic tertiary amine, MPTP, by brain monoamine oxidase, *Biochem. Biophys. Res. Commun.* **120**:574.

Chieuh, C. C., Burns, R. S., Markey, S., Jakobowitz, D., Ebert, M. H., and Kopin, I. J., 1983, Effects of *N*-methyl-4-phenyl-1,2,3,6-tetrahydropyridine, a cause of an extrapyramidal syndrome, on the nigrostriatal dopaminergic system in the rat, guinea pig, and monkey. 5th Catecholamine Symposium, Gothenburg (Abstr.), 74.

Cohen, G., Pasik, P., Cohen, B., Leist, A., Mytilineou, C., and Yahr, M. D. 1985, Pargyline and deprenyl prevent the neurotoxicity of 1-methyl-4-phenyl-1,2,3,6-tetrahydropyridine (MPTP) in monkeys, *Eur. J. Pharmacol.* **106**:209.

Cotzias, G. C., Papvasiliov, P. S., and Gellene, R., 1969, Modification of parkinsonism—Chronic treatment with L-dopa, *N. Engl. J. Med.* **280**:337.

Csanda, E., and Tarczy, M., 1983, Clinical evaluation of deprenyl (selegiline) in the treatment of Parkinson's disease, *Acta Neurol. Scand.* **95**(Suppl.):117.

Csanda, E., Antal, J., Anthony, M., and Csanaky, A., 1978, Experiences with L-deprenyl in parkinsonism, *J. Neural. Transm.* **43**:263.

Csanda, E., Antal, J., and Fornadi, F., 1980, Clinical experience in extrapyramidal disease with selective MAO-B inhibitor, deprenyl, in: *Monoamine Oxidases and Their Selective Inhibition* (K. Magyar, ed.), pp. 127–132, Pergamon Press-Akademiai Kiado, Budapest.

Davis, G. C., Williams, A. C., Markey, S. P., Ebert, M. H., Caine, E. D., Reichart, C. M., and Kopin, I. J., 1979, Chronic parkinsonism secondary to intravenous injection of meperidine analogues, *Psychiatr. Res.* **1**:249.

Dupont, E., 1977, Epidemiology of Parkinsonism, in: *Symposium on Parkinsonism* (J. Worm-Petersen and J. Bottcher, eds.), pp. 65–75, M. S. D., Denmark.

Ehringer, H., and Hornykiewicz, O., 1960, Verteilung von Noradrenalin und Dopamin (3-Hydroxytyramin) im Gehirn des Menschen und ihr Verhalten bei Erkrankungen des extrapyramidalen Systems, *Klin. Wschr.* **38**:1336.

Ekstedt, B., Magyar, K., and Knoll, J., 1979, Does the B form selective monoamine oxidase inhibitor lose selectivity by long term treatment? *Biochem. Pharmacol.* **28**:919.

Elsworth, J. D., Glover, V., Reynolds, G. P., Sandler, M., Lees, A. J., Phuapradit, P., Shaw, K. M., Stern, G. M., and Kumar, P., 1978, Deprenyl administration in man: a selective monoamine oxidase B inhibitor without the "cheese effect," *Psychopharmacology* **57**:33.

Gerstenbrand, F., Ransmayr, G., and Poewe, W., 1983, Deprenyl (selegiline) in combination treatment of Parkinson's disease. *Acta Neurol. Scand.* **95**(Suppl.):123.

Ghose, K., Turner, P., and Coppen, A., 1975, Intravenous tyramine pressor response in depression, *Lancet* **1**:1317.

Gilman, A., Goodman, L. S., and Gilman, A., 1980, *Goodman and Gilman's The Pharmacological Basis of Therapeutics*, p. 429, Macmillan, New York.

Glover, V., Sandler, M., Owen, V., and Riley, G., 1977, Dopamine is a monoamine oxidase B substrate in man, *Nature* **265**:80.

Glowinski, J., and Iversen, L. L., 1966, Regional studies of catecholamines in the rat brain. I. The disposition of ^{3}H-norepinephrine, ^{3}H-dopamine and ^{3}H-DOPA in various regions of the brain, *J. Neurochem.* **13**:655.

Hallman, H., Olson, L., and Jonsson, G., 1984, Neurotoxicity of the meperidine analogue, *N*-methyl-4-phenyl-1,2,3,6-tetrahydropyridine on brain catecholamine neurons in the mouse, *Eur. J. Pharmacol.* **97**:133.

Harsing, L. G., Jr., Magyar, K., Tekes, K., Vizi, E. S., and Knoll, J., 1979, Inhibition by deprenyl of dopamine uptake in rat striatum: A possible correlation between dopamine uptake and acetylcholine release inhibition, *Pol. J. Pharmacol. Pharm.* **31**:297.

Hassler, R., 1938, Zur Pathologie der Paralysis agitans und des postencephalitischen Parkinsonismus, *J. Psychol. Neurol.* (*Lpz.*) **48**:387.

Heikkila, R. E., Hess, A., and Duvoisin, R. C. 1984, Dopaminergic neurotoxicity of 1-methyl-4-phenyl-1,2,5,6-tetrahydropyridine in mice, *Science* **224**:1451.

Hunter, K. R., Boakes, A. J., Laurence, D. R., and Stern, G. M., 1970, Monoamine oxidase inhibitors and 1-dopa, *Br. Med. J.* **3**:388.

Johnston, J. P., 1968, Some observations upon a new inhibitor of monoamine oxidase brain tissue, *Biochem. Pharmacol.* **17**:1285.

Karoum, F., Chuang, L.-W., Eisler, T., Calne, D. B., Leibowitz, M. R., Quitkin, F. M., Klein, D. F., and Wyatt, R. J., 1982, Metabolism of (−)deprenyl to amphetamine and metamphetamine may be responsible for deprenyl's therapeutic benefit: A biochemical assessment, *Neurology* **32**:503.

Kerecsen, L., Kalasz, H., Tarcali, J., Fekete, J., and Knoll, J., 1985, Measurement of DA and DOPAC release from rat striatal preparations "in vitro" using HPLC with electrochemical detection, in: *Chromatography, the State of the Art* (H. Kalasz and L. S. Ettre, eds.), pp. 195–203, Akademiai Kiado, Budapest.

Kissinger, P. T., Craig, S. B., and Shoup, R. E., 1981, Neurochemical applications of liquid chromatography with electrochemical detection, *Life Sci.* **28**:455.

Knoll, J., 1976, Analysis of the pharmacological effects of selective monoamine oxidase inhibitors, in: *Monoamine Oxidase and Its Inhibition* G. E. W. Wolstenholme and J. Knight, eds.), pp. 135–161, Ciba Foundation Symposium 39 (new series) Elsevier-Excerpta Medica, North-Holland, Amsterdam.

Knoll, J., 1978a, The pharmacology of selective irreversible monoamine oxidase inhibitors, in: *Enzyme-Activated Irreversible Inhibitors* (N. Seiler, M. J. Jung and J. Koch-Weser, eds.) pp. 253–269, Elsevier North Holland Biomedical Press, Amsterdam, New York, Oxford.

Knoll, J., 1978b, On the dual nature of monoamine oxidase, *Horizons Biochem. Biophys.* **5**:37.

Knoll, J., 1978c, The possible mechanism of action of (−)deprenyl in Parkinson's disease, *J. Neural Transm.* **43**:117.

Knoll, J., 1979a, (−)Deprenyl, the MAO inhibitor without the "cheese effect," *Trends in Neurosciences* **2**:111.

Knoll, J., 1979b, Structure-activity relationship of the selective inhibitors of MAO-B, in: *Monoamine Oxidase: Structure, Function and Altered Functions* (T. P. Singer, R. W. von Korff, and D. L. Murphy, eds.) pp. 431–446, Academic Press, New York.

Knoll, J., 1980, Monoamine oxidase inhibitors: Chemistry and pharmacology, in: *Enzyme Inhibitors as Drugs* (M. Sandler, ed.), pp. 151–272, The Macmillan Press Ltd, London.

Knoll, J., 1981, The pharmacology of selective MAO inhibitors, in: *Monoamine Oxidase Inhibitors—The State of the Art* (M. B. H. Youdim and E. S. Paykel, eds.), pp. 45–61, Wiley, New York.

Knoll, J., 1982, Selective inhibition of B type monoamine oxidase in the brain: a drug strategy to improve the quality of life in senescence, in: *Strategy in Drug Research* (J. A., Keverling-Buisman, ed.) pp. 107–135, Elsevier, Amsterdam.

Knoll, J., 1983, Deprenyl (selegiline): The history of its development and pharmacological action, *Acta Neurol. Scand.* **95**(Suppl.):57.

Knoll, J., and Magyar, K., 1972, Some puzzling effects of monoamine oxidase inhibitors, *Adv. Biochem. Psychopharmacol.* **5**:393.

Knoll, J., Ecsery, Z., Nievel, J. G., and Knoll, B., 1964, Phenylisopropil-methylpropynilamin HCl, E-250, egy uj hataspektrumu pszichoenergetikum, *MTA V. Oszt. Kozl.* **15**:231.

Knoll, J., Ecsery, Z., Kelemen, K., Nievel, J. G., and Knoll, B., 1965, Phenylisopropyl-methylpropinylamine (E-250): A new spectrum psychic energizer, *Arch. Int. Pharmacodyn. Ther.*, **155**:154.

Knoll, J., Vizi, E. S., and Somogyi, G., 1967, A phenylisopropylmethylpropynilamine (E-250) tyraminantagonista hatasa, *MTA V. Oszt Kozl.* **18**:31.

Knoll, J., Vizi, E. S., and Somogyi, G., 1968, Phenylisopropyl-methylpropinylamine (E-250): A monoamine oxidase inhibitor antagonizing the effects of tyramine, *Arzneimittel-Forsch.* **18**:109.

Knoll, J., Yen, T. T., and Dallo, J., 1983, Long-lasting, true aphrodisiac effect of (−)deprenyl in sexually sluggish old male rats, *Mod. Probl. Pharmacopsychiatry* **19**:135.

Kolata, G., 1983, Monkey model of Parkinson's disease, *Science* **220**:705.

Langston, J. M., and Ballard, P. Jr., 1983, Parkinson's disease in a chemist working with 1-methyl-4-phenyl-1,2,5,6-tetrahydropyridine, *N. Engl. J. Med.* **309**:310.

Langston, J. M., Ballard, P. Jr., Tetrud, J. W., and Irwin, I., 1983, Chronic parkinsonism in humans due to a product of meperidineanalog synthesis, *Science* **219**:979.

Langston, J. M., Irwin, I., and Langston, E. B., 1984, Pargyline prevents MPTP-induced parkinsonism in primates, *Science* **225**:1480.

Lees, A. J., Shaw, K. M., Kohout, L. J., Stern, G. M., Elsworth, J. D., Sandler, M., and Youdim, M. B. H., 1977, Deprenyl in Parkinson's disease, *Lancet* **2**:791.

Magyar, K., Skolnik, J., and Knoll, J., 1971, Radiopharmacological analytic studies with deprenyl-C^{14}, in: *V. Conferentia Hungaric pro Therapia et Investigatione in Pharmacologia* (E. P. Leszkovszky, ed.), pp. 103–109, Akademiai Kiado, Budapest.

Martilla, R., 1974, Epidemiological, Clinical and Virus-Serological Studies of Parkinson's Disease, Thesis, Reports from the Department of Neurology, University of Turku, Finland.

Martilla, R. J., 1980, Etiology of Parkinson's disease, in: *Parkinson's Disease. Current Progress, Problems and Management* (U. K. Rinne, M. Klinger, and G. Stamm, eds.), pp. 3–15, Elsevier North Holland Biomedical Press, Amsterdam-New York.

Mendis, N., Pare, C. M. P., Sandler, M., Glover, V., and Stern, G., 1981, Failure of the selective monoamine oxidase B inhibitor, (−)deprenyl to alleviate depression: relationship to tyramine insensitivity? in: *Monoamine Oxidase Inhibitors. The State of the Art* (M. B. H. Youdim and E. S. Paykel, eds.), pp. 171–176, John Wiley and Sons Ltd., Chichester-New York-Brisbane-Toronto.

Mytilineou, C., and Cohen, G., 1984, 1-Methyl-4-phenyl-1,2,3,6-tetrahydropyridine destroys dopamine neurons in explants of rat embryo mesencephalon, *Science* **225**:529.

Pare, C. M. B., Sandler, M., and Stern, G., 1978, Clinical usefulness of deprenyl-A new M.A.O.I., 11th CINP Congress. Vienna, abstract, p. 286.

Parsons, B., and Rainbow, T. C., 1984, High-affinity binding sites for ^{3}H-MPTP may correspond to monoamine oxidase, *Eur. J. Pharmacol.* **102**:375.

Pollock, M., and Hornabrook, R. W., 1966, The prevalence, natural history and dementia of Parkinson's disease, *Brain* **89**:429.

Portin, R., and Rinne, U. K., 1980, Neuropsychological responses of parkinsonian patients to long-term levodopa treatment, in: *Parkinson's Disease. Current Progress, Problems, and Management* (U. K. Rinne, M. Klinger, and G. Stamm, eds.), pp. 271–304, Elsevier/North Holland Biomedical Press, Amsterdam, New York.

Portin, R., and Rinne, U. K., 1983, The effect of deprenyl (Selegiline) on cognition and emotion in parkinsonian patients undergoing long-term levodopa treatment, *Acta Neurol. Scand.* **95**(Suppl.):135.

Presthus, J., and Hajba, A., 1983, Deprenyl (selegiline) combined with levodopa and a decarboxylase inhibitor in the treatment of Parkinson's disease, *Acta Neurol. Scand.* **95**(Suppl.):127.

Reynolds, G. P., Elsworth, J. D., Blau, K., Sandler, M., Lees, A. J., and Stern, G. M., 1978, Deprenyl is metabolized to methamphetamine and amphetamine in man, *Br. J. Clin. Pharmacol.* **6**:542.

Riederer, P., Youdim, M. B. H., Birkmayer, W., and Jellinger, K., 1978, Monoamine oxidase activity during (−)deprenyl therapy: Human brain post mortem studies, in: *Advances in Biochemistry and Pharmacology*, 19 (P. J. Roberts and L. L. Iversen, Eds.), pp. 377–382, Raven Press, New York.

Rinne, U. K., 1980a, Levodopa in combination with an extracerebral decarboxylase inhibitor, in: *Parkinson's Disease. Current Progress, Problems and Management* (U. K. Rinne, M. Klinger, and G. Stamm, eds.), pp. 323–334, Elsevier/North Holland Biomedical Press, Amsterdam, New York.

Rinne, U. K., 1980b, Long-term L-deprenyl treatment of on-off phenomena in Parkinson's disease, in: *Monoamine Oxidases and Their Selective Inhibition* (K. Magyar ed.), pp. 145–149, Pergamon Press-Akademiai Kiado, Budapest.

Rinne, U. K. (ed.), 1983, A New Approach to the Treatment of Parkinson's Disease, *Acta Neurol. Scand.* **95**(Suppl.):144.

Rinne, U. K., Sonninen, V., and Siirtola, T., 1970, L-Dopa treatment in Parkinson's disease, *Eur. Neurol.* **4**:348.

Rinne, U. K., Siirtola, T., and Sonninen, V., 1978, L-deprenyl treatment of on-off phenomena in Parkinson's disease, *J. Neural Transm.* **43**:253.

Sandler, M., Glover, V., Ashford, A., and Stern, G. M., 1978, Absence of "cheese effect" during deprenyl therapy: Some recent studies, *J. Neural Transm.* **43**:209.

Simpson, L. L., 1978, Evidence that deprenyl a type B monoamine oxidase inhibitor, is an indirectly acting sympathomimetic amine, *Biochem. Pharmacol.* **27**:1591.

Squires, R. F., 1972, Multiple forms of monoamine oxidase in intact mitochondria as characterized by selective inhibitors and thermal stability: A comparison of eight mammalian species. *Adv. Biochem. Psychopharmacol.* **5**:355.

Stern, G. M., 1980, Current adjuvants to levodopa therapy, in: *Parkinson's Disease. Current Progress, Problems and Management* (U. K. Rinne, M. Klinger, and G. Stamm, eds.), pp. 357–361, Elsevier/North Holland Biomedical Press, Amsterdam, New York.

Stern, G. M., Lees, A. J., and Sandler, M., 1978, Recent observations on the clinical pharmacology of (−)deprenyl, *J. Neural Transm.* **43**:245.

Stern, G. M., Lander, C. M., Lees, A. J., and Ward, C., 1982, Deprenyl, in: *Rose and Capildeo Research Progress in Parkinson's Disease*, pp. 324–344, Pitman Medical, London.

Tozer, T. N., Neff, N. H., and Brodie, B. B., 1966, Application of steady-state kinetics to the synthesis rate and turnover time of serotonin in the brain of normal and reserpine treated rats, *J. Pharmacol. Exp. Ther.* **153**:177.

Tretiakoff, C., 1919, Contribution a l'etude de l'anatomie pathologique du locus niger de Sommering avec quelques deductions relatives a la pathogenie des troubles du tonus musculaire et de la maladie de Parkinson. These med. no. 293, Paris.

Varga, A., and Tringer, L., 1967, Clinical trial of a new type of promptly acting psychoenergetic agent (phenyl-isopropyl-methyl-propinylamine HCl) (E-250), *Acta Med. Acad. Sci. Hung.* **23**:289.

Vizi, E. S., Harsing, L. G., and Knoll, J., 1977, Presynaptic inhibition leading to disinhibition of acetylcholine release from interneurons of the caudate nucleus: effects of dopamine, β-endorphin and D-Ala^{2}-Pro^{5}-enkephalinamide, *Neuroscience* **2**:953.

Wajsbort, J., and Youdim, M. B. H., 1980, The action of L-deprenyl in L-dopa treated parkinsonian patients, with special reference to the "on-off" effect, in: *Monoamine Oxidases and Their Selective Inhibition* (K. Magyar, ed.), pp. 139–144, Pergamon Press-Akademiai Kiado, Budapest.

Yahr, M. D., 1978, Overview of present day treatment of Parkinson's disease, *J. Neural Transm.* **43**:227.

Yahr, M. D., 1980, Pharmaco-therapy of parkinsonism, in: *Monoamine Oxidases and Their Selective Inhibition* (K. Magyar, ed.), pp. 117–125, Pergamon Press-Akademiai Kiado, Budapest.

Yahr, M. D., Duvoisin, R. C., and Shear, M. J., 1969, Treatment of parkinsonism with levodopa, *Arch. Neurol. (Chicago)* **21**:343.

Yahr, M. D., Mendoza, M. R., Moros, D., Bergmann, K. J., 1983, Treatment of Parkinson's disease in early and late phases. Use of pharmacological agents with special reference to deprenyl/selegiline, *Acta Neurol. Scand.* **95**(Suppl.):95.

Ziering, A., Berger, L., Heineman, S. D., and Lee, J., 1947, Piperidine derivatives. Part III. 4-Arylpiperidines, *J. Org. Chem.* **12**:894.

Zsilla, G., and Knoll, J., 1982, The action of (−)deprenyl on monoamine turnover rate in rat brain, in: *Typical and Atypical Antidepressants* (E. Costa and G. Racagni, eds.), pp. 211–217, Raven Press, New York.

Zsilla, G., Barbaccia, M. L., Gandolfi, O., Knoll, J., and Costa, E., 1983, (−)Deprenyl a selective MAO-B inhibitor increases H-imipramine binding and decreases β-adrenergic receptor function, *Eur. J. Pharmacol.* **89**:111.

Zsilla, G., Foldi, P., Held, G., Szekely, A. M., and Knoll, J., 1986, The effect of repeated doses of (−)deprenyl on the dynamics of monoaminergic transmission. Comparison with clorgyline, *Pol. J. Pharmacol. Pharm.* **86**.

CHAPTER 3

Parkinson's Disease

Current Concepts

ABRAHAM N. LIEBERMAN and MENEK GOLDSTEIN

1. INTRODUCTION

After more than a decade levodopa (combined with a peripheral decarboxylase inhibitor) remains the standard treatment for Parkinson's disease (PD) with up to 90% of patients responding. However, levodopa does not halt the advance of the underlying disease, but by improving mobility, it delays the onset of fatal complications (Marsden and Parkes, 1977; Yahr *et al.*, 1972). Thus, after 2–5 years many PD patients have increased disability. The increased disability may manifest itself by the reappearance of old symptoms, i.e., symptoms present before levodopa treatment, or by new symptoms, symptoms not present before levodopa treatment. Most of the new symptoms are disease manifestations seen only because patients live longer (postural instability, dementia), and some of the symptoms may result from the chronic effects of levodopa itself (dyskinesias, diurnal oscillations in performance).

Parkinson's disease is characterized by a dopamine (DA) deficiency resulting from degeneration of pigmented neurons that have their cell bodies in the substantia nigra (SN) and terminate in the striatum (Bernheimer *et al.*, 1973). Within the striatum, these terminals synapse with at least two different populations of receptors (Kebabian *et al.*, 1972; Schwartz *et al.*, 1978). One population of receptors is linked to adenyl cyclase and is mainly localized on striatal interneurons. The other population of receptors is not linked to adenyl cyclase and is localized mainly on the axons of a descending corticostriatal tract. In addition, there are presynaptic DA receptors on the nigrostriatal neurons themselves (autoreceptors). Stimulation of the autoreceptors results in feedback inhibition of DA synthesis. While it is known that the major efferent pathway from the striatum is to the globus pallidus and thence to the thalamus, events after stimulation of the postsynaptic receptors are still not clear. In

ABRAHAM N. LIEBERMAN • Department of Neurology, New York University School of Medicine, New York, New York 10016. MENEK GOLDSTEIN • Department of Neurochemistry, New York University School of Medicine, New York, New York 10016.

addition to the pigmented DA neurons in the SN, there are pigmented DA neurons in the mesocortex, hypothalamus, and spinal cord, and pigmented noradrenergic neurons in the locus ceruleus, the dorsal vagal nucleus, and the sympathetic ganglia (Hartog Jager and Bethlem, 1960). PD may also involve nonpigmented neurons which contain other neurotransmitters including acetylcholine, γ-aminobutyric acid, glutamic acid, glycine, serotonin, and the enkephalins. In many patients dementia and/or depression may be associated with PD. In one study of 520 patients with PD, 168 (32%) had a dementia (Liebermen *et al.*, 1979a). Although the demented patients were older than the nondemented patients (70.4 versus 65.5 years), the incidence of dementia in PD was 10-fold higher than among controls (the age-matched spouses of the patients). In a second study of 36 patients, the prevalence of dementia and pathologically established Alzheimer's changes were six times higher among PD patients (33%) than in age-matched controls (5.1%) (Boller *et al.*, 1980). In a third study, 19 of 34 cases of PD (56%) had pathologically established Alzheimer's changes (Hakim and Mathieson, 1979), while in a fourth study, 26 of 55 PD patients (47%) without dementia were significantly depressed compared to only 12% of their spouses (Mayeux *et al.*, 1981).

2. EVALUATION OF PD

To evaluate the efficacy of treatment it is necessary to have a standardized means of comparing patients. To accomplish this we devised and subsequently modified an examination of the major motor symptoms of the disease: rigidity, tremor, bradykinesia, and the gait disorder (Lieberman *et al.*, 1980a) (Table 1). Rigidity is assessed at the neck and the major joints of the upper and lower extremities and is graded on a scale of 0 (absent) to 3, 4 (marked) at the joint of the extremity where it is maximal. A total score of 15 is derived for rigidity reflecting its relative importance. Tremor is similarly assessed at the head, face, neck, and the upper and lower extremities. Tremor is graded on the basis of amplitude. Notation is made whether tremor is present. Bradykinesia, which may manifest itself as a slowness in initiating and/or completing movements and/or as a decrease in the amplitude and/or power of movements, is assessed at the extremities by having the patient perform movements involving all four limbs simultaneously. Bradykinesia of whole-body movements is assessed by observing the patient rise from a chair. The gait disorder may manifest itself as a disturbance in stance (postural stability) and/or locomotion (gait). Postural stability and gait are assessed separately. The scores for rigidity, tremor, bradykinesia, postural stability, and gait are then summed and expressed on the N.Y.U. Disability Scale, where 0 represents no disability and 100% represents complete disability (Table 1). Dyskinesias are assessed at the head, face, neck, and all four extremities and are graded on a scale where 0 represents no dyskinesias, and 100% represents very marked dyskinesias. Dyskinesias are reported separately (Table 1). Patients are also staged on the N.Y.U. Functional

Disability Scale (Table 2) and on the Hoehn and Yahr Scale (Hoehn and Yahr, 1967). Because of their importance, postural stability and gait are evaluated further through an examination which assesses 20 different aspects of stance and locomotion (Table 3).Patients who have no difficulty with a particular aspect of stance or locomotion are rated as 0 on that aspect; patients who have some difficulty are rated as 3; and patients who cannot perform the particular aspect are rated as 5. The scores for the 20 aspects are summed, with 0 representing no disability and 100% complete disability (Table 3). Many levodopa-treated patients eventually experience diurnal oscillations in performance, both "wearing off" and/or "on–off" phenomena (Marsden and Parkes, 1976). Wearing off refers to the gradual waning of the benefit from a dose of levodopa, usually 2–4 hr after the dose, with the off period lasting until the next dose. On–off refers to the abrupt loss of effectiveness of a dose of levodopa that may last minutes to hours and that is followed by an equally abrupt return of effectiveness. On each visit an attempt is made to examine each patient while the patient is in both an on and an off period in the hospital. The duration of the diurnal oscillations in performance is assessed by having a trained nurse or a neurologist examine the patients each waking hour and note whether they are on (can walk without difficulty) or whether they are off (can walk with difficulty or cannot walk). Patients and/or their family members are also trained to assess themselves and periodically perform such assessments at home.

At each visit patients are questioned about the presence of symptoms that may be attributed to the striatal DA deficiency. These symptoms are those which were present before levodopa treatment, were ameliorated by levodopa, and then reappeared. These symptoms include difficulty in turning in bed; difficulty arising from a chair; difficulty in initiating gait; tendency to walk with short steps and/or shuffle; difficulty in turning or pivoting while walking; postural deformity with the assumption of a simian posture; and difficulty with hygiene, dressing, feeding, speaking, and writing. Patients are also questioned about the presence of symptoms that may be attributed to the effects of chronic levodopa treatment. These symptoms were not present before levodopa treatment, appeared during treatment, and decreased when levodopa was decreased or stopped. They include diurnal oscillations in performance and dyskinesias which may be associated with each other to varying degrees. The dyskinesias consist of a mixture of choreoathetosis, dystonic posturing, and painful leg spasms. The dyskinesias usually appear after several months of levodopa treatment. Patients are also questioned about the presence of symptoms which probably result from disease progression, but which may be aggravated by levodopa. These symptoms include confusion, depression, agitation, paranoia, delusions, hallucinations, and sleep disturbances. Patients are then questioned about the presence of symptoms that may be attributed to the involvment of nondopaminergic motor systems. These symptoms may have been present before levodopa treatment and are only partly ameliorated by levodopa. The symptoms include stress- and movement-induced freezing spells, falls, and propulsions (Narabayashi *et al.,* 1981). Finally, patients are questioned about the

TABLE 1
NYU Disability Scale

	Turning in bed
	Arising from chair
	Postural deformity
	Start hesitation
	Short steps
	Shuffling gait
	Freezing
	Decreased arm swing
	Turning/pivoting
	Falling
	Propulsions
	Hygiene
	Dressing
	Feeding
	Swallowing/cooking
	Drooling
	Speaking
	Facial masking
	Tremor
	Handwriting

ON–OFF STATUS AT TIME OF ASSESSMENT

☐ **ON** ☐ **OFF**

☐ STAGE 1 Mild, unilateral disease
☐ STAGE 2 Bilateral involvement
☐ STAGE 3 Mild-moderate gait
☐ STAGE 4 Marked gait disturbance
☐ STAGE 5 Confined to bed or wheelchair

Rigidity [MAXIMUM 15]
0 = Absent
1 = Minimal
2 = Moderate
3, 4 = Marked

Right upper limb	Left upper limb	Neck	Right lower limb	Left lower limb
☐ 0	☐ 0	☐ 0	☐ 0	☐ 0
☐ 1	☐ 1	☐ 1	☐ 1	☐ 1
☐ 2	☐ 2	☐ 2	☐ 2	☐ 2
☐ 4	☐ 4	☐ 3		

Bradykinesia [30]

Extremity	Body
☐ 0 = None	☐ 0 = None
☐ 5 = Minimal	☐ 5 = Minimal
☐ 10 = Moderate	☐ 10 = Moderate
☐ 15 = Marked	☐ 15 = Marked

[10] Tremor
0 = Absent
1 = Present
4 = Marked

BP Sit
BP Stand
PR
WGT.

Right upper limb	Left upper limb		Right lower limb	Left lower limb
☐ 0	☐ 0	☐ Rest	☐ 0	☐ 0
☐ 1	☐ 1	☐ Sust. action	☐ 1	☐ 1
☐ 4	☐ 4			

Gait [30]
- ☐ 0 = Freely ambulatory, good stepping
- ☐ 10 = Minimal difficulty
- ☐ 20 = Moderate difficulty, no assistance
- ☐ 30 = Marked difficulty

Posture stability [15]
- ☐ 0 = Normal
- ☐ 5 = Retropulsion but recovers unaided
- ☐ 10 = Would fall if not caught
- ☐ 15 = Unable to stand

Dyskinesia [100]

RUE	0		0: None			LUE	0	
	1	5	1: Present				1	5
	2	20	2: Marked				2	20
RLE	0		Face, head, and neck	0		LLE	0	
	1	5		1	20		1	5
	2	20					2	
							20	

TABLE 2
N.Y.U. Parkinson's Disease Functional Scale[a]

Stage	Status	Description
0	No symptoms	
1	Fully employed, symptoms	Transporation: goes alone Work schedule: full Work assistance: none
2	Partially employed, reduced schedule, assistance at work	Transportation: may go alone Walking: goes out alone ADL: independent
3	Independent: (1) retired non-PD related (2) retired, PD related	Transportation: may/may not go alone Walking: goes out alone ADL: independent (but slow)
4	Semidependent	Walking: May/may not go out alone ADL: mostly independent No aide or part-time aide
5	Dependent: (1) living at home (2) institutionalized	Walking: does not go out alone ADL: mostly dependent Full-time aide

[a] ADL = activities of daily life.

presence of symptoms that indicate autonomic nervous system involvement. These symptoms include blurring of vision, tearing, increased salivation, dysphagia, weight loss, respiratory distress (stridor, dyspnea), constipation, urinary difficulty, change in libido, postural hypotension, altered sweating, and abnormal temperature sensations.

3. LEVODOPA DECREASED RESPONSE

When a patient with parkinsonian features fails to benefit at all from levodopa, a condition defined as primary levodopa failure, then the diagnosis of PD should be questioned (Lieberman *et al.*, 1973). There are several disorders which resemble PD, but which do not respond to levodopa (Lieberman, 1974). They include (1) the "lacunar state," where multiple small, often clinically inapparent, infarcts damage the striatonigral region, leaving it honeycombed with tiny holes (or lacunae); (2) brain tumors in the same region; and (3) obstruction to the flow of cerebrospinal fluid without increased intracranial pressure (normal pressure hydrocephalus). These three conditions may be diagnosed by a computed tomogram (CT scan) of the head. Other (rarer) disorders which also resemble PD, which do not respond to levodopa, but which cannot be diagnosed through a CT scan, include the rigid form of Huntington's disease, Wilson's disease, Hallervorden–Spatz disease, progressive supranuclear palsy, and striatonigral degeneration. Prior to levodopa treatment some of these dis-

TABLE 3
NYU Gait Postural Stability

Gait and posture	0 Without Difficulty	3 Difficulty	5 Marked Difficulty
Getting in/out chair ×3			
Postural stability			
Heel standing			
Toe standing			
Stand right foot 10 sec			
Standing left foot 10 sec			
Start hesitation			
Arm swing			
Short steps			
Shuffling			
Turning right foot (moving)			
Turning left foot			
Freezing			
Heel walking			
Toe walking			
Tandem walking			
Backward walking			
Anteropulsion			
Retropulsion			
Falling			

orders may be suspected on clinical examination and others on the basis of a positive family history (PD is rarely inherited). However, many of these disorders are not suspected until after a therapeutic trial of levodopa. In these patients, especially those with progressive supranuclear palsy and striatonigral degeneration, levodopa not only does not result in alleviation of their parkinsonian symptoms, it does not result in dyskinesias. Presumably, in these patients, degeneration of the striatal DA receptors has occurred and this precludes a response to levodopa.

A subgroup of patients who are considered as being levodopa failures are those patients whose parkinsonism is complicated by cerebellar dysfunction, pyramidal tract dysfunction, and/or amyotrophy (Lieberman *et al.*, 1973). In these patients, although the parkinsonism may respond, in part, to treatment, the major disability results from cerebellar dysfunction, pyramidal tract abnormalities, and/or amyotrophy, and these features do not respond to levodopa.

Another group of patients who are considered as being levodopa failures are patients who are unable to tolerate levodopa because of gastrointestinal symptoms, e.g., nausea, vomiting, and anorexia. Between 2 and 8 g of levodopa a day is required to achieve an antiparkinsonian effect. Up to 90% of the administered levodopa is converted in the periphery outside the brain into DA, and it is this DA which is responsible for the nausea, vomiting, and anorexia. Only 10% of the administered levodopa reaches the brain where it is also converted into DA. To inhibit the conversion of levodopa to DA in the periphery so that more levodopa gets into the brain, but at the same time not inhibit the conversion of levodopa to DA in the brain, a decarboxylase inhibitor which itself does not cross into the brain is combined with levodopa. Two such peripherally acting (extracerebral) decarboxylase inhibitors (PDIs), carbidopa and benserazide, are available for clinical use. In the average patient, between 75 and 200 mg of carbidopa or benserazide is required to inhibit the extracerebral conversion of levodopa to DA. Carbidopa, which is available in the United States and England, is combined with levodopa as Sinemet in a ratio of 10 mg of levodopa to 1 mg of carbidopa (10:1), or in a ratio of 4 mg of levodopa to 1 mg of carbidopa (4:1). The 4:1 tablet contains the same amount of PDI as the 10:1 tablet and thus inhibits the extracerebral conversion of levodopa to DA to the same degree. However, since the new tablet contains less levodopa, the minimum amount of levodopa required for therapeutic efficacy can be delivered, thus sparing the patient some of the symptoms of DA excess. Trials comparing levodopa/carbidopa or levodopa/benserazide versus levodopa alone have consistently demonstrated the superiority of levodopa combined with a PDI (Lieberman *et al.*, 1975). Thus, when levodopa and a PDI are used together, only one-fifth of the dose of levodopa is required, and the antiparkinsonian effect is greater, is achieved sooner, and symptoms of extracerebral DA excess (nausea, vomiting, and anorexia) are reduced. When, despite the coadministration of a PDI, a patient experiences symptoms of extracerebral DA excess, then one may administer a peripheral DA antagonist. The most widely used DA antagonists in the United States are the phenothiazines. However, as these drugs cross the blood–brain barrier, they also act as central DA antagonists and thus may exacerbate PD. Domperidone is a new peripheral DA antagonist which is effective against the symptoms of extracerebral DA excess, and as it does not cross the blood–brain barrier, domperidone does not exaccerbate PD. However, domperidone is currently not available in the United States. On those occasions when, despite the coadministration of a PDI, we have had to control the symptoms of extracerebral DA excess, principally nausea and vomiting, we have given our patients cimetidine. It may be noteworthy that drugs which act as antinauseants (the phenothiazines, domperidone) elevate serum prolactin, while antiparkinsonian drugs, levodopa, and the DA agonists, which may act as nauseants, lower serum prolactin. The association between a drug's ability to act as an antinauseant and to raise serum prolactin may be related, and the DA receptors that mediate nausea may share some properties with the receptors that relese prolactin. Certain drugs may have the ability to interact with both sets of receptors. Thus, cimetidine, which in our patients acts as an antinauseant, also elevates serum prolactin.

4. TREATMENT OF MILD LEVODOPA DECREASED RESPONSE

When a patient's benefits on levodopa begin to decline, after an initial good response, the condition is referred to as secondary levodopa failure (Marsden and Parkes, 1977). These patients exhibit symptoms of increasing parkinsonism often accompanied by diurnal oscillations in performance. Among those patients whose response to levodopa is just starting to decline, several strategies may be employed including increasing the dose of levodopa, changing the ratio of levodopa/carbidopa, giving the same amount of levodopa, but giving it more frequently or giving it on "demand": as soon as the patient begins to go "off." In these patients it may also be helpful to add another antiparkinsonian drug, even if the drug had already been used in the past and discontinued because of a lack of efficacy. Amantadine (Symmetrel), which acts by releasing DA from presynaptic nerve terminals, is effective in many such patients, but unfortunately, the benefits are usually only temporary. When patients do not respond to amantadine, when their response decreases, or when they develop side effects (confusion, urinary retention, livedo reticularis), then an anticholinergic drug may be used. At one stage, early in the disease, a balance between the two neurotransmitters acetylcholine and DA may be required for control of some symptoms of PD. Decreasing the activity of acetylcholine through the use of an anticholinergic agent may, early in PD, be almost as useful as increasing DA. The anticholinergic drugs are effective against rigidity and tremor, but rarely against bradykinesia. The anticholinergic drugs result in drying of the mouth, blurring of vision, mental changes, and difficulty in urinating. Often these side effects are more troubling than the parkinsonian symptoms they alleviate. There are two classes of anticholinergics: (1) the piperidyl derivatives, which include trihexyphenidyl (Artane), biperiden (Akineton), and procyclidine (Kemadrin), and (2) the tropanol derivatives, which include benztropine (Cogentin). A poor response or side effects with one class should not exclude a trial with the other class. Other drugs that may be used are the antihistamines, orphenadrine (Disipal), diphenhydramine (Benadryl), and the phenothiazine ethopropazine (Parsidol). Ethopropazine is particularly effective against tremor. If patients are depressed, then antidepressants which improve mood may often also improve some parkinsonian symptoms. Several antidepressants are available, each having specific advantages and disadvantages.

Bromocriptine (Parlodel), the only dopamine agonist available on a nonexperimental basis in the United States, has been successfully used as a first treatment for PD. Patients may be started on low doses (5–50 mg/day); later, many patients require higher doses (31–100 mg/day); and eventually all patients require the addition of levodopa. However, patients treated with bromocriptine alone do not develop diurnal oscillations in performance, and few patients experience dyskinesias. It is thought, by many investigators, that the duration of the optimal response to levodopa (the "levodopa honeymoon"), prior to the decline in response to levodopa and before the appearance of diurnal oscillations in performance, is a function of the duration of levodopa treatment itself regardless of disease severity. It is thought, by these same investigators, that levodopa treatment should be delayed as long as possible so that the optimal levodopa response will occur late in the disease (when the patient needs levodopa the most), rather than early in the disease (when the patient's requirement for levodopa is least demanding), thus delaying the levodopa honeymoon. Bromocriptine, anticholinergics, and amantadine may all be considered as an initial treatment for PD. Many investigators now advocate adding bromocriptine to levodopa during the "levodopa honeymoon," soon after starting levodopa. Bromocriptine is initially administered in low doses (1.0–2.5 mg/day) and built up gradually (in increments of 1.0–2.5 mg/day each week) to a maximum of 10–20 mg/day). It is thought that such combined treatment may prolong the levodopa honeymoon. Such combination therapy in PD, using two drugs with different antiparkinsonian effects, is becoming as important as combination therapy in other conditions including hypertension, infectious diseases, and cancer.

5. TREATMENT OF MODERATE DECREASED LEVODOPA RESPONSE

Different strategies are required for those patients whose benefit from levodopa has declined further, whose parkinsonian symptoms are more severe, and who are experiencing disabling oscillations in performance, both wearing-off and on–off phenomena. On the assumption that the increasing parkinsonian symptoms and some of the oscillations are, in part, related to inadequate delivery of levodopa to presynaptic neurons, several attempts have been made to augment this delivery. They include prescribing a low-protein diet (to maximize the amount of levodopa absorbed through the gut); increasing the dose of and/or the frequency of administering levodopa; and varying the ratio of levodopa to carbidopa. On the assumption that some symptoms (dyskinesias, mental changes) and some of the oscillations may be related to the effects of chronic levodopa administration with accumulation of DA and of toxic metabolites and/or alteration of the DA receptors by these metabolites, attempts have been made to lessen the effects of such a "DA excess." These attempts include, paradoxically, decreasing the dose of levodopa and even a period of complete levodopa abstinence or "holiday" (Direnfeld *et al.*, 1980; Weiner *et al.*, 1980). After a drug holiday of 3–21 days, symptoms of DA excess lessen while the baseline parkinsonian picture emerges. However, with reinstitution of levodopa, often a lower dose than before the holiday, the parkinsonian signs may improve. The dyskinesias and mental changes lessen, and often there are fewer oscillations. Unfortunately, many patients cannot tolerate a holiday, many patients do not benefit, and in the remainder the improvement is only temporary. In addition, the holiday may be accompanied by serious complications of immobility including aspiration, pneumonia, phlebitis, pulmonary embolus, and depression.

Another approach toward augmenting the effect of levodopa with fewer symptoms of DA excess is through the use of a monoamine oxidase (MAO) inhibitor. Levodopa is catabolized by catechol-*O*-methyl transferase and by MAO. Several early investigators combined levodopa with a MAO inhibitor, but this combination often resulted in hypertensive crises. Subsequently it was shown that there are two forms of MAO: "A" and "B" (Knoll, 1978). MAO-A, present mainly outside the brain, is specialized for metabolizing amines such as serotonin and tyramine which are attached to a 5-hydroxy-indole ring. MAO-B, present mainly inside the brain, is specialized for metabolizing amines such as DA and norepinephrine which are attached to a phenol ring. The MAO inhibitors that were available to the early investigators and the only ones that are still available in the United States are nonselective: they inhibit both MAO-A and MAO-B. Inhibition of MAO-A results in the dietary amine tyramine gaining access to the circulation. As an average meal of some foods (especially cheese) contains enough tyramine to provoke a marked pressor response, ingestion of these foods by patients who are on MAO inhibitors can lead to hypertensive crisis with serious consequences.

Deprenyl (Jumex), in doses of up to 10 mg/day, selectively results in inhibition of MAO-B, thus avoiding the pressor responses that may occur with dietary indiscretion. We administered deprenyl to 25 patients with PD whose response to levodopa had diminished and who were experiencing wearing-off, but not on–off phenomena. Improvement occurred in many of our patients, usually at a 10% lower dose of levodopa. Our findings are similar to those reported by others (Birkmayer *et al.*, 1977; Lieberman *et al.*, 1984a). There are several theories as to why deprenyl may be effective in these patients. One theory is that deprenyl's antiparkinsonian effect is related to the drug's conversion in the body to amphetamine and metamphetamine. These metabolites of deprenyl may, in turn, augment the activity of the patient's endogenous serotonin and norepinephrine, which, in turn, may be responsible for the antiparkinsonian effect. As deprenyl has a greater antiparkinsonian effect than amphetamine, it is unlikely that the metabolites of deprenyl are responsible for the antiparkinsonian effect. Another theory is that deprenyl's antiparkinsonian effect is related to the drug's antidepressant activity. However, deprenyl's antiparkinsonian efficacy is greater than its antidepressant activity. This contrasts with the effects of the tricyclic compounds, whose antidepressant activities are greater than that of deprenyl, but whose antiparkinsonian activity is less than that of deprenyl. We believe that the reason that deprenyl, combined with a lower dose of levodopa, is more beneficial than a higher dose of levodopa without deprenyl is that deprenyl and a lower dose of levodopa result in a similar increase in intracerebral DA as the higher dose of levodopa but with less 3-*O*-methoxydopa (3-*O*-MD). The combination of increased intracerebral DA and decreased 3-*O*-MD may be beneficial because recent evidence suggests that 3-*O*-MD itself lowers intracerebral DA and thus negates much of the benefit of a higher dose of levodopa (Reches and Fahn, 1982).

Another approach, in these patients, is to combine levodopa/carbidopa (Sinemet) with levodopa/benserazide (Madopar). It has been observed that there are some patients who will respond better to Madopar than to Sinemet and vice versa (Diamond *et al.*, 1978). However, in addition to these differences in response, there are differences in the pharmacokinetics of levodopa after it is combined with benserazide (as Madopar) from when it is combined with carbidopa (as Sinemet) (Lieberman *et al.*, 1978; Rinne and Lölsä, 1979). Peak levodopa levels are higher and occur sooner, but decline more rapidly with Madopar. In order to achieve more uniform blood dopa levels, we sought to exploit these differences in levodopa pharmacokinetics by combining Madopar with Sinemet. We treated 38 patients, all of whom exhibited symptoms of increasing parkinsonism, including 22 who also experienced diurnal oscillations in performance, both wearing-off and on–off phenomena. Previous attempts to change the dose, sequence, or ratio of levodopa to carbidopa in these patients had been unrewarding. Ten of the patients improved on the combination of Madopar and Sinemet with a 30% decline in disability, while many other patients experienced a more uniform antiparkinsonian effect. The dose of levodopa in Sinement before adding Madopar was 910 mg, and the ratio of levodopa/carbidopa was 9:1. The dose of levodopa after adding Madopar was approximately the same, but the ratio of levodopa/benserazide + carbidopa was less, 6:1. This suggests that the combination of benserazide + carbidopa may result in a more complete inhibition of extracerebral dopa decarboxylase and thus allow more levodopa to reach the brain. Adverse effects of the combination of benserazide + carbidopa were those of intracerebral DA excess (dyskinesias, mental changes). These adverse effects also suggest that the combination of Madopar and Sinemet may allow more levodopa to reach the brain. The amelioration of diurnal oscillations in performance, predominantly wearing-off phenomena, suggests that the combination does result in more uniform daily blood levodopa levels.

6. TREATMENT OF MARKED DECREASED LEVODOPA RESPONSE

Another approach in patients who exhibit a decreased response to levodopa, but whose symptoms are even more severe, is to use the DA agonists. Use of the DA agonists in these patients is based on the following assumptions. Initially the administration of levodopa compensates for the striatal DA deficiency mainly through an increased conversion of levodopa to DA by the remaining nigrostriatal neurons, and to a lesser extent by the development of supersensitivity of the postsynaptic striatal DA receptors, and through the conversion of levodopa to DA in striatal interneurons that, under ordinary circumstances, do not synthesize DA (Hefti *et al.*, 1980; Hornykiewicz, 1974; Melamed *et al.*, 1980; Ungerstedt, 1971). However, as PD progresses, levodopa no longer compensates for the DA deficiency as the remaining nigrostriatal neurons are too few in number to generate enough DA to stimulate the striatal receptors. Supersensitivity of the postsynaptic striatal DA receptors and the conversion of levodopa to DA in striatal neurons are not enough to compensate for the loss of nigrostriatal neurons. Although there is no proof that the loss of nigrostriatal neurons is paramount, there is a correlation between the degree of degeneration of the nigrostriatal neurons, the striatal DA deficiency, and the severity of PD symptoms (Horynkiewicz, 1974). In refutation of this, it has been argued that some severely incapacitated patients (with extensive loss of nigrostriatal neurons) respond as well as less incapacitated patients (with less extensive loss of nigrostriatal neurons). Furthermore, it has been postulated that the critical factor affecting the decreased response to levodopa is not the progressive loss of nigrostriatal neurons, but the duration of levodopa therapy with the accumulation of excessive DA and other metabolites which then alter the striatal DA receptors (Lesser *et al.*, 1979). Proponents of this theory have also had their assumptions questioned. Thus, it has been noted that while many severely disabled patients do indeed respond to levodopa, their response is usually not as good as nor as long as that of less disabled patients (Markham and Diamond, 1981).

The DA agonists bypass the degenerating presynaptic nigrostriatal neurons and stimulate the postsynaptic DA receptors directly. The most useful of the DA agonists are the ergot alkaloids. Fundamental to our understanding of how these drugs work and to our selecting effective compounds has been the development of experimental models of PD: rats with nigrostriatal lesions and monkeys with ventral tegmental mesencephalic lesions (Goldstein *et al.*, 1973). With these models, a large number of compounds can be tested. Such screening is essential as only a few of the many available compounds can be tested in man. For example, lergotrile, a semisynthetic ergot, was shown to have antiparkinsonian activity but was found to be hepatotoxic in 25% of patients (Lieberman *et al.*, 1979b). Several congeners of lergotrile were available, but all of them could not be subjected to a long and expensive clinical trial. It was reasoned that the cyano group and the halogen of the ergoline portion of the lergotrile molecule were likely to be responsible for the drug's hepatotoxicity. Thus, compounds without the cyano group and the halogen were tested in the models to see whether they had anti-PD activity. Three such compounds with alkyl side chains at the N6 position of the ergoline molecule were found to have such activity (Goldstein *et al.*, 1980). It was then found that the anti-PD activity of these three compounds correlated with the length of the alkyl side chain at the N6 position, with the propyl compound (pergolide) having more activity than the ethyl or methyl compound.

Studies using experimental models have also enabled us to determine that some agonists require the presence of presynaptic DA synthesis and stores, while others do not. Thus, the activity of bromocriptine may be blocked by pretreatment with α-methyl paratyrosine, which inhibits DA synthesis, and by reserpine, which depletes DA stores. Because the activity of bromocriptine is dependent on presynaptic DA synthesis and stores, bromocriptine may not be effective in some severely disabled patients with extensive loss of presynaptic nigrostriatal neurons and depleted DA synthesis and stores. In contrast, the activity of lisuride, another semisynthetic ergoline, is independent of presynaptic DA synthesis and storage, and thus, lisuride may be effective in some of these patients with extensive loss of presynaptic nigrostriatal neurons (Horowski, 1978). Studies with the experimental models have also shown that the various agonists have different affinities for the pre- and postsynaptic DA receptors and for both the agonist and the antagonist binding sites on the post-synaptic striatal membranes. Finally, studies with the experimental models have shown that the ergots act on different striatal DA receptors. Thus, pergolide may be referred to as a "cyclase-linked" drug because it predominantly stimulates the adenyl-cyclase-linked receptors; whereas lisuride, bromocriptine, and lergotrile may be referred to as "noncyclase-linked" drugs because they predominantly stimulate the non-adenyl-cyclase-linked receptors. There are limitations to the experimental model. The most apparent limitation is that the models create a fixed lesion in the mesencephalon which does not progress, unlike the lesion of PD that does.

Bromocriptine was the first ergot alkaloid to be successfully tested in both high (greater than 30 mg/day) and low doses (less than 5–30 mg/day) in PD (Calne *et al.*, 1978; Lieberman *et al.*, 1980b). Bromocriptine combined with levodopa is effective in patients who are responding to levodopa and whose response to levodopa has diminished. Bromocriptine's half-life (3 hr) is longer than that of levodopa (2 hr). This is one reason that it is effective in many patients with diurnal oscillations in performance, particularly patients with wearing-off phenomena. Bromocriptine alone, in low doses (5–30 mg/day) is also effective in newly diagnosed patients with mild or moderate disease who have never been treated with levodopa or in patients who cannot tolerate levodopa (Lees *et al.*, 1978). Bromocriptine alone in high doses (30–100 mg/day) is effective in patients with advanced disease. Patients who are so treated with bromocriptine alone develop dyskinesias or diurnal oscillations in performance. We combined bromocriptine with levodopa in 106 patients with advanced PD and increasing disability despite optimal treatment with levodopa alone (Lieberman *et al.*, 1980b). Fifty-one of these patients had diurnal oscillations in performance, both wearing-off and/or on–off phenomena; 55 of the 106 patients (54%) had, upon addition of bromocriptine to levodopa, an at least one-stage decrease in disease severity; 24 of the patients (44%) experienced a decrease in disease severity in their on period, and 19 in both their on and off periods. Mean dose of bromocriptine was 51.0 mg, permitting a 26% reduction in the dose of levodopa. Bromocriptine was discontinued in 54 patients because of adverse effects including mental changes (29 patients). All adverse effects were reversible on stopping the drug.

Pergolide is a semisynthetic ergot alkaloid that is more potent on a mg/mg basis and longer-acting than bromocriptine (Lemberger and Crabtree, 1979). In the first clinical study in PD, pergolide was added to levodopa in 13 patients with advanced disease and diurnal oscillations in performance (Lieberman *et al.*, 1981). Among the nine patients who completed the trial, there was a significant reduction in disability. Pergolide also resulted in a significant increase in the number of hours the patients were on. The mean daily dose of pergolide was 2.4 mg. Pergolide was subsequently investigated in 56 patients with advanced PD, who were no longer satisfactorily responding to levodopa (Lieberman *et al.*, 1982). The group included 45 patients with diurnal oscillations. Pergolide, when combined with levodopa, resulted in a 44% decrease in disability as assessed in the on period, a 15% decrease in disability as assessed in the off period, and a 148% increase in the number of hours in which patients were on (from 4.6 ± 0.3 hr to 11.4 ± 0.6 hr) (mean ± SEM). All these changes were significant ($p \leq 0.05$, Student's paired t test). Mean dose of pergolide was 2.5 mg (range, 0.2–10.0 mg). Mean duration of the study was 13 months (range, 1 day–34 months). Maximum improvement occurred within 2 months and began to decline within 6 months. The major adverse effects necessitating discontinuing pergolide were the occurrence of an organic confusional syndrome (six patients) and increased dyskinesias (four patients). Nine patients discontinued pergolide because of a lack of effect or declining effect. The long-term effects of pergolide were further studied in 17 patients including 15 with on–off phenomena. Mean duration of the study was 27.8 months (range, 24–38 months). Mean disability score, which had decreased by 60% at the time of pergolide's peak effect, was decreased by only 19.6% after 2 years of treatment. On–off phenomena, which had improved initially, again became prominent. Four patients maintained some of their improvement; nine patients lost much of their improvement; and four lost all their improvement and were now worse than before beginning pergolide. Mean dose of pergolide was 2.2 mg (range, 0.8–5.0 mg); mean dose of levodopa (in Sinemet) was reduced from 998 to 800 mg (range, 150–1500 mg).

Lisuride hydrogen maleate is another semisynthetic ergot that is effective in experimental models of PD. Lisuride is more potent but shorter acting than bromocriptine, and its effects are independent of presynaptic DA synthesis or stores (Horowski, 1978). Lisuride is also a central serotonin agonist. Lisuride was administered to 63 patients with advanced PD, who were no longer satisfactorily responding to levodopa (Lieberman *et al.*, 1983b). The group included 40 patients with diurnal oscillations in performance. Lisuride combined with levodopa resulted in a 34% decrease in PD disability as assessed in the on period, a 16% decrease in disability as assessed in the off period, and a 96% increase in the number of hours during which patients were on (from 5.5 to 10.8 hours). All of these changes were significant ($p \leq 0.001$). The major adverse effect limiting the use of lisuride was the occurrence of an organic confusional syndrome. This was related, in part, to the presence of an underlying dementia and to the concurrent use of anticholinergic drugs.

Treatment with pergolide was compared with bromocriptine in 25 patients, all of whom were also receiving levodopa (Lieberman *et al.*, 1983c). All 25 patients had on–off phenomena. At the time bromocriptine was added to levodopa, the mean age of the patients was 61.8 years; mean PD duration was 9.0 years; and mean levodopa treatment duration was 6.1 years. For the group as a whole, disability as determined in the on period decreased by 36% (from 28.7 to 18.5). Disability, as determined in the off period decreased by 25% (from 59.5 to 44.4). The number of hours in which patients were on increased by 62% (from 7.1 to 11.5). All of these changes were significant ($p \leq 0.05$). Bromocriptine had to be discontinued in nine patients because of adverse effects. Mean dose of bromocriptine was 50 mg (range, 10–100 mg), and mean duration of treatment was 23 months (range, 2–65 months). At the time of their treatment with pergolide the patients were older: 65.5 years; had the disease longer: 12.7 years; and were more disabled. Disability as determined in the on period decreased significantly by 40% (from 43.5 to 26.3). Disability, as determined in the off period decreased significantly by 21% (from 69.0 to 54.8). The number of hours in which patients were on increased significantly by 224% (from 3.4 to 11.0 hr). The mean dose of pergolide was 2.1 mg (range, 0.1–10.0 mg), and the mean duration of treatment was 6.2 months (range, 0.5–20 months). Pergolide was discontinued in eight patients because of adverse effects. In our study, both bromocriptine and pergolide appeared to be equally effective in reducing parkinsonian symptoms. Pergolide was more effective in reducing diurnal oscillations in performance. In another study, in patients at the same stage of the disease, the drugs were comparable (Lewitt *et al.*, 1983).

Treatment with lisuride was compared with bromocriptine in 25 patients, all of whom were also receiving levodopa (Lieberman *et al.*, 1983a). Nineteen patients had "on–off" phenomena. At the time bromocriptine was added to levodopa, the mean age of the patients was 62.7 years; mean PD duration was 8.9 years; and mean levodopa treatment duration was 6.2 years. For the group, disability as determined in the on period decreased by 34%. Disability in the off period decreased by 20%, and the number of hours during which patients were on increased from 9.6 to 12.8. All of these changes were significant ($p \leq 0.01$–0.05). Bromocriptine had to be discontinued in 11 patients because of adverse effects. Mean dose of bromocriptine was 55 mg (range, 20–100 mg), and mean duration of treatment was 21 months (range, 1–65 months). At the time lisuride was added to levodopa, the patients were older: 65.4 years; had the disease longer: 11.4 years; and were more disabled. Nonetheless, disability decreased in the on period by 34%. Disability in the off period decreased by 17%, and the number of hours during which patients were on increased from 3.9 to 8.9. All of these changes were significant ($p \leq 0.01$–0.05). The mean dose of lisuride was 2.8 mg (range, 0.6–5.0 mg); and the mean duration of treatment was 7.4 months (range, 1.0–22.0 months). Lisuride was discontinued in eight patients because of adverse effects. Both bromocriptine and lisuride were found to be equally useful in managing patients with advanced PD.

Treatment with lisuride was compared with pergolide in 25 patients in whom the response to levodopa had diminished (Lieberman *et al.*, 1984b). Sixteen of the patients had on–off phenomena. At the time lisuride or pergolide was added to levodopa, the mean age of the patients was 64.0 years; mean duration of disease was 12.4 years; and mean duration of levodopa treatment was 9.0 years. On lisuride, disability in the on period decreased significantly from 41.9 to 26.3 ($p \leq 0.01$). Disability in the off period also decreased significantly from 72.3% to 61.5% ($p \leq 0.05$). The number of hours during which patients were on increased significantly from 4.0 to 8.6 ($p \leq 0.01$). Lisuride was discontinued in nine patients because of adverse effects, and in eight because of a decrease in efficacy. Mean dose of lisuride was 2.6 mg (range, 0.6–5.0 mg), and mean duration of treatment was 7 months (range, 1–22 months).

On pergolide, disability in the on period decreased significantly from 45.0 to 20.6 ($p \leq 0.01$). Disability in the off period decreased significantly from 71.3 to 54.4 ($p \leq 0.05$). The number of hours in which patients were on increased significantly from 3.8 to 10.4 h ($p \leq 0.01$). The mean dose of pergolide was 2.5 mg (range, 0.1–10.0 mg), and the mean duration of treatment was 7.5 months (range, 0.5–22 months). Pergolide was discontinued in six patients because of adverse effects and in three because of a decrease in efficacy.

Bromocriptine, lisuride, and pergolide were useful in most patients. The differential antiparkinsonian effects of these drugs and the differential spectrum of adverse reactions may reflect the differences in the ability of these drugs to stimulate the different DA receptors.

7. MPTP-INDUCED PARKINSONISM

PD has always been considered an idiopathic degeneration of the central nervous system. Recent work suggests that the premature death of the pigmented DA neurons in the substantia nigra might result from the effect of an environmental toxin (Langston *et al.*, 1983). Thus, 1-methyl-4-phenyl, 1,2,5,6-tetrahydropyridine (MPTP), a by-product of meperidine synthesis admixed with varying amounts of meperidine, was sold as a "synthetic heroin" in California. The admixture was used intravenously by a number of young individuals four of whom then developed severe parkinsonism. All four patients responded to levodopa. Two also required bromocriptine. Two years later all four patients have shown disease progression, and all continue to require levodopa or bromocriptine. One patient is experiencing a decreased response to levodopa. A similar case of MPTP-induced parkinsonism was reported earlier in a patient who then died. Postmortem examination of this patient revealed nerve cell loss in the substantia nigra comparable to that seen in idiopathic parkinsonism. Subsequent studies with monkeys have shown that intravenous or intraperitoneal administration of MPTP results in destruction of the DA neurons in the substantia nigra with loss of DA innervation in the striatum. The cellular destruction is specific for the nigral DA neurons. Pretreatment of DA neurons *in vitro* with a MAO inhibitor, either pargyline (a nonspecific Type A and Type B inhibitor) or deprenyl (a specific Type B inhibitor), protects the DA neurons. Speculation has centered about the possibility that human PD might result from the inadvertant production of an MPTP-like substance by the brain itself or from inadvertent exposure to an environmental MPTP-like substance. Identification of such an endogenous or exogenous substance and possible prevention of its activity by treatment with MAO inhibitors is now central to PD research.

8. SUMMARY

PD is a progressive disorder of the central nervous system (CNS). Most of the symptoms of PD result from degeneration of pigmented neurons that have their cell bodies in the substantia nigra and terminate in the striatum. Initially, the symptoms can be treated by the use of levodopa alone or combined with a peripheral decarboxylase inhibitor (carbidopa, benserazide). However, levodopa does not halt the advance of PD, and in time, disability increases. The increased disability is frequently accompanied by diurnal oscillations in performance: wearing-off and/or on–off phenomena; mental changes (dementia, depression); and autonomic insufficiency. Some of the symptoms arise from progressive loss of neurons in the substantia nigra with the remaining neurons unable to convert the administered levodopa to DA in sufficient quantities to stimulate the striatal receptors. Some of the symptoms arise from an excessive accumulation of toxic levodopa metabolites, and some of the symptoms arise from the degeneration of pigmented dopaminergic, noradrenergic, and nonpigmented neurons in other regions of the CNS and outside the CNS in the sympathetic ganglia.

Treatment of patients whose response to levodopa is beginning to decline may consist of rearranging the dose and scheduling of levodopa, changing the ratio of levodopa to carbidopa, or supplementing levodopa with amantadine or an anticholinergic agent or with bromocriptine (in low doses, 5–30 mg). Treatment of patients whose response to levodopa has declined further may consist of adding deprenyl, a type-B MAO inhibitor, or adding another decarboxylase inhibitor, benserazide, to carbidopa. The simultaneous combination of levodopa with two separate decarboxylase inhibitors may result in a more uniform delivery of levodopa to the striatal receptors. Treatment of patients whose response to levodopa has declined even further may also consist of adding a dopamine agonist. Therapy with three separate dopamine agonists, bromocriptine, pergolide, and lisuride, is discussed and compared.

9. REFERENCES

Bernheimer, H., Birkmayer, W., Hornykiewicz, O., Jellinger, K., and Seitelberger, F., 1973, Brain dopamine and the syndromes of Parkinson and Huntington. Clinical, morphological and neurochemical correlations, *J. Neurol. Sci.* **20**:415–455.

Birkmayer, W., Riederer, P., Ambrozi, L., and Youdim, M. B. H., 1977, Implications of combined treatment with "Madopar" and L-deprenil in Parkinson's disease: A long-term study, *Lancet* **1**:439–443.

Boller, F., Mizutani, T., Roessmann, U., and Gambetti, P., 1980, Parkinson disease, dementia, and Alzheimer disease. Clinicopathological correlations, *Ann. Neurol.* **7**:329–335.

Calne, D. B., Williams, A. C., Neophytides, A., Plotkin, C., Nutt, J. C., and Teychene, P. F., 1978, Long-term treatment of parkinsonism with bromocriptine, *Lancet* **1**:735–738.

Diamond, S. G., Markham, C. H., and Treciokas, L. J., 1978, A double-blind comparison of levodopa, Madopar, and Sinemet in Parkinson disease, *Ann. Neurol.* **3**:267–272.

Direnfeld, L. K., Feldman, R. G., Alexander, M. P., and Kelly-Hayes, M., 1980, Is L-DOPA drug holiday useful? *Neurology* **30**:785–788.

Goldstein, M., Battista, A. F., Ohmoto, T., Anagnoste, B., and Fuxe, K., 1973, Tremor and involuntary movements in monkeys: Effect of L-DOPA and of a dopamine receptor stimulating agent, *Science* **179**:816–187.

Goldstein, M., Lieberman, A., Lew, J. Y., Asano, T., Rosenfeld, M. R., and Markman, M. H.,

1980, Interaction of pergolide with central dopaminergic receptors, *Proc. Natl. Acad. Sci. USA* **77**:3725–3728.

Hakim, A. M., and Mathieson, G., 1979, Dementia in Parkinson disease: A neuropathologic study, *Neurology* **29**:1209–1214.

Hartog Jager, W. A. den, and Bethlem, J., 1960, The distribution of lewy bodies in the central and autonomic nervous system in idiopathic paralysis agitans, *J. Neurol., Neurosurg. Psychiatry* **23**:283–289.

Hefti, F., Melamed, E., and Wurtman, R. J., 1980, Partial lesions of the dopaminergic nigrostriatal system in rat brain: Biochemical characterization, *Brain Res.* **195**:123–137.

Hoehn, M. M., and Yahr, M. D., 1967, Parkinsonism: Onset, progression, and mortality, *Neurology* **17**:427–442.

Hornykiewicz, O., 1974, The mechanisms of action of L-dopa in Parkinson's disease, *Life Sci.* **15**:1249–1259.

Horowski, R., 1978, Differences in the dopaminergic effects of the ergot derivatives bromocriptine, lisuride and d-LSD as compared with apomorphine, *Eur. J. Pharmacol.* **51**:157–166.

Kebabian, J. W., Petzold, G. L., and Greegard, P., 1972, Dopamine-sensitive adenylate cyclase in caudate nucleus of rat brain, and its similarity to the "dopamine receptor," *Proc. Natl. Acad. Sci. USA* **69**:2145–2149.

Knoll, J., 1978, The possible mechanisms of action of (-) deprenyl in Parkinson's disease, *J. Neural. Transmis.* **43**:177–198.

Langston, J. W., Ballard, P., and Tetrud, J. W., 1983, Chronic parkinsonism in humans due to a product of meperidine analog synthesis, *Science* **219**:979–980.

Lees, A. J., Haddad, S., Shaw, K. M., Kohout, L. J., and Stern, G. M., 1978, Bromocriptine in parkinsonism: A long-term study, *Arch. Neurol.* **35**:503–505.

Lemberger, L., and Crabtree, R. E., 1979, Pharmacologic effects in man of a potent, long-acting dopamine receptor agonist, *Science* **205**:1151–1153.

Lesser, R. P., Fahn, S., Snider, S., Cote, L. J., Isgreen, W. P., and Barrett, R. E., 1979, Analysis of the clinical problems in parkinsonism and the complications of long-term levodopa therapy, *Neurology* **29**:1253–1260.

Lewitt, P. A., Ward, C. D., Larsen, T. A., Raphaelson, M. I., Newman, R. P., Foster, N., Dambrosia, J. M., and Calne, D. M., 1983, Comparison of pergolide and bromocriptine therapy in parkinsonism, *Neurology* **33**:1009–1014.

Lieberman, A. N., 1974, Parkinson's disease: A clinical review, *Am. J. Med. Sci.* **267**:66–80.

Lieberman, A. N., Goodgold, A. L., and Goldstein, M., 1973, Treatment failures with levodopa in parkinsonism, *Neurology* **22**:1205–1210.

Lieberman, A., Goodgold, A., Jonas, S., and Liebowitz, M., 1975, Comparison of dopa decarboxylase inhibitor (carbidopa) combined with levodopa and levodopa alone in Parkinson's disease, *Neurology* **25**:911–916.

Lieberman, A., Estey, E., Gopinathan, G., Ohashi, T., Sauter, A., and Goldstein, M., 1978, Comparative effectiveness of two extracerebral DOPA decarboxylase inhibitors in Parkinson disease, *Neurology* **28**:964–968.

Lieberman, A., Dziatelowski, M., Kupersmith, M., Serby, M., Goodgold, A., Korein, J., and Goldstein, M., 1979a, Dementia in Parkinson disease, *Ann. Neurol.* **6**:355–359.

Lieberman, A.N., Gopinathan, G., Estey, E., Kupersmith, M., Goodgold, A., and Goldstein, M., 1979b, Lergotrile in Parkinson disease: Further studies, *Neurology* **29**:267–272.

Lieberman, A. N., Dziatelowski, M., Gopinathan, G., Kupersmith, M., Neophytides, A., and Konein, J., 1980a, The evaluation of Parkinson's disease, in: *Advances in Biochemical Psychopharmacology,* Vol. 23 (D. Calne, M. Goldstein, A. Lieberman, and M. Thorner, eds.), pp. 227–286, Raven Press, New York.

Lieberman, A. N., Kupersmith, M., and Neophytides, A., 1980b,Bromocriptine in Parkinson's disease: Report on 106 patients treated for up to 5 years, in: *Ergot Compounds and Brain Function: Neuroendocrine and Neuropsychiatric Aspects* (D. Calne, M. Goldstein, A. Lieberman, and M. Thorner, eds.), pp. 245–253, Raven Press, New York.

Lieberman, A., Goldstein, M., Leibowitz, M., Neophytides, A., Kupersmith, M., Pact, V., and

Kleinberg, D., 11981, Treatment of advanced Parkinson disease with pergolide, *Neurology* **31**:675–682.

Lieberman, A. N., Goldstein, M., Gopinathan, G., Leibowitz, M., Neophytides, A., Walker, R., Hiesigen, E., and Nelson, J., 1982, Further studies with pergolide in Parkinson disease, *Neurology* **32**:1181–1184.

Lieberman, A. N., Gopinathan, G., Neophytides, A., Leibowtiz, M., Walker, R., and Hiesigen, E., 1983a, Bromocriptine and lisuride in Parkinson disease, *Ann. Neurol.* **13**:44–47.

Lieberman, A. N., Goldstein, M., Gopinathan, G., Leibowitz, M., Neophytides, A., Walker, R., and Hiesigen, E., 1983b, Further studies with lisuride in Parkinson's disease, *Eur. Neurol.* **23**:119–123.

Lieberman, A. N., Neophytides, A. N., Leibowitz, M., Gopinathan, G., Pact, V., Walker, R., Goodgold, A., and Goldstein, M., The comparative efficacy of pergolide and bromocriptine in patients with Parkinson disease, 1983c, in: *Experimental Therapeutics of Movement Disorders,* Vol. 37 (S. Fahn, D. B. Calne, and I. Shoulson, eds.), pp. 95–108, Raven Press, New York.

Lieberman, A. N., Gopinathan, G., and Neophytides, A., 1984a, The antiparkinsonian efficacy of deprenyl: A specific type "B" monoamine oxidase inhibitor, *NY State J. Med.* **84**:13–16.

Lieberman, A. N., Leibowitz, M., Neophytides, A. N., Gopinathan, G., Walker, R., Hiesigen, E., Collins, M., and Goldstein, M., 1984b, Pergolide and lisuride in advanced Parkinson's disease, *Adv. Neurol.* **40**:503–507.

Markham, C. H., and Diamond, S. G., 1981, Evidence to support early levodopa therapy in Parkinson disease, *Neurology* **31**:125–131.

Marsden, C. D., and Parkes, J. D., 1976, "On–off" effects in patients with Parkinson's disease on chronic levodopa therapy, *Lancet* **1**:292–296.

Marsden, C. D., and Parkes, J. D., 1977, Success and problems of long-term levodopa therapy in Parkinson's disease, *Lancet* **1**:345–349.

Mayeux, R., Stern, Y., Rosen, J., and Leventhal, J., 1981, Depression, intellectual impairment, and Parkinson disease, *Neurology* **31**:645–650.

Melamed, E., Hefti, F., and Wurtman, R. J., 1980, Nonaminergic striatal neurons convert exogenous L-dopa to dopamine in parkinsonism, *Ann. Neurol.* **8**:558–563.

Narabayashi, H., Kondo, T., and Hayashi, A., 1981, L-Threo 3,4-dihydroxy-phenylserine treatment for akinesia and freezing of parkinsonism, *Proc. Japan Acad.* **57**(ser B):351–354.

Reches, A., and Fahn, S., 1982, 3-*O*-methyldopa blocks dopa metabolism in rat corpus striatum, *Ann. Neurol.* **12**:267–271.

Rinne, U. K., and Mölsä, P., 1979, Levodopa with benserazide or carbidopa in Parkinson disease, *Neurology* **29**:1584–1589.

Schwartz, R., Creece, J., and Coyle, J. T., 1978, Dopamine receptors localized on cerebral cortical afferents to rat corpus straitum, *Nature* **271**:766–768.

Ungerstedt, U., 1971, Postsynaptic supersensitivity after 6-hydroxydopamine induced degeneration of the nigro-striatal dopamine system, *Acta Physiol. Scandinavia.,* **367**(Suppl.):69–93.

Weiner, W. J., Koller, W. C., Perlik, S., Nausieda, P. A., and Klawans, H. L., 1980, Drug holiday and management of Parkinson disease, *Neurology* **30**:1257–1261.

Yahr, M. D., Wolf, A., and Antuner, J., 1972, Autopsy findings in parkinsonism following treatment with levodopa, *Neurology* **22**(Suppl.):56–71.

CHAPTER 4

The Clinical Evaluation of Drug Therapy in Parkinsonism and Models of Dysfunction of Brain Dopamine Systems in Animals

A Review

J. R. BIANCHINE, R. D. SCHWARTZ, and C. W. RICHARD III

1. INTRODUCTION

The observation by Ehringer and Hornykiewicz (1960) that patients who had died with Parkinson's disease showed cellular degeneration of basal ganglia structures and a deficiency of dopamine (DA) within this area was a key link in the concept that a deficiency in a specific substance was intimately involved in the pathogenesis of a neurological disorder. The efficacy of levodopa treatment in Parkinson's disease obviously reinforced this idea. At about the same time as these discoveries, both biochemical and histological techniques became available for researchers to examine brain catecholamine systems that appeared to be involved in neurological illnesses (Carlsson, 1959; Dahlström and Fuxe, 1964).

Coincidental to these very important basic observations with exciting practical clinical utility in movement disorders, it became essential to develop methods for evaluation of the clinical efficiacy of drug therapy in Parkinson's disease and to develop animal models for evaluation of dopaminergic drugs. The objective of this chapter is to summarize both the basic and clinical evaluation of drugs impacting the dopaminergic system.

J. R. BIANCHINE, R. D. SCHWARTZ, and C. W. RICHARD III • Department of Pharmacology, College of Medicine, The Ohio State University, Columbus, Ohio 43210. *Present address for J. R. B.:* Medical Affairs, American Critical Care, McGaw Park, Illinois 60085.

1.1. Clinical Evaluation of Parkinsonism

Parkinson's disease, a form of movement disorder that appears in the later decades of life, produces progressively profound disability in movement (Bianchine, 1974, 1980; Rose and Capildeo, 1981). Parkinsonism is manifested clinically by a combination of four major clinical features: tremor, rigidity, bradykinesia, and disturbance of posture.

A brief description of each of the major clinical features of parkinsonism follows:

Tremor represents a rhythmical alternating contraction (3–5/sec) of a given muscle group and its antagonist. Distal muscles are more commonly involved than proximal muscles. Tremor may occur unilaterally for years before becoming generalized. Tremor present during rest usually disappears on purposeful movement or during sleep but becomes more intense with anxiety or stress.

Rigidity is a clinical manifestation of increased muscle tone. On examination, there is resistance to passive movement of an extremity in the direction of displacement. The terms "cogwheel" and "ratchet" rigidity vividly describe the type of resistance offered to passive movement in parkinsonism.

Bradykinesia actually represents a clinical composite of three factors: marked poverty of spontaneous movement, loss of normal associated movements, and slowness in initiation of all voluntary movements. "Masked" facies classically demonstrate all these features.

Postural fixation is usually a late development in parkinsonism. The patient is unable to maintain an upright position of the trunk while standing or walking. The patient then may find himself "chasing" his own center of gravity. This obviously will seriously compromise ambulation.

Several other aspects of parkinsonism may become important and difficult clinical problems for individual patients. These factors include swallowing defects, delayed gastric emptying, severe constipation, seborrhea, difficulty in speech, and emotional depression.

Of course, since parkinsonism commonly begins in the fifth decade of life or later, a whole raft of diseases associated with aging may further complicate the evaluation and management of these patients.

Each of these major and minor features of Parkinsonism is highly variable and labile. Certain parkinsonians manifest only one or two of these features, sometimes in severe form. Others unfortunately manifest many, if not all, of these major and minor features. In addition to great clinical differences between individual Parkinsonian patients, there may be significant day-to-day and even within-day variations of individual patients.

1.2. Clinical Scaling Assessments

In summary, parkinsonism is a highly variable, multifaceted manifestation of basal ganglia disease which appears to defy quantitation and evaluation.

Nonetheless, several clinical rating systems have been devised to measure the clinical status or disability of individual patients with parkinsonism (Alba *et al.*, 1968; Canter *et al.*, 1961; Langrall and Joseph, 1972).

The Northwestern University Disability Scale is a good example of a clinical scaling system. It assesses each of the following parameters:

1. Rigidity
2. Tremor
3. Bradykinesia
4. Gait
5. Posture
6. Postural stability
7. Swallowing
8. Sialorrhea
9. Seborrhea
10. Walking
11. Dressing
12. Eating
13. Feeding
14. Hygiene
15. Speech

Each of these parameters is assessed on an arbitrary basis by means of scores ranging from 0 to 4. A score of 0 represents no clinical involvement while a score of 4 represents maximal involvement. Upon repeated evaluation on a regular basis, a baseline score of each of these parameters may be established. Some investigators prefer to summate the diverse scores from each of these parameters to arrive at a total "global score." While this is a highly artificial and contrived score, it may give some estimation of overall status of the patient.

The limitations of such scoring systems are great. These rating scales are obviously highly subjective, and significant interobserver variations exist. It is readily apparent that global scores might remain unchanged even though significant shifts occurred in two or more of the individual subparameters that contribute to the total global score. Unfortunately, a plethora of disability scales have been formulated. It appears as if each major neurology center has its own scaling system. It would be very helpful if an ad hoc committee of eminent authorities in parkinsonism would devise a single system that might be acceptable to most clinics nationwide and then strongly recommend its general use. Despite the shortcomings of these rather crude scaling systems, they are sufficiently sensitive to detect the pharmacological efficacy of standard antiparkinsonian drugs such as anticholinergic drugs and levodopa.

There is a great inclination among investigators to consider the nominal values derived from these arbitrary scaling procedures as quantitative measurements. Hence, some investigators cannot resist the temptation to carry out standard statistical manipulations with these scores. Of course, this is an improper and meaningless manipulation of purely nominal data. A wide variety of nonparametric methods may be applied to evaluate this sort of information. These methods do not make the assumption that the data are continuous or directly comparable.

1.3. Quantitative Scaling Assessments

Considerable effort has been directed to devising specific, accurate, and reproducible techniques that quantitate each of the major parameters of Parkinson's disease. Some of these special techniques are very elaborate and time-consuming. They may employ expensive recording and monitoring equipment and simply are very impractical for general use. Other techniques are relatively simple but unfortunately are not very discriminating. A brief description of some of these techniques follows.

1.4. Rigidity

Rigidity may be the most easily quantified phenomena in parkinsonism. While various methods have been described for rigidity measurements, it appears that a very practical method is based on the principle of torque measurement about a joint (Boman and Meurman, 1970; Johns and Draper, 1964; Knutsson and Martensson, 1971; Schwab, 1964). For example, muscle tone in the thigh can be accurately quantitated by measuring the swinging characteristics of the lower limb as a function of time after eliciting a knee jerk. The "swing time" then can be recorded and analyzed with a variety of devices. A typical study would establish a baseline measurement of rigidity for the individual patient, and then a treatment would be administered and repeat measurement made. Each patient thus serves as his own control. Numbers derived from such procedures are real measurements and hence may be subjected to statistical analysis.

1.5. Gait

Goniometers attached to adjustable knee-cage-type braces and attached to both legs may be used to quantitate gait in the parkinsonian patient. Electronic signal reducers, analogue recording devices, and computer programs for data analysis can reduce each stride into meaningful data. Knutsson and Martensson (1971) and Knutsson (1975) have subdivided the stride of Parkinsonian patients into distinctly measureable components, which include speed of forward progression, duration of walking cycles, stride length, swing-to-support-time ratio, duration of double-limb support, and symmetry of steps. They made these analyses by means of intermittent light photography and filming, a very time-consuming approach but one that yields precise data of gait.

1.6. Bradykinesia

Bradykinesia in specific limbs may be evaluated with slight variations of the techniques noted previously. Of course, as bradykinesia becomes severe or incapacitating, these techniques, which demand significant cooperation and movements by the patient, lose applicability.

1.7. Tremor

The involuntary tremors seen in patients with Parkinson's disease remain a poorly understood phenomenon and one which is difficult to quantitate.

Tremor is probably the most labile and variable component of Parkinson's disease. Any sort of stress, sometimes stress of a very minor nature, can precipitously exacerbate tremor for certain patients.

Therefore, one might elicit maximal amplitude of tremor by giving a patient a standardized stress. This might be a standardized difficult mental task or calculation that must be completed in a brief time interval.

Several techniques have been utilized to record tremor (Velasco and Velasco, 1973). These techniques vary from very simple (but effective) draw-a-circle tests to very intricate electronic measurements of tremor coupled with a computer. For example, needle electrodes inserted into a selected set of flexor and extensor muscles in the hand can nicely record muscle tremor. An integrator can be used to quantitate this sort of information.

Tremulousness in the voice is difficult to demonstrate by acoustical measurements. Tremor of the lips can be nicely documented with properly recorded electromyography (Leanderson *et al.*, 1972).

A definitive study is needed to compare the relative merits of a general clinical disability scaling system with selected elaborate analytical systems as described previously.

1.8. Placebo

It is clear that no matter which sort of system is used, either a gross clinical scale or a precise analytical system, major extrinsic variables must be adequately controlled.

A therapeutically beneficial placebo effect of 20–30% has been shown repeatedly in parkinsonian patients. Recent work with a new experimental inhibitor of dopa decarboxylase confirmed a mean placebo effect of 27% among a total of 48 parkinsonian patients studied. This, of course, is a factor of significance and must be dealt with squarely.

It is well known among clinicians that any abrupt cessation or significant alteration in the established therapeutic regime for an individual parkinsonian patient may markedly worsen the clinical status of such patients. Therefore, gradual withdrawal of medication to be deleted or to be replaced by a test medication is essential. Usually this withdrawal may require at least 2 weeks. Despite this precaution, the investigator may be faced with significant deterioration in clinical status of the patient.

Most clinical investigators, expert in management of parkinsonism, are not excessively reluctant to slowly withdraw an anticholinergic, antihistaminic, or sedative drug in order to evaluate another similar drug. However, as the fundamental value of levodopa for the majority of parkinsonians is now clearly established, many investigators are reluctant to discontinue levodopa therapy for study purposes. This raises the question, should future studies of ancillary or secondary antiparkinsonian drugs be studied using patients on "optimal" or "baseline" levodopa therapy?

While hard data that would help answer this question are not available as yet, many investigators, including these authors, have the "gut feeling" that indeed, whenever possible, levodopa should form the basic therapy for parkinsonians around which the total therapeutic regime is constructed. Similarly, evaluation of new "secondary" drugs in parkinsonism might be accomplished with levodopa as a preexisting standard or baseline therapy.

The utilization of levodopa as baseline therapy during evaluation of a second drug introduces a highly complex variable into the experimental design. For example, levodopa has a short but variable serum half-life. Therefore, time of dosing with levodopa may be critical to ensure maintenance of a stable baseline. Levodopa is absorbed primarily in the small intestine, and its rate of absorption is highly dependent on gastric juice pH and gastric emptying time. Since levodopa is an amino acid, it must complete for absorption sites with the amino acid load derived from a meal. Therefore, amount and timing of meals may be especially critical during the study (Peaston and Bianchine, 1970; Bianchine *et al.*, 1971).

An interesting possible drug–drug interaction may appear in such a testing system. It is well known that anticholinergic drugs useful in parkinsonians also may cause considerable slowing of the gut with resultant delay in gastric emptying. This, of course, may significantly minimize the therapeutic benefit derived from the concomitantly administered levodopa. The ingestion of vitamin B_6, pyridoxine, an important cofactor for dopa decarboxylase, must be controlled as it may alter the rate of extracerebral metabolism of levodopa (Bianchine *et al.*, 1978).

1.9. Conclusion

Despite remarkable clinical variations among parkinsonian patients and even more troublesome day-to-day and within-day variations for individual parkinsonians, the currently available techniques described in Section 1 serve reasonably well and can discriminate between active antiparkinsonian drugs and placebo. While the available clinical disability scaling systems are adequate, a single universally accepted scaling technique should be established and adhered to so that some sort of comparison may be made between studies performed in the different major clinics.

2. ANIMAL MODELS IN DRUG-INDUCED DYSFUNCTION OF BRAIN DA SYSTEMS

In order to construct animal models to test drugs that will affect brain DA, sites must be identified for pharmacological manipulation. Moore and Bloom (1978), Lindvall and Björklund (1978), and Lloyd *et al.* (1981) present review articles on the anatomical mapping of central afferent and efferent DA pathways. In brief, there are at least seven DA pathways with the majority of brain DA contained in only several of these paths. The mesostriatal system has two major divisions: the nigrostriatal path (A_9 designation by Dahlström and Fuxe), which originates in the substantia nigra and terminates in the striatum, and the mesolimbic (A_{10}) path, which projects to areas such as the nucleus accumbens, olfactory tubercle, and stria terminalis from cell bodies in the ventral tegmentum. Associated with these pathways are interneurons and fiber systems which utilize such substances as acetylcholine (ACh), norepinephrine (NE), serotonin (5-HT), GABA, substance P, glutamate, and β-endorphin. These pathways and substances have been shown to be intimately involved in drug-induced brain dysfunction.

Recent evidence using radioligand receptor binding techniques has suggested that there may be multiple DA receptor types (Seeman, 1980; Creese *et al.*, 1983; Leff and Creese, 1983). It is currently unclear which receptor is involved in a specific neurological disoder, but it may someday be possible to design drugs that preferentially affect a particular receptor which has been linked to a specific disorder.

Proposed animal models are constructed by combining the anatomical, behavioral, and pharmacological data and correlating these results with clinical reports. Of the proposed animal models currently in use for studying DA-dependent behavior, we will examine locomotor behavior, stereotype behavior, drug-induced circling behavior, and neuroleptic-induced dyskinetic movements in primates and rodents.

2.1. Locomotor Behavior

Behavior has been broadly classified into two general categories: conditioned and unconditioned. Spontaneous motor activity, an unconditioned behavior, appears to be necessary for specific motivated behaviors such as feeding, drinking, and mating, and also for conditioned or learned behaviors (e.g., avoidance, self-stimulation). The spontaneous motor behavior of rats has been shown to be dependent on intact forebrain DA systems. If forebrain DA is depleted, this type of behavior is markedly reduced, while excessive stimulation of this same system results in an increase in locomotion (Iversen, 1977).

Depletion of forebrain DA has been accomplished by pharmacological and mechanical methods. Reserpine and tetrabenazine have been shown to deplete amine storage granules, while α-methyl-*p*-tyrosine blocks tyrosine hydroxylase, the rate-limiting enzyme in catecholamine synthesis. Although each of these agents affects NE as well as DA, it is the effect on DA transmission which has best correlated with changes in locomotor behavior. Restoration of newly synthesized DA stores is required for normal behavior. Blockade of DA receptors by neuroleptic drugs also markedly reduced spontaneous motor activity in rats. Drugs such as α-flupenthixol, which are more selective for DA receptors than other neuroleptics affecting both NE and DA receptors, give further specificity to the idea of DA control of locomotion (Iversen, 1977; Kelly, 1977).

Selective lesioning of the mesolimbic DA system with 6-hydroxydopamine, but not the nigrostriatal system, severely diminishes locomotor activity. In addition, bilateral lesions to the ventral or dorsal NE bundles do not affect locomotor behavior (Kelly, 1977; Creese and Iversen, 1975). Taken together, these pieces of evidence lend further support to the idea that brain DA systems control locomotion.

Release of DA by amphetamine, direct stimulation of the DA receptor by apomorphine, or blockade of the DA uptake system by uptake inhibitors such as nomifensine all increase locomotor activity. Stimulation of supersensitive DA receptors seen after chronic neuroleptic treatment also results in large increases in animal activity. The supersensitivity of DA receptors has been suggested to be responsible for tardive dyskinesia (Klawans, 1973; Baldessarini and Tarsy, 1980; Tarsy and Baldessarini, 1984).

Although measurement of locomotor behavior may be useful as an indicator of dopaminergic activity, many different types of drugs can influence this behavior. Agents that have been shown to stimulate activity include amphetamine, apomorphine, serotonergic agonists, cocaine, opiates, muscarinic antagonists, nicotinic agonists, and GABA antagonists. Depressants such as barbiturates, alcohol, neuroleptics, cholinergic agonists, serotonergic antagonists, and GABA agonists decrease motor activity. The dose of a specific drug may also be important since drugs like amphetamine and morphine stimulate motor activity at low doses, but as the dose is increased, the appearance of stereotyped behaviors block motor activity (Kelly *et al.*, 1975).

At best, measurement of locomotor activity is a crude measurement of dopaminergic activity.

2.2. Stereotypic Behavior

It has repeatedly been observed that amphetamine, which releases DA from nerve terminals, and apomorphine, which acts directly on DA receptors, both produce intense stereotyped behavior. At high doses this behavior is seen as sniffing, biting, gnawing, and licking with increased locomotion seen at low doses. Since this syndrome is produced by DA agonists and blocked by DA antagonists, it has been strongly suggested that stereotyped behavior is produced through DA pathways (Kelly *et al.*, 1975).

The neuronal pathways mediating stereotypy have been examined by a variety of techniques. Bilateral lesions of the striatum markedly reduced both amphetamine- and apomorphine-induced stereotypy (Naylor and Olley, 1972; Costall and Naylor, 1973). Bilateral electrolytic lesions of the globus pallidus, an area receiving striatal efferent projections, also blocked both apomorphine and amphetamine stereotypy (Costall and Naylor, 1973). Stereotypy caused by amphetamine, which appears to release newly synthesized DA, can be blocked by the prior administration of α-methyl-*p*-tyrosine but not by prior reserpine administration (Weissman *et al.*, 1966). Direct injection of DA into the striatum elicits sniffing and biting in rats, while similar injections into the cortex were ineffective. In contrast, antagonists injected directly into the striatum abolish drug-induced stereotypy and produce catalepsy (Fog *et al.*, 1967, 1968).

As with locomotor activity, brain systems other than those mediated by DA have been shown to alter stereotypic behavior. Anatomical studies have shown that DA neurons synapse with and are inhibitory on cholinergic neurons. Thus, the final output of striatum is influenced by cholinergic activity. The data showing that anticholinergics potentiate the effects of amphetamine while cholinergic agonists augment α-methyl-*p*-tyrosine blockade of amphetamine-induced stereotypy are consistant with the anatomical studies (Arnfred and Randrup, 1968). Effects on stereotypy by GABA, serotonin (5-HT), and NE are less clear with conflicting reports clouding the question of their activities on stereotyped behavior (Kelly, 1977).

Measurement of drug-induced stereotyped behavior is a more sensitive indicator of dopaminergic activity than measurement of only motor activity. However, most studies rely on visually scoring an animal's behavior and are therefore highly subjective and differ from laboratory to laboratory. They do offer a rapid procedure that can grossly measure the activation and blockade of brain DA system.

2.3. Drug-Induced Circling Behavior

Circling behavior controlled by brain DA pathways appears to be the result of an imbalance of dopaminergic activity between the two corpora striata (Ungerstedt, 1971). This asymmetry is usually the result of lesioning procedures, electrical stimulation, or drug administration directly into brain sites.

Many circling experiments unilaterally lesioning a striatum do so by electrolytic or chemical means. Electrolytic lesions are dependent on current strength and duration and nonspecifically destroy all cells in the area of the electrode tip. Neurotoxins, such as 6-OHDA, which destroys only catecholamine neurons, appear to be more selective as to the type of cells destroyed. Once lesions create the DA imbalance, circling away from the side of higher DA activity is usually seen. The administration of amphetamine results in circling toward the lesions (ipsilateral direction) since DA is released from intact nerve terminals on the unlesioned side. Apomorphine, a direct-acting agonist, causes ipsilateral circling in electrolytically lesioned animals, but circling away from the lesion (contralateral direction) in 6-OHDA-treated animals owing to the development of supersensitivity on the lesioned side (Ungerstedt, 1971). The DA precursor, levodopa, would show a profile similar to that of apomorphine. Administration of neuroleptic drugs, which block DA receptors, blocks circling behavior induced by DA agonists.

Unilateral electrical stimulation of brain DA-containing regions also results in contralateral circling which can be blocked by DA antagonists (Arbuthnott and Ungerstedt, 1975). However, variation in electrode placement has resulted in confusion concerning direction and type of circling observed.

Direct injection of various drugs into discrete brain sites also results in circling behavior. Injection of DA or DA agonists into one striatum will result in circling which is similar to electrical stimulation at the same site (Costall and Naylow, 1974). In contrast, DA receptor blockers injected intrastriatally will result in ipsilateral circling (Costall *et al.*, 1972).

The circling model has been widely used for examining DA agonists which would be of benefit of Parkinson's disease and DA receptor blockers used in the treatment of schizophrenia. However, the model suffers from at least two criticisms. Nondopaminergic agents affect circling, and circling can be elicited at sites that do not involve DA pathways. In general, cholinergic drugs reduce DA-dependent circling while anticholinergics potentiate it. Increased 5-HT activity appears to decrease circling, with the converse true for blockade of 5-HT receptors associated with specific DA pathways. GABA neurons associated with the nigrostriatal pathway also modulate DA-dependent circling, but the action is site dependent. In addition, opiate peptides and substance P have also been found to alter circling. It appears, then, that many substances are involved either directly or in modulatory capacities in what initially appeared to be a simple behavior. In regard to sites associated with circling, many brain sites have been shown to produce postural assymmetries and circling. However, it is important to remember that it is an imbalance of the basal ganglia which is primarily responsible for this motor activity (Pycock, 1980).

For review articles concerning these points on circling behavior, see Glick *et al.* (1976), Dankova *et al.* (1978), and Pycock (1980).

2.4. Dyskinetic Movements in Primates and Rodents following Neuroleptic Treatment

One of the primary goals of pharmacologists and clinicians working on schizophrenia is the development of an antipsychotic agent that would not induce extrapyramidal side effects (EPS) or tardive dyskinesia as a result of DA receptor blockade. Although it was once believed that the clinical efficacy of antipsychotics was inseparable from the neurological side effects, more recent work suggests that separation of activity may be possible.

One animal model that appears to predict the ability of an antipsychotic to produce extrapyramidal dysfunction is the acute dyskinetic syndrome in primates following neuroleptic treatment (Weiss *et al.*, 1977; Liebman and Neale, 1980; Neale *et al.*, 1984; Gerlach *et al.*, 1984). In the model of Liebman and Neale, haloperidol was administered to squirrel monkeys at intervals of 7 days, and within 7–25 weeks of treatment, the monkeys would display dystonia (bizarre posture) and dyskinesia (abnormal movements) upon acute neuroleptic challenge. Once acutely challenged, the animals are observed for behavioral changes, and in some of the animals that had been pretrained, performance in Sidman avoidance task was measured. The monkeys developed dystonias and dyskinesia which was correlated to the blockade of avoidance performance. Drugs scoring high EPS liability also score high in humans (Neale *et al.*, 1981). Using a similar model to evaluate the roles of DA, ACh, and GABA in dystonias, Casey *et al.* (1980) found that DA agonists and ACh antagonists appeared to reduce dystonic symptoms while DA antagonists and ACh agonists intensified reactions. The role of GABA, however, remains unclear with a modulatory role possible.

The primate model appears to be an excellent preclinical screen for the extrapyramidal potential of prospective antipsychotic agents. The major drawback appears to be cost since few institutions have the necessary facilities for large colonies of monkeys. In addition to the cost, the model does differ from the human condition since the movements are transient and self-resolving as well as enhanced following each dose of neuroleptic, whereas in man they are not (Goetz *et al.*, 1983). The high correlation with human clinical studies may, however, outweigh these factors.

Because of the high cost of the primate model, a second approach in studying abnormal motor movements has been to utilize rodents. Rats or guinea pigs are chronically administered neuroleptics and then challenged with either apomorphine or amphetamine. Characteristic gnawing and mouthing movements develop which appear to correlate with increased dopaminergic striatal activity (Glassman and Glassman, 1980; Waddington *et al.*, 1982; Goetz *et al.*, 1983). However, unlike man, no actual motor movements are elicited, and after drug withdrawal, the mouthing movements may disappear.

3. CONCLUSION

The remarkable clinical advances in the management of Parkinson's disease associated with the development of dopaminergic drugs have been equally paralleled by excellent animal models for the study of drug action on brain dopamine systems. We have contrasted the rather simple but effective clinical methods that are useful in the evaluation of dopaminergic drugs in Parkinson's disease and the specific techniques used to evaluate these same compounds in animal models of dopamine-mediated behavior.

It is obvious that these techniques will allow even more accelerated advances in this important area of movement disorders.

4. REFERENCES

Alba, A., Trainor, Frieda, S., Ritter, W., and Dacso, M. M., 1968, A clinical disability rating for Parkinson patients, *J. Chron. Dis.* **21**:507–522.

Arbuthnott, G. W., and Ungerstedt, U., 1975, Turning behavior induced by electrical stimulation of the nigro-neostiratal system of the rat, *Exp. Neurol.* **47**:162–172.

Arnfred, T., and Randrup, A., 1968, Cholinergic mechanisms in brain inhibiting amphetamine-induced stereotyped behavior, *Acta. Pharmacol. Toxicol.* **26**:384–394.

Baldessarini, R. J., and Tarsy, D., 1980, Dopamine and the pathophysiology of dyskinesias induced by antipsychotic drugs, *Ann. Rev. Neurosci.* **3**:23–41.

Bianchine, J. R., 1974, Evaluation of drug therapy in Parkinson's disease, in: *Principles and Techniques of Human Research and Therapeutics: Psychopharmacological Agents*, Vol. 8 (F. G. McMahon, ed.), pp. 171–179, Futura Publishing Company, New York.

Bianchine, J. R., 1980, Drugs for Parkinson's disease; Centrally acting muscle relaxants, in: *The*

Pharmacologic Basis of Therapeutics (A. G. Gilman, L. S. Goodman, and A. Gilman, eds.), pp. 475–493, Macmillan, New York.

Bianchine, J. R., Calimlim, L. R., Morgan, J. P., Dujovne, C. A., and Lasagna, L., 1971, Metabolism and absorption of L-3,4 dihydroxyphenylalanine in patients with Parkinson's disease, *Ann. NY Acad. Sci.* **179:**126–140.

Bianchine, J. R., Shaw, G. M., Greenwald, J. E., and Dandalides, S. M., 1978, Clinical aspects of dopamine agonists and antagonists, *Fed. Proc.* **37:**2434–2439.

Boman, K., and Meurman, T., 1970, Investigation on the effect of some drugs on the Parkinsonian rigidity, *Acta. Neurol. Scand.* **46:**71–84.

Canter, G. J., DeLatorre, R., and Mier, M., 1961, A method for evaluating disability in patients with Parkinson's disease, *J. Nerv. Ment. Dis.* **133:**143–147.

Carlsson, A., 1959, The occurance, distribution, and physiological role of catecholamines in the nervous system, *Pharmacol. Rev.* **11:**490–493.

Casey, D. E., Gerlach, J., and Christensson, E., 1980, Dopamine, acetylcholine, and GABA effects in acute dystonia in primates, *Psychopharmacology* **70:**83–87.

Costall, B., and Naylor, R. J., 1973, The role of telencephalic dopaminergic systems in the mediation of apomorphine-stereotyped behaviour, *Eur. J. Pharmacol.* **24:**8–24.

Costall, B., and Naylor, R. J., 1974, Specific asymmetric behavior induced by the direct chemical stimulation of neostriatal dopaminergic mechanisms, *Naunyn-Schmeideberg's Arch. Pharmacol.* **285:**83–98.

Costall, B., Naylor, R. J., and Olley, J. E., 1972, Catalepsy and circling behaviour after intracerebral injections of neuroleptic, cholinergic and anticholinergic agents into the caudate-putamen, globus pallidus and substantia nigra of rat brain, *Neuropharmacology* **11:**645–663.

Creese, I., and Iversen, S. D., 1975, The pharmacological and anatomical substrates of the amphetamine response in the rat, *Brain Res.* **83:**419–436.

Creese, I., Sibley, D. R., Hamblin, M. W., and Leff, S. E., 1983, The classification of dopamine receptors, *Annu. Rev. Neurosci.* **6:**43–71.

Dahlström, A., and Fuxe, K., 1964, Evidence for the existence of monoamine-containing neurons in the central nervous system. I. Demonstration of monoamines in the cell bodies of brain stem neurons, *Acta. Physiol. Scand.* **232**(Suppl.):1–55.

Dankova, J., Bedard, P,. Langelier, P., and Poirier, L. J., 1978, Dopaminergic agents and circling behavior, *Gen. Pharmacol.* **9:**295–302.

Erhinger, H., and Hornykiewicz, O., 1960, Verteilung von noradrenalin und dopamin (3-hydroxytyramine) in geshirn des menschen und iht verhalten bei erkrangungen des extrapyramidalen systems, *Klin. Wochenschir.* **38:**1236–1239.

Fog, R., Randrup, A., and Parkkenberg, H., 1967, Aminergic mechanisms in corpus striatum and amphetamine-induced stereotyped behaviour, *Psychopharmacol.* **11:**179–193.

Fog, R., Randrup, A., and Parkkenberg, H., 1968, Neuroleptic action of quaternary chlorpromazine and related drugs injected into various brain areas in rats, *Psychopharmacology* **12:**428–432.

Gerlach, J., Bjorndal, N., and Christensson, E., 1984, Methylphenidate, apomorphine, THIP, and diazepam in monkeys: Dopamine-GABA behavior related to psychoses and tardive dyskinesia, *Psychopharmacology* **82:**131–134.

Glassman, R. B., and Glassman, H. N., 1980, Oral dyskinesia in brain-damaged rats withdrawn from a neuroleptic: Implication for models of tardive dyskinesia, *Psychopharmacology* **69:**19–25.

Glick, S. D., Jerussi, T. P., and Fleisher, L. N., 1976, Turning in circles: The neuropharmacology of rotation, *Life. Sci.* **18:**889–896.

Goetz, C. G., Klawans, H. L., and Carvey, P., 1983, Animals models of tardive dyskinesia: Their use in the search for new treatment methods, *Mod. Prob. Pharmacopsychiatry.* **21:**5–20.

Iversen, S. D., 1977, Brain dopamine systems and behavior, in: *Handbook of Psychopharmacology*, Vol. 1 (L. L. Iversen, S. D. Iversen, and S. H. Snyder, eds.), pp. 333–374, Plenum Press, New York.

Johns, R. J., and Draper, I. T., 1964, The control of movement in normal subjects, *Bull. Johns Hopkins Hosp.* **115:**447–464.

Kelly, P. H., 1977, Drug-induced motor behavior, in: *Handbook of Psychopharmacology*, Vol. 8 (L. L. Iversen, S. D. Iversen, and S. H. Snyder, eds.), pp. 295–320, Plenum Press, New York.
Kelly, P. H., Seviour, P. W., and Iversen, S. D., 1975, Amphetamine and apomorphine responses in the rat following 6-OHDA lesions of the nucleus accumbens septi and corpus striatum, *Brain Res.* **94**:507–522.
Klawans, H. L., Jr., 1973, The pharmacology of tardive dyskinesias, *Am. J. Psychiatry.* **130**:82–86.
Knutsson, E., 1975, An analysis of Parkinsonian gait, *Brain* **95**:475–486.
Knutsson, E., and Martensson, A., 1971, Quantitative effects of L-dopa on different types of movements and muscle tone in Parkinsonian patients, *Scand. J. Rehab. Med.* **3**:121–130.
Langrall, H. M., and Joseph, C., 1972, Evaluation of safety and efficacy of levodopa in Parkinson's disease and syndrome, *Neurology* **22**:1–14.
Leanderson, R., Meyerson, B. A., and Persson, A., 1972, Lip muscle function in parkinsonian dysarthria, *Acta. Otolaryng.* **74**:350–357.
Leff, S. E., and Creese, I., 1983, Dopamine receptors re-examined, *Trends in Pharmacological Science* **4**(11):463–467.
Liebman, J., and Neale, R., 1980, Neuroleptic-induced acute dyskinesias in squirrel monkeys: Correlation with propensity to cause extra-pyramidal side effects, *Psychopharmacology* **68**:25–29.
Lindvall, O., and Björklund, A., 1978, Anatomy of the dopaminergic neuron systems in the rat brain, *Adv. Biochem. Psychopharmacol.* **19**:1–23.
Lloyd, K. G., Broekkamp, C. L. E., Cathala, F., Worms, P., Goldstein, M., and Asano, T., 1981, Animal models for the prediction and prevention of dyskinesias induced by dopaminergic drugs, in: *Apomorphine and Other Dopaminomimetics: Clinical Pharmacology*, Vol. 2 (G. U. Corsini and G. L. Gessa, eds.), pp. 123–133, Raven Press, New York.
Moore, R. Y., and Bloom, F. E., 1978, Central catecholamine neuron systems: Anatomy and physiology of the dopamine systems, *Ann. Rev. Neurosci.* **1**:129–169.
Naylor, R. J., and Olley, J. E., 1972, Modification of the behavioural changes induced by amphetamine in the rat lesions in the caudate nucleus, the caudate-putamen and globus pallidus, *Neuropharmacology* **11**:91–99.
Neale, R., Fallon, S., Gerhardt, S., and Liebman, J. M., 1981, Acute dyskinesias in monkeys elicited by halopemide, mezilamine, and the antidyskinetic drugs, oxiperomide and tiapride, *Psychopharmacology* **75**:254–257.
Neale, R., Gerhardt, S., and Liebman, J. M., 1984, Effects of dopamine agonists, catecholamine depletors, and cholinergic and GABAergic drugs in acute dyskinesias in squirrel monkeys, *Psychopharmacology* **82**:20–26.
Peaston, M. J. T., and Bianchine, J. R., 1970, Metabolic studies and clinical observations during L-dopa treatment of Parkinson's disease, *Br. Med. J.* **1**:400.
Pycock, C. J., 1980, Turning behaviour in animals, *Neuroscience* **5**:461–514.
Rose, F. C., and Capildeo, R. (eds.), 1981, *Research Progress in Parkinson's Disease*, Pitman Books Limited, London.
Schwab, R. S., 1964, Problems in clinical estimation of rigidity, *Clin. Pharmacol. Ther.* **5**:942–946.
Seeman, P., 1980, Brain dopamine receptors, *Pharmacol. Rev.* **32**:229–313.
Tarsy, D., and Baldessarini, R. J., 1984, Tardive dyskinesia, *Annu. Rev. Med.* **35**:605–623.
Ungerstedt, U., 1971, Striatal dopamine release after amphetamine or nerve degeneration revealed by rotational behaviour, *Acta. Physiol. Scand* **367**(Suppl.):49–68.
Velasco, F., and Velasco, M., 1973, A quantitative evaluation of the effects of L-DOPA on Parkinson's disease, *Neuropharmacology* **12**:89–99.
Waddington, J. L., Cross, A. J., Gamble, S. J., and Bourne, R. C., 1982, Spontaneous orofacial dyskinesia and dopaminergic function in rats after six months of neuroleptic treatment, *Science* **220**:530–532.
Weiss, B., Santelli, S., and Lusink, G., 1977, Movement disorders induced in monkeys by chronic haloperidol treatment, *Psychopharmacology* **53**:289–293.
Weissman, A., Koe, B. K., and Tenen, S. S., 1966, Anti-amphetamine effects following inhibition of tyrosine hydroxylase, *J. Pharmacol. Exp. Ther.* **151**:339–352.

PART II
Tardive Dyskinesia

CHAPTER 5

Tardive Dyskinesia

Epidemiology, Pathophysiology, and Pharmacology

JES GERLACH, DANIEL E. CASEY, and SØREN KORSGAARD

Two decades ago, Faurbye and co-workers at Saint Hans Hospital introduced the term tardive dyskinesia (TD) to designate the now well-known syndrome of involuntary hyperkinetic movements that may develop during or following long-term treatment with neuroleptic drugs (Uhrbrand and Faurbye, 1960; Faurbye *et al.*, 1964). In recent years, TD has attracted increasing attention, not only from psychiatrists faced with this serious side effect of an otherwise beneficial antipsychotic treatment, but also from pharmacologists and neurochemists who recognized TD as a valid clinical correlate to the behavioral and biochemical effects of neuroleptic drugs in animals. The syndrome has therefore been the subject of a large number of clinical and animal investigations (for book reviews, see Baldessarini *et al.*, 1980; Fann *et al.*, 1980; DeVeaugh-Geiss, 1982; Jeste and Wyatt, 1982; Bannet and Belmaker, 1983; Klawans, 1983; Casey and Gerlach, 1984; Casey *et al.*, 1985).

1. CLINICAL CHARACTERISTICS OF TARDIVE DYSKINESIA

TD is characterized by involuntary hyperkinetic movements, varying in localization and appearance, occurring most often in the oral region as the so-called buccolinguomasticatory syndrome. Choreoathetoid movements of the extremities can be observed in half of the cases, while facial grimacing and movements of the head and trunk occur more rarely. The pattern of movements

JES GERLACH and SØREN KORSGAARD • Department H, Saint Hans Hospital, DK-4000 Roskilde, Denmark. DANIEL E. CASEY • Departments of Psychiatry, Research, and Neurology, Veterans Administration Medical Center and Oregon Health Sciences University, Portland, Oregon 97207.

in TD varies from patient to patient. In younger patients, it often consists of regular, relatively simple tongue protrusions with synchronous jaw movements. In older patients, the movements are more complex and irregular, the tongue rotating and licking in all directions, both in and out of the mouth. The individual patient often shows a characteristic movement pattern which can be accentuated or weakened by pharmacological treatment, but which still retains its individual features.

Pharmacologically, TD is a heterogeneous syndrome. It may be subdivided into at least two groups according to the response to pharmacological probes. Most cases are unmasked by discontinuance of the neuroleptic treatment or a supplementary anticholinergic treatment and are suppressed by an increased neuroleptic dose. This is the typical TD. A minority of patients (about 10%), however, react oppositely to such pharmacological manipulation. This group may best be termed paradoxical TD, although other terms, such as initial hyperkinesia and acute dyskinesia, have been proposed (Gerlach, 1979).

While the typical TD occurs mainly in elderly patients and usually involves the oral region and the peripheral part of the extremities, paradoxical TD is more likely to develop in younger patients (although all age groups may be affected) and has a more widespread localization. Often, but not always, paradoxical TD is associated with elements of dystonia and akathisia, while TD in its typical form does not have these other neurological features.

Generally, patients are unaffected or only slightly disturbed by the TD, but the movements are particularly striking to those in their surroundings and can therefore become a considerable problem, especially for people who work. Some cases of TD are complicated by difficulties of speech and swallowing and by irregular respiration. In extremely severe cases, the movements can be painful and even life-threatening because of respiratory and gastrointestinal complications (Casey and Rabins, 1978).

In this chapter, we shall concentrate on the typical TD syndrome, evaluate its prevalence and long-term course and its predisposing and etiological factors, and discuss pathogenetic mechanisms and the influence of pharmacological manipulation of various transmitter systems [dopamine (DA), acetylcholine, serotonin, γ-aminobutyric acid (GABA), and neuropeptides].

2. PREVALENCE OF TD

The literature suggests a prevalence ranging from 0.5% to 65% (for reviews, see Baldessarini *et al.*, 1980; Jeste and Wyatt, 1982; Kane *et al.*, 1985), reflecting such variables as heterogeneous populations (especially with respect to age and neuroleptic/anticholinergic treatment), varying definitions of the syndrome, and varying assessment methods. During 1960–1980, the prevalence among neuroleptic-treated psychiatric *inpatients* rose progressively and reached a mean level of 25% in 1976–1980 (Jeste and Wyatt, 1982), which may

TABLE 1
Prevalence of Tardive Dyskinesia among 202 Chronic Institutionalized Patients

	<40 years	40–60 years	>60 years
+ Neuroleptics − Anticholinergics	19%	39%	56%
− Neuroleptics (2 days) + Anticholinergics	38%	65%	75%

be due to an increased interest and vigilance toward the syndrome (Casey, 1985a). The average prevalence of TD across *various populations* is 15–20% (Baldessarini *et al.*, 1980; Kane and Smith, 1982; Casey, 1985a).

To better understand the true prevalence of TD, it is also necessary to evaluate the prevalence of spontaneous dyskinesia (SD) as a control group (true prevalence = TD − SD). Recent reviews calculated SD rates to be 4–7% (Baldessarini *et al.*, 1980; Kane and Smith, 1982), suggesting a true TD prevalence of 10–15%. A new evaluation of 18 studies which reported prevalence rates in both TD and SD patients throws more light on these figures (Casey and Hansen, 1984). These studies met the criteria that treated and untreated patient groups were (1) evaluated by the same investigators (2) using the same assessment techniques to determine dyskinesia rates in (3) similarly aged patients residing in institutional settings. Mean TD prevalence was 19.8% in more than 7200 patients, and SD prevalence was 5.9% in more than 6600 patients. This gives a net difference of 14.9%, which is in agreement with the above-mentioned 10–15% level.

To illustrate the importance of age and psychopharmacological treatment, we shall briefly report the results of a recent epidemiological–pharmacological study (Gerlach and Korsgaard, 1981). The evaluated population consisted of 202 long-term hospitalized and long-term treated schizophrenics, with thus a very "chronic" course of illness. These patients were evaluated twice: (1) during treatment with their neuroleptic drug (unchanged for at least 3 months) and without anticholinergic drugs (for at least 2 days); (2) during anticholinergic treatment (biperiden 6–12 mg/day for 2–4 weeks) and without neuroleptics (for the last 2 days). As shown in Table 1, the prevalence of TD increased with age, from 19% (below 40 years) to 56% (above 60 years) in patients receiving neuroleptic treatment and no anticholinergic drugs, and from 38% to 75% in the same patients receiving anticholinergic treatment and no neuroleptic drugs for 2 days. Thus the prevalence increased by 34–100% following anticholinergic treatment + 2 days without neuroleptics. This result shows the difficulties in evaluating a true frequency of TD in patients when they are on psychotropic treatment.

3. LONG-TERM COURSE OF TARDIVE DYSKINESIA

Though TD has been studied for more than two decades, little is known about the long-term outcome of this syndrome or about the appropriate role of continuing neuroleptic treatment in chronically psychotic patients with TD. Reversibility of TD has been recognized since the initial publications identifying this syndrome (Schönecker, 1957; Sigwald *et al.*, 1959; Uhrbrand and Faurbye, 1960), but for many years, the irreversibility of this syndrome has been more commonly emphasized and is still widely believed to be the expected outcome.

Though it is difficult to compare much of the earlier literature about the outcome of TD because of substantial methodological differences, some trends are evident. In many patients, symptoms stabilize or slowly improve when neuroleptics are discontinued or decreased, though in some patients, symptoms increase (Smith and Baldessarini, 1980; Kane and Smith, 1982). A recent prospective 5-year evaluation of patients with TD who discontinued, decreased, or increased their neuroleptic medications found that TD improved the most in those who were able to discontinue drugs, particularly in the younger patients, but those who remained on neuroleptic drugs at low doses [approximately 300 mg/day chlorpromazine (CPZ) equivalents] also improved. Furthermore, in the patients who received higher neuroleptic doses (500–600 mg/day CPZ equivalents), the TD symptoms remained stable (Casey and Toenniessen, 1983). Thus, for many patients there is an acceptable tradeoff between the risks of TD and benefits of neuroleptics. Even in the presence of continued neuroleptic treatment, TD may gradually decrease and does not inevitably become worse, as has often been feared.

What is the risk of TD becoming irreversible? If irreversibility is defined as persistent TD occurring more than 5 years after discontinuation of neuroleptic treatment, the risk of irreversibility appears to be considerably lower than originally thought. Marsden (1985) suggests that 60% of TD's will recover within such a 5-year drug-free period. With an average TD prevalence of 20%, this implies that only 8% become irreversible. From this, the 5% of spontaneous dyskinesia should be deducted, leaving only 3% as neuroleptic-induced irreversible TD. These figures are tentative, but the assumption that the number of irreversible TD caused by neuroleptic drugs is relatively low appears to hold (see also Itoh and Yagi, 1979; Glazer *et al.*, 1984).

4. PREDISPOSING FACTORS

Predisposing factors are a necessary precondition for development of TD. Reliable data concerning the relative importance of these factors are, however, limited. The following points should be emphasized:

1. *Age* is the most important predisposing factor (Crane, 1973; Jeste and Wyatt, 1982; Kane and Smith, 1982; Toenniessen *et al.*, 1985; see also

Table 1). Increasing age increases both risk of developing TD and severity and persistence of the syndrome.

2. *Sex*. Females appear to be slightly more predisposed to TD than males, perhaps only in patients older than 70 (Bell and Smith, 1978; Kane and Smith, 1982).
3. *Psychiatric diagnosis* may be related to development of TD. Patients with affective or schizoaffective disorder appear to have a significantly greater incidence of TD than schizophrenic patients (Kane *et al.*, 1985; for review, see Casey, 1984). Severely disturbed schizophrenic patients often display various movement disturbances including TD (Gerlach and Korsgaard, 1981; Owens, 1985). Anxiety, agitation, and stress may temporarily precipitate or aggravate the TD syndrome.
4. *Brain lesions*. Some reports indicate that TD is related to brain damage (Edwards, 1970). As yet, however, it has not been possible to identify such a defect.
5. Finally, *an unknown constitutional factor*, perhaps genetic vulnerability, has been assumed to explain why some, but not all, patients receiving neuroleptic treatment develop TD. Some patients develop TD within a few months, while others can be treated with high doses for life without showing the slightest sign of TD.

5. ETIOLOGICAL FACTORS

5.1. Neuroleptics

Many retrospective epidemiological studies have attempted to establish the relationship between neuroleptic treatment and development of TD, but the results have been divergent and inconsistent. However, some studies in humans and monkeys suggest that both dosage and duration of treatment have to be considered as etiological factors in TD (Gardos *et al.*, 1977; Gunne and Bárány, 1979; Domino, 1985; Owens, 1985; Toenniessen *et al.*, 1985). In patients 55 years and older, a significant correlation has been found between prevalence of TD and length as well as total dose of neuroleptic treatment, only during the first 2–3 years of treatment (Toenniessen and Casey, 1982; see also Toenniessen *et al.*, 1985). Thereafter, over a 3- to 20-year period, no correlation was found. These findings suggest that, at least for this age group, the period of maximum vulnerability for developing TD is early in the treatment course (see also Gardos *et al.*, 1983; Kane *et al.*, 1985). These results help also to explain why previous studies failed to find a significant relationship between TD and measures of drug treatment.

The crucial question whether some neuroleptics are more liable to induce TD than others, when antipsychotic equipotent doses are compared, will be discussed in Section 7.

5.2. Anticholinergic Drugs

Anticholinergic drugs (antiparkinsonian or antidepressant drugs) can uncover latent TD and accentuate existing TD by increasing hypermotility and amplitude (Gerlach *et al.*, 1974b; Klawans *et al.*, 1980). On the other hand, it has not been finally clarified whether anticholinergics contribute to the development of TD symptoms which might emerge following withdrawal of a combined neuroleptic–anticholinergic treatment. Until now, most clinical and animal studies suggest that only the neuroleptic (antidopaminergic) treatment, and not the possible concomitant anticholinergic treatment, plays an etiological role for withdrawal TD (for further discussion, see Section 7).

Other drugs such as *antihistamines* (Thach *et al.*, 1975), *phenytoin* (Lühdorf and Lund, 1977), and *lithium* (Beitman, 1978) can temporarily induce symptoms that resemble TD. Electroconvulsive treatment can unmask latent TD (Uhrbrand and Faurbye, 1960), but in other cases it improves the syndrome (Price and Levin, 1978; Rosenbaum *et al.*, 1977) or leaves it unchanged (Asnis and Leopold, 1978).

6. PATHOGENETIC MECHANISMS

Figure 1 shows the well-known diagram of the DA synapse. DA is synthesized in the presynaptic nerve terminal and is stored in the synaptic vesicles. When the nerve impulse arrives at the terminal, it triggers an influx of calcium ions leading to a release of DA from the vesicles into the synaptic space. The released DA interacts with specific receptor sites located postsynaptically on different target neurons as well as presynaptically on the DA terminal (Carlsson, 1975; Iversen, 1979; Seeman, 1980).

6.1. The DA Receptor Supersensitivity Theory

The traditional hypothesis to explain the development of TD is based on the DA supersensitivity theory. Neuroleptics block postsynaptic DA receptors and cause an increased acetylcholine turnover. Clinically, this leads to parkinsonism (Seeman, 1980). During long-term treatment, however, various phenomena of adaptation occur: the DA receptor blockade induces DA receptor supersensitivity, which may be seen as an increased behavioral response to DA agonists (Baldessarini *et al.*, 1980; Seeman, 1980; Christensen *et al.*, 1985). The supersensitivity appears to be related to a proliferation of DA receptors, a decreased acetylcholine turnover (Marsden and Jenner, 1981), and probably adaptive phenomena in other neurons, e.g., GABA neurons. This DA supersensitivity theory constitutes the commonly accepted background for the spontaneous decrease in parkinsonism during long-term neuroleptic treatment and the possibility for gradually developing TD.

To explain the sometimes concomitant occurrence of TD and parkinson-

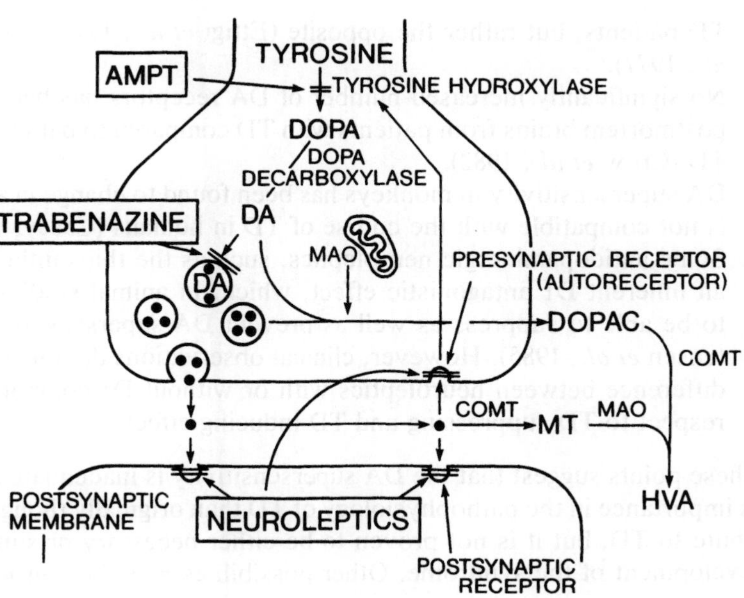

Figure 1. The dopamine synapse. For further explanation, see text. The figure shows the site of action of the three types of antidopaminergic drugs: α-methyl-para-tyrosine (AMPT), tetrabenazine, and traditional neuroleptics.

ism, it has been proposed that TD depends on an increased sensitivity of certain receptors facilitated by DA, while parkinsonism depends on blocking of other receptors inhibited by DA (Klawans *et al.*, 1980; Cools and Van Rossum, 1976).

This DA supersensitivity, however, is not in agreement with various clinical, pharmacological, and biochemical observations in TD:

1. During neuroleptic treatment, no correlation can be found between the time course of development of DA supersensitivity in animals and TD in patients. The DA supersensitivity in animals develops in all cases after a few days of treatment and may even be seen after a single injection (Christensen, 1981), whereas TD develops in only some patients and usually after treatment for months or years.
2. TD in psychiatric patients shows no or only slight aggravation during treatment with DA agonists such as L-dopa (Gerlach and Casey, 1983) and may just be due to a nonspecific activation, in contrast to the effect of DA agonists in the animal supersensitivity model.
3. If DA supersensitivity plays an essential role in TD, DA agonists should, as in animal models, counteract the effect of neuroleptics and make the symptoms disappear after withdrawal. This does not seem to be the case (see Section 7).
4. Endocrinological studies do not support an increased DA sensitivity in

TD patients, but rather the opposite (Ettigi *et al.*, 1976; Tamminga *et al.*, 1977).
5. No significantly increased number of DA receptors has been found in postmortem brains from patients with TD compared to patients without TD (Crow *et al.*, 1982).
6. DA supersensitivity in monkeys has been found to change in a way that is not compatible with the course of TD in humans (Casey, 1985b).
7. Some antidopaminergic neuroleptics, such as the thioxanthenes, have an inherent D1 antagonistic effect, which, in animal studies, appears to be able to suppress as well as prevent DA supersensitivity (Christensen *et al.*, 1985). However, clinical observations do not suggest any difference between neuroleptics with or without D1 components with respect to TD-suppressing and TD-inducing effect.

These points suggest that the DA supersensitivity is inadequate or at least of less importance in the pathophysiology of TD than originally thought. It may contribute to TD, but it is not proven to be either necessary or sufficient for the development of the syndrome. Other possibilites must be considered.

6.2. The GABA Hypothesis

TD may depend on decreased activity in a subgroup of GABA neurons originating from striatum (Scheel-Krüger, 1983; Gunne *et al.*, 1984; Fibiger and Lloyd, 1984). Animal studies have shown that different GABA neurons project from striatum: one type (from the anterior part of striatum to globus pallidus, lateral segment) is inhibited by DA, while another type (from the posterior part of striatum to substantia nigra, zona reticulata, and to globus pallidus, medial segment) is stimulated by DA (see Fig. 2) (Scheel-Krüger, 1983). This means that neuroleptics disinhibit (stimulate) the GABA neurons projecting to the lateral segment of globus pallidus, while they inhibit the GABA neurons to substantia nigra and medial globus pallidus (illustrated in Fig. 2).

From experiments with intracerebral injection techniques, it is known that increased GABA activity in lateral globus pallidus in animals induces symptoms related to parkinsonism, while decreased GABA function in the medial segment of pallidus and in the zona reticulata of substantia nigra is associated with hyperkinetic movements (Scheel-Krüger, 1983; Scheel-Krüger and Arnt, 1985). This implies that neuroleptics affect GABA neurons in the anterior part of striatum to produce parkinsonlike symptoms while, at the same time, they may facilitate hyperkinetic movements by effect on other GABA neurons in the posterior part of striatum. The net result may depend on a balance between these two, partly counteracting effects. Initially in neuroleptic treatment, the Parkinson effect may be dominating, later the hyperkinetic effect (TD).

The observation of a neuroleptic-induced decrease of GABA activity in striatonigral neurons is in accordance with the finding of supersensitive GABA

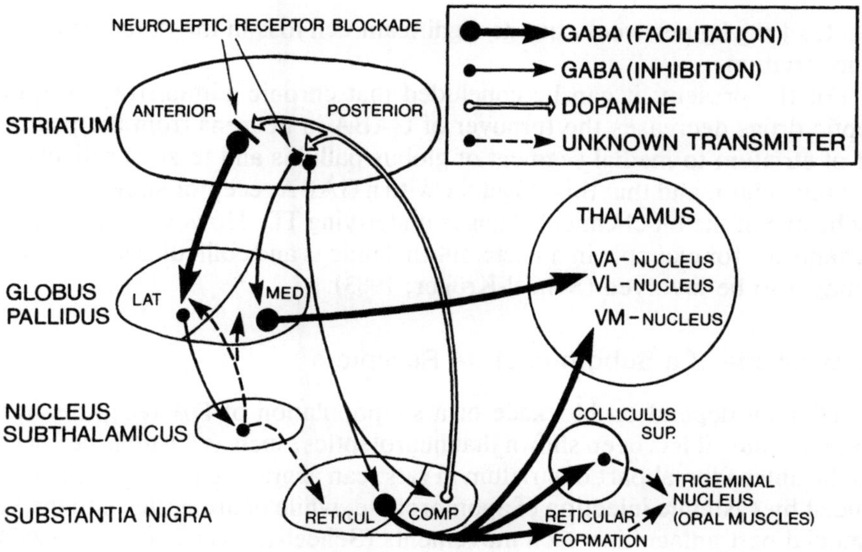

Figure 2. Diagram of some dopamine (DA)–GABA–GABA interactions in the basal ganglia which participate in the regulation of oral movements. Blockade of DA receptors in striatum leads to GABA facilitation in some neurons and inhibition in others, with the opposite reaction in the subsequent GABA neuron (Scheel-Krüger and Arnt, 1985).

receptors in nigra after neuroleptic treatment (Gale and Casu, 1981; Coward, 1982).

More direct evidence for the GABA hypothesis in TD comes from a monkey study (Gunne *et al.*, 1984), in which it was shown that the concentration of glutamic acid decarboxylase (GAD, the GABA-synthesizing enzyme) is decreased in substantia nigra (57%, $p < 0.001$), in medial pallidal segment (26%, $p < 0.05$), and in nucleus subthalamicus (32%, $p < 0.01$) in postmortem brains from *Cebus* monkeys with persistent dyskinesias. Furthermore, one monkey with unilateral dyskinesia showed a GAD decrease only in the affected side. This supports the relationship between a decreased GABA function in certain striatofugal neurons and the presence of TD.

Whether these biochemical changes include loss of neurons cannot be answered at present. Owing to methodological difficulties, no consistent results have been obtained from neuropathological examinations of human brains in TD patients. These difficulties can be reduced in animal experiments. In 1973, Pakkenberg *et al.* reported that a count of neurons in the striatum of rats treated for 12 months with perphenazine, 3.4 mg/kg per 14 days, showed a cell loss of 19% ($p < 0.005$). Later studies, however, with the same neuroleptic in a 10-fold dose (40 mg/kg per 14 days) for 6 months showed no significant changes in cell count, neither in the striatum nor in the cortex (Fog *et al.*, 1976). In a later study, Nielsen and Lyon (1978) found a cell loss of 10% ($p < 0.05$) in the ventrolateral part of the striatum in rats after treatment with flupenthixol decanoate, 4 mg/kg per week, but no significant cell loss in the dorsomedial part of the striatum.

For the present, it can be concluded that chronic administration of neuroleptic drugs decreases the turnover of GABA in neurons from the posterior part of striatum to medial segment of globus pallidus and to zona reticulata of substantia nigra, and that this, together with a GABA receptor supersensitivity, may be one of the biochemical changes underlying TD. However, other GABA mechanisms, for example in nucleus subthalamicus and colliculus superior (Fig. 2), may also be involved (Scheel-Krüger, 1983).

6.3. Blockade of a Subgroup of DA Receptors

TD may depend on blockade of a subpopulation of DA receptors in the striatum. Thus, it has been shown that neuroleptics, such as (-)sulpiride injected into the anterodorsal part of striatum in rats, can aggravate licking and gnawing induced by systemic injection of apomorphine, while neuroleptics into the ventromedial part antagonize such movements (Scheel-Krüger and Arnt, 1985).

The hypothesis of a DA receptor blockade as a pathogenetic factor underlying TD is supported by observation of hyperkinetic movements including oral dyskinesia and choreoathetoid movements of trunk and extremities directly in relation to administration of neuroleptic drugs. In monkeys, a single dose of haloperidol can produce TD-like mouth openings and tongue protrusions and writhing movements of the extremities, often mixed with dystonic features (Weiss and Santelli, 1978; Casey *et al.*, 1980a). In some humans, similar movements (paradoxical TD) can be induced by antidopaminergic treatment (Casey and Denney, 1977; Gerlach and Korsgaard, 1981). Even a combination of tetrabenazine and pimozide, a strong specific antidopaminergic drug combination, has been found to maintain hyperkinetic movements indistinguishable from typical TD (our own unpublished observation). These antidopaminergic dyskinesias diminish with decreasing neuroleptic dose and may be antagonized by anticholinergics.

These observations suggest that antidopaminergic treatment via distinct DA receptors may produce parkinsonism as well as dyskinesias. The clinical picture may depend on a balance between these two tendencies, a balance that may change over time, e.g., from parkinsonism to dyskinesia during long-term neuroleptic treatment. As can be seen, there are similarities between this and the above-mentioned GABA hypothesis, but whether there is a direct causal relationship between them remains to be determined.

6.4. An Unspecific Lowering of the Threshold for Spontaneous Dyskinesia

While mentally and neurologically healthy subjects usually tolerate stress, anxiety, and pharmacological influence on DA and cholinergic receptors without developing movement disturbances, such influences may result in obvious changes in aged people. It may be said that such people have a low threshold for development of movement disturbances, an increased vulnerability that may result in spontaneous dyskinesia.

As TD and spontaneous dyskinesia are indistinguishable syndromes, it may be hypothesized that neuroleptics, like old age, may facilitate or precipitate the development of spontaneous dyskinesia. A neuroleptic drug (during treatment as well as following withdrawal) may produce a state of distress or increased drive (including akathisia and tardive akathisia) which may unspecifically lead to dyskinesia, the nature and localization of which depend on the individual predisposition (Gerlach and Korsgaard, 1981; Marsden, 1985). Neuroleptics may promote or exacerbate a tendency to developing spontaneous motor disorders inherent in at least some forms of the illness (Owens, 1985).

The biochemical mechanisms underlying such an unspecific effect of neuroleptics are still unclear, but it might be reasonable to suggest that the biochemical alterations underlying TD resemble the brain processes of old age, i.e., a decreased activity in several neurotransmitter systems, including DA, GABA, and acetylcholine (Gottfries, 1981).

In concluding these observations on the pathogenetic mechanisms of TD, it should be emphasized that TD is a heterogeneous syndrome, not dependent on a simple defect within a single transmitter system, but rather a disturbed balance in a complex interplay between different transmitter systems and different subgroups of neurons, an imbalance including reduced DA neurotransmission and DA receptor supersensitivity as well as reduced GABA turnover with GABA receptor supersensitivity.

7. PHARMACOLOGICAL MANIPULATION OF TD

To improve our knowledge about the pathophysiological mechanisms discussed in Section 6 and to evaluate the treatment possibilities, a series of pharmacological trials were conducted in patients with TD. The strategy has been to systematically investigate the effect of drugs that influence the central nervous system's (CNS) neurotransmitters, DA, acetylcholine, GABA, serotonin, and peptide fragments of the β-lipotropin hormone (β-LPH_{61-91}).

7.1. Antidopaminergic Drugs

All CNS antidopaminergic drugs have an antihyperkinetic effect. These influences, however, vary from drug to drug, depending on (1) striatal antidopaminergic potency, (2) anticholinergic effect, (3) mechanisms of action (pre- and postsynaptic receptor blockade, presynaptic catecholamine depletion or synthesis inhibition), (4) dosage, and (5) sedative effect. Therefore, it is not possible to characterize in a specific, simple formula the antihyperkinetic effect of individual drugs.

To illustrate the antihyperkinetic potency of various antidopaminergic

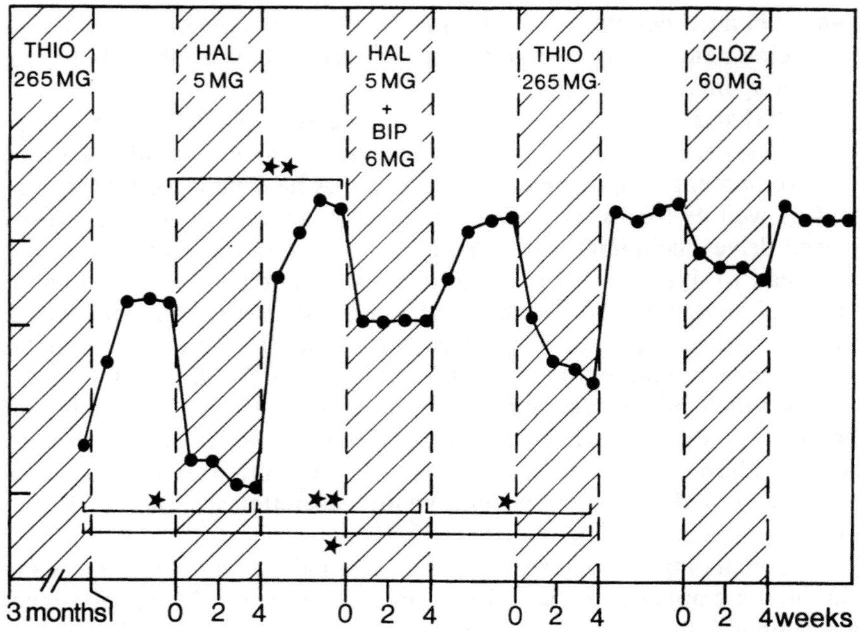

Figure 3. Mean hyperkinesia scores of 16 psychiatric patients with tardive dyskinesia. Each patient was first treated with thioridazine (THIO) for 3 months. After 4 weeks without medicine, the patients were treated with four different types of neuroleptic treatment: haloperidol (HAL), haloperidol + biperiden (HAL + BIP), THIO, and clozapine (CLOZ), all of 4 weeks duration and all interrupted by 4 weeks without drugs. The antihyperkinetic effects of all drugs given (except CLOZ) were significant ($p < 0.01$). Other p- values are shown in the figure ($*, p < 0.05$; $**, p < 0.01$).

drugs, Fig. 3 shows the TD intensity before, during, and after treatment with various receptor-blocking neuroleptics. Haloperidol, a potent DA receptor blocker, strongly suppresses TD. This observation is in accordance with other clinical studies showing that haloperidol (Kazamatsuri *et al.*, 1972) and another specific antidopaminergic drug, pimozide (Claveria *et al.*, 1975), markedly decrease TD. Also sulpiride, the most selective D2 receptor blocker (Jenner and Marsden, 1981), has such a marked antihyperkinetic effect when administered in doses up to 2100 mg/day (Gerlach and Casey, 1984).

From Fig. 3, it can also be seen that the weaker DA receptor blocking agent, thioridazine, has antihyperkinetic effects, although less than haloperidol ($p < 0.05$). However, the antihyperkinetic effect of thioridazine and other high-dose neuroleptics may slightly increase in the course of prolonged treatment. This may also apply to clozapine, which appears to have only minimal or no effect in short-term studies (Gerlach and Simmelsgaard, 1978), but some effect in long-term studies (Simpson *et al.*, 1978). In general, however, high-dose neuroleptics have only weak antihyperkinetic effect unless given in high doses.

Antidopaminergic treatment can also be performed with drugs acting on the presynaptic side, such as the synthesis inhibitor α-methyl-para-tyrosine (AMPT) and the storage inhibitors reserpine and tetrabenazine (see Fig. 1). AMPT (3–4 g daily) decreases TD approximately 50% (Gerlach *et al.*, 1974b; Gerlach and Thorsen, 1976). The effect of AMPT, however, is in general weaker than the effect of postsynaptic DA receptor blockers, not only in TD, but also in schizophrenia (Wålinder *et al.*, 1976; Magelund *et al.*, 1979), and the antihyperkinetic effect of AMPT appears to diminish or disappear completely during prolonged treatment (Nasrallah *et al.*, 1977). Therefore, and because of the nephrotoxic effects of AMPT, this drug will be of no practical value in the treatment of TD.

More useful is the DA storage depletor tetrabenazine. Tetrabenazine, 50–150 mg daily, has been shown to have a marked antihyperkinetic effect (Brandrup, 1961), although also in this case, some tolerance may develop during prolonged treatment. Unfortunately, this type of treatment also produces parkinsonism and may aggravate TD in the long run.

The most serious drawback associated with neuroleptic treatment of TD is the risk of aggravating the syndrome. Retrospective studies have suggested that there is no obvious difference between the ability of particular neuroleptics to induce TD. However, the prospective study summarized in Fig. 3 showed that TD was more intensive after treatment with haloperidol, 5 mg/day (and a strong TD suppression), than after the initial treatment with thioridazine, 250 mg/day (and a weaker TD suppression) ($p < 0.01$). This result could suggest that haloperidol is more potent than thioridazine in its ability to suppress TD as well as to produce rebound aggravation of TD, a suggestion that is supported by studies in monkeys showing that the exacerbation of TD following a challenge dose of haloperidol (0.05 mg/kg) is more prolonged than the corresponding reaction to thioridazine (1 mg/kg) (Gunne and Bárány, 1979). However, the study reported in Fig. 3 has certain limitations: the treatment periods lasted only 4 weeks and were not randomized (except haloperidol/haloperidol + biperiden). Furthermore, the second thioridazine treatment resulted in rebound hyperkinesia of the same magnitude as haloperidol treatment.

Consequently, the hypothesis about a correlation between the TD-suppressing effect and the following rebound aggravation has been reevaluated in a recent Nordic multicenter study. Thirty-four chronic psychiatric patients with TD have been exposed to chlorprothixene (a high-dose neurolepticlike thioridazine), perphenazine (a medium-dose neuroleptic), and haloperidol (a low-dose neuroleptic) in periods of 6 months, each followed by drug-free periods of 6 weeks' duration in a randomized crossover study. The study has been completed, but the statistical analysis of the videotaped alterations in TD and parkinsonism during and following the different treatment periods has not yet been finished. The preliminary evaluations suggest the following:

1. Chlorprothixene had no significant TD-suppressing effect, while perphenazine and haloperidol both reduced TD comparable to the effect of haloperidol shown in Fig. 3.

2. Following discontinuation of each of the three neuroleptics, TD returned to the pretreatment level. No rebound aggravation was found in any of the three groups.
3. The parkinsonian scores were unchanged during chlorprothixene therapy, but increased during treatment with perphenazine and haloperidol, an increase that corresponded to the concomitant decrease in TD.

Thus, this study did not confirm the hypothesis of a rebound aggravation correlated to a prior neuroleptic-induced TD suppression. Two explanations can be given for this:

1. Neuroleptic treatment (dose and duration) may be correlated to the induction of TD only during the first 1–3 years of treatment, not later (see Section 3). In our study, all patients were long-term treated (>10 years), which may explain why no rebound aggravation could be observed. The hypothesis may be correct, but can be established only in studies of TD in its initial phase.
2. The other possibility is that rebound aggravation is of no or minor importance in the development of TD. This possibility is in agreement with the suggestion discussed earlier (Section 6) that DA supersensitivity apparently plays a limited role in TD, and that other biochemical disturbances, e.g., within the GABA system, are of greater importance. If so, the conclusion may be that the degree of antidopaminergic effect is unrelated to development of TD, and that the DA potency of neuroleptics is not a useful parameter for differentiating between neuroleptics with respect to TD-inducing capacity. This would be in agreement with most studies showing that no difference can be found between the ability of different neuroleptics to induce TD.

Is it at all possible to administer neuroleptic drugs without risk of inducing TD? The dibenzoazepines clozapine and fluperlapine may be promising possibilities. Clozapine is a nonselective DA antagonist with weak antidopaminergic properties, but strong antinoradrenergic, antihistaminergic, antiserotoninergic, and anticholinergic effect. In a clinical study, it was found that clozapine in medium doses (225 mg/day) even reduced the homovanillic acid concentration in cerebrospinal fluid (Gerlach *et al.*, 1975), suggesting an atypical, perhaps presynaptic effect on the DA system and, at this dose level, practically no effect on postsynaptic DA receptors. In several studies it has been found that clozapine does not induce acute extrapyramidal side effects or TD (Gerlach *et al.*, 1974a; Claghorn *et al.*, 1983; Pi and Simpson, 1983), and worldwide no TD has been reported with this drug—except one case, in which an intensification of TD was found during treatment with clozapine, 250–450 mg/day (Doepp and Buddeberg, 1975).

Another dibenzoazepine derivative, fluperlapine, has been introduced. This drug is characterized by (1) a weak D1 and D2 receptor blockade, (2) no noradrenaline receptor-blocking effect, but a slight inhibition of noradrenaline

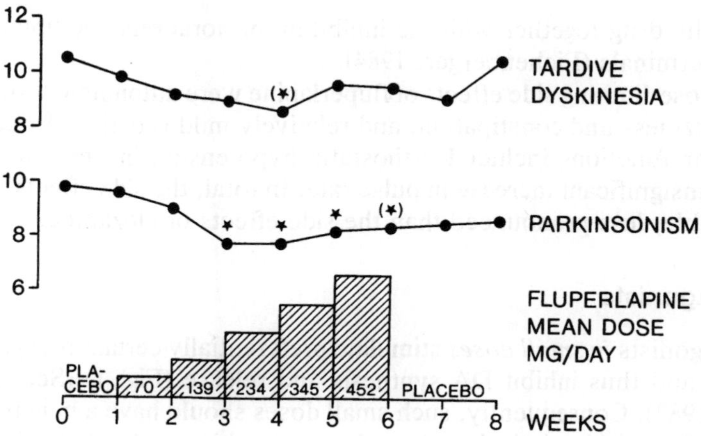

Figure 4. Effects of increasing doses of fluperlapine on hyperkinesia and parkinsonism. Values are mean scores ± SEM. *, $p < 0.05$ compared with pretreatment scores.

reuptake, like antidepressant drugs, (3) no cataleptic and only weak antistereotypic effect, (4) no production of DA supersensitivity (in one study, hyposensitivity to apomorphine has been found), (5) a marked anticholinergic effect, (6) sedative effect, although less than clozapine, and (7) a low prolactin increase (Eichenberger, 1984).

These pharmacological properties suggest that fluperlapine has an antipsychotic effect (like clozapine), and perhaps an antidepressant effect. It may not induce parkinsonism or TD. On the other hand, the drug should induce autonomic side effects and sedation like other high-dose neuroleptics. These suggestions have been largely confirmed in preliminary clinical trials (Woggon *et al.*, 1984; Fischer-Cornelssen, 1984).

We have studied the effect of fluperlapine in TD and parkinsonism in a placebo-controlled trial (Korsgaard *et al.*, 1984). The drug was given to 11 psychiatric patients (mean age 60 years) in doses gradually increasing to 200–600 mg/day. The effect was evaluated blindly by means of videotapes.

As shown in Fig. 4, the dyskinesia score tended to decrease during fluperlapine treatment, but compared to pre- as well as posttreatment placebo scores, only minimal differences were found. Surprisingly, the parkinsonian scores were slightly, but not significantly reduced (except during the highest fluperlapine dose). If this observation of an antiparkinsonian effect of fluperlapine can be confirmed in other studies, it might be of considerable clinical importance, for example, in the treatment of psychoses in patients with Parkinson's disease and in psychiatric patients vulnerable to extrapyramidal side effects of conventional neuroleptics. Such a potential antiparkinsonian effect of fluperlapine is in agreement with its ability to counteract tetrabenazine-induced catalepsy in rats, and it may be explained by the strong anticholinergic effect of the drug together with the inhibition of noradrenaline uptake in presynaptic terminals (Eichenberger, 1984).

The dose-limiting side effects of fluperlapine were autonomic disturbances, mainly dizziness and constipation, and relatively mild sedation. Effect on cardiovascular functions included orthostatic hypotension in one patient and a clinically insignificant increase in pulse rate. In total, the side effects were mild and tolerable, less pronounced than the side effects of clozapine.

7.2. DA Agonists

DA agonists *in small doses* stimulate preferentially certain presynaptic DA receptors and thus inhibit DA synthesis and release (Fig. 1) (Seeman, 1980; Meltzer, 1982). Consequently, such small doses should have a beneficial effect in TD. In fact, early clinical trials with apomorphine and L-dopa showed that these drugs in small doses may reduce TD (Smith *et al.*, 1977; Carroll *et al.*, 1977). However, the effect was limited, and studies with other DA agonists, such as bromocriptine and piribedil, do not suggest any significant effect (Chase and Tamminga, 1980). Hopefully, more selective presynaptic DA agonists, such as *N-n*-propyl-3-(3-hydroxyphenyl)-piperidine (Hjorth *et al.*, 1981), will clarify the therapeutic possibilities of this presynaptic treatment approach.

DA agonists *in higher doses* may counteract the postsynaptic DA receptor supersensitivity induced by neuroleptic drugs. Findings of reduced supersensitivity in animals after prolonged treatment with DA agonists support this proposal (Friedhoff *et al.*, 1977). We have examined this treatment approach more closely (Casey *et al.*, 1982). Thirteen patients with TD were treated for 4–8 weeks with levodopa + benserazide, a peripheral decarboxylase inhibitor (Madopar), over a wide dose range, corresponding to 3.0–9.0 g/day of levodopa. TD scores increased slightly in some patients during levodopa. After discontinuation of the drug, TD scores returned to pretreatment baseline levels without further improvement in those patients receiving concurrent neuroleptic medications ($n = 9$), but in the neuroleptic-free patients, TD scores decreased 25% in three patients and resolved in one younger patient. Psychological effects of depression or increased psychotic symptoms occurred at higher drug doses. Similar negative results have been found by Hardie *et al.* (1983) and Nasrallah *et al.* (1983). These results do not support the proposal that receptor sensitivity modification with levodopa is an effective therapeutic approach to TD (for further discussion, see Gerlach and Casey, 1983), and they do not support the concept of a DA supersensitivity in the pathophysiology of TD.

7.3. Anticholinergic Drugs

Cholinergic hypofunction is a purported pathophysiological factor underlying TD, possibly interacting with DA receptor supersensitivity. Consequently, anticholinergic drugs can uncover latent TD or aggravate existing TD (Gerlach *et al.*, 1974b; Gerlach and Thorsen, 1976; Klawans and Rubovits,

1974). This effect can be seen in Fig. 3 in which biperiden, an anticholinergic drug, given together with haloperidol, markedly reduced the antihyperkinetic effect of haloperidol ($p < 0.01$).

Since anticholinergics aggravate TD, it has been suggested that these drugs may also increase the risk of developing TD. This theoretical problem was also investigated in the previously mentioned study in which 16 elderly psychiatric patients with TD were treated with haloperidol and haloperidol + biperiden in randomized sequence (Fig. 3). No significant difference was found between the hyperkinesias following the two treatment periods. This result, however, does not exclude the possibility of a differential effect following more prolonged treatment.

Though we do not yet know the role of anticholinergics in the pathogenesis of TD, there is ample reason to justify the use of anticholinergic antiparkinsonian drugs when signs of neuroleptic-induced dystonia, parkinsonism, akathisia, or paradoxical TD are present and a reduction in the antidopaminergic treatment is not feasible, owing to psychotic symptoms. On the other hand, routine prophylactic administration of anticholinergics should be avoided, as this will lead to prescription of these drugs for many patients who do not need them.

7.4. Cholinomimetics

It has been a logical approach to increase cholinergic activity as a treatment of TD. Physostigmine, an acetylcholine esterase inhibitor, may decrease TD, but may also produce no change, or an aggravation of symptoms (Gerlach *et al.*, 1974b; Klawans and Rubovits, 1974; Tarsy *et al.*, 1974; Casey and Denney, 1977). The first attempt to increase the cholinergic activity was to use precursor loading with deanol, whose effect on cholinergic mechanisms is unclear, but later investigations found this drug to be generally ineffective (Casey, 1977). Choline and lecithin have been reported to have slight antihyperkinetic effect, but side effects limit their practical use (Jeste and Wyatt, 1982), and in a recent study, RS 86, a new cholinomimetic drug, showed no effect in TD (Noring *et al.*, 1984). Taken together, these findings indicate that cholinomimetics are not useful for treating TD.

7.5. Serotonin

Several reports indicate a complex interaction between serotoninergic neurons and the nigrostriatal dopaminergic system and suggest that serotonin may have a role in the pathophysiology of TD.

In a monkey model of TD, antihyperkinetic effects have been demonstrated by both the serotonin precursor 5-hydrotryptophan and the purported serotonin antagonist methysergide, while cyproheptadine, another presumed serotonin antagonist, both reduced (0.40 mg/kg) and intensified (0.20 mg/kg) TD (Gunne and Bárány, 1980). In monkeys previously treated with long-term haloperidol, oral hyperkinesias were induced by the serotonin uptake inhibitor paroxetine, as well as by the serotonin antagonist mianserin (Korsgaard *et al.*, 1982c).

The clinical studies of serotoninergic influences in hyperkinetic movement disorders till now are few. Enhanced central serotoninergic activity during treatment with 5-hydroxytryptophan aggravates Huntington's chorea (Lee *et al.*, 1969), while treatment with parachlor-phenylalanin, a serotonin synthesis inhibitor, does not affect the disease (Chase *et al.*, 1972). L-Dopa-induced hyperkinesias in parkinsonian patients are intensified by L-tryptophan, but patients with TD showed no consistent change during treatment with L-tryptophan, D-L-tryptophan, 5-hydroxytryptophan combined with carbidopa or the specific serotonin uptake inhibitor citalopram (Jus *et al.*, 1974; Nasrallah *et al.*, 1982), or the antiserotoninergic drug cyproheptadine (Gardos and Cole, 1978).

Although animal data indicate that serotonin is involved—in a highly interactive manner with DA, acetylcholine, and other transmitters—in motor functioning and movement disorders, no therapeutically useful effects have been observed with serotonin drugs in TD in humans.

7.6. GABA Agonists

Studies of cerebrospinal fluid from patients with TD, Parkinson's disease, and Huntington's chorea have shown that the concentration of GABA and GAD is decreased, suggesting a decreased GABA activity in these neurological diseases (Chase and Tamminga, 1979; Meldrum, 1982). In Section 6.2 it was hypothesized that a decreased GABA turnover in projections from striatum to substantia nigra and globus pallidus (medial part) plays a role in the pathophysiology of TD. Therefore, it is relevant to evaluate the effect of GABA agonists in TD.

In three placebo-controlled, video-evaluated clinical trials, we have tested three GABA-mimetics [γ-acetylenic GABA (GAG), γ-vinyl GABA (GVG), and tetrahydroxyisoxazolopyridinol (THIP)] in psychiatric patients with TD (see Table 2).

GAG is an inhibitor of GABA transaminase (GABA-T) that raises brain GABA severalfold and also decreases striatal DA turnover. Figure 5 shows the effect of increasing doses of GAG (from 75 to 225 mg/day) at 2-week intervals in nine patients concurrently taking neuroleptics. TD scores were significantly reduced ($p < 0.01$) at GAG doses of 110–225 mg/day when compared with pre- and posttreatment placebo scores (for details, see Casey *et al.*, 1980b). Interestingly, there was a significant correlation between the decrease in TD scores during GAG treatment and the strength of neuroleptic dose. In other words, those patients who were taking a high amount of neuroleptic medication had substantial improvement with GAG, whereas those patients taking little or no neuroleptic medication had minimal effects during GAG treatment. Parkinsonian scores increased in the four oldest patients, three of whom had mild parkinsonian symptoms prior to taking GAG. These scores returned to pre-

TABLE 2
Three Studies Evaluating the Effect of GABA Mimetics in Tardive Dyskinesia

Number of patients	GABA drugs	Dose	Effect on tardive dyskinesia	Parkinsonism	Psychic symptoms	Side effects	
10	γ-Acetylenic GABA (GAG)	75–225 mg/day (mean 195)	8 patients improved (correlated to neuroleptic dose)	Aggravation in 4 patients	No effect	Sedation	5
						Confusion	2
						Myoclonia	1
10	γ-Vinyl GABA (GVG)	2–6 g/day (mean 4.4)	8 patients improved (correlated to parkinsonism)	Aggravation in 6 patients	1 schizophrenia patient deteriorated	Sedation	4
						Confusion	2
						Dizziness	1
						Myoclonia	0
13	THIP	20–120 mg/day (mean 67)	No effect	Aggravation in 7 patients	No significant effect (tension and depression tended to decrease)	Sedation	5
						Confusion	2
						Dizziness	5
						Vomiting	2
						Myoclonia	1

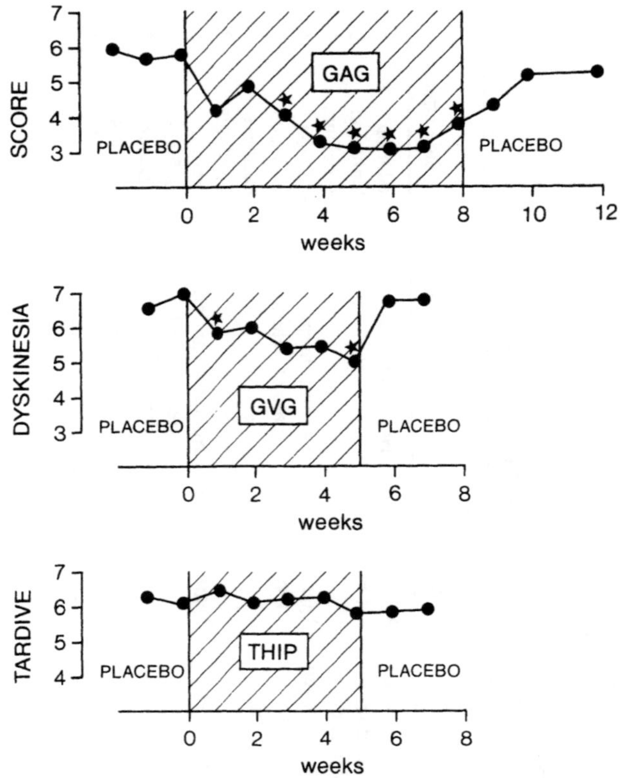

Figure 5. Mean hyperkinesia scores for psychiatric patients with tardive dyskinesia during treatment with γ-acetylenic GABA (GAG) (75–225 mg/day), γ-vinyl GABA (GVG) (2–6 g/day), and THIP (20–120 mg/day). For further explanation, see text and Table 2. *, $p < 0.05$.

treatment levels during the final placebo period. For other side effects, see Table 2.

In a second similar study with another GABA-T inhibitor, GVG, similar results were found (Korsgaard *et al.*, 1983). The TD score decreased slightly ($p < 0.05$) during treatment with GVG, 2–6 g/day, when compared with pre- and posttreatment scores (Fig. 5). The parkinsonism increased in six patients, and this increase was correlated with the antihyperkinetic effect ($p < 0.05$). As in the GAG study, the side effects were sedation and confusion, and in a single case dizziness (Table 2). These side effects disappeared after dose reduction, but then the antihyperkinetic effect also disappeared.

In a third study on the effect of GABA agonists in TD, we have used a direct GABA receptor stimulant, THIP (Korsgaard *et al.*, 1982a). The study revealed that THIP, 20–120 mg/day, had no significant effect on TD, although preexistent parkinsonism increased ($p < 0.05$). Side effects of THIP were more pronounced than those of the two GABA-T inhibitors and included sedation, confusion, dizziness, vomiting, and myoclonic jerks (Table 2).

Muscimol, another direct GABA agonist, has revealed antihyperkinetic effect together with serious toxic side effects (Tamminga *et al.*, 1979). And preliminary studies with progabide, a third GABA agonist, has shown some beneficial effect in TD, although no effect in L-dopa-induced dyskinesia (Morselli *et al.*, 1985).

The main conclusion from these studies is that GABA agonists may have some antihyperkinetic effect (see also Tamminga *et al.*, 1985), but this effect may be secondary to the side effects of parkinsonism and sedation. These relatively negative results may be explained by the complexity of GABA functions in the brain and the existence of functionally counteracting subgroups of GABA pathways, some facilitating dyskinesias, others inducing parkinsonianlike symptoms (see Section 6). The question remains open until more selective GABA agonists (e.g., with specific affinity to GABA receptors in substantia nigra, zona reticulata) become available for further clinical–pharmacological evaluation of the GABA hypothesis in TD.

7.7. Neuropeptides

In recent years, β-endorphin (β-lipotropin$_{61-91}$) and its peptide fragments have been found to affect DA-mediated mechanisms and appear to be involved in the regulation of neuropsychiatric functions. β-endorphin, as well as the smaller peptide fragments enkephalin (β-lipotropin$_{61-65}$) and des-tyr^1-γ-endorphin (DTγE, β-lipotropin$_{62-77}$), have been used in the treatment of schizophrenia and depression, with inconsistent results (for review, see Shah and Donald, 1982).

Figure 6 shows the effect of the met-enkephalin synthetic analogue (FK 33-824) in TD in comparison with morphine, 10 mg, and the opiate antagonist naloxone, 0.8 mg. The suppression of hyperkinesia scores with FK 33-824 was statistically significant ($p < 0.05$) 1 hr after i.m. injection of 3 mg, though the overall clinical impact was only modest. The effects of 1 mg and 2 mg of FK 33-824 were nearly identical to the effects of 3 mg. In two patients, the hyperkinetic movements disappeared completely (the only patients who were concurrently receiving a high neuroleptic dose), whereas the hyperkinesias were only slightly affected in the other six patients on minimal or no neuroleptic medication. FK 33-824 slightly increased parkinsonism in four patients who had parkinsonian symptoms prior to this drug trial. The altered parkinsonian symptom was increased bradykinesia, but there was no change in tremor or rigidity. Six hours after the injection of FK 33-824, the hyperkinesia and parkinsonian scores had returned to the pretreatment level.

No change in the psychiatric state of the patients was observed. Side effects were dizziness, a feeling of heaviness in the extremities, slurred speech, pursed lips, injection of conjunctivae, and dryness of the mouth.

Morphine, 10 mg, tended to reduce TD (Fig. 6). A substantial decrease was seen in the two patients who were taking a high neuroleptic dose and who also improved during FK 33-824. Three of the patients with reduced hyper-

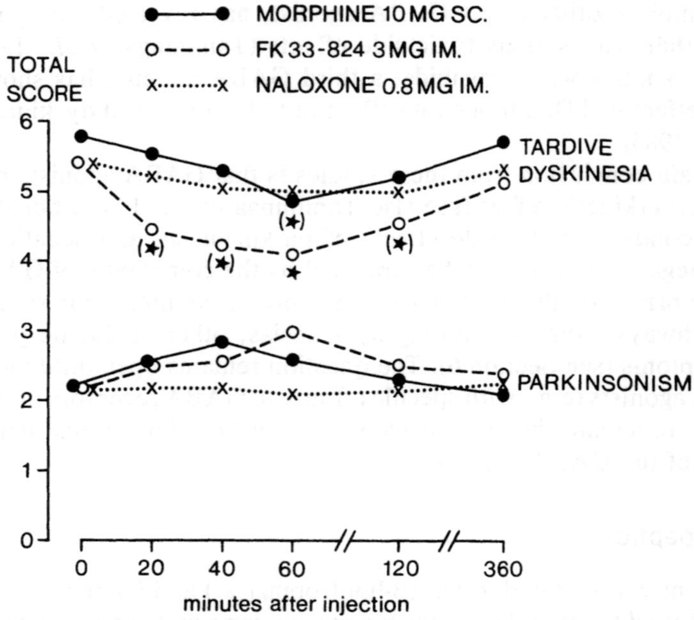

Figure 6. Mean hyperkinesia scores and mean parkinsonism scores before and following injection of morphine, FK 33-824, and naloxone. n = 8. *, $p < 0.05$; (*), $0.05 < p < 0.10$.

kinesia scores also showed increased parkinsonism. The response to morphine was complicated by the side effects of drowsiness, dizziness, ataxia, dysarthria, and nausea.

Naloxone, 0.8 mg, had no consistent effect on the hyperkinesias or the parkinsonian scores (Fig. 6). No other neurological or mental changes were seen following naloxone.

Recent reports have shown that the β-LPH_{62-77} fragment, des-tyr[1]-γ-endorphin (DTγE), produces behavioral effects in animal models which are similar to the effects of neuroleptic drugs, but does not show opiatelike activity (De Wied *et al.*, 1978). DTγE was reported to have antipsychotic activity in schizophrenic patients. In light of these findings, the role of DTγE was investigated in TD and parkinsonism (Casey *et al.*, 1981; Korsgaard *et al.*, 1982b). It was shown that a wide range of DTγE (0.5–120 mg) given as single injections i.m. had no significant effect in TD or parkinsonism, and no side effects.

These early reports do not identify a primary role for these specific neuropeptide fragments in the pathophysiology or treatment of drug-induced movement disorders, particularly TD or parkinsonism. However, this area of drug development is still in the early stages, and new molecular subdivisions of the β-lipotropin hormone will undoubtedly be considered in future research of drug-induced and idiopathic dyskinesia.

8. SUMMARY

TD is a syndrome of abnormal involuntary movements which develops during or following neuroleptic treatment. The average prevalence of the syndrome in various patient populations is 15–20% of which approximately 5% may be considered as SD. The syndrome appears only in predisposed individuals, at least in older patients, within the first 2–3 years of treatment.

The pathogenetic mechanism underlying TD is still unknown. The earlier widely accepted theory of DA supersensitivity is not compatible with several clinical and biochemical features of TD. Therefore, three new hypotheses are proposed: (1) TD may depend on a decreased GABA turnover in a subgroup of neurons projecting from striatum to globus pallidus and substantia nigra. This hypothesis is based on pharmacological and biochemical studies in rats and monkeys, but has not yet been adequately tested in humans, mainly owing to lack of sufficiently selective GABA agonists. (2) TD may depend on a blockade of a subgroup of DA receptors in striatum. Neuroleptics injected directly into the anterodorsal region of striatum in rats produce an intensification of apomorphine licking and gnawing, and in predisposed humans and monkeys, acute antidopaminergic treatment can produce dyskinetic movements which—although usually mixed with dystonic features and akathisia—have all the characteristics of TD. (3) TD may be related to an unspecific lowering of the threshold for spontaneous dyskinesia, a phenomenon which, like advanced age, may be associated with a reduced activity in several transmitter systems.

Pharmacological manipulation of TD has been performed with DA antagonists and agonists, cholinergic antagonists and agonists, serotoninergic antagonists and agonists, GABA agonists, and neuropeptides:

1. *DA antagonists* have a varying antihyperkinetic effect, low-dose neuroleptics like haloperidol and pimozide (especially in combination with tetrabenazine) being more effective than high-dose neuroleptics like thioridazine and clozapine. With the exception of clozapine, and maybe fluperlapine, all neuroleptics available have been associated with the induction of TD. Based on the DA supersensitivity theory and observation of rebound aggravation in monkeys, it has been proposed that neuroleptics with a strong antidopaminergic effect (and the ability to markedly suppress TD) have a relatively higher capacity to induce TD than weaker antidopaminergic drugs, but this theory has not yet been proved clinically.
2. *DA agonists* may, in small doses, via presynaptic DA receptors diminish DA turnover and thereby attenuate TD. However, no selective presynaptic DA agonists are yet available for clinical use. In higher doses, DA agonists may reverse DA receptor supersensitivity, but no beneficial effect has been found in human TD.
3. *Anticholinergics* uncover/aggravate TD, probably without influencing the primary, pathophysiological process, perhaps indirectly by counteracting parkinsonism.

4. *Cholinomimetics* have shown inconsistent effect in TD and produce disturbing side effects. They appear to be without therapeutic value in TD.
5. *Serotonin antagonists and agonists* may be involved in the regulation of motor dysfunctions, but no consistent effect in TD has been observed so far.
6. *GABA agonists*. The transaminase inhibitors γ-acetylenic GABA and γ-vinyl GABA can decrease TD slightly, but aggravate parkinsonism. THIP, a direct GABA receptor agonist, appears to have no significant effect in TD, although it may increase preexisting parkinsonism. Muscimol and progabide, two other GABA agonists, have been found to have some beneficial effect in TD, muscimol, however, with toxic side effects. In general, the antidyskinetic effect of GABA agonists available until now has been limited and complicated with side effects. Clinicians are eagerly awaiting new GABA agonists with selective affinity to subgroups of GABA receptors.
7. Among the *neuropeptides*, the met-enkephalin analogue FK 33-824 has slight antihyperkinetic and parkinsonian aggravating effect when added to a relatively intense neuroleptic treatment, while DTγE in single doses up to 120 mg/day is without effect in TD.

9. REFERENCES

Asnis, G. M., and Leopold, M. A., 1978, A single-blind study of ECT in patients with tardive dyskinesia, *Am. J. Psychiatry*. **135:**1235.

Baldessarini, R. J., Cole, J. O., Davis, J. M., Gardos, G., Preskorn, S. H., Simpson, G. M., and Tarsy, D., 1980, Tardive dyskinesia: A task force report, American Psychiatric Association, American Psychiatric Press, Washington D.C.

Bannet, J., and Belmaker, R. H., 1983, *New Directions in Tardive Dyskinesia Research*, Karger, Basel.

Beitman, B. D., 1978, Tardive dyskinesia reinduced by lithium carbonate, *Am. J. Psychiatry* **135:**1229.

Bell, R. C. H., and Smith, R. C., 1978, Tardive dyskinesia: Characterization and prevalence in a statewide system, *J. Clin. Psychiatry* **39:**39.

Brandrup, E., 1961, Tetrabenazine treatment in persisting dyskinesia caused by psychopharmaca, *Am. J. Pyschiatry* **118:**551.

Carlsson, A., 1975, Receptor-mediated control of dopamine metabolism, in: *Pre- and Postsynpatic Receptors* (E. Usdin and W. E. Bunney, eds.), pp. 49–63, Marcel Dekker Inc., New York.

Carroll, B. J., Curtis, C. C., and Kokmen, E., 1977, Paradoxical response to dopamine agonists in tardive dyskinesia, *Am. J. Psychiatry* **134:**785.

Casey, D. E., 1977, Deanol in the management of involuntary movement disorders: A review, *Dis. Nerv. Syst.* **38:**7.

Casey, D. E., 1984, Tardive dyskinesia and affective disorders, in: *Tardive Dyskinesia and Affective Disorders* (G. Gardos and D. E. Casey, eds.), pp. 2–17, American Psychiatric Press, Washington, DC.

Casey, D. E., 1985a, Spontaneous and tardive dyskinesia: Clinical and laboratory studies, *J. Clin. Psychiatry* **46:**42–47.

Casey, D. E., 1985b, Behavioral effects of long-term neuroleptic treatment in *Cebus* monkeys, in: *Dyskinesia—Research and Treatment* (*Psychopharmacoloy*, Suppl. 2) (D. E. Casey, T. Chase, A. V. Christensen, and J. Gerlach, eds.), pp. 211–216, Springer-Verlag, Berlin, Heidelberg.

Casey, D. E., and Denney, D., 1977, Pharmacological characterization of tardive dyskinesia, *Psychopharmacology* **54:**1.

Casey, D. E., and Gerlach, J., 1984, Tardive dyskinesia: management and new treatment, in: *Guidelines for the Use of Psychotropic Drugs* (H. C. Stancer, P. E. Garfinkel, and V. M. Rakoff, eds.), pp. 183–203, Spectrum, New York.

Casey, D. E., and Hansen, T. E., 1984, Spontaneous dyskinesia, in: *Neuropsychiatric Movement Disorders* (D. V. Jeste and R. J. Wyatt, eds.), pp. 68–95, American Psychiatric Press, Washington, DC.

Casey, D. E., and Rabins, P., 1978, Tardive dyskinesia as a life-threatening illness, *Am. J. Psychiatry* **135:**486.

Casey, D. E., and Toenniessen, L. M., 1983, Neuroleptic treatment in tardive dyskinesia; can it be developed into a clinical strategy for long-term treatment? in: *New Directions in Tardive Dyskinesia Research* (J. Bannet and R. H. Belmaker, eds.), pp. 65–79, Karger, Basel.

Casey, D. E., Gerlach, J., and Christensson, E., 1980, Dopamine, acetylcholine, and GABA effects in acute dystonia in primates, *Psychopharmacology* **70:**83.

Casey, D. E., Gerlach, J., Magelund, G., and Rosted Christensen, T., 1980b, γ-Acetylenic GABA in tardive dyskinesia, *Arch. Gen. Psychiatry* **37:**1376.

Casey, D. E., Korsgaard, S., Gerlach, J., Jørgensen, A., and Simmelsgaard, H., 1981, Effect of des-tyrosine-γ-endorphin in tardive dyskinesia, *Arch. Gen. Psychiatry* **38:**158.

Casey, D. E., Gerlach, J., and Bjørndal, N., 1982, Levodopa and receptor sensitivity modification in tardive dyskinesia, *Psychopharmacology* **78:**89.

Casey, D. E., Chase, T., Christensen, A. V., and Gerlach, J., 1985, *Dyskinesia—Research and Treatment* (*Psychopharmacology*, Suppl. 2), Springer-Verlag, Berlin, Heidelberg.

Chase, T. M., and Tamminga, C. A., 1979, GABA system participation in motor, cognitive and endocrine function in man, in: *GABA-Neurotransmitters. Pharmacochemical, Biochemical and Pharmacological Aspects* (P. Krogsgaard-Larsen, J. Scheel-Krüger, and H. Kofod, eds.), pp. 283–296, Academic Press, New York.

Chase, T. N., and Tamminga, C. A., 1980, Pharmacological studies of tardive dyskinesia, in: *Long-Term effects of neuroleptics, Advances in Biochemical Psychopharmacology* Vol. 24 (F. Cattabeni, G. Racagni, P. F. Spano, and E. Costa, eds.), pp. 457–461, Raven Press, New York.

Chase, T. N., Watanabe, A. M., Brodie, K. H., and Donnelly, E. F., 1972, Huntington's chorea. Effect of serotonin depletion, *Arch. Neurol.* **26:**282.

Christensen, A. V., 1981, Dopamine hyperactivity: Effects of neuroleptics alone and in combination with GABA agonists, in: *Biological Psychiatry 1981* (C. Perris, G. Struwe, and B. Jansson, eds.), pp. 828–832, Elsevier/North Holland, Amsterdam.

Christensen, A. V., Arnt, J., and Svendsen, O., 1985, Pharmacological differentiation of dopamine D-1 and D-2 antagonists after single and repeated administration, in: *Dyskinesia—Research and Treatment* (*Psychopharmacology*, Suppl. 2) (D. E. Casey, T. Chase, A. V. Christensen, and J. Gerlach, eds.), pp. 182–190, Springer-Verlag, Berlin, Heidelberg.

Claghorn, J. L., Abuzzahab, F. S., Wang, R., Larson, C., Gelenberg, A. J., Klerman, G. L., Tuason, V., and Steinbook, R., 1983, The current status of clozapine, *Psychopharmacol. Bull.* **19:**138.

Claveria, L. E., Teychenne, P. F., Calne, D. B., Haskayne, L., Petrie, A., Pallis, C. A., and Lodge-Patch, I. C., 1975, Tardive dyskinesia treated with pimozide, *J. Neurol. Sci.* **24:**393.

Cools, A. R., and Van Rossum, J. M., 1976, Excitation-mediating and inhibition-mediating dopamine-receptors: A new concept towards a better understanding of electrophysiological, biochemical, pharmacological, functional and clinical data, *Psychopharmacologia* **45:**243.

Coward, D. M., 1982, Classical and non-classical neuroleptics induce supersensitivity of nigral GABA-ergic mechanisms in the rat, *Psychopharmacology* **78:**180.

Crane, G. E., 1973, Persistent dyskinesia, *Br. J. Psychiatry* **122:**395.

Crow, T. J., Cross, A. J., Johnstone, E. C., Owen, F., Owens, D. G. C., and Waddington, J. L.,

1982, Abnormal involuntary movements in schizophrenia: Are they related to the disease process or its treatment? Are they associated with changes in dopamine receptors? *J. Clin. Psychopharmacol.* **2**:336.
DeVeaugh-Geiss, J., 1982, *Tardive Dyskinesia and Related Involuntary Movement Disorders*, John Wright, PSG Inc., Boston-Bristol-London.
De Wied, D., Kovács, G. L., Bohus, B., Van Ree, J. M., and Greven, H. M., 1978, Neuroleptic activity of the neuropeptide β-LPH_{62-77}(des-tyr-^{1}-γ-endorphin (DTγE), *Eur. J. Pharmacol.* **49**:427.
Doepp, S., and Buddeberg, C., 1975, Extrapyramidale Symptome unter Clozapin, *Nervenarzt* **46**:589.
Domino, E. F., 1985, Induction of tardive dyskinesia in *Cebus* apella and *Macaca speciosa* monkeys: A review, in: *Dyskinesia—Research and Treatment* (*Psychopharmacology*, Suppl. 2) (D. E. Casey, T. Chase, A. V. Christensen, and J. Gerlach, eds.), pp. 217–223, Springer-Verlag, Berlin, Heidelberg.
Edwards, H., 1970, The significance of brain damage in persistent oral dyskinesia, *Br. J. Psychiatry* **116**:271.
Eichenberger, E., 1984, Pharmacology of fluperlapine compared with clozapine. *Arzneim.-Forsch./Drug Res.* **34**:110.
Ettigi, P., Nair, N. P. V., Lal, S., Cervantes, P., and Guyda, H., 1976, Effect of apomorphine on growth hormone and prolactin secretion in schizophrenic patients, with or without oral dyskinesia, withdrawn from chronic neuroleptic therapy, *J. Neurol. Neurosurg. Psychiatry* **39**:870.
Fann, W. E., Smith, R. C., Davis, J. M., and Domino, E. F., 1980, *Tardive Dyskinesia. Research and Treatment*, Spectrum, Jamaica, New York.
Faurbye, A., Rasch, P. J., Petersen, P. B., Brandborg, G., and Pakkenberg, H., 1964, Neurological symptoms in pharmacotherapy in psychoses, *Acta Psychiatr. Scand.* **40**:10.
Fibiger, H. C., and Lloyd, K. G., 1984, Neurobiological substrates of tardive dyskinesia: The GABA hypothesis, *Trends In Neuroscience*, p. 462.
Fischer-Cornelssen, K. A., 1984, Fluperlapine in 104 schizophrenic patients. Open multicenter trial, *Arzneim.-Forsch./Drug. Res.* **34**:125.
Fog, R., Pakkenberg, H., Juul, P., Bock, E., Jørgensen, O. S., and Andersen, J., 1976, High-dose treatment of rats with perphenazine enanthate, *Psychopharmacology* **50**:305.
Friedhoff, A. J., Bonnet, K., and Rosengarten, H., 1977, Reversal of two manifestations of dopamine receptor supersensitivity by administration of L-dopa, *Res. Commun. Chem. Pathol. Pharmacol.* **16**:411.
Gale, K., and Casu, M., 1981, Dynamic utilization of GABA in substantia nigra: Regulation by dopamine and GABA in the striatum, and its clinical and behavioral implications, *Mol. Cell Biochem.* **39**:369.
Gardos, G., and Cole, J. O., 1978, Pilot study of cyproheptadine (Periactin) in tardive dyskinesia, *Psychopharmacol. Bull.* **14**:18.
Gardos, G., Cole, J., and La Brie, R., 1977, Drug variables in the etiology of tardive dyskinesia. Application of discriminant function analyses, *Prog. Neurol. Psychopharmacol.* **1**:147.
Gardos, G., Perenyi, A., and Cole, J. O., 1983, Tardive dyskinesia: Changes after three years, *J. Clin. Psychopharmacol.* **3**:315.
Gerlach, J., 1979, Tardive dyskinesia, *Dan. Med Bull.* **46**:209.
Gerlach, J., and Casey, D. E., 1983, Dopamine agonists in clinical research for new tardive dyskinesia treatment, in: *New Directions in Tardive Dyskinesia Research* (J. Bannet and R. H. Belmaker, eds.), pp. 97–110, Karger, Basel.
Gerlach, J., and Casey, D. E., 1984, Sulpiride in tardive dyskinesia, *Acta Psychiatr. Scand.* **69**(Suppl. 311):93.
Gerlach, J., and Korsgaard, S., 1981, Classification and prevalence of neuroleptic-induced hyperkinetic disorders, in: *Biological Psychiatry 1981* (C. Perris, G. Struwe, and B. Jansson, eds.), pp. 844–851, Elsevier/North Holland, Amsterdam.
Gerlach, J., and Simmelsgaard, H., 1978, Tardive dyskinesia during and following treatment with

haloperidol, haloperidol + biperiden, thioridazine and clozapine, *Psychopharmacology* **59**:105.

Gerlach, J., and Thorsen, K., 1976, The movement pattern of oral tardive dyskinesia in relation to anticholinergic and antidopaminergic treatment, *Int. J. Pharmacopsychiatr.* **11**:1.

Gerlach, J., Koppelhus, P., Jelweg, E., and Monrad, A., 1974a, Clozapine and haloperidol in a single-blind cross-over trial: Therapeutic and biochemical aspects in the treatment of schizophrenia, *Acta Psychiatr. Scand.* **50**:410.

Gerlach, J., Reisby, N., and Randrup, A., 1974b, Dopaminergic hypersensitivity and cholinergic hypofunction in the pathophysiology of tardive dyskinesia, *Psychopharmacologia* **34**:21.

Gerlach, J., Thorsen, K., and Fog, R., 1975, Extrapyramidal reactions and amine metabolites in cerebrospinal fluid during haloperidol and clozapine treatment of schizophrenic patients, *Psychopharmacologia* **40**:341.

Glazer, W. M., Moore, D. C., Schooler, N. R., Brenner, L. M., and Morgenstern, H., 1984, Tardive dyskinesia. A discontinuation study, *Arch. Gen. Psychiatry* **41**:623.

Gottfries, C. G., 1981, Levels of monoamines, monoamine metabolites, and activity in related enzyme systems correlated to normal aging and in patients with dementia of Alzheimer type, in: *Apomorphine and Other Dopaminomimetics*, vol. 2, *Clinical Pharmacology* (G. U. Corsini and G. L. Gessa, eds.), pp. 243–249, Raven Press, New York.

Gunne, L-M., and Bárány, S., 1979, A monitoring test for the liability of neuroleptic drugs to induce tardive dyskinesia, *Psychopharmacology* **63**:195.

Gunne, L. M., and Bárány, S., 1980, A primate model for tardive dyskinesia, in: *Tardive Dyskinesia: Research and Treatment* (W. E. Fann, R. C. Smith, and J. M. Davis, eds.), pp. 1–26, Spectrum, New York.

Gunne, L.-M., Häggström, J.-E., and Sjöquist, B., 1984, Persistent neuroleptic-induced dyskinesia associated with regional changes within brain GABA and dopamine systems, *Nature (London)* **309**:347.

Hardie, R. J., Lees, A. J., and Stern, G. M., 1983, Sustained levodopa therapy in tardive dyskinesia, *J. Neurol. Neurosurg. Psychiatry* **46**:685.

Hjorth, S., Carlsson, A., Wikström, H., Lindberg, P., Sanchez, D., Hacksell, U., Arvidsson, L. E., Svensson, U., and Nilsson, J. L. G., 1981, 3-PPP, a new acting DA-receptor agonist with selectivity for autoreceptors, *Life Sci.* **28**:1225.

Itoh, H., and Yagi, G., 1979, Reversibility of tardive dyskinesia, *Folia Psychiat. Neurolog. Jap.* **33**:43.

Iversen, L. L., 1979, The chemistry of the brain, *Sci. Am.* **241**:134.

Jenner, P., and Marsden, C. D., 1981, Substituted benzamide drugs as selective neuroleptic agents, *Neuropharmacology* **20**:1285.

Jeste, D. V., and Wyatt, R. J., 1982, *Understanding and Treating Tardive Dyskinesia*, The Guilford Press, New York.

Jus, K., Jus, A., Gautier, J., Villeneuve, A., Pires, P., Pineau, R., and Villeneuve, R., 1974, Studies on the action of certain pharmacological agents on tardive dyskinesia and on the rabbit syndrome, *Int. J. Clin. Pharmacol.* **9**:138.

Kane, J. M., and Smith, J. M., 1982, Tardive dyskinesia: Prevalence and risk factors, 1959 to 1979, *Arch. Gen. Psychiatry* **39**:473.

Kane, J. M., Woerner, M., and Lieberman, J., 1985, Tardive dyskinesia: Prevalence, incidence, and risk factors, in: *Dyskinesia—Research and Treatment* (*Psychopharmacology*, Suppl. 2) (D. E. Casey, T. Chase, A. V. Christensen, and J. Gerlach, eds.), pp. 72–78, Springer-Verlag, Berlin, Heidelberg.

Kazamatsuri, H., Chien, C. P., and Cole, J., 1972, Treatment of tardive dyskinesia. II. Short-term efficacy of dopamine-blocking agents haloperidol and thiopropazate, *Arch. Gen. Psychiatry* **27**:100.

Klawans, H. L. (ed.), 1983, Proceedings of a symposium on tardive dyskinesia. *Clin. Neuropharmacol.* **6**:75.

Klawans, H. L., and Rubovits, R., 1974, Effect of cholinergic and anticholinergic agents on tardive dyskinesia, *J. Neurol. Neurosurg. Psychiatry* **27**:941.

Klawans, H. L., Goetz, C. G., and Perlik, S., 1980, Tardive dyskinesia: Review and update, *Am. J. Psychiatry* **137:**900.

Korsgaard, S., Casey, D. E., Gerlach, J., Hetmar, O., Kaldan, B., and Mikkelsen, L. B., 1982a, Effect of THIP, a new GABA agonist in tardive dyskinesia, *Arch. Gen. Psychiatry* **39:**1017.

Korsgaard, S., Casey, D. E., and Gerlach, J., 1982b, High-dose des-tyrosine-γ-endorphin in tardive dyskinesia, *Psychopharmacology* **78:**285.

Korsgaard, S., Gerlach, J., and Christensson, E., 1982c, Behavioural aspects of serotonin-dopamine interactions in monkeys, in: *CINP Congress Abstracts*, Jerusalem, p. 400.

Korsgaard, S., Casey, D. E., and Gerlach, J., 1983, Effect of γ-vinyl GABA in tardive dyskinesia, *Psychiatr. Res.* **8:**261.

Korsgaard, S., Noring, U., and Gerlach, J., 1984, Fluperlapine in tardive dyskinesia and parkinsonism, *Psychopharmacology* **84:**76.

Lee, D. K., Markham, C. H., and Clark, W. G., 1969, Serotonin (5-HT) in Huntington's chorea, in: *Progress in Neurogenetics* (A. Barbeau and J. R. Brunette, eds.), Excerpta Medica Foundation, Amsterdam.

Lühdorf, K., and Lund, M., 1977, Phenytoin induced hyperkinesia, *Epilepsia* **18:**409.

Magelund, G., Gerlach, J., and Casey, D. E., 1979, Neuroleptic-potentiating effect of α-methyl-p-tyrosine compared with haloperidol and placebo in a double-blind cross-over trial, *Acta Psychiatr. Scand.* **60:**185.

Marsden, C. D., 1985, Is tardive dyskinesia a unique disorder? in: *Dyskinesia—Research and Treatment* (*Psychopharmacology*, Suppl. 2) (D. E. Casey, T. Chase, A. V. Christensen, and J. Gerlach, eds.), pp. 64–71, Springer-Verlag, Berlin, Heidelberg.

Marsden, C. D., and Jenner, P., 1981, The nature and selectivity of changes in cerebral neuronal receptor function induced by continuous long term neuroleptic administration to rats, in: *Biological Psychiatry 1981* (C. Perris, G. Struwe, and B. Jansson, eds.), pp. 820–823, Elsevier/North Holland, Amsterdam.

Meldrum, B., 1982, Pharmacology of GABA, *Clin. Neuropharmacol.* **5:**293.

Meltzer, H. Y., 1982, Dopamine autoreceptor stimulation: Clinical significance, *Pharmacol. Biochem. Behavior* **17** (suppl. 1):1.

Morselli, P. L., Fournier, V., Bossi, L., and Musch, B., 1985, Clinical activity of GABA agonists in neuroleptic- and L-dopa-induced dyskinesia, in: *Dyskinesia—Research and Treatment* (*Psychopharmacology*, Suppl. 2) (D. E. Casey, T. Chase, A. V. Christensen, and J. Gerlach, eds.), pp. 128–136, Springer-Verlag, Berlin, Heidelberg.

Nasrallah, H. A., Donnelly, E. F., Bigelow, L. B., Rivera-Calimlim, L., Rogol, A., Potkin, S., Rauscher, F. P., Wyatt, R. J., and Gillin, J. C., 1977, Inhibition of dopamine synthesis in chronic schizophrenia, *Arch Gen. Psychiatry* **34:**649.

Nasrallah, H. A., Smith, R. E., Dunner, F. J., and McCalley-Whitters, M., 1982, Serotonin precursor effects in tardive dyskinesia, *Psychopharmacology* **77:**234.

Nasrallah, H. A., Dunner, F. J., Smith, R. E., McCalley-Whitters, M., 1983, A placebo-controlled trial of L-dopa in tardive dyskinesia, *Psychopharmacol. Bull.* **19:**122.

Nielsen, E. B., and Lyon, M., 1978, Evidence for cell loss in corpus striatum after longterm treatment with a neuroleptic drug (fluopenthixol) in rats, *Psychopharmacology* **59:**85.

Noring, U., Juul Povlsen, U., Casey, D. E., and Gerlach, J., 1984, Effect of a cholinomimetic drug (RS 86) in tardive dyskinesia and drug-related parkinsonism, *Psychopharmacology* **84:**569.

Owens, D. G. C., 1985, Involuntary disorders of movement in chronic schizophrenia—The role of the illness and its treatment, in: *Dyskinesia—Research and Treatment* (*Psychopharmacology*, Suppl. 2) (D. E. Casey, T. Chase, A. V. Christensen, and J. Gerlach, eds.), pp. 79–87, Springer-Verlag, Berlin, Heidelberg.

Pakkenberg, H., Fog, R., and Nilakantan, B., 1973, The long-term effects of perphenazine enanthate on the rat brain, *Psychopharmacologia* **29:**329.

Pi, E. H., and Simpson, G. M., 1983, Atypical neuroleptics Clozapine and the benzamides in the prevention and treatment of tardive dyskinesia, in: *New Directions in Tardive Dyskinesia Research* (J. Bannet and R. H. Belmaker, eds.), pp. 80–86, Karger, Basel.

Price, T. R. P., and Levin, R., 1978, The effects of electroconvulsive therapy on tardive dyskinesia, *Am. J. Psychiatry* **135**:991.

Rosenbaum, A. H., Niven, R. G., Hanson, N. P., *et al.*, 1977, Tardive dyskinesia: Relationship with a primary affective disorder, *Dis. Nerv. Syst.* **38**:423.

Scheel-Krüger J., 1983, The GABA receptor and animal behaviour evidence that GABA transmits and mediates dopaminergic functions in the basal ganglia and the limbic system, in: *GABA Receptors* (S. J. Enna, ed.), pp. 215–255, Humana Press, Clifton, NJ.

Scheel-Krüger, J., and Arnt, J., 1985, New aspects on the role of dopamine, acetylcholine, and GABA in the development of tardive dyskinesia, in: *Dyskinesia—Research and Treatment* (D. E. Casey, T. Chase, A. V. Christensen, and J. Gerlach, eds.) (*Psychopharmacology*, Suppl. 2), pp. 46–71, Springer-Verlag, Berlin, Heidelberg.

Schönecker, M., 1957, Ein eigentümliches Syndrom im oralen Bereich bei Megaphenapplikation, *Nervenarzt* **28**:35.

Seeman, P., 1980, Brain dopamine receptors, *Pharmacol. Rev.* **32**:229.

Shah, N. S., and Donald, A. G., 1982, *Endorphins and Opiate Antagonists in Psychiatric Research*, Plenum Medical, New York, London.

Sigwald, J., Bouttier, D., Raymondeaud, C., *et al.*, 1959, Quatre cas de dyskinésie facio-bucco-linguo-masticatrice a l'evolution prolongée secondaire à un traitement par les neuroleptiques, *Rev. Neurol.* **100**:751.

Simpson, G., Lee, J., and Shrivastava, R., 1978, Clozapine in tardive dyskinesia, *Psychopharmacology* **56**:75.

Smith, J. M., and Baldessarini, R. J., 1980, Changes in prevalence, severity and recovery in tardive dyskinesia with old age, *Arch. Gen. Psychiatry* **37**:1368.

Smith, R. C. Tamminga, C. A., Haraszti, J., Pandey, G. N., and Davis, J. M., 1977, Effects of dopamine agonists in tardive dyskinesia, *Am. J. Psychiatry* **134**:763.

Tamminga, C. A., Smith, R. C., Pandey, G., Frohman, L. A., and Davis, J. M., 1977, A neuroendocrine study of supersensitivity in tardive dyskinesia, *Arch. Gen. Psychiatry* **34**:1199.

Tamminga, C. A., Crayton, J. W., and Chase, T. N., 1979, Improvement in tardive dyskinesia after muscimol therapy, *Arch. Gen. Psychiatry* **36**:595.

Tamminga, C. A., Thaker, G. K., and Chase, T. N., 1985, GABA dysfunction in the pathophysiology of tardive dyskinesia, in: *Dyskinesia—Research and Treatment* (*Psychopharmacology*, Suppl. 2) (D. E. Casey, T. Chase, A. V. Christensen, and J. Gerlach, eds.), pp. 122–127, Springer-Verlag, Berlin, Heidelberg.

Tarsy, D., Leopold, N., and Sax, D., 1974, Physostigmine in choreiform movement disorders, *Neurology* **24**:28.

Thach, B. T., Chase, T. N., and Bosma, J. F., 1975, Oral facial dyskinesia associated with prolonged use of antihistaminic decongestants, *N. Engl. J. Med.* **293**:486.

Toenniessen, L. M., and Casey, D. E., 1982, Duration of neuroleptic exposure and risk of tardive dyskinesia, *Proc. Soc. Biol. Psychiatry* **69**:100.

Toenniessen, L. M., Casey, D. E., and McFarland, B. H., 1985, Tardive dyskinesia in the aged: Duration of treatment relationships, *Arch. Gen. Psychiatry* **42**:278–284.

Uhrbrand, L., and Faurbye, A., 1960, Reversible and irreversible dyskinesia after treatment with perphenazine, chlorpromazine, reserpine and electroconvulsive therapy, *Psychopharmacologia* **1**:408.

Wålinder, J., Skott, A., Carlsson, A., and Ross, B. E., 1976, Potentiation by metyrosine of thioridazine effects in chronic schizophrenics, *Arch. Gen. Psychiatry* **33**:501.

Weiss, B., and Santelli, S., 1978, Dyskinesias evoked in monkeys by weekly administration of haloperidol, *Science* **200**:799.

Woggon, B., Heinrich, K., Küfferle, B., Müller-Oerlinghausen, B., Pöldinger, W., Rüther, E., and Schied, H. W., 1984, Results of a multicenter AMDP study with fluperlapine in schizophrenic patients, *Arzneim.-Forsch./Drug Res.* **34**:122.

CHAPTER 6

Tardive Dyskinesia and Subtyping of Schizophrenia

DILIP V. JESTE, CHARLES A. KAUFMANN, and RICHARD JED WYATT

1. INTRODUCTION

The history of psychiatry, especially that of the "functional psychoses," is replete with systems of classification proposed by "splitters" and "lumpers." Kraepelin (1907) split manic depressive illness from dementia precox and lumped under the latter category different conditions described by Morel, Kahlbaum, and Hecker. Bleuler (1911) clearly recognized the heterogeneous nature of schizophrenia and split the syndrome into paranoid, catatonic, simple and hebephrenic subtypes. Since then a number of attempts have been made to subdivide the schizophrenic syndrome into rational subgroups. Most of the traditional classifications have been based on "naturalistic" variables such as symptomatology and course. In recent years, several investigators have suggested clinical utility and potential biological validity of subtyping disorders according to patterns of response to "extraneous" variables, such as drug treatment (Klein, 1965). Since the most effective treatment of chronic schizophrenia is neuroleptic administration, it would seem logical to attempt subdividing schizophrenic patients on the basis of their response to neuroleptic treatment. While therapeutic response to neuroleptics has been used to characterize patient subgroups (neuroleptic responders versus nonresponders), we believe that it might also be useful to subtype neuroleptic-treated chronic schizophrenic patients according to the development of tardive dyskinesia (TD). In a recent study we classified schizophrenic patients into different subgroups using a number of different dimensions (Jeste *et al.*, 1982). This paper deals with the study using the TD dimension.

DILIP V. JESTE, CHARLES A. KAUFMANN, and RICHARD JED WYATT • Neuropsychiatry Branch, National Institute of Mental Health, Saint Elizabeths Hospital, Washington, D.C. 20032.

2. RATIONALE

1. TD is a serious side effect of long-term neuroleptic treatment. A review of the literature on the epidemiology of TD shows that the reported mean prevalence of dyskinesia among neuroleptic-treated inpatients and outpatients since 1976 has been about 26% (Jeste and Wyatt, 1982). According to Crane (1973), TD is likely to be the major limiting factor in the use of neuroleptics. TD is no longer a phenomenon of innocuous curiosity but a serious public health problem in the psychopharmacology of schizophrenia, with clinical, theoretical, and medicolegal implications. Spontaneously occurring abnormal involuntary movements are not rare in chronic schizophrenic patients. Nevertheless, the reported prevalence of dyskinesia is three times greater among neuroleptic-treated patients than among non-neuroleptic-treated controls. The epidemiological data, as well as experimental work with monkeys, strongly favor a causal association between neuroleptics and TD (Kovacic and Domino, 1982).

2. TD is not a momentary reaction to drugs. The symptoms tend to be persistent even after neuroleptic withdrawal in nearly two-thirds of patients who develop TD (Jeste and Wyatt, 1982).

3. TD can be operationally defined (Jeste and Wyatt, 1982). This may also help in the somewhat difficult task of separating spontaneously occurring abnormal movements in schizophrenic patients (Casey and Hansen, 1984) from drug-induced dyskinesias.

4. Although TD is an adverse effect of neuroleptics, there are several reasons to believe that predisposing constitutional factors are also necessary for its development (Kane *et al.*, 1984). Only a certain proportion of patients treated with neuroleptics are afflicted with dyskinesia. Some patients develop persistent dyskinesia with a relatively brief course of treatment, while others are free of dyskinesia despite high-dose intake of neuroleptics for a number of years. Aging and certain types of brain damage have been thought to be two of the factors predisposing to persistent TD. Also favoring the contribution of patient-related variables to the pathogenesis of TD is the fact that a number of investigators have failed to find significant differences between dyskinetic and nondyskinetic groups in treatment-related variables such as length of neuroleptic treatment (Demars, 1966; Kennedy *et al.*, 1971; Brandon *et al.*, 1971; Jus *et al.*, 1976; Mallya *et al.*, 1979; Smith *et al.*, 1978; Simpson *et al.*, 1978), type of neuroleptics administered (Faurbye *et al.*, 1964; Heinrich *et al.*, 1968; Kennedy *et al.*, 1971; Jus *et al.*, 1976; Mallya *et al.*, 1979; Pandurangi *et al.*, 1978; Perris *et al.*, 1979; Simpson *et al.*, 1978; Jeste *et al.*, 1979), polypharmacy (Demars, 1966; Simpson *et al.*, 1978), prescription of antiparkinsonian drugs (Jus *et al.*, 1976; Gardos *et al.*, 1977; Bell and Smith, 1978; Simpson *et al.*, 1978; Jeste *et al.*, 1979), and physical treatments including electroconvulsive therapy, leukotomy, and insulin coma (Pryce and Edwards, 1966; Demars, 1966; Brandon *et al.*, 1971; Simpson *et al.*, 1978; Pandurangi *et al.*, 1978).

We are not suggesting that variables associated with neuroleptic treatment are unrelated to TD. Indeed, we define TD as a syndrome in the etiology of which neuroleptics are a necessary factor. We believe, however, that neuroleptics are not a sufficient factor and that constitutional factors are also necessary in the pathogenesis of TD. It is, therefore, permissible to assume that there may be some biological differences between patients who are predisposed to develop TD with neuroleptics and those who are not so predisposed.

5. There is generally little correspondence between TD and therapeutic response to neuroleptics (Jeste and Wyatt, 1982). It seems that the antipsychotic and dyskinesia-producing effects of neuroleptics may not be totally interdependent and, therefore, need to be studied separately.

3. PATIENTS AND METHODS

Over the past 6 years, 109 inpatients at Saint Elizabeths Hospital with presumed schizophrenia, including 102 meeting research diagnostic criteria (RDC) for schizophrenia (Spitzer *et al.*, 1975), were studied. A preliminary analysis of dopamine-β-hydroxylase (DBH) activity in some of these patients was reported earlier (Jeste *et al.*, 1982). An effort was made to include a diverse group of patients, varying in age (19–95 years), duration of illness (1–63 years), and diagnostic subtype (primarily, chronic paranoid and undifferentiated schizophrenia). Patients were examined by two psychiatrists for the presence of TD. Thirty-nine patients met specific diagnostic criteria for TD (Jeste and Wyatt, 1982), based on phenomenology, duration, relationship to neuroleptic treatment, and exclusion of other movement disorders. The severity of abnormal movements was rated in all patients with the abnormal involuntary movement scale (National Institute of Mental Health, 1976); a global score of 2 or greater (mild, but definitely abnormal) was also required for the diagnosis of TD.

Subgroups of patients were further evaluated with the brief psychiatric rating scale (BPRS) (Overall and Gorham, 1962) and assayed for plasma DBH activity (with a photometric assay) (Nagatsu and Udenfriend, 1972). In addition, 65 patients underwent CT scans, which were assessed for ventricle/brain ratio (VBR) (Weinberger *et al.*, 1979).

All patients were studied in the steady state, primarily when neuroleptic-free for more than 2 weeks or when on steady-dose neuroleptic treatment for more than 4 weeks. All biochemical assays and neuroradiological assessments were performed "blind" to diagnostic and TD status.

Data on plasma DBH activity and VBR in some of the earlier patients have been published previously (Jeste *et al.*, 1982; Jeste and Wyatt, 1982). Futher details of this study will be reported elsewhere (Kaufmann, Jeste, Linnoila *et. al.*, submitted for publication).

4. RESULTS

As shown in Table 1, patients with and without TD were comparable for gender, race, and diagnosis. TD patients were older (53.1 ± 23.1 versus

TABLE 1
Description of Patients

	TD	Non-TD
n	39	70
Age (years)[a]	53.1 ± 23.1[b]	42.4 ± 19.9
Gender (% men/women)	68/32	70/30
Race (% black/white)	15/85	16/84
Diagnosis[c] (% CPS/CUS/other)	23/64/13	44/54/2
Duration (years)[a]	26.7 + 17.8[b]	19.0 ± 14.0

[a] Values are means ± SD.
[b] $p < 0.02$
[c] CPS, chronic paranoid schizophrenia; CUS, chronic undifferentiated schizophrenia (RDC).

42.4 ± 19.9 years, T = −2.55, $p < 0.02$, *t* test) and had longer durations of illness (26.7 ± 17.8 versus 19.0 ± 14.0 years, T = −2.49, $p < 0.02$, *t* test).

Patients with TD had significantly greater plasma DBH activity than those without TD (43.5 ± 23.1 versus 26.8 ± 15.6 nmole/ml per min, T = −3.61, $p < 0.001$, *t* test) (Fig. 1). This difference was not accounted for by differences in age or duration: neither variable correlated with DBH activity. Might differences in DBH be explained by differences in treatment between TD and non-TD patients? While comparable numbers of TD and non-TD patients were off-and-on neuroleptics (TD-off 18, TD-on 9, non-TD-off 36, non-TD-on 11, $p < 0.35$, chi square), TD patients were on significantly lower doses (64 versus 229 mg chlorpromazine equivalents, $p < 0.05$, *t* test). Thus, the elevation in

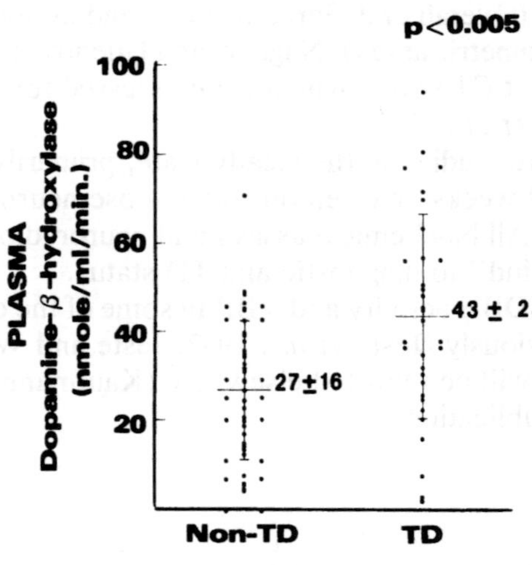

FIGURE 1. Plasma dopamine-β-hydroxylase activities in patients with and without tardive dyskinesia (TD). T = 3.61, $p < 0.001$, *t* test.

DBH activity in TD patients could conceivably be due to their relatively lower neuroleptic dose (DeLisi *et al.*, 1981). This would appear unlikely as patients on and off neuroleptics had almost identical values for DBH, and DBH did not correlate with neuroleptic dose (in chlorpromazine equivalents).

Patients with TD also had significantly larger VBR's than those without TD (11.0 ± 6.1 versus 6.0 ± 5.2); however, VBR *was* highly correlated with age ($r_p = 0.68, p < 0.0001$), and the difference in VBR was lost when age was considered as a covariate ($F = 0.02, p < 0.90$, analysis of covariance).

We did not observe a significant correlation between DBH and VBR (in all patients studied together, $-r_p = 0.07$, or in TD and non-TD patients considered individually $-r_p = -0.22$ and $r_p = 0.18$, respectively).

As a group, patients with TD have a greater DBH activity than those without. Yet, as is apparent from Fig. 1, not all patients with TD have elevated DBH activity. It seems that other factor(s) besides enzyme activity may contribute to the development of TD. To further delineate those factors, we divided all subjects (both TD and non-TD) into those with DBH activities above and below the mean (33.2 nmole/ml per min). TD patients with "high" and "low" DBH activities, so defined, did not differ in age, duration, gender, race, or mean neuroleptic dose (in chlorpromazine equivalents). They did not differ in abnormal movement localization or severity. They did not differ in clinical psychopathology rating scores, although there was a trend toward higher BPRS total score in TD patients with "high" DBH activity (2.16 ± 0.42 versus 1.09 ± 0.85, $p < 0.08$, *t* test). While TD patients with "high" and "low" DBH activity could not be differentiated, patients with "low" DBH activity and TD *could* be separated from those with "low" DBH activity and no TD. The former were more likely to demonstrate enlarged VBR's on CT scan (14.54 ± 7.41 versus 7.07 ± 6.02, $F = 2.88, p < 0.05$, analysis of variance; $p < 0.05$ Scheffé posthoc test). As noted, VBR was highly correlated with age, and patients with TD were older than those without. Might "low" DBH patients with TD be older than those without TD accounting for the observed differences in VBR? This would appear unlikely: on average, "low"-DBH TD patients were older than their non-TD counterparts (56.7 versus 43.5 years), but this did not achieve statistical significance. Moreover, the mean ages of "high"-DBH TD and non-TD patients were even more discrepant (59.4 versus 41.8 years, $p < 0.02$, *t* test), but these patients demonstrated almost identical VBR's (9.61 ± 3.92 and 9.84 ± 5.73, respectively).

Structural brain abnormalities in "low"-DBH TD patients appeared limited to VBR enlargement. None of eight patients with TD demonstrated cortical atrophy, while 10 of 25 patients without TD demonstrated such atrophy ($p < 0.08$, Fisher's exact probability test). If anything, this is in a direction opposite to the VBR findings.

Finally, both DBH activity and VBR were used to classify all subjects in a discriminant function analysis. As within-group covariance matrices were nonhomogeneous, a quadratic disciminant analysis was performed. Seventy percent of tardive and 79.3% of nontardive subjects were correctly classified, yielding an overall success rate of 75.5%. Using a more stringent jackknifed quadratic classification, 65.0% of tardive and 75.9% of nontardive subjects were correctly classified, yielding an overall success rate of 70.4%

5. DISCUSSION

We found that 75.5% of the schizophrenic patients could be classified into dyskinesia and nondyskinesia groups on the basis of plasma DBH activity and VBR.

We do not know whether the high plasma DBH activity in our patients suggested a constitutional predisposition to TD or was a result of long-term neuroleptic treatment. The effect of neuroleptics on plasma DBH activity is uncertain (Meltzer *et al.*, 1980; DeLisi *et al.*, 1981). It is worth noting that in our study the nondyskinesia patients too had received neuroleptic treatment for a similar period and, yet, had significantly lower plasma DBH activity than the dyskinesia patients. It is, of course, conceivable that neuroleptics might have had different effects on enzyme activities in different patients or that there might have been subtle differences in the treatment patterns for the two groups that were not obvious from chart reviews. (For example, the records of drug interruptions were often unsatisfactory.)

There was no significant difference (*t* test) in the plasma DBH activities of dyskinesia patients on neuroleptics and those of dyskinesia patients off neuroleptics. A similar subdivision of nondyskinesia patients into those on neuroleptics and those off neuroleptics also revealed no significant difference in the enzyme activities of these two subgroups (*t* test). This suggests that the difference between dyskinetic and nondyskinetic groups in DBH activities was not likely to have been an artifact of current medication.

We do not know whether the high plasma DBH activity in TD reflected similar trends in DBH activities in certain areas of the brain. If it did, our findings would conform with the theory of central (primarily nigrostriatal) catecholaminergic hyperactivity in TD (Baldessarini and Tarsy, 1978). The elevated plasma DBH activity in TD parallels the finding in Huntington's disease (Lieberman *et al.*, 1972), which is also presumed to be associated with relatively increased striatal catecholaminergic activity. Lieberman *et al.* (1972) noted reduced plasma DBH activity in untreated patients with Parkinson's disease. Parkinsonism and TD are believed to be "biochemical opposites" (Klawans, 1973). Furthermore, there are reports of successful use of DBH inhibitors in the treatment of iatrogenic dyskinesias. For example, Viukari and Linnoila (1977) found that fusaric acid reduced neuroleptic-induced TD, while Birket-Smith and Anderson (1973) obtained improvement in L-dopa-induced dyskinesia with disulfiram.

It remains to be studied whether the association of TD with high plasma DBH activity is restricted to schizophrenic patients or is also seen in nonschizophrenic patients (e.g., those with affective disorders). It is possible that even among schizophrenic patients, there may be biochemical subtypes of TD, and that only certain of these subtypes are associated with high plasma DBH activities. Indeed, whereas Perenyi *et al.* (1981) found that TD patients had significantly greater plasma DBH activity, Glazer *et al.* (unpublished data) failed to find such a difference in their outpatients.

We should stress that all our dyskinetic patients had persistent TD. Patients with reversible TD were not included. Elsewhere we have reviewed the literature on differences between reversible and persistent forms of TD (Jeste *et al.*, 1983). However, our present study was not addressed to this issue. The relevance of reversible TD to a subtyping of schizophrenia remains to be investigated.

It should be mentioned that neuroradiological abnormalities in TD patients with low DBH were confined to ventricular enlargement. There was no evidence of greater cortical atrophy in low-DBH patients with TD compared to those without TD, suggesting that structural brain abnormalities might be limited to subcortical structures (such as basal ganglia). Of note, rats subjected to "bifrontal cortical" ablations (which often extend into subcortical structures) develop spontaneous vacuous chewing movements which are intensified after chronic neuroleptic administration and exceed those of unoperated rats receiving chronic neuroleptics (Glassman and Glassman, 1980; Gunne and Growdon, 1982). Abnormal movements continue for up to 7 weeks after neuroleptic withdrawal and may provide an animal model of TD. It is particularly interesting that abnormal oral movements are seen most frequently in rats that have sustained considerable damage to the basal ganglia; some rats with frontal cortical damage alone show no abnormal movements (Glassman and Glassman, 1980). These animal studies may provide preclinical support for our clinical observation of selective VBR enlargement in low-DBH patients with TD.

TD is probably not a unitary disorder, but a syndrome with heterogeneous symptomatology and, undoubtedly, heterogeneous pathophysiology. While drug (primarily neuroleptic) treatment is a necessary factor in the development of TD, it is not the only factor. Noradrenergic (elevated DBH) and neuroanatomical (enlarged VBR) abnormalities may also be among the risk factors and suggest a rational approach to subtyping and pharmacological treatment.

In conclusion, our study supported the concept of classifying neuroleptic-treated chronic schizophrenic patients into those with versus those without TD. Some possibly relevant biological differences between the two groups were found.

6. SUMMARY

Over the past 6 years, 109 inpatients at Saint Elizabeths Hospital (102 meeting RDC for schizophrenia) were studied with a battery of neurological, behavioral, biochemical, and neuroradiological measures. Thirty-nine patients met specific diagnostic criteria for TD.

Patients with TD had significantly greater plasma DBH activity (TD: 43.5 ± 23.1, non-TD: 26.8 ± 15.7 nmole/ml per min, T = −3.61, $p < 0.001$). TD patients with low DBH activities (below the mean) had significantly larger ventricles than non-TD patients with low DBH activities (VBR 14.54 ± 7.41 and 7.07 ± 6.02, respectively, F = 2.88, $p < 0.05$).

These results extend earlier findings of elevated plasma DBH activity in TD. Moreover, the presence of larger ventricles in TD patients with low DBH activity suggests the possible contribution of heterogeneous factors to the pathophysiology of this disorder.

7. REFERENCES

Baldessarini, R. J., and Tarsy, D., 1978, Relationship of the actions of neuroleptic drugs to the pathophysiology of tardive dyskinesia, *Int. Rev. Neurobiol.* **21:**1–45.

Bell, R. C. H., and Smith, R. C., 1978, Tardive dyskinesia: Characterization and prevalence in a statewide system, *J. Clin. Psychiatry* **39:**39–47.

Birket-Smith, E., and Anderson, J. V., 1973, Treatment of side effects of levodopa, *Lancet* **1:**431.

Bleuler, E., 1911, *Dementia Praecox or the Group of Schizophrenias* (J. Zinkin, translator), International University Press, New York, 1950.

Brandon, S., McClelland, H. A., and Protheroe, C., 1971, A study of facial dyskinesia in a mental hospital population, *Br. J. Psychiatry* **118:**171–184.

Casey, D. E., and Hansen, T. E., 1984, Spontaneous dyskinesias, in: *Neuropsychiatric Movement Disorders* (D. V. Jeste and R. J. Wyatt, eds.), pp. 68–95, American Psychiatric Press, Washington, DC.

Crane, G. E., 1973, Clinical psychopharmacology in its 20th year, *Science* **181:**124–128.

DeLisi, L. E., Phelps, B. H., Wise, C. D., Apostles, P. S., Rosenblatt, J. E., Bigelow, L., and Wyatt, R. J., 1981, An effect of neuroleptic medication on plasma dopamine-β-hydroxylase activity, *Biol. Psychiatry* **16:**873–878.

Demars, J. P. C. A., 1966, Neuromuscular effects of long-term phenothiazine medication, electroconvulsive therapy and leucotomy, *J. Nerv. Ment. Dis.* **143:**73–79.

Faurbye, A., Rasch, P-J., Petersen, P. B., Brandborg, G., and Pakkenberg, H., 1964, Neurological symptoms in pharmacotherapy of psychoses, *Acta Psychiatr. Scand.* **40:**10–27.

Gardos, G., Cole, J. O., and La Brie, R. A., 1977, Drug variables in the etiology of tardive dyskinesia: Application of discriminant function analysis, *Prog. Neuropsychopharmacol.* **1:**147–154.

Glassman, R. B., and Glassman, H. N., 1980, Oral dyskinesias in brain-damaged rats withdrawn from a neuroleptic: Implication for models of tardive dyskinesia *Psychopharmacology* **69:**19–25.

Gunne, L. M., and Growdon, J. H., 1982, A model for oral dyskinesia in rats, *J. Clin. Psychopharmacol.* **2:**308–311.

Heinrich, K., Wagener, I., and Bender, H-J., 1968, Spate extrapyramidal hyperkinesen bei neuroleptischer langzelttherapie, *Pharmakopsychiat. Neuropsychopharmakol.* **1:**169–195.

Jeste, D. V., and Wyatt, R. J., 1982, *Understanding and Treating Tardive Dyskinesia*, Guilford Press, New York.

Jeste, D. V., Potkin, S. G., Sinha, S., Feder, S. L., and Wyatt, R. J., 1979, Tardive dyskinesia—Reversible and persistent, *Arch. Gen. Psychiatry* **36:**585–590.

Jeste, D. V., Kleinman, J. E., Potkin, S. G., Luchins, D. J., and Weinberger, D. R., 1982, Ex Uno Multi: Subtyping the schizophrenic syndrome, *Biol. Psychiatry* **17:**199–222.

Jeste, D. V., Jeste, S. D., and Wyatt, R. J., 1983, Reversible tardive dyskinesia: Implications for therapeutic strategy and prevention of tardive dyskinesia, in: *New Research in Tardive Dyskinesia* (J Bannet and R. Belmaker, eds.), pp. 34–48, Plenum Press, New York.

Jus, A., Pineau, R., Lachance, R., Pelchat, G., Jus, K., Pires, P., and Villeneuve, R., 1976, Epidemiology of tardive dyskinesia, Part II, *Dis. Nerv. Syst.* **37**:257–261.

Kane, J. M., Woerner, M., Lieberman, J., *et al.*, 1984, Tardive dyskinesia, in: *Neuropsychiatric Movement Disorders* (D. V. Jeste and R. J. Wyatt, eds.), pp. 98–118, American Psychiatric Press, Washington, DC.

Kennedy, P. F., Hershon, H. I., and McGuire, R. J., 1971, Extrapyramidal disorders after prolonged phenothiazine therapy, *Br. J. Psychiatry* **118**:509–518.

Klawans, H. L., Jr., 1973, The pharmacology of tardive dyskinesia, *Am. J. Psychiatry* **130**:82–86.

Klein, D. F., 1965, Diagnosis and pattern of reaction to drug treatment: Clinically derived formulations, in: *Classification in Psychiatry and Psychopathology* (M. M. Katz, J. O. Cole and W. E. Barton, eds.), US Government Printing Office, Washington, DC.

Kovacic, B., and Domino, E. F., 1982, A monkey model of tardive dyskinesia (TD): Evidence that reversible TD may turn into irreversible TD, *J. Clin. Psychopharmacol.* **2**:305–307.

Kraepelin, E., 1907, *Dementia Praecox and Paraphrenia*, (R. M. Barclay, and G. M. Robertson, translators), Robert E. Krieger, New York, 1971.

Lieberman, A. N., Freeman, L. S., and Goldstein, M., 1972, Serum dopamine-β-hydroxylase activity in patients with Huntington's chorea and Parkinson's disease *Lancet* **1**:153–154.

Mallya, A., Jose, C., Baig, M., Williams, R., Cho, D., Mehta, D., and Volavka, J., 1979, Antiparkinsonics, neuroleptics and tardive dyskinesia, *Biol. Psychiatry* **14**:645–649.

Meltzer, H. Y., Nasr, S. J., and Tong, C., 1980, Serum dopamine-β-hydroxylase activity in schizophrenia, *Biol. Psychiatry* **15**:781.

Nagatsu, T., and Udenfriend, S., 1972, Photmetric assay of dopamine-β-hydroxylase activity in human blood, *Clin. Chem.* **18**:980–983.

National Institute of Mental Health, 1976, Abnormal involuntary movement scale, in: *ECDEU Assessment Manual* (W. Guy, ed.), pp. 534–537, US Department of Health, Education and Welfare, Rockville, Maryland.

Overall, J. E., and Gorham, D. R., 1962, The brief psychiatric rating scale, *Psychol. Rep.* **10**:799–812.

Pandurangi, A. K., Ananth, J., and Channabasavanna, S. M., 1978, Dyskinesia in an Indian mental hospital, *Int. J. Psychiatry* **20**:339–342.

Perenyi, A., Arato, M., Bagdy, G., *et al.*, 1981, Long-acting neuroleptics and tardive dyskinesia. Is the high activity of serum dopamine-β-hydroxylase indicative of "individual sensitivity"? Abstracts of the III World Congress of Biological Psychiatry, Stockholm, Abstract #F368.

Pryce, I. G., and Edwards, H., 1966, Persistent oral dyskinesia in female mental hospital patients, *Br. J. Psychiatry* **112**:983–987.

Simpson, G. M., Varga, E., Lee, J. H., and Zoubok, B., 1978, Tardive dyskinesia and psychotropic drug history, *Psychopharmacology* **58**:117–124.

Smith, J. M., Oswald, W. T., Kucharski, L. T., and Waterman, L. J., 1978, Tardive dyskinesia: Age and sex differences in hospitalized schizophrenics, *Psychopharmacology* **58**:207–211.

Spitzer, R. L., Endicott, J., and Robins, E., 1975, Clinical criteria for psychiatric diagnosis and DSM-II, *Am. J. Psychiatry* **132**:1187–1192.

Viukari, M., and Linnoila, M., 1977, Effect of fusaric acid on tardive dyskinesia and mental state in psychogeriatric patients, *Acta Psychiatr. Scand.* **56**:57–61.

Weinberger, D. R., Torrey, E. F., Neophytides, A. N., *et al.*, 1979, Lateral cerebral ventricular enlargement in chronic schizophrenia, *Arch. Gen. Psychiatry* **36**:735–739.

CHAPTER 7

Brain Peptides, Neuroleptic-Induced Tolerance, and Dopamine Receptor Supersensitivity

Implications in Tardive Dyskinesia

HEMENDRA N. BHARGAVA

1. INTRODUCTION

The number of prescriptions written in an average community in the United States includes 20% for medication intended to affect mental processes (Baldessarini, 1980). Among these medications are several classes of drugs that are effective in the symptomatic treatment of psychoses. The phenothiazines as a class, and especially fluphenazine, the prototype, are the most widely used in the treatment of psychotic patients. Another drug used to treat psychosis is haloperidol, which is a butyrophenone derivative. Although structurally different from phenothiazines, haloperidol shares many of their pharmacological properties. These agents have been shown to be effective in the treatment of the manic phase of manic–depressive illnesss and in schizophrenia.

1.1. Biochemical Effects of Neuroleptic Agents

Because the extrapyramidal effects of most neuroleptic drugs are prominent, a great deal of interest has centered on the specific actions of these compounds in the basal ganglia. With the exception of reserpine, all neuroleptic drugs block the dopamine (DA) receptors and increase the turnover rate of DA in the corpus striatum. A DA-sensitive adenylate cyclase has been demonstrated in the caudate nucleus of the rat brain. The neuroleptic drugs chlorpromazine and haloperidol are potent competitive inhibitors of the stimulatory effect of DA on this enzyme (Kebabian *et al.*, 1972). The locus of action on

HEMENDRA N. BHARGAVA • Department of Pharmacodynamics, College of Pharmacy, The University of Illinois at Chicago, Health Sciences Center, Chicago, Illinois 60612.

the neuroleptic drugs, at least in the rat caudate nucleus, may be this presumed "DA receptor." Similar findings in the mesolimbic dopaminergic system support this concept. Furthermore, the administration of L-dopa to rats results in a rapid increase in the concentration of cyclic adenosine 3′,5′-monophosphate (cAMP), owing to the conversion of L-dopa to DA and subsequent activation of adenylate cyclase (Garelis and Neff, 1974).

Histochemical techniques have demonstrated that neurons in the mesolimbic system, particularly in the nucleus accumbens, contain DA. Moreover, DA turnover rate in the limbic system is increased by neuroleptic drugs. This increase in turnover rate, while blocked by trihexyphenidyl, an anticholinergic–antiparkinsonian agent, in the striatum, is not blocked in the limbic system (Anden, 1972). The obvious relationship of the emotional disturbances in schizophrenia to the functions of the limbic system has led to the speculation that the site of neuroleptic drug activity resides in these areas.

Like the caudate nucleus, the limbic system also contains a DA-stimulated adenylate cyclase. The activation of this enzyme is also blocked by phenothiazines and other neuroleptic drugs (Clement-Cormier *et al.*, 1974). These authors postulated that the relevant dopamine receptors in both the caudate nucleus and the limbic system may be linked to adenylate cyclase and proposed that the extrapyramidal symptoms (caudate) and the neuroleptic effects (limbic) could be explained by one basic mechanism of action.

The neuroleptic drugs are believed to act by blocking DA receptors in the nervous system (van Rossum, 1966; Janssen and Allewijn, 1969; Horn and Snyder, 1971; Iversen, 1975). The indirect evidence includes the increase in the firing rate of DA neurons (Bunney *et al.*, 1973), decrease of the stimulated release of DA from dopaminergic neurons (Seeman and Kee, 1975), acceleration of turnover rate of DA (Carlsson and Lindquist, 1963), and blockade of the effect of dopamine-mimetic drugs such as amphetamine and apomorphine (Snyder *et al.*, 1974; Matthysse, 1973). The potency of the neuroleptic drugs in the above test is not correlated with their potencies in affecting schizophrenia (Iversen, 1975; Stawarz *et al.*, 1974). For instance, haloperidol and spiroperidol are 20 to 100 times more potent clinically than chlorpromazine, yet they are equal to or weaker than chlorpromazine in blocking the DA-sensitive adenylate cyclase (Iversen, 1975; Clement-Cormier *et al.*, 1974; Karobath and Leitich, 1974). The presynaptic coupling–blocking actions of the neuroleptics (Seeman and Lee, 1975) correlate better with the clinical neuroleptic potencies, but even here the correlation is not linear. Seeman *et al.*, (1975), by using ligand binding studies, showed that the clinical potencies of neuroleptic drugs could be related to their potency in displacing stereospecific DA receptor binding of [^{3}H]haloperidol and [^{3}H]dopamine to rat brain striata and their subcellular fractions.

1.2. Extrapyramidal Reactions—Tardive Dyskinesia

Even though phenothiazines and butyrophenones are commonly used drugs in the treatment of psychoses, they produce neurological syndromes involving the extrapyramidal system. These drug-induced reactions are particularly prominent during treatment with the piperazine phenothiazines and with haloperidol. Four varieties of extrapyramidal syndromes are associated with the use of neuroleptic drugs (Baldessarini, 1980). Three of them usually appear concomitantly with the administration of the drug, and one is the late-appearing syndrome that occurs either after the drug is discontinued or during prolonged treatment. The first three are the parkinsonian syndrome, akathisia, and acute dystonic reactions. The late-appearing syndrome is called tardive dyskinesia.

Tardive dyskinesia is characterized by stereotyped involuntary movements consisting in sucking and smacking of the lips, lateral jaw movements, and fly-catching dartings of the tongue. There may be choreiform and purposeless quick movement of the extremities. Tardive dyskinesia is an undesirable consequence for many patients maintained on long-term neuroleptic medication (Crane, 1968; Faurbye *et al.*, 1964; Hunter *et al.*, 1964). Such dyskinesias may seriously limit social rehabilitation and in some cases may impair speech, eating, and breathing (Schmidt and Jarcho, 1966).

1.3. Prevalence of Tardive Dyskinesias

In recent years, numerous estimates of the prevalence of tardive dyskinesia have been reported. In 23 studies, prevalence estimates varied from 0.5 to 41.3% (Kazamatsuri *et al.*, 1972a). More recently, Smith *et al.* (1978) examined 213 schizophrenic outpatients treated with neuroleptic medications and found that the incidence of tardive dyskinesia could be related to age, but not to the sex of the patients. Thus, both sexes showed significant linear increases of tardive dyskinesia with increasing age. A comparison of prevalence values between the outpatient sample and the inpatient sample indicated increased levels of mild symptomatology in the outpatients. The latter may represent withdrawal dyskinesias in the process of remission. Similar additional studies on the prevalence have appeared in the literature (Gardos *et al.*, 1977; Kane *et al.*, 1980).

1.4. Pathophysiology of Tardive Dyskinesias

Although the pathophysiology of tardive dyskinesia is not known, several studies indicate the involvement of brain receptors for the neurotransmitters. Both behavioral and biochemical evidence has been presented for receptor involvement. Schelkunov (1967) reported that following chronic administration of a variety of classical neuroleptics (drugs that induce catalepsy, antagonize apomorphine-induced stereotypies, and accelerate striatal DA turnover in animals), mice and rats showed increased and prolonged behavioral responses to DA agonists, amphetamine, and apomorphine. The enhanced sensitivity to dopaminergic stimulation persisted for several weeks after withdrawal of the neuroleptic agent. These findings have since been confirmed in guinea pigs, as well as in mice and rats (Tarsy and Baldessarini, 1973) and have been suggested to be due, in all probability, to "chemical denervation" suspersensitivity of the striatal DA receptors (Fjalland and Moller-Nielsen, 1974). Furthermore, it has been suggested that this mechanism could be responsible for the persistent and irreversible tardive dyskinesias appearing during prolonged administration of classical neuroleptic drugs, on reduction of dosage, or on cessation of treatment (Klawans and Rubovits, 1974b). Indeed, lesions in the substantia nigra have been seen at autopsy in patients with tardive dyskinesia (Christiansen *et al.*, 1970).

Further evidence for the involvement of supersensitivity of DA receptors in tardive dyskinesia stems from the fact that clozapine, a potent neuroleptic drug without cataleptogenic or apomorphine antagonistic properties (Stille *et al.*, (1971), has been found to be free from extrapyramidal side effects in man (Ionescu *et al.*, 1973; Angst *et al.*, 1971), and to date no cases of tardive dyskinesias have been reported. If a relationship exists between the proposed neuroleptic-induced supersensitivity of DA receptors and the appearance of tardive dyskinesias, then it would be expected that clozapine would fail to induce hypersensitivity in the animal model. Indeed, Sayers *et al.*, (1975), using apomorphine-induced responses, showed that clozapine failed to induce supersensitivity of DA receptors.

In addition to the behavioral evidence, biochemical evidence has also been provided for the existence of DA receptor supersensitivity after long-term neuroleptic treatment. Thus, the behavioral supersensitivity is accompanied by a lowering of K_a of DA for the activation of striatal adenylate cyclase and an increase in the striatal content of Ca^{2+}-dependent protein that activates cAMP phosphodiesterase (Gnegy *et al.*, 1977a,b). The activator protein is stored in striatal membranes and can be released by membrane phospharylation in cytosol. The protein increases the activity of the high-K_m phosphodiesterase, but when it is bound to striatal membranes, it facilitates the activation of striatal adeynlate cyclase by DA (Gnegy *et al.*, 1976). The increase in protein activator content of striatal membranes caused by haloperidol could be a primary factor in causing supersensitivity to the biochemical and behavioral effects of dopamine receptor agonists.

Using DA receptor binding assays, it has been shown that chronic administration of haloperidol to rodents results in an increase in the number of DA receptors in the striatum (increased B_{max} value); however, the affinity (dissociation constant, K_d) of radioligands to DA receptor remains unchanged (Burt *et al.*, 1977; Muller and Seeman, 1977). Most of these experiments used 3 weeks of parenteral administration of haloperidol. In one study, where haloperidol was given for 10 weeks, orally mixed with food, it was found that not only was there an increase in the number (B_{max}) of postsynaptic DA receptors labeled by [^{3}H]spiroperidol but also an increase in the dissociation constant (K_d) (Ebstein *et al.*, 1979). The role of enhanced K_d values or lowered affinity in tardive dyskinesia is not yet clear.

In addition to dopaminergic hypersensitivity, a cholinergic hypofunction has been suggested in the pathophysiology of tardive dyskinesia (Fann *et al.*, 1974, 1975; Gerlach *et al.*, 1974; Christian and Moller-Nielsen, 1979). It is clear from the preceding discussion that supersensitivity of DA-dependent mechanisms is the reason for the late-appearing dyskinesias after prolonged neuroleptic drug treatment.

1.5. Pharmacotherapy of Tardive Dyskinesia and Its Potential Pitfalls

Baldessarini (1980) stated that tardive dyskinesia, which appears after months or years of treatment with neuroleptic drugs and becomes worse on withdrawal, is related to excessive function of DA, the prevention of which is *crucial* but the *treatment* of which is *unsatisfactory*. Several treatment strategies for tardive dyskinesia have been attempted with varying degree of success. Most of them included drugs that modified dopaminergic and cholinergic transmission in the central nervous system.

Gerlach (1977) studied the effect of α-methylparatyrosine (AMPT) (an inhibitor of catecholamine synthesis) (4 g/day for 3 days) and biperiden (an anticholinergic drug) (12 mg/day for 3 weeks) in 24 psychiatric patients with neuroleptic-induced tardive dyskinesia. During the whole study, patients continued with their psychopharmacological treatment. The author found that AMPT, primarily through inhibition of DA neurotransmission, left the amplitude unchanged or slightly reduced the frequency of tardive dyskinesia. Furthermore, AMPT often increased the duration of each separate tongue protrusion or mouth opening. He also found that anticholinergic–antiparkinsonian drugs increase both the frequency and the amplitude of tardive dyskinesia. In some patients, baclofen (a GABAergic drug) was helpful, but it could not be used in older patients because it caused sedation, muscular weakness, and confusion.

Growdon *et al.* (1977) used oral choline to increase brain acetylcholine in 20 patients with tardive dyskinesia. Choreic movements decreased in nine patients, worsened in one, and were unchanged in 10. These authors concluded that oral doses of choline can be useful in neurological diseases in which an increase in acetylcholine is desired. However, patients receiving choline chloride experience unpleasant side effects such as development of a "dead fish" body odor. Oral doses of lecithin, the major source of dietary choline, may be an alternate way to treat patients with tardive dyskinesia. In fact, prolonged increase in serum choline after lecithin administration was observed in humans (Wurtman *et al.*, 1977). When lecithin was given to eight chronic schizophrenic patients in doses that increased acetylcholine availability, no difference between the severity of tardive dyskinesia as assessed by rating scales, and by frequency counts of abnormal movements, was observed during lecithin or placebo administration (Branchey *et al.*, 1979). Furthermore, anticholinergic–antiparkinsonian drugs have been reported to worsen the severity of tardive dyskinesia in schizophrenic patients (Burnett *et al.*, 1980; Gerlach, 1977).

Gardos *et al.* (1979) have studied the efficacy of papaverine, a nonnarcotic antispasmodic agent, in the treatment of tardive dyskinesia in 23 psychogeriatric and 18 chronic schizophrenic patients. They found that the oral–facial dyskinesia was significantly reduced by papaverine in the psychogeriatric group during the first 6 weeks. Only a few patients showed at least 50% improvement of dyskinesia scores. Overall, the drug effects were modest.

The symptoms of tardive dyskinesia can also be suppressed by drugs that block DA receptors, like sulpiride, but often at the cost of a concomitant increase in parkinsonism (Casey *et al.*, 1979).

In summary, drugs that either block catecholamine synthesis (AMPT) (Gerlach, 1977), deplete the brain of monoamines (reserpine, tetrabenazine) (Kazamatsuri *et al.*, 1972b), or antagonize actions of DA on postsynaptic receptors (phenothiazines, haloperidol) (Kazamatsuri *et al.*, 1972c) often suppress tardive dyskinesia, whereas drugs that indirectly stimulate DA receptor activity (amphetamine, L-dopa) often exacerbate the abnormal movements (Gerlach 1977; Klawans and McKendall, 1971). Drugs that increase the concentration of acetylcholine within brain synapses (physostigmine, deanol, lecithin, choline) tend to suppress the chorea of tardive dyskinesia, whereas anticholinergics (scopolamine) make it worse (Klawans and Rubovits, 1974a; Gerlach, 1977).

1.6. Brain Peptides and Tardive Dyskinesia

It is well known that the secretion of many pituitary hormones is controlled by peptide factors present in the hypothalamus. In mammals, the release of melanocyte-stimulating hormone (MSH) from the pituitary is controlled by a hypothalamic factor known as MSH-release inhibiting factor (MRIF or MIF). This factor has been isolated and characterized as Pro-Leu-Gly-NH_2 (Nair *et al.*, 1971). In addition to its endocrine activity, MIF exerts effects on the central nervous system which are independent of its action on the pituitary. Thus, MIF can potentiate the behavioral effects of L-dopa in both intact and hypophysectomized animals (Plotnikoff *et al.*, 1971; Huidobro-Toro *et al.*, 1974). MIF has been found to be useful in man as an antiparkinsonian agent (Barbeau *et al.*, 1976).

The results of clinical studies indicated that orally administered MIF was relatively effective in alleviating, at least temporarily, the symptoms of tardive dyskinesia (Ehrensing *et al.*, 1977). Similarly, parkinsonian patients given MIF orally showed a decrease in dopa-induced dyskinesias, a predominantly lingual–facial–buccal dyskinesia similar to tardive dyskinesia (Kastin and Barbeau, 1972). The temporary remission of tardive dyskinesia symptoms by MIF may be related to its entry into the brain.

Disposition studies by Greenberg *et al.* (1976), indicated that peptides like MIF do penetrate the brain by a passive transport process, but the amount of injected radiolabeled MIF reaching the brain was only 0.5%/g brain weight and 0.1%/g brain weight at 15 sec and 10 min, respectively, after the injection. Furthermore, the binding of MIF to synaptosomes derived from brain was extremely small. Recently, several analogues of MIF have been synthesized, and the cyclic peptides like cyclo(Leu-Gly), which is theoretically derived from MIF, have been shown to have a long biological half-life (Rainbow *et al.*, 1979), whereas the half-life of MIF in plasma has been found to be only 9 min (Redding *et al.*, 1973). Therefore, it is possible that the limited success with MIF in tardive dyskinesia may be related to its rapid inactivation and short half-life in the body. The longer-acting analogues like cyclo(Leu-Gly) or related agents may show promise in the treatment of tardive dyskinesia.

Since the development of tardive dyskinesia following chronic administration of neuroleptics is probably related to the development of DA receptor supersensitivity induced as a result of denervation of dopaminergic terminals, the effect of MIF and cyclo(Leu-Gly) on DA receptor supersensitivity induced by various drugs was explored in the author's laboratory. The denervation supersensitivity has been known to be induced by 6-hydroxydopamine, as well as by the neuroleptic drugs.

2. EXPERIMENTAL PROCEDURES AND RESULTS

2.1. Brain Peptides and DA Receptor Sensitivity Animal Experiments

2.1.1. Effect of Cyclo(Leu-Gly) on Supersensitivity of Brain DA Receptors Induced by Intracerebroventricular Injection of 6-Hydroxydopamine

Male albino Swiss Webster mice (20–25 g) were used. Selective depletion of brain DA was achieved by using a combination of pargyline (75 mg/kg), a monoamine oxidase inhibitor, and desmethylimipramine (DMI) (40 mg/kg), an inhibitor of monoamine uptake into the noradrenergic neurons. Thirty minutes after the DMI injection, mice were divided into two groups. One group of mice received the vehicle (i.c.v.) (0.01 N HCl containing 0.1% ascorbic acid), whereas the other received 6-hydroxytryptamine (6-OHDA) (75 μg/mouse). Twenty-four hours later mice from each group were divided into two subgroups. One subgroup of mice was injected s.c. with water (vehicle), while the other subgroup of mice received cyclo (Leu-Gly) (50 μg/mouse). The injections of water and cyclo(Leu-Gly) were repeated in their respective groups for an additional 13 days. Forty-eight hours after the last injection of water or cyclo(Leu-Gly), brain DA levels and apomorphine-induced changes in locomotor activity and body temperature were determined in separate groups of mice.

Mice in all the four groups—vehicle–6-OHDA, cyclo(Leu-Gly)–6-OHDA, vehicle–vehicle, and cyclo(Leu-Gly)–vehicle—were injected with apomorphine (1 mg/kg i.p.). Fifteen minutes after the apomorphine injection, the motor activity of mice was recorded for 15 min subsequent to a 5-min preambulatory period in a Stoelting activity monitor (The Stoelting Co., Chicago, IL). The activity was expressed as mean counts ± SEM. Mice were used only once. The rectal temperatures of mice were measured immediately prior to and at 30 min after apomorphine (1 mg/kg) injection using a rectal probe and a telethermometer (Yellow Springs Instrument Co., Yellow Springs, OH). The data

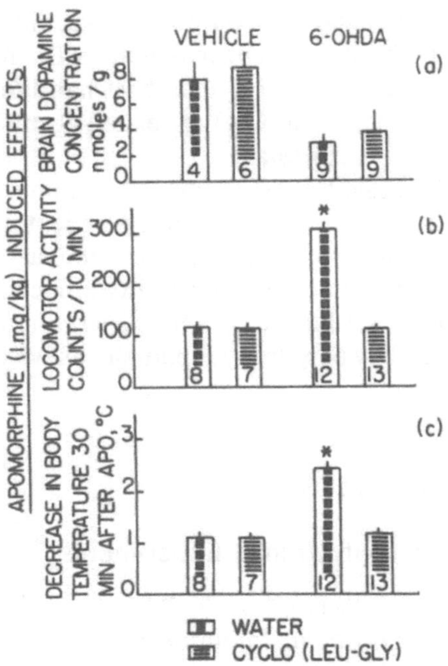

FIGURE 1. Effect of cyclo(Leu-Gly) (CLG) on brain levels of dopamine (a), and apomorphine (1.0 mg/kg i.p.) induced changes in locomotor activity (b) and hypothermia (c) in 6-hydroxydopamine (6-OHDA) treated mice. Mice were injected with pargyline, desipramine, 6-OHDA, and CLG as described in the text. The injections of water (vehicle) and CLG were given for 14 days. Forty-eight hours after the last injection of water or CLG, brain DA levels and apomorphine-induced responses were determined. The numbers within the bars represent the number of mice used. The vertical lines on the bars represent the SEM. Asterisk represents $p < 0.05$ versus all other groups.

were expressed as the difference between the temperature readings at 0 and 30 min. The activity and temperature data were analyzed by using Student's *t* test (two tailed).

Intracerebral administration of 6-OHDA depleted brain DA concentration (from 7.81 to 1.5 nmole/g brain) (Fig. 1a) and induced supersensitivity of brain DA receptors (Fig. 1 b,c). The latter was evidenced by enhanced responses to apomorphine (1 mg/kg i.p.) on locomotor activity (Fig. 1b) and on body temperature (Fig. 1c) (Ritzmann and Bhargava, 1980; Bhargava, 1983b). Daily treatment of mice with cyclo(Leu-Gly) prevented the development of supersensitivity of DA receptors without altering the brain DA concentration. Thus, it indicates that the receptor proliferation (receptor supersensitivity) induced by 6-OHDA can be blocked by cyclo(Leu-Gly), a longer-acting analogue of MIF.

2.1.2. Effect of Cyclo(Leu-Gly) on Haloperidol-Induced Supersensitivity of Brain DA Receptors

Using a similar behavioral profile as above, i.e., employing apomorphine-induced changes in locomotor activity and body temperature, the effect of cyclo(Leu-Gly) on supersensitivity of brain DA receptors induced by chronic administration of haloperidol was determined (Bhargava and Ritzmann, 1980; Bhargava, 1983b).

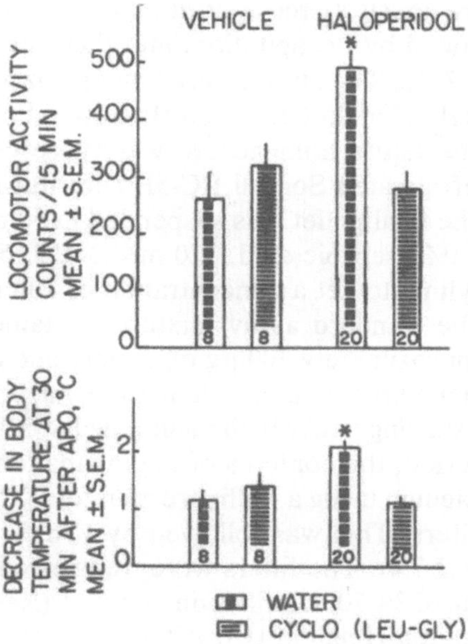

FIGURE 2. Effect of cyclo(Leu-Gly) (CLG) on haloperidol-induced dopamine receptor supersensitivity. Mice were injected with water (vehicle) or CLG (50 g/mouse) followed by either vehicle (0.01 μM tartaric acid) or haloperidol (1 mg/kg i.p.) daily for 21 days. Responses to apomorphine (1 mg/kg) were determined 48 hr after the last injection. The numbers within the bar indicate the number of mice used. The vertical lines on the lines represent the SEM. Asterisk represents $p < 0.05$ versus all other groups.

Male Swiss Webster mice (20–25 g) were injected s.c. with either water (vehicle) or cyclo(Leu-Gly) (50 μg/mouse). One hour later each group of mice was divided into two subgroups. One subgroup of mice received the vehicle (0.01 M tartaric acid), and the other subgroup of mice received haloperidol (1 mg/kg, i.p.). This treatment was repeated every day for 21 days. Forty-eight hours after the last injection, behavioral responses to apomorphine (1 mg/kg, i.p.) were determined. As shown in Fig. 2, chronic administration of haloperidol to mice resulted in the development of DA receptor supersensitivity as evidenced by greater locomotor activity and hypothermic responses to apomorphine when compared with similar responses in vehicle-treated mice. The latter were blocked by concurrent administration of cyclo(Leu-Gly).

The effect of MIF and cyclo(Leu-Gly) on Da receptor supersensitivity induced by chronic administration of haloperidol to rats was also investigated (Bhargava, 1984a). Male Sprague–Dawley rats (approximately 115 g) were divided into three groups and injected s.c. with water (vehicle of the peptides), MIF (2 mg/kg), or cyclo(Leu-Gly) (2 mg/kg), respectively. Each group of rats was further divided into two groups. Rats in one subgroup were administered haloperidol (1.5 mg/kg p.o.) whereas those in the other subgroup received 1 ml/kg of the vehicle (0.3 N tartaric acid). This treatment was repeated every day for 21 days. Four days after the last drug or vehicle administration rats were tested for the dopamine receptor function by two methods, one involving the motor activity response to apomorphine as described previously for the mouse and the other using [^{3}H]spiroperidol binding to the striatal membranes.

The binding of [^{3}H]spiroperidol was carried out essentially as described previously (Creese *et al.*, 1979) with slight modifications. The rats were sacrificed by decapitation, and their striata were quickly removed and frozen at −70°C. The striata were homogenized in 30 volumes of 50 mM Tris-HCl buffer (pH 7.7 at 25°C) using a Brinkman Polytron homogenizer (setting 6 for 20 sec). The tissue homogenate was twice centrifuged at 49,000X g for 15 min in a refrigerated Sorvall RC-5B centrifuge after resuspending in fresh Tris buffer. The final pellet was suspended in 50 nM Tris buffer (pH 7.4 at 29°C containing 0.1% ascorbic acid, 120 mM NaCl, 5 mM KCl, 2mM $CaCl_2$, and 10 μM pargyline) to get a concentration of 10 mg wet tissue per ml of incubation buffer. The standard assay mixture contained 0.2 ml of the homogenate containing approximately 200 μg of protein and 0.1 ml of [^{3}H] spiroperiodol and buffer to make up the total volume to 1 ml. Incubation was carried out in triplicate in a shaking water bath maintained at 37°C for 15 min. At the end of the incubation period, the contents of the incubation tubes were rapidly filtered under partial vacuum using a Millipore manifold filtration unit and Whatman GF/F glass fiber filters. This was followed by two 5-ml washes of ice-cold 50 mM Tris buffer (pH 7.4). The filters were transferred to liquid scintillation vials containing 10 ml of 3a 70 scintillation cocktail (Research Products International Corp., Arlington Heights, IL). After an overnight equilibration period, the radioactivity in the samples was determined in a Packard Tricarb liquid scintillation spectrometer. The stereospecific binding of [^{3}H] spiroperidol was defined as the diffference in binding in the absence and presence of 1 μM *d*-butaclamol. The concentration of protein in the homogenates was determined according to the method of Lowry *et al.* (1951).

The apparent dissociation constant (K_d) and the maximal binding capacity (B_{max}) were calculated from saturation curves generated by using six to seven concentrations of [^{3}H]spiroperidol. The Scatchard plots were generated by the least-square regression analyses. Four rats were used for each treatment group. The means of B_{max} and K_d values in different groups were analyzed by one-way analysis of variance followed by Schaffe's S test.

The study was designed to develop agents that could be used concurrently or prophylactically to prevent the appearance of the symptoms of tardive dyskinesia. However, an issue of greater importance is whether the potential agents work when given after the symptoms of tardive dyskinesia develop. Since our studies indicated that concurrent treatment with MIF or cyclo(Leu-Gly) prevented haloperidol-induced proliferation of [^{3}H]spiroperidol binding sites in the striatum (Bhargava, 1984a), studies were undertaken in which the peptide was given after the termination of haloperidol treatment (Bhargava, 1984b). In this design, the rats were divided into two groups. One group of rats was given haloperiodol (1.5 mg/kg) intragastrically (i.g.) for 21 days. Rats in the other group received the vehicle (i.g.). On days 22 to 24 the rats from each group were divided into three subgroups and received water (vehicle), MIF (2 mg/kg), and cyclo(Leu-Gly) (2 mg/kg) s.c., respectively, every day. On day 25, dopamine receptor sensitivity was determined by behavioral and biochemical methods described previously, using the apomorphine-induced locomotor ac-

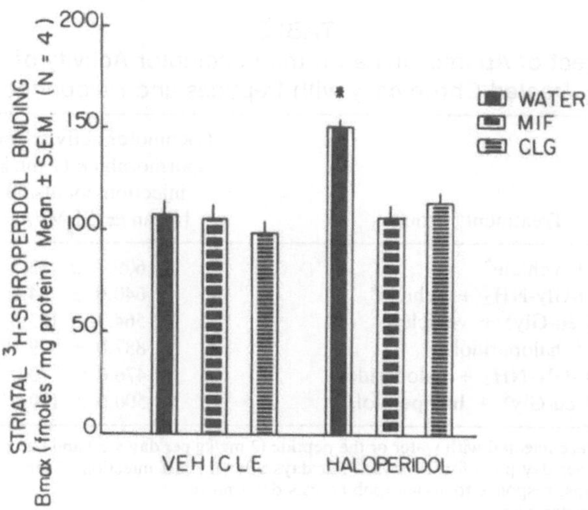

FIGURE 3. Effects of Pro-Leu-GlyNH$_2$ (MIF) and cyclo(Leu-Gly) (CLG) on the binding of [^{3}H]spiroperidol to striatal membranes of rats treated chronically with haloperidol. The vertical lines on the bars (means) represent the SEM ($n = 4$). Asterisk represents $p < 0.05$ versus all other groups.

tivity changes and determining the binding of [^{3}H]spiroperidol to the striatal membranes.

Similar blockage of DA receptor supersensitivity was observed when the peptides were given after the termination of haloperidol treatment to rats. As shown in Fig. 3, chronic intragastric administration of haloperidol for 21 days followed by a 3-day withdrawal resulted in a 30% increase in the number of [^{3}H]spiroperidol binding sites in the rat striatum. Administration of peptides for 3 days by themselves had no effect on [^{3}H]spiroperidol binding. However, when MIF or cyclo(Leu-Gly) was given for 3 days after the termination of haloperidol treatment, the increaase in the B_{max} value of [^{3}H]spiroperidol was completely blocked. Similar blockage of DA receptor supersensitivity was also observed using apomorphine-induced motor activity as the behavioral paradigm (data not shown) (Bhargava, 1984b).

Chronic administration of haloperidol to rats resulted in the development of supersensitivity of DA receptors. As shown in Table 1, a greater locomotor activity response to apomorphine was observed in haloperidol-treated rats when compared with vehicle-treated rats. Chronic administration of MIF or cyclo(Leu-Gly) did not modify apomorphine-induced changes in locomotor activity response in vehicle-injected rats but antagonized the increased response to apomorphine in chronically haloperidol-treated rats. This effect of peptides was very similar to that observed in the mouse with cyclo(Leu-Gly) (Bhargava and Ritzmann, 1980).

The presence of supersensitivity of DA receptors in rats treated chronically

TABLE 1
Effect of Apomorphine on the Locomotor Activity of Rats Treated Chronically with Peptides and Haloperidol

Treatment[a] group	Locomotor activity following apormorphine (1 mg/kg i.p.) injection counts/30 min [mean ± SEM (n = 6)]
Water + vehicle[b]	609.8 ± 83.5
Pro-Leu-Gly-NH_2 + vehicle[b]	640.0 ± 84.0
Cyclo(Leu-Gly) ± vehicle[b]	564.3 ± 103.6
Water + haloperidol	887.0 ± 79.9[c]
Pro-Leu-Gly-NH_2 + haloperidol	476.0 ± 70.0[d]
Cyclo(Leu-Gly) + haloperidol	500.0 ± 120.0[d]

[a] Rats were injected with water or the peptide (2 mg/kg per day s.c.) and haloperidol (1.5 mg/kg per day p.o.) for 21 days. Four days after the last injection of thhe vehicles or the drugs, response to apomorpohine was determined.
[b] 0.3 N tartaric acid.
[c] $p < 0.05$ versus all other groups.
[d] $p < 0.05$ versus water + haloperidol group.

with haloperidol was also evidenced by increased number of [^{3}H]spiroperidol binding sites. As shown in Table 2, a 32% increase in the B_{max} values over the controls was noted in haloperidol-treated rats; however, the dissociation constant, K_d, values remained unchanged. Concurrent administration of MIF or cyclo(Leu-Gly) blocked the increases in B_{max} values for [^{3}H]spiroperidol binding in haloperidol-treated rats.

TABLE 2
Effect of Peptides on [^{3}H]Spiroperidol Binding in Striatum of Rats Treated Chronically with Haloperidol

Treatment[a] group	[^{3}H]Spiroperidol binding parameters [mean ± SEM (n = 4)]	
	B_{max} (fmoles/mg protein)	K_d (pM)
Water + vehicle[b]	135 ± 3	115 ∓ 9
Pro-Leu-Gly-NH_2 + vehicle[b]	128 ± 8	113 ± 4
Cyclo(Leu-Gly) + vehicle[b]	112 ± 13	117 ± 6
Water + haloperidol	178 ± 8[c]	114 ± 4
Pro-Leu-Gly-NH_2 ± haloperidol	121 ± 12[d]	113 ± 5
Cyclo(Leu-Gly) + haloperidol	129 ± 17[d]	112 ± 6

[a] Rats were treated with water, or the peptide (2 mg/kg per day s.c.) and haloperidol (1.5 mg/ per day p.o.) for 21 days. Four days after the last injection of the vehicle or the drugs, the binding of [^{3}H]spiroperidol to striatal membranes was determined.
[b] 0.3 N tartaric acid.
[c] $p < 0.05$ versus all other groups.
[d] $p < 0.05$ versus water + haloperidol group.

TABLE 3
The Effect of MIF or CLG on Tolerance to the Cateleptic Effect of Haloperidol

Treatment[a]	Catalepsy (bar time in seconds) [mean ± SEM (n = 6)] 1 hr after haloperidol[b] injection	2 hr after haloperidol[b] injection
Water + vehicle[c]	22.7 ± 5.2	16.8 ± 4.6
MIF + vehicle[c]	24.7 ± 3.9	16.2 ± 2.5
CLG + vehicle[c]	20.8 ± 3.7	12.5 ± 2.1
Water + haloperidol	1.7 ± 0.3[d]	1.3 ± 0.3[d]
MIF + haloperidol	11.8 ± 0.3[e]	4.7 ± 2.0[e]
CLG + haloperidol	19.3 ± 6.7[e]	8.7 ± 1.9[e]

[a] Rats were injected daily with water, MIF, or CLG (2 mg/kg s.c.) 1 hr before vehicle, or haloperidol (1.5 mg/kg p.o.) for 21 days. The catalepsy was measured 24 hr after the last injection of haloperidol.
[b] 3 mg/kg i.p.
[c] 0.3 N tartaric acid.
[d] $p < 0.05$ versus water + vehicle group.
[e] $p < 0.05$ versus water + haloperidol group.

2.1.3. *Effect of MIF and Cyclo(Leu-Gly) on the Development of Tolerance to the Pharmacological Effects of Haloperidol in the Rat*

Since it is known that tolerance to the pharmacological effects of haloperidol-like drugs is associated with the development of supersensitivity of brain DA receptors, the effects of MIF and cyclo(Leu-Gly) on the two processes were examined (Bhargava, 1981). Male Sprague–Dawley rats were injected with water, MIF (2 mg/kg s.c.), or cyclo(Leu-Gly) (2 mg/kg s.c.). One hour later rats from each group were further divided into two subgroups. Rats from one subgroup received the vehicle (0.3 N tartaric acid p.o.), and the other subgroups received haloperidol (1.5 mg/kg p.o.). This treatment was carried out for 21 days. Twenty-four hours after the last treatment, the effect of haloperidol (3 mg/kg i.p.) on catalepsy (measured by the bar time) and body temperature was determined.

As shown in Tables 3 and 4, chronic administration of haloperidol to rats resulted in the development of tolerance to the cataleptic and the hypothermic effects. Chronic injection of either of the two peptides did not affect the cataleptic or hypothermic effect of haloperidol. However, when the peptides were injected concurrently with haloperidol, the development of tolerance to the pharmacological effects of haloperidol was blocked. Furthermore, cyclo(Leu-Gly) appeared to be more potent than MIF in inhibiting the development of tolerance to haloperidol. Thus, MIF and cyclo(Leu-Gly) were effective in

TABLE 4
The Effect of MIF or CLG on Tolerance to the Hypothermic Effect of Haloperidol

Treatment[a]	Body temperature (°C) [mean ± SEM (n = 6)] Basal temperature	1 hr after haloperidol[b] injection
Water + vehicle[c]	37.41 ± 0.15	36.55 ± 0.07
MIF + vehicle[c]	37.35 ± 0.07	36.72 ± 0.10
CLG + vehicle[c]	37.37 ± 0.12	36.72 ± 0.11
Water + haloperidol	37.25 ± 0.10	37.14 ± 0.10[d]
MIF + haloperidol	37.21 ± 0.10[e]	36.81 ± 0.10[e]
CLG + haloperidol	37.28 ± 0.10[e]	36.65 ± 0.16[e]

[a] Rats were injected daily with water, MIF, or CLG (2 mg/kg s.c.) 1 hr before vehicle, or haloperidol (1.5 mg/kg p.o.) for 21 days. The body temperature was measured after haloperidol which was given on day 22.
[b] 3 mg/kg i.p.
[c] 0.3 N tartaric acid.
[d] $p < 0.05$ versus water + vehicle group.
[e] $p < 0.05$ versus water + haloperidol group.

blocking both the development of tolerance to the pharmacological effects and the supersensitivity of brain DA receptors induced by haloperidol.

2.1.4. Studies on the Mechanism of Action of Brain Peptides on Haloperidol-Induced Effects

Since the preceding studies indicated that MIF and cyclo(Leu-Gly) blocked DA receptor supersensitivity, it was important to assess whether these peptides modified the activity of haloperidol. Although the behavioral data presented previously showed that chronic administration of peptides did not influence the haloperidol effects, biochemical evidence was also necessary. Experiments were undertaken to study the interaction of MIF and cyclo(Leu-Gly) with [^{3}H]spiroperidol (antagonist) and [^{3}H]apomorphine (agonist) binding sites in the brain (Bhargava, 1982, 1983a).

The binding of [^{3}H]apomorphine was carried out essentially as described previously (Creese *et al.*, 1979) for [^{3}H]spiroperidol except that the incubation was carried out for 10 min. The standard assay mixture contained 0.2 ml of the homogenate, containing approximately 200 μg of protein, 0.1 ml of [^{3}H]apomorphine, 0.1 ml of competing agent (appropriate peptide), and buffer to make up the total volume of 1 ml. The specific binding of [^{3}H]apomorphine was defined as the difference in binding observed in the absence and in the presence of 10 μM unlabeled apomorphine. For interaction studies, concentrations of peptides ranged from 10^{-8} to 10^{-4} M. The concentration of

TABLE 5
Effect of Pro-Leu-Gly-NH_2(MIF) and Cyclo(Leu-Gly)(CLG) on the Binding of [^{3}H]Apomorphine and [^{3}H]Spiroperidol in the Rat Striatum

Concentration of peptide (M)	Amount of [^{3}H]radioligand bound (fmoles/mg protein)(mean ± SEM)			
	[^{3}H]Spiroperidol binding[a] (n = 4)		[^{3}H]Apomorphine binding[b] (n = 8)	
	MIF	CLG	MIF	CLG
0	72.5 ± 3.3	73.0 ± 5.0	121.4 ± 3.1	115.5 ± 5.3
10^{-8}	77.7 ± 3.0	73.4 ± 3.4	128.8 ± 5.1[c]	127.7 ± 6.0
10^{-7}	72.5 ± 3.0	70.7 ± 2.8	139.6 ± 5.3[d]	132.3 ± 6.0[e]
10^{-6}	74.0 ± 3.0	71.6 ± 2.4	135.5 ± 4.3[c]	136.3 ± 3.7[e]
10^{-5}	76.2 ± 3.0	70.8 ± 2.3	137.2 ± 3.7[d]	144.4 ± 2.6[e]
10^{-4}	77.0 ± 3.0	71.0 ± 2.3	127.1 ± 5.0	137.7 ± 4.0[e]

[a] The incubations were carried out with 0.5-nM concentration of [^{3}H]spiroperidol.
[b] The incubations were carried out with 3-nM concentration of [^{3}H]apomorphine.
[c] $p < 0.05$ versus the control.
[d] $p < 0.02$ versus the control.
[e] $p < 0.01$ versus the control.

[^{3}H]apomorphine was 3 nM. Protein concentration in the homogenate was determined according to the method of Lowry *et al.* (1951). The specific binding was expressed as fmoles of [^{3}H]apomorphine bound per mg of protein.

The assay procedure to determine [^{3}H]spiroperidol binding was identical to that for [^{3}H]apomorphine except that the incubation was carried out for 15 min. The specific binding of [^{3}H]spiroperidol was defined as the difference in binding in the absence and presence of 1 μM *d*-butaclamol. For interaction studies, the concentration of [^{3}H]spiroperidol used was 0.5 nM, and that of the test peptides ranged from 10^{-8} to 10^{-4} M.

In order to determine whether the modification in binding of apomorphine by peptides was related to changes in the affinity or the number of receptors, the specific binding of [^{3}H]apomorphine (five to six concentrations) was determined in the striatum in the presence of 1 μM concentration of MIF or cyclo(Leu-Gly). This was followed by Scatchard analysis to determine the dissociation constant (K_d) and the number of binding sites (B_{max}).

As can be seen from Table 5, both MIF and cyclo(Leu-Gly) (10^{-8} to 10^{-4} M) failed to affect the [^{3}H]spiroperidol binding. [^{3}H]apomorphine, on the other hand labeled the DA receptors with B_{max} of 352.0 fmoles/mg of protein and a K_d of 5.45 nM. However, both peptides enhanced the binding of [^{3}H]apomorphine in the striatum (Table 5) The Scatchard analysis revealed that in the presence of 1 μM of each of the peptides, the number of maximal binding sites was unaffected. However, the affinity to the receptors was enhanced as evidenced by decreased K_d values (Table 6).

TABLE 6
Effects of Peptides on the Affinity and Density of Binding Sites for [^{3}H]Apomorphine in the Rat Striatum

		Specific [^{3}H]apomorphine binding [mean ± SEM (n = 5)]	
Peptide	Concentration (μM)	B_{max} (fmoles/mg protein)	K_d (nM)
Control	—	352.0 ± 13.2	5.45 ± 0.21
MIF	1.0	326.2 ± 13.7	4.44 ± 0.15[a]
Cyclo(Leu-Gly)	1.0	324.6 ± 13.4	4.56 ± 0.10[a]

[a] $p < 0.05$ versus the control.

3. DISCUSSION

It is clear from the preceding studies that chronic administration of neuroleptic drugs like haloperidol results in the development of tolerance to their pharmacological effects. For instance, tolerance development has been observed with respect to the cataleptic effect (Asper *et al.*, 1973; Gessa and Tagliamonte, 1975; Ezrin-Waters and Seeman, 1977; Bhargava, 1981) and the hypothermic effect (Bhargava, 1981) of haloperidol. The development of tolerance to both the cataleptic and the hypothermic effects was inhibited by MIF and its analogue.

Chronic administration of haloperidol also induced supersensitivity of brain DA receptors, particularly of the striatum. Both behavioral and biochemical evidence has been presented in favor of this concept. The behavioral studies include mainly the increased apomorphine-induced responses (e.g., locomotor activity, stereotypies, hypothermia) (Gianutsos *et al.*, 1974; Gnegy *et al.*, 1977a,b; Bhargava and Ritzmann, 1980; Sayers *et al.*, 1975; Tarsy and Baldessarini, 1973; Yarbrough, 1975). The biochemical evidence includes the metabolism of DA in striatal tissue (Smith *et al.*, 1978; Scatton, 1977; Stanley and Wilk, 1980) and increased binding of [^{3}H]spiroperidol or haloperidol to striatal membranes (Burt *et al.*, 1977; Muller and Seeman, 1977; Ebstein *et al.*, 1979; Owen *et al.*, 1980). Not only the number of maximal binding sites, B_{max}, was increased after chronic haloperidol administration, but in two studies (Ebstein *et al.*, 1979; Owen *et al.*, 1980) the apparent dissociation constant, K_d, was also found to be increased. The importance of increased affinity of radioligands to Da receptors in relation to tardive dyskinesia is not known at present. The increased number of receptors has been suggested to be due to an increase in the amount of a CA^{2+}-dependent protein activator of adenylate cyclase and phosphodiesterase in the striatum (Gnegy *et al.*, 1977a,b). The supersensitivity of striatal DA receptors may be associated with this protein.

Multiple classes of brain DA receptors have been suggested (Kebabian and Calne, 1979). One class, D1 DA receptors, appears to be located on neuronal cell bodies intrinsic to the striatum and is thought to be associated functionally with DA-sensitive adenylate cyclase. The second class, D2 DA receptors, is located on axons of the glutaminergic corticostriate pathway. Although neuroleptic drugs bind to both receptors, chronic administration of the neuroleptic haloperidol preferentially increases [^{3}H]spiroperidol binding to D2 DA receptors (Rosenblatt *et al.*, 1979).

The studies in our laboratories suggest that MIF and related peptides derived from the hypothalamus can inhibit the development of tolerance to the cataleptic and hypothermic effects of haloperidol (Bhargava, 1981), as well as the behavioral supersensitivity of brain DA receptors (Bhargava and Ritzmann, 1980; Bhargava, 1983b). Our studies indicate that the increased binding of [^{3}H]spiroperidol to striatal membranes in chronic haloperidol-treated rats can be inhibited by MIF and cyclo(Leu-Gly) whether the peptides are given concurrently with haloperidol or after the termination of haloperidol treatment (Bhargava, 1984a,b). These peptides do not alter the binding of [^{3}H]spiroperidol (neuroleptic receptors) to DA receptors *in vitro* but enhance the binding of [^{3}H]apomorphine to DA receptors (Bhargava, 1982; Bhargava, 1983a). Therefore, it is possible to use agents that are DA receptor agonists to inhibit DA receptor supersensitivity induced by neuroleptic drugs. Indeed, dopaminergic agonist amantadine (Allen *et al.*, 1980) and bromocryptine or the combination of L-dopa and carbidopa (List and Seeman, 1979) decrease the supersensitivity of DA receptors in the striatum induced by haloperidol.

The effects of long-term treatment with cyclo(Leu-Gly) and haloperidol on biochemical parameters indicative of striatal DA target cell supersensitivity in the rat have been investigated (LeDouarin *et al.*, 1984). Acute administration of cyclo(Leu-Gly) (2 mg/kg s.c.) was found not to affect striatal homovanillic acid, dihydroxyphenylacetic acid, and acetylcholine (ACh) levels in vehicle or haloperidol (1 mg/kg i.p.) treated rats. Cyclo(Leu-Gly) (2 mg/kg per day s.c.) given concurrently with haloperidol for 14 days also did not prevent the decreases in striatal DA metabolites observed 2 days after withdrawal and the tolerance to the elevation of DA metabolites that occurs following administration of haloperidol during withdrawal. Similarly, the tolerance to the lowering of striatal ACh levels seen after chronic haloperidol treatment was unaffected by cyclo(Leu-Gly). The authors concluded that the haloperidol-induced supersensitivity of DA receptors by cyclo(Leu-Gly) does not involve its action on nigrostriatal DA and striatal cholinergic neurons and is probably exerted distally to both DA and ACh synapses. Studies from this laboratory also indicate that MIF and cyclo(Leu-Gly) do not interact with striatal cholinergic muscarinic receptors labeled with [^{3}H]quinuclidinyl benzilate in the rat (Das and Bhargava, 1986).

4. SUMMARY

In summary, peptides derived from hypothalamus and their analogues, which can enhance dopaminergic transmission, may prove to be useful in combating tardive dyskinesia symptoms induced by long-term administration of neuroleptic drugs.

ACKNOWLEDGMENTS. The studies described here were supported by a United States Public Health Service Grant No. DA-02598 from the National Institute on Drug Abuse and a grant from the Campus Research Board, University of Illinois at Chicago, Health Sciences Center, Chicago. The author thanks Mr. George A. Matwyshyn for providing technical assistance and Mr. Paulose Pathrose for his help in preparation of this manuscript.

5. REFERENCES

Allen, R. M., Lane, J. D., and Brauchi, J. T., 1980, Amantadine reduces haloperidol-induced dopamine receptor hypersensitivity in the striatum, *Eur. J. Pharmacol.* **65**:313–315.

Anden, N. E., 1972, Dopamine turnover in the corpus striatum and the limbic system after treatment with neuroleptic and antiacetylcholine drugs, *J. Pharm. Pharmacol.* **24**:905–906.

Angst, J., Bente, D., Berner, P., Heimann, H., Helmchen, H., and Hippius, H., 1971, Das Klinische Wirkungs bild von clozapine (Unterschung mit dem AMP-system), *Pharmacopsychiatria.* **4**:201–211.

Asper, H., Baggiolini, M., Burki, H. R., Lauener, H., Ruch, W., and Stille, G., 1973, Tolerance phenomena with neuroleptics: Catalepsy, apomorphine stereotypies and strital dopamine metabolism in the rat after single and repeated administration of loxapine and haloperidol, *Eur. J. Pharmacol.* **22**:287–294.

Baldessarini, R. J., 1980, Drugs and the treatment of psychiatric disorders, in: *The Pharmacological Basis of Therapeutics* A. G. Gilman, L. S. Goodman, and A. Gilman, eds., pp. 391–447, Macmillan, New York.

Barbeau, A., Roy, M., and Kastin, A. J., 1976, Double-blind evaluation of oral L-prolyl-L-leucylglycinamide in Parkinson's disease, *Can. Med. Assoc. J.* **24**:120–122.

Bhargava, H. N., 1981, The effects of hypothalamic peptide factor MIF, and its cyclic analog on tolerance to haloperidol in the rat, *Life Sci.* **29**:45–51.

Bhargava, H. N., 1982, Effects of melanotropin release inhibiting factor, and related compounds, on ^{3}H-spiroperidol and ^{3}H-apomorphine binding to rat striatal and hypothalamic dopamine receptors, *Pharmacologist* **24**:121.

Bhargava, H. N., 1983a, The effect of melanotropin release inhibiting factor, its metabolites and analogs on ^{3}H-spiroperidol and ^{3}H-apomorphine binding sites, *Gen. Pharmacol.* **14**:609–614.

Bhargava, H. N., 1983b, Cyclo(Leu-Gly): A possible treatment for tardive dyskinesia? in: *Modern Problems of Pharmacopsychiatry, "New Directions in Tardive Dyskinesia Research,"* Vol. 21 (J. Bannet and R. H. Belmaker, eds.), pp. 196–205, Karger, Basel.

Bhargava, H. N., 1984a, Effects of prolyl-leucyl-glycinamide and cyclo(leucyl-glycine) on the supersensitivity of brain dopamine receptors induced by chronic administration of haloperidol to rats, *Neuropharmacology* **23**:439–444.

Bhargava, H. N., 1984b, Enhanced ^{3}H-spiroperidol binding induced by chronic haloperidol treatment inhibited by peptides administered during the withdrawal phase, *Life Sci.* **34**:887–879.

Bhargava, H. N., and Ritzmann, R. F., 1980, Inhibition of neuroleptic-induced dopamine receptor supersensitivity by cyclo(Leu-Gly), *Pharmacol. Biochem. Behav.* **13**:633–636.

Branchey, M. H., Branchey, L. B., Bark, N. M., and Richardson, M. A., 1979, Lecithin in the treatment of tardive dyskinesia, *Commun. Psychopharmacol.* **3**:303–307.

Bunney, B. S., Walters, J. R., Roth, R. H., and Aghajanian, G. K., 1973, Dopaminergic neurons: Effect of antipsychotic drugs and amphetamine on single cell activity, *J. Pharmacol. Exp. Ther.* **185**:560–571.

Burnett, G. B., Prange, A. J., Wilson, I. C., Joliff, L. A., Creese, I., and Snyder, S. H., 1980, Adverse effect of anticholinergic-antiparkinsonian drugs in tardive dyskinesia: An investigation of mechanism, *Neuropsychobiology* **6**:109–120.

Burt, D. R., Creese, I., and Snyder, S. H., 1977, Antischizophrenic drugs: Chronic treatment elevates dopamine receptor binding in brain, *Science* **197**:326–328.

Carlsson, A., and Lindquist, M., 1963, Effect of chlorpromazine and haloperidol on formation of 3-methoxytyramine and normetanephrine in mouse brain, *Acta Pharmacol. Toxicol.* **20**:140–144.

Casey, D. E., Gerlach, J., and Simmelsgaard, H., 1979, Sulpiride in tardive dyskinesia, *Psychopharmacology* **66**:73–77.

Christian, A. V., and Moller-Nielsen, I., 1979, Dopaminergic supersensitivity: Influence of dopamine agonists, cholinergics,, anticholinergics, and drugs used for the treatment of tardive dyskinesia, *Psychopharmacology* **62**:111–116.

Christiansen, E., Moller, J. E., and Fourbye, A., 1970, Neuropathological investigation of 28 brains from patients with dyskinesia, *Acta Psychiatr. Scand.* **46**:14–23.

Clement-Cormier, Y. C., Kebabian, J. W., Petzoid, G. L., and Greengard, P., 1974, Dopamine sensitive adenylate cyclase in mammalian brain: a possible site of action of antipsychotic drugs, *Proc. Natl. Acad. Sci. USA* **71**:1113–1117.

Crane, G. E., 1968, Tardive dyskinesia in patients treated with major neuroleptics: A review of the literature, *Am. J. Psychiatry* **124**:40–48.

Creese, I., Usdin, T. B., and Snyder, S. H., 1979, Dopamine receptor binding regulated by guanine nucleotides, *Mol. Pharmacol.* **16**:69–76.

Das, S., and Bhargava, H. N., 1986, Effects of Pro-Leu-Gly-NH_2 and cyclo(Leu-Gly) on the binding of ^{3}H-quinuclidinyl benzilate to striatal cholinergic muscarinic receptors, *Peptides* (in press).

Ebstein, R. P. Pickholz, D., and Belmaker, R. H., 1979, Dopamine receptor changes after long-term haloperidol treatment in rats, *J. Pharm. Pharmacol.* **31**:558–559.

Ehrensing, R. H., Kastin, A. J. Larsons, P. F., and Bishop, G. A., 1977, Melanocyte stimulating hormone release inhibiting factor-1 and tardive dyskinesia, *Dis. Nerv. Syst.* **38**:303–307.

Ezrin-Waters, C., and Seeman, P., 1977, Tolerance to haloperidol catalepsy, *Eur J. Pharmacol.* **41**:321–327.

Fann, W. E., Lake, C. R., Gerber, C. J., and McKenzie, G. M., 1974, Cholinergic suppression of tardive dyskinesia, *Psychopharmacologia* **37**:101–107.

Fann, W. E. Sullivan, J. L. III, Miller, R. D., and McKenzie, G. M., 1975, Deanol in tardive dyskinesia: a preliminary report, *Psychopharmacologia* **42**:135–137.

Fourbye, A., Rasche, P. J., and Peterson, B., 1964, Neurological symptoms in pharmacotherapy of psychoses, *Acta Psychiatr. Scand.* **40**:10–27.

Fjalland, B., and Moller-Nielsen, I., 1974, Enhancement of methylphenidate-induced stereotypies by repeated administration of neuroleptics, *Psychopharmacologia (Berlin)* **34**:105–109.

Gardos, G., Cole, J. O., and LaBrie, R. L., 1977, The assessment of tardive dyskinesia, *Arch. Gen. Psychiatry* **34**:1206–1212.

Gardos, G., Granacher, R. P., Cole, J. O., and Sniffin, C., 1979, The effects of papaverine in tardive dyskinesia, *Prog. Neuropsychopharmacol.* **3**:543–550.

Garelis, E., and Neff, N. H., 1974, Cyclic adenosine monophosphate: Selective increase in caudate nucleus after administration of L-dopa, *Science* **183**:532–533.

Gerlach, J., 1977, The relationship between parkinsonism and tardive dyskinesia, *Am. J. Psychiatry* **134**:781–784.

Gerlach, J., Reisby, N., and Randrup, A., 1974, Dopaminergic hypersensitivity and cholinergic hypofunction in the pathophysiology of tardive dyskinesia, *Psychopharmacologia* **34**:21–35.

Gessa, G. L., and Tagliamonte, A., 1975, Effect of methadone and dextromoramide on dopamine metabolism: Comparison with haloperidol and amphetamine, *Neuropharmacology* **14**:913–920.

Gianutsos, G., Drawbaugh, R. B., Hynes, M. D., and Lal, H., 1974, Behavioral evidence for dopaminergic supersensitivity after chronic haloperidol, *Life Sci.* **14**:887–898.

Gnegy, M. E., Uzunov, P., and Costa, E., 1976, Regulations of the dopamine stimulation of striatal adenylate cyclase by an endogenous Ca^{++}-binding protein, *Proc. Natl. Acad. Sci. USA* **73**:3887–3890.

Gnegy, M. E. Uzunov, P., and Costa, E., 1977a, Participation of an endogenous Ca^{++}-binding protein activator in the development of drug-induced supersensitivity of striatal dopamine receptors, *J. Pharmacol. Exp. Ther.* **202**:558–564.

Gnegy, M. E., Lucchelli, A., and Costa, E., 1977b, Correlation between drug-induced supersensitivity of dopamine dependent striatal mechanisms and the increase in striatal content of the Ca^{++}-regulated protein activator of cAMP phosphodiesterase, *Naunyn-Schmiedb. Arch. Pharmacol.* **301**:121–127.

Greenberg, R., Whalley, C. E., Jourdikian, F., Mendelson, I. S., and Walter, R., 1976, Peptides readily penetrate the blood brain barrier: Uptake of peptides by synaptosomes is passive, *Pharmacol. Biochem. Behav.* **5**:151–158.

Growdon, J. H. Hirsch, M. J. Wurtman, R. J., and Weiner, W., 1977, Oral choline administration to patients with tardive dyskinesia, *N. Engl. J. Med.* **297**:524–527.

Horn, A. S., and Snyder, S. H., 1971, Chlorpromazine and dopamine: Conformational similarity that correlate with the antischizophrenic activity of phenothiazine drugs, *Proc. Natl. Acad. Sci. USA* **68**:2325–2328.

Huidobro-Toro, J. P. deCarolis, A. S., and Longo, V. G., 1974, Action of two hypothalamic factors (TRH,MIF) and of angiotensin II on the behavioral effects of L-dopa and 5-hydroxytryptophan in mice, *Pharmacol. Biochem. Behav.* **2**:105–109.

Hunter, R., Earl., C. J., and Janz, D., 1964, A syndrome of abnormal movements and dementia in leucotomized patients treated withh phenothiazines, *J. Neurol. Neurosurg. Psychiatry* **27**:219–223.

Ionescu, R., Nica, S. U., Oproiu, L., Niturad, A., and Tudoarche, B., 1973, Double blind study in psychopathic behavioral disorders (clozapine and pericyazine), *Pharmacopsychiatria* **6**:294–299.

Iversen, L., 1975, Dopamine receptors in the brain: A dopamine sensitive adenylate cyclase models synaptic receptors, illuminating antipsychotic drug action, *Science* **188**:1084–1089.

Janssen, P.A. J., and Allewijn, T. F. M., 1969, The distribution of the butyrophenones, haloperidol, trifluperidol, moperone, and clofluperiol in rats, and its relationships with their neuroleptic activity, *Arzneim. Forsch.* **19**:199–208.

Kane, J. Wegner, J., Stenzler, S., and Ramsey, P., 1980, The prevalence of the presumed tardive dyskinesia in psychiatric inpatients and outpatients, *Psychopharmacology* **69**:247–251.

Karobath, M., and Leitich, H., 1974, Antipsychotic drugs and dopamine stimulated adenylate cyclase prepared from corpus striatum of rat brain, *Proc. Natl. Acad. Sci. USA* **71**:2915–2918.

Kastin, A. J., and Barbeau, A., 1972, Prelininary clinical studies with L-prolyl-L-leucyl-glycineamide in Parkinson's disease, *Can. Med. Assoc. J.* **107**:1079–1081.

Kazamatsuri, H., Chien, C. P., and Cole, J. O., 1972a, Therapeutic approaches to tardive dyskinesia: a review of the literature, *Arch. Gen. Psychiatry* **27**:491–499.

Kazamatsuri, H., Chien, C., and Cole, J. O., 1972b, Treatment of tardive dyskinesia I. Clinical efficacy of a dopamine-depleting agent tetrabenazine, *Arch. Gen. Psychiatry* **27**:95–99.

Kazamatsuri, H., Chien, C., and Cole, J. O., 1972c, Treatment of tardive dyskinesia III. Short-term efficacy of dopamine-blocking agents, haloperidol and thiopropazate, *Arch. Gen. Psychiatry* **27**:100–103.

Kebabian, J. W., and Calne, D. W., 1979, Multiple receptors for dopamine, *Nature* (*London*) **277**:93–96.

Kebabian, J. W., Petzold, G. L., and Greengard, P., 1972, Dopamine sensitive adenylate cyclase in the caudate nucleus of rat brain and its similarity to the "dopamine receptor," *Proc. Natl. Acad. Sci. USA* **69**:2145–2149.

Klawans, H. L., Jr., and McKendall, R. R., 1971, Observations on the effect of levodopa on tardive lingual-facial-buccal dyskinesia, *J. Neurol. Sci.* **14**:189–192.

Klawans, H. L., and Rubovits, R., 1974a, Effect of cholinergic and anticholinergic agents on tardive dyskinesia, *J. Neurol. Neurosurg. Psychiatry* **37**:941–947.

Klawans, H. L., and Rubovits, R., 1974b, An experimental model of tardive dyskinesia, *J. Neural Transm.* **33**:235–246.

LeDouarin, C., Fage, D., and Scatton, B., 1984, Effects of cyclo(Leu-Gly) on neurochemical indices of dopaminergic supersensitivity induced by prolonged haloperidol treatment, *Life Sci.* **34**:393–399.

List, S. J., and Seeman, P., 1979, Dopamine agonists reverse the elevated ^{3}H-neuroleptic binding in neuroleptic-pretreated rats, *Life. Sci.* **24**:1447–1452.

Lowry, O. H., Rosenbrough, N. J., Farr, A. L., and Randall, R. J., 1951, Protein measurement with the Folin phenol reagent, *J. Biol. Chem.* **193**:265–275.

Matthysse, S., 1973, Antipsychotic drug actions: A cue to the neuropathology of schizophenia?, *Fed. Proc.* **32**:200–204.

Muller, P., and Seeman, P., 1977, Brain neurotransmitter receptors after long term haloperidol: dopamine, acetylcholine, serotonin, alpha-noradrenergic and naloxone receptors, *Life Sci.* **21**:1751–1758.

Nair, R. M. G., Kastin, A. J., and Schally, A. V., 1971, Isolation and structure of hypothalamic MSH release-inhibiting hormone, *Biochem. Biophys. Res. Commun.* **43**:1376–1381.

Owen, F., Cross, A. J. Waddinton, J. L., Poulter, M., Gamble, S. J., and Crow, T. J., 1980, Dopamine mediated behavior and ^{3}H-spiroperone binding to striatal membranes in rats after nine months haloperidol administration, *Life Sci.* **26**:55–59.

Plotnikoff, N. P., Kastin, A. J., Anderson, M. S., and Schally, A. V., 1971, Dopa potentiation by a hypothalamic factor, MSH release-inhibiting hormone (MIF), *Life Sci.* **10**:1279–1283.

Rainbow, T. C., Flexner, J. B. Flexner, L. B., Hoffman, P. L., and Walter R., 1979, Distribution survival and biological effects in mice of a behaviorally active enzymatically stable peptides, pharmacokinetics of cyclo(Leu-Gly) and puromycine induced amnesia, *Pharmacol. Biochem. Behav.* **10**:787–793.

Redding, T. W., Kastin, A. J., Nair, R. M. G., and Schally, A. V., 1973, Distribution, half-life and excretion of ^{14}C- and ^{3}H-labeled L-prolyl-L-leucyl-glycinamide in the rat, *Neuroendocrinology* **11**:92–100.

Ritzmann, R. F., and Bhargava, H. N., 1980, The effect of cyclo(Leu-Gly) on chemical denervation supersensitivity of dopamine receptors-induced by intracerebroventricular injection of 6-hydorxydopamine in mice, *Life Sci.* **27**:2075–2080.

Rosenblatt, J. E., Shore, D., Neckers, L. M., Perlow, M. J., Freed, W. J., and Wyatt, R. J., 1979, Effects of chronic haloperidol on caudate ^{3}H-spiroperidol binding in lesioned rats, *Eur. J. Pharmacol.* **60**:387–388.

Sayers, A. C., Burki, H. R., Ruch, W., and Asper, H., 1975, Neuroleptic induced hypersensitivity of striatal dopamine receptors in the rat as a model of tardive dyskinesias. Effects of clozapine, haloperidol, loxapine and chlorpromazine, *Psychopharmacologia* **41**:97–104.

Scatton, B., 1977, Differential regional development of tolerance to increase in dopamine turnover upon repeated neuroleptic administration, *Eur. J. Pharmacol.* **46**:363–369.

Schelkunov, E. L., 1967, Adrenergic effect of chronic administration of neuroleptics, *Nature* (*London*) **214**:1210–1213.

Schmidt, W. R., and Jarcho, L. W., 1966, Persistent dyskinesias following phenothiazine therapy, *Arch. Neurol.* (*Chicago*) **14**:369–377.

Seeman, P., and Lee, T., 1975, Antipsychotic drugs: Direct correlation between clinical potency and presynaptic action on dopamine neurons, *Science* **188**:1217–1219.

Seeman, P., Chau-Wong, M., Tedesco, J., and Wong, K., 1975, Brain receptors for antipsychotic drugs and dopamine: Direct binding assays, *Proc. Natl. Acad. Sci. USA* **72**:4376–4380.

Smith, R. C., Narsimhachari, N., and Davis, J. M., 1978, Increased effect of apomorphine on homovanillic acid in rats terminated from chronic haloperidol, *J. Neural Transm.* **42**:159–162.

Snyder, S. H., Banerjee, S. P., Yamamura, H. I., and Greenberg, D., 1974, Drugs, neurotransmitters and schizophrenia, *Science* **184**:1243–1253.

Stanley, M., and Wilk, S., 1980, Acute and chronic effects of haloperidol and clozapine on dopamine metabolism in two dopamine rich areas of the rat brain, *Res. Commun. Psychol. Psychiat. Behav.* **5**:37–47.

Stawarz, R. J., Robinson, S., Sulser, F., and Dingell, J. V., 1974, On the significance of the increase of homovanilic acid (HVA) caused by antipsychotics in the corpus striatum and limbic forebrain, *Fed. Proc.* **33**:246.

Stille, G., Lauener, H., and Eichenberger, E., 1971, The pharmacology of 8-chloro-11-(4-methyl-1-piperazinyl)-5-H-dibenzo (b,e) (1,4) diazepine (Clozapine), *Il Farmaco* **26**:603–625.

Tarsy, D., and Baldessarini, R. J., 1973, Pharmacologically induced behavioral supersensitivity to apomorphine, *Nature New Biol.* **245**:262–263.

van Rossum, J. M., 1966, The significance of dopamine receptor blockade for the mechanism of neuroleptic drugs, *Arch. Int. Pharmacodyn. Ther.* **160**:492–494.

Wurtman, R. J., Hirsch, M. J., and Growdon, J. H., 1977, Lecithin consumption elevates serum free choline levels, *Lancet* **2**:68–69.

Yarbrough, G. C., 1975, Supersensitivity of caudate neurons after repeated administration of haloperidol, *Eur. J. Pharmacol.* **31**:367–369.

CHAPTER 8

Prevention of Tardive Dyskinesia

EDMOND H. PI and GEORGE M. SIMPSON

1. INTRODUCTION

Neuroleptics produce many unwanted effects including extrapyramidal side effects. The latter usually occur early in treatment and are seen frequently in psychiatric practice. Recently, much attention has been given to the late-onset condition, tardive dyskinesia (TD). Despite the fact that this syndrome was described in the literature soon after the introduction of neuroleptics in the early 1950s (Schonecker, 1957: Sigwald *et al.*, 1959), there is still wide disagreement among investigators regarding its prevalence (estimates have ranged from 0.5% to 56%). This reflects the need for a precise definition of TD (Simpson *et al.*, 1982) and has hampered clinicians' efforts to make an early diagnosis of TD. The availability of standardized, objectively defined rating scales (NIMH, 1975; Simpson *et al.*, 1979) and other diagnostic devices represents a significant advance in clarifying these problems (Schooler and Kane, 1982). These developments provide clinicians with a useful tool to objectively document the evolution and severity of TD. This represents the first step toward the goal of preventing the debilitating and sometimes irreversible effects of this disorder.

This chapter discusses strategies regarding "prevention," the primary aim in dealing with TD, since no single successful treatment has been found for the disorder. However, the concept of preventive measures needs to be extended beyond that of primary prevention measures to avoid its occurrence. Secondary prevention measures to limit its progress and tertiary prevention measures to ameliorate its disabling effects are also discussed.

2. PRIMARY PREVENTION OF TARDIVE DYSKINESIA

2.1. Criteria for the Use of Neuroleptics

Neuroleptics should be prescribed only if there are no other effective alternatives available or if the proven benefits of neuroleptics are superior to

---

EDMOND H. PI • Department of Psychiatry, University of Southern California School of Medicine, Los Angeles, California 90033. GEORGE M. SIMPSON • Department of Psychiatry, The Medical College of Pennsylvania, Philadelphia, Pennsylvania 19129.

those of other treatments for a given psychiatric disorder. For example, patients suffering from affective disorders are reported to be more prone to TD than are schizophrenics when neuroleptics are prescribed (Davis *et al.*, 1976). Routine use of combined pharmacotherapy, i.e., neuroleptic and antidepressant or neuroleptics alone, should be avoided if at all possible in treating affective disorders. Personality disorders and other nonpsychotic disorders are other examples where the use of neuroleptics significantly increases the risk of developing TD and alternative treatments exist without such risk.

In treating chronic schizophrenia, clinicians often have the dilemma that without neuroleptics, the patient's psychotic symptoms relapse, while with neuroleptics, TD may develop or worsen. In many instances, high doses of neuroleptics are prescribed for months, for years, or even for life. Therefore, it is crucial to reassess such a patient's diagnosis using diagnostic criteria with well-established validity and reliability. It is also necessary to reevaluate other factors that might affect the treatment. Psychosocial stressors may exaggerate patients' behavioral problems and result in an increase in neuroleptic dose which could be avoided with environmental interventions. Also, many "refractory" schizophrenics suffer from a defect state marked by negative symptoms (i.e., withdrawal, blunted affect) but few or no positive symptoms (i.e., delusions, hallucinations, and overactivity); such patients may obtain little benefit from further neuroleptic administration (Pi and Simpson, 1981a).

Although the use of megadoses of neuroleptics can be considered heroic treatment for "refractory" schizophrenia (McClelland *et al.*, 1976), it should be kept in mind that such patients may already be getting too much neuroleptic (Curry 1970; Simpson *et al.*, 1970; Simpson, 1975). If little or no therapeutic benefit is observed within a reasonable time period, it is preferable to change neuroleptics or take the patient off medication entirely for a trial period.

Much effort has been made during the past two decades to find a neuroleptic that would not cause TD. One promising drug, clozapine, unfortunately produced a high incidence of agranulocytosis, which limited its use (Senn *et al.*, 1977; Simpson and Lee, 1978; Singer, 1983; Jeste and Wyatt, 1983).

2.2. Patterns of Neuroleptic Usage

2.2.1. Specific Neuroleptics

Whether certain neuroleptics are more prone than others to produce TD remains controversial (Kane and Smith, 1982; Mukherjee *et al.*, 1982; Perris *et al.*, 1979; Crane, 1968). Virtually all currently prescribed neuroleptics have been associated with TD. Fluphenazine decanoate has been discussed as perhaps being more likely to produce TD. However, patients who receive this drug may be more severely ill, more vulnerable in terms of individual susceptibility to TD, and more drug compliant since this is given parenterally to such subjects.

2.2.2. *Dosages of Neuroleptics*

It has been reported that higher total cumulative dosage of neuroleptics administered relates to severity of TD (Kane *et al.*, 1982; Simpson *et al.*, 1978; Smith *et al.*, 1978; Crane, 1973). There seems to be a much lower prevalence of TD in countries where lower dosages are prescribed, although no significant differences in prevalence were found in transcultural studies comparing high- and low-dosage countries (Ogita *et al.*, 1975). On the other hand, such studies are often poorly controlled for other variables, such as age, sex, and body weight of the patients.

Neuroleptics have proven profoundly effective in the treatment of schizophrenia. However, in many cases their dosages could be substantially reduced, without affecting the general climate of inpatient wards or long-term prognosis of psychosis (Crane, 1976; Pi *et al.*, 1983). For maintenance treatment, 400 mg of chlorpromazine a day or less or equivalent doses of other neuroleptics will suffice for most schizophrenics. In a recent study, plasma levels of chlorpromazine and dosages were much lower than the therapeutic range usually suggested in the literature (Falloon *et al.*, 1985). Lower plasma levels of neuroleptics may lower the risk of developing TD. Jeste *et al.* (1979) found higher plasma levels of neuroleptics in TD than non-TD patients. However, it is premature to make general conclusions about the pharmacokinetics, pharmacodynamics, and plasma levels of neuroleptics (Simpson and Pi, 1981). The hypothesis that patients with TD may not metabolize neuroleptics as efficiently as patients without TD has not yet been clarified (Fairbairn *et al.*, 1983).

2.2.3. *Length of Time of Neuroleptic Administration*

Preliminary results of a prospective study suggest that TD cases have had significantly longer exposure to neuroleptics than non-TD cases (Kane *et al.*, 1982). This is generally assumed to be a risk factor for TD, although more data are needed.

2.2.4. *Extrapyramidal Side Effects and Antiparkinson Agents*

The question of extrapyramidal side effects (EPS) predisposing to TD also needs more data for clarification (Crane, 1972; Mukherjee *et al.*, 1982). Some researchers have questioned whether antiparkinson agents worsen TD (Gerlach, 1977; Defraites *et al.*, 1977; Chouinard *et al.*, 1979). Recently, Kane *et al.* (1982) reported a possible correlation between TD and length of exposure to antiparkinson agents. While awaiting definitive findings, the clinician should probably avoid routine use of antiparkinson agents for prophylactic reasons (Kiloh *et al.*, 1973; Simpson *et al.*, 1981). A careful periodic evaluation (e.g., every 3 months) as to the need for continuation of antiparkinson agents is also recommended, since at least two-thirds of patients do not develop EPS when antiparkinson agents are gradually reduced and eventually discontinued (McClelland *et al.*, 1974).

2.3. Nonpharmacological Risk Factors

Most of the information regarding risk factors is based on retrospective studies. Though limited information is available to aid in the prediction of subjects at risk, it is important for the clinician to be aware of factors that have been incriminated.

2.3.1. Age

It appears that older patients show greater likelihood of developing TD and greater severity when it occurs (Simpson *et al.*, 1978; Smith *et al.*, 1978; Johnson *et al.*, 1982). Lower doses of neuroleptics should usually be prescribed for the elderly. This recommendation, of course, is in accordance with general pharmacokinetic and pharmacodynamic concepts (Simpson and Pi, 1981).

2.3.2. Sex

Females appear more susceptible than males for the development of severe TD. It has been suggested that estrogen may play a role (Gordon *et al.*, 1980), but one must also consider that female patients may receive higher doses of neuroleptics than male patients (Laska *et al.*, 1973). Whatever the reasons, extra caution seems in order when prescribing neuroleptics for older females—the most vulnerable risk group (Yassa and Nair, 1984).

2.3.3. Organicity

Whether this is a factor is debatable. Recently, Varga *et al.* (1982) found a high prevalence of spontaneous involuntary movement disorders in elderly nursing home patients who had never been treated with neuroleptics. This suggests that elderly people are indeed at risk for abnormal movements whether TD or not, but whether organicity is a factor in developing these movements is not clear. Movement disorders were described in the literature prior to the introduction of neuroleptics (Kraepelin, 1919; Turek, 1975); a more recent study found no difference in the incidence of TD in chronic schizophrenics who never received neuroleptics compared with those who did (Owens *et al.*, 1982). This suggests that chronic schizophrenia itself may predispose to movement disorders similar to TD.

Individual susceptibility, alcohol abuse, and organic mental syndrome, e.g., mental retardation, have been suggested as other risk factors for TD (Simpson and Kline, 1976; Gerlach, 1979; Yassa *et al.*, 1984) which need further study to confirm.

2.3.4. Electroconvulsive Therapy

Retrospective reports that prior treatment with electroconvulsive therapy (ECT) correlates with TD may relate to the use of ECT primarily for affective disorder patients, who may be at greater risk for TD, or for chronic patients treated with ECT as a last resort (Davis *et al.*, 1976; Rosenbaum *et al.*, 1977; Harma *et al.*, 1983). The relationship of TD to prior treatment with ECT or insulin has not been a consistent finding (Simpson *et al.*, 1978).

In summary, special consideration should be given when neuroleptics are prescribed for patients who are old or female, who show evidence of organicity, a history of alcohol abuse, ECT, or insulin coma therapy, or who have a history of affective disorder.

2.4. Drug Holidays

Recently, Kane and Smith (1982) pointed out that there are insufficient data to support the role of drug holidays, as affecting the risk of TD, but that drug holidays may help the clinician to diagnose the masked or covert type of tardive dyskinesia.

2.5. Conclusion

In dealing with tardive dyskinesia, the importance of primary prevention needs to be reemphasized. We should prescribe neuroleptics only when they are indicated and use the lowest possible dosage and for the shortest duration of time. Familiarity with factors that may predispose to the development of TD is essential.

3. SECONDARY PREVENTION OF TARDIVE DYSKINESIA

3.1. Early Detection of Tardive Dyskinesia

Periodic side effect monitoring for patients who must take neuroleptics is another important part of prevention. Subjective complaints are rarely made by the TD subjects (Alexopoulos, 1979; Rosen *et al.*, 1982), which complicates the early detection of TD. The difficulty is increased by the lack of clarity in defining the disorder and by the subtlety of "characteristic signs."

All clinicians who prescribe neuroleptics should know the typical presentation of TD, including the classic buccolingual–masticatory triad and choreoathetoid movements of the extremities and other parts of the body (Simpson *et al.*, 1979, 1981; Pi and Simpson, 1981b). They should be able to perform a routine neurological examination and should be acquainted with TD rating scales, which are sensitive clinical tools for monitoring changes in the syndrome.

Education and training sessions to familiarize clinicians with TD could be a valuable means of preventing the development of the disorder. However, the impact of some educational efforts on the prescribing habits of physicians—including monitoring side effects and systematic reduction of neuroleptic doses—may be disappointing (Crane, 1977).

If the neuroleptic-treated patient shows any sign of TD during a periodic examination, careful assessment of the benefit-to-risk ratio should be carried out. If the benefit of neuroleptics clearly outweighs the risk, the neuroleptic dosage should be gradually titrated to the lowest possible dose, with periodic assessments of both psychosis and TD. A detailed recommendation for management of TD is shown in Fig. 1.

3.2. Recognition of Reversible TD

In many cases (up to 90%) TD disappears or at least improves over time—even on a continuing stable dose of neuroleptics (Hunter *el al.*, 1964; Heinrich *et al.*, 1968; Quitkin *et al.*, 1977; Casey, 1985). A recent discontinuation study of neuroleptic therapy found an extremely low reversibility rate (1/33) in an outpatient population with TD. The estimated probability of showing a 50% reduction in movement after a drug-free period of 18 months is 87.2% (Glazer *et al.*, 1984). Another study found complete remission of abnormal movement in six patients with TD after a neuroleptic-free period of more than 2 years (Klawans *et al.*, 1984). Although it is impossible to predict reversibility in each individual subject, or to distinguish immediately between TD as a transient sign of neuroleptic withdrawal versus a persistent disorder, early detection can prevent or at least reduce the risk of permanent neurological disability.

3.3. Consideration of Differential Diagnosis

After a comprehensive history and neurological evaluation, if no organic etiological factor such as Huntington's chorea, Wilson's disease, or Tourette syndrome can be determined, and the subject gives a positive history of neuroleptic ingestion, then a diagnosis of TD can be established (Granacher, 1981). Other conditions, including dyskinesias induced by tricyclic antidepressants, phenytoin, amphetamines, antihistamines, and steriods, also need to be considered (Simpson *et al.*, 1983). Some of these disorders are treatable and reversible.

3.4. Clinical Management

In order to limit progression and worsening of TD, a decision to continue or discontinue neuroleptics must be made. Periodic assessment of psychiatric status will help clinicians make this decision. When it is decided to reduce the dosage of neuroleptics, we recommend the regimen described by Simpson *et al.* (1983). (See Fig. 1.)

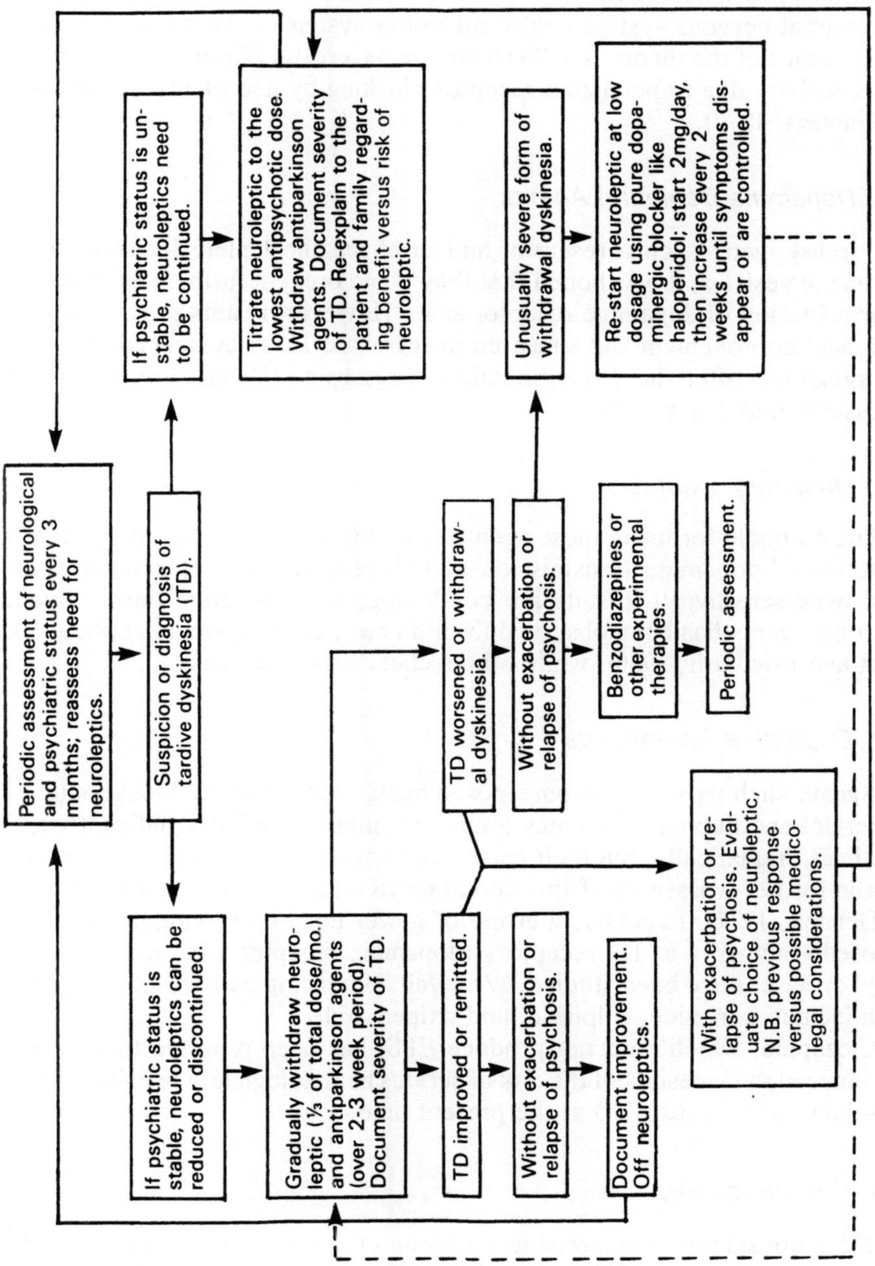

FIGURE 1. Scheme for clinical management of tardive dyskinesia (TD). (From Simpson *et al.*, 1983.)

3.5. Experimental Pharmacotherapy

Various pharmacological agents including agonists and antagonists of different central nervous system neurotransmitter systems have been studied as a consequence of the theory that TD develops as a result of dopamine receptor supersensitivity due to persistent receptor blocking by neuroleptics (Rubovits and Klawans, 1972).

3.5.1. Dopamine-Depleting Agents

Because agents such as reserpine and tetrabenazine deplete dopamine from presynaptic vesicles, it was hoped that they could prevent further sensitization and proliferation of dopamine receptor sites. However, results proved limited to transient improvement and appeared to aggravate TD as evidenced by worsened symptoms after the discontinuation of reserpine (Donatelli *et al.*, 1983; Baldessarini and Tarsy, 1978).

3.5.2. Dopamine Agonists

The rationale for using these agents (e.g., levodopa, apomorphine) lies in facilitation of dopamine transmission and thereby reversal of antagonist-induced hypersensitivity. Results are conflicting; only moderate improvement with some agents has been observed (Smith *et al.*, 1977; Bjorndal *et al.*, 1980; Alpert and Friedhoff, 1980), while others claim greater success.

3.5.3. Dopamine Antagonists

Agents such as so-called pure postsynaptic dopaminergic blockers (e.g., haloperidol and fluphenazine) may improve symptoms of TD significantly and sometimes dramatically, but their causative relationship to TD may eventually result in further worsening of this condition (Kazamatsuri *et al.*, 1973; Casey and Denney, 1977). Recently, a group of newer dopamine antagonists which supposedly act only at D2 receptors (dopamine receptor sites not linked to adenyl cyclase) have been studied. We await convincing data for these agents, which include pimozide, sulpiride, and oxiperomide.

Clozapine, which dose not produce EPS, has been reported to suppress drug-induced dyskinesia, but the risk of serious hematological side effects limits its usefulness in treating TD at the present time.

3.5.4. Cholinergic Agents

TD is considered a heterogeneous group of disorders which may involve cholinergic and/or dopaminergic pathways (Tarsy and Baldessarini, 1976). Although agents like physostigimine, deanol, choline, and lecithin improved some TD cases, results need confirmation (Simpson *et al.*, 1977; Growden, 1979).

3.5.5. *Anticholinergic Agents*

These agents (e.g., benztropine, trihexyphenidyl) worsen TD (Gerlach, 1977), usually reversibly (Gardos and Cole, 1983).

3.5.6. *γ-Aminobutyric Acid, Serotonin, and Norepinephrine Agents*

Conflicting results have been generated from the few controlled studies (Baldessarini and Tarsy, 1978; Gerlach *et al.*, 1978).

3.5.7. *Anxiolytics*

Anxiety can aggravate dyskinetic movements. Benzodiazepines may be helpful in removing anxiety and subsequently reducing the TD symptoms, but the possibility of abuse needs to be considered. Benzodiazepines also affect the GABA system and could have a beneficial effect via this mechanism.

3.5.8. *Miscellaneous Agents*

These include endorphin, naloxone, propranolol, pyridoxine, clonidine, cyproheptadine, manganese, fusaric acid, and estrogens. None of them has been shown to produce any consistent therapeutic effects (Simpson *et al.*, 1976, 1982; Schrodt *et al.*, 1982; Jeste and Wyatt, 1982). While lithium may not directly benefit TD, it may exert a prophylactic effect against the development of TD (Pert *et al.*, 1978).

3.6. Nonpharmacological Therapy

Certain degrees of symptomatic improvement has been reported in some TD patients with nonpharmacological therapy, including transcutaneous nerve stimulation, prosthodontia, electroconvulsive therapy, and surgery (Jeste and Wyatt, 1982).

4. TERTIARY PREVENTION OF TARDIVE DYSKINESIA

Rehabilitation

Although most TD patients are chronically and severely mentally ill, they may still benefit from certain rehabilitation programs.

Physiotherapy may improve some neuromuscular dysfunctions (e.g., facial or buccal pain). Special vocational skills can be learned to maximize personal potential with the goal of eventual self-support. Finally, appropriate psychotherapeutic intervention (e.g., relaxation techniques, biofeedback, supportive psychotherapy, family therapy) may ease some of the suffering caused by the physical and psychological impairment of TD. An educational program for the patients and their relatives may also be helpful, but this intervention must be carried out carefully lest it provoke an excessive fear and resulting noncompliance with essential treatment.

5. CONCLUSION

It is clear that prevention is more important than cure in dealing with TD, despite encouraging findings regarding the reversibility of TD (Quitkin *et al.*, 1977; Wegner and Kane, 1982). Although at present it is not possible to eliminate TD, it is possible to lower the risk if clinicians practice early detection and follow basic principles of rational pharmacotherapy. This includes indications for use and risk versus benefit of neuroleptics when neuroleptics are used for maintenance purposes. Side effects and psychopathology should be monitored at reasonable intervals, e.g., every 3 months.

6. REFERENCES

Alexopoulos, G. S., 1979, Lack of complaints in schizophrenics with tardive dyskinesia, *J. Nerv. Ment. Dis.* **167**:125–127.

Alpert, M., and Friedhoff, A. J., 1980, Clinical application of receptor modification treatment, in: *Tardive Dyskinesia—Research and Treatment* (W. E., Fann, R. C. Smith, J. M. Davis, and E. F. Domino, eds.), pp. 471–473, SP Medical and Scientific Books, New York.

Baldessarini, R. J., and Tarsy, D., 1978, Tardive dyskinesia, in: *Psychopharmacology: A Generation of Progress* (M. A. Lipton, A. DiMascio, and K. F. Killam, eds.), pp. 993–1004, Raven Press, New York.

Bjorndal, N., Casey, D. E., Gerlach, J., Hansen, H., Korsgaard, S., and Rasmussen, J. 1980, The effects of levodopa in tardive dyskinesia, in: *Phenothiazines and Structurally Related Drugs: Basic and Clinical Studies* (E. Usdin, ed.), pp. 321–324, Elsevier North Holland, Amsterdam.

Casey, D. E., 1985, Tardive dyskinesia: Reversible and irreversible, in: *Dyskinesia Research and Treatment* (D. E. Casey *et al.*, eds.), pp. 88–97, Springer-Verlag, New York.

Casey, D. E., and Denney, D., 1977, Pharmacological characterization of tardive dyskinesia, *Psychopharmacology* **54**:1–8.

Chouinard, G., deMoutigny, C., and Annable, L., 1979, Tardive dyskinesia and antiparkinsonian medication, *Am. J. Psychiatry* **136**:228–229.

Crane, G. E., 1968, Tardive dyskinesia in patients treated with major neuroleptics: A review of the literature, *Am. J. Psychiatry* **124**(Suppl.):40–48.

Crane, G. E., 1972, Psuedoparkinsonism and tardive dyskinesia, *Arch. Neurol.* **27**:426–430.

Crane, G. E., 1973, Persistent dyskinesia, *Br. J. Psychiatry* **122**:395–405.

Crane, G. E., 1976, Risks of long-term therapy with neuroleptic drugs, in: *Antipsychotic Drugs: Pharmacodynamics and Pharmacokinetics* (G. Sedvall, B. Uvnas, and G. Y. Zotterman, Eds.), pp. 441–419, Pergamon Press, New York.

Crane, G. E., 1977, The prevention of tardive dyskinesia, *Am. J. Psychiatry* **134**:756–758.

Curry, S. H., Davis, J. M., Janowsky, D. S., *et al.*, 1970, Factors affecting chlorpromazine plasma levels in psychiatric patients, *Arch. Gen. Psychiatry* **22**:209–215.

Davis, K. L., Berger, P. A., and Hollister, L. E., 1976, Tardive dyskinesia and depressive illness, *Psychopharmacol. Commun.* **2**:125–130.

Defraites, E. G., Davis, K. L., and Berger, P. A., 1977, Coexisting tardive dyskinesia and parkinsonism: A case report, *Biol. Psych.* **12**:267–272.

Donatelli, A., Geisen, L., and Feuer, E., 1983. Case report of adverse effect of reserpine on tardive dyskinesia, *Am. J. Psychiatry* **140**:239–240.

Fairbairn, A. F., Rowell, F. J., Hui, S. H., Hassanyeh, F., Robinson, A. J., and Decdeston, D., 1983. Serum concentration of depot neuroleptics in tardive dyskinesia, *Br. J. Psychiatry* 142:579–583.

Fallon, I. R. H., Boyd, J. L., McGill, C. W., *et al.*, 1985, Family management in the prevention of morbidity of schizophrenia, *Arch. Gen. Psychiatry* **42**:887–896.

Gardos, G., and Cole, J. O., 1983. Tardive dyskinesia and anticholinergic drugs, *Am. J. Psychiatry* **140**:200–202.

Gerlach, J., 1977, The relationship between parkinsonism and tardive dyskinesia, *Am. J. Psychiatry* **134**:781–783.

Gerlach, J., 1979, Tardive dyskinesia, *Dan. Med. Bull.* **26**:209–245.

Gerlach, J., Rye, T., and Kristjansen, P., 1978, Effect of Baclofen on tardive dyskinesia, *Psychopharmacology* **56**:145–151.

Glazer, W. M., Moore, D. C., Schooler, N. R., Brenner, L. M., and Morgenstern, H., 1984, Tardive dyskinesia, a discontinuation study. *Arch. Gen. Psychiatry* **41**:623–629.

Gordon, J. H., Borison, R. L., and Diamond, B. T., 1980, Modulation of dopamine receptor sensitivity by estrogen, *Biol. Psychiatry* **15**:389–396.

Granacher, R. P., 1981, Differential diagnoses of tardive dyskinesia: An overview, *Am. J. Psychiatry* **138**:1288–1297.

Growdon, J., 1979, Choline and lecithin administration to patients with tardive dyskinesia, *Can. J. Neurol. Sci.* **6**:80.

Harma, B. J., Nasrallah, H. A., Clancy, J., and Finn, R., 1983, Psychiatric diagnosis and risk for tardive dyskinesia, *Arch. Gen. Psychiatry* **40**:346–347.

Heinrich, K., Wegener, I., and Bender, H. J., 1968, Spate extrapyramidal hyperkinesen bei neuroleptischer Langzeit-Therapie, *Pharmakopsychiatry Neuropsychopharmakol.* **1**:169–195.

Hunter, R., Earl, C. J., and Thornicroft, S., 1964, An apparently irreversible syndrome of abnormal movements following phenothiazine medication, *Proc. Roy. Soc. Med.* **57**:758–762.

Jeste, D. V., and Wyatt, R. J., 1982, Therapeutic strategies against tardive dyskinesia. Two decades of experience, *Arch. Gen. Psychiatry* **39**:803–816.

Jeste, D. V., and Wyatt, R. J., 1983, Clozapine and tardive dyskinesia, *Arch. Gen. Psychiatry* **40**:347–348.

Jeste, D. V., Rosenblatt, J. E., Wagner, R. L., and Wyatt, R. J., 1979, High serum neuroleptic levels in tardive dyskinesia? *N. Engl. J. Med.* **301**:1184.

Johnson, G. F. S., Hunt, G. E., and Rey, J. M., 1982, Incidence and severity of tardive dyskinesia increase with age, *Arch. Gen. Psychiatry* **39**:486.

Kane, J. M., and Smith, J. M., 1982, Tardive dyskinesia: Prevalence and risk factors, 1959 to 1979, *Arch. Gen. Psychiatry* **39**:473–481.

Kane, J. M., Woerner, M., Weinhold, P., Wegner, J., and Kinon, B., 1982, A prospective study of tardive dyskinesia development: Preliminary results, *J. Clin. Pharmacol.* **2**:345–349.

Kazamatsuri, H., Chien, C. P., and Cole, J. O., 1973, Long-term treatment of tardive dyskinesia with haloperidol and tetrabenazine, *Am. J. Psychiatry* **130**:479–483.

Kiloh, L. G., Smith, J. S., and Williams, S. E., 1973, Antiparkinson drugs as causal agents in tardive dyskinesia, *Med. J. Aust.* **2**:591–593.

Klawans, H. L., Tanner, C. M., and Barr, A., 1984, The reversibility of "permanent" tardive dyskinesia, *Clin. Neuropharmacol.* **7**:153–159.

Kraepelin, E., 1919, *Dementia Praecox and Paraphrenia* (R. M. Barclay, Translator), Livingstone, Edinburgh.

Laska, E., Varga, E., Wanderling, J., Simpson, G., Logemann, G. W., and Shah, B. K., 1973, Patterns of psychotropic drug use for schizophrenia, *Dis. Nerv. Syst.* **34**:294–305.

McClelland, H. A., Blessed, G., Bhata, S., Ali, N., and Clarke, P., 1974, Abrupt withdrawal of antiparkinsonian drugs in schizophrenic patients, *Br. J. Psychiatry* **124**:151–159.

McClelland, H. A., Farquharson, R. G., Leyburn, P., Furness, J. A., and Schiff, A. A., 1976, Very high dose fluphenazine decanoate: A controlled trial in chronic schizophrenia, *Arch. Gen. Psychiatry* **33**:1435–1439.

Mukherjee, S., Rosen, A. M., Cardenas, C., Varia, V., and Olarte, S., 1982, Tardive dyskinesia in psychiatric outpatients: A study of prevalence and association with demographic, clinical and drug history variables, *Arch. Gen. Psychiatry* **36**:466–469.

National Institute of Mental Health, Psychopharmacology Research Branch, 1975, Development of a dyskinetic movement scale, *Early Clinical Drug Evaluation Unit* **4**:3–6.

Ogita, K., Yagi, G., and Itoh, H., 1975, Comparative analysis of persistent dyskinesia of long-term usage with neuroleptics in France and in Japan, *Folia Psychiatr. Neurol. Jpn.* **29**:315–320.

Owens, D. G. C., Johnstone, E. C., and Frith, C. D., 1982, Spontaneous involuntary disorders of movement, their prevalence, severity, and distribution in chronic schizophrenics with and without treatment with neuroleptics, *Arch. Gen. Psychiatry* **39**:452–461.

Perris, C., Dimitrijevic, P., Jacobsson, L., Paulsson, P., Rapp, W., and Froberg, H., 1979, Tardive dyskinesia in psychiatric patients treated with neuroleptics, *Br. J. Psychiatry* **135**:509–514.

Pert, A., Rosenblatt, J. E., Sivit, C., Pert, C. B., and Bunney, W. E., Jr., 1978, Long-term treatment with lithium prevents the development of dopamine receptor supersensitivity, *Science* **201**:171–173.

Pi, E. H., and Simpson, G. M., 1981a, The treatment of refractory schizophrenia: Pharmacotherapy and clinical implications of blood level measurement of neuroleptics, *Int. J. Pharmacopsychiatry* **16**:154–161.

Pi, E. H., and Simpson, G. M., 1981b, Tardive dyskinesia and abnormal tongue movements, *Br. J. Psychiatry* **139**:526–528.

Pi, E. H., Miller, A., Rosenberg, M. R., Schultz, C., and Simpson, G. M., 1983, Therapeutic effects of psychiatric hospitalization, *J. Psychiatr. Eval. Treatment* **5**:135–142.

Quitkin, F., Rifkin, A., Gochfeld, L., and Klein, D. F., 1977, Tardive dyskinesia: Are first signs reversible? *Am. J. Psychiatry* **134**:84–87.

Rosen, A. M., Mukherjee, S. Olarte, S., Varia, V., and Cardenas, C., 1982, Perception of tardive dyskinesia in outpatients receiving maintenance neuroleptics, *Am. J. Psychiatry* **139**:372–373.

Rosenbaum, A. H., Niven, R. G., Hanson, N. P., and Swanson, D. W., 1977, Tardive dyskinesia: Relationship with a primary affective disorder, *Dis. Nerv. Syst.* **38**:423–427.

Rubovits, R., and Klawans, H. L., 1972, Implications of amphetamine-induced stereotyped behavior as a model for tardive dyskinesia, *Arch. Gen. Psychiatry* **27**:502–507.

Schonecker, M., 1957, Ein eigentumlichen syndrom im oralem bereich bei Megaphen applikation, *Nervenarzt* **28**:35.

Schooler, N. R., and Kane, J. M., 1982, Research diagnoses for tardive dyskinesia, *Arch. Gen. Psychiatry* **39**:486–487.

Schrodt, G. R., Wright, J. H., Simpson, R., Moore, D. P., and Chase, S., 1982, Treatment of tardive dyskinesia with propranolol, *J. Clin. Psychiatry* **42**:328–331.

Senn, H. J., Jungi, W. F., Kunz, H., and Poldinger, W., 1977, Clozapine and agranulocytosis, *Lancet* **1**:547.

Sigwald, J., Bouttier, D., Raymond, C., and Piot, C., 1959, Quatre cas de dyskinesie famio-bucio-linguo-masticatrice a evolution prolongee secondaire a un traitment par les neuroleptiques, *Rev. Neurol.* **100**:751–755.

Simpson, G. M., 1975, CNS effects of neuroleptic agents, *Psychiatr. Ann.* **5**:53–60.

Simpson, G. M., and Kline, N. S., 1976, Tardive dyskinesia: Manifestation, incidence, etiology and treatment, in: *The Basal Ganglion* (M. D. Yahr, ed.), pp. 427–432, Raven Press, New York.

Simpson, G. M., and Lee, H., 1978, A ten-year Review of antipsychotics, in: *Psychopharmacology, A Generation of Progress* (M. A. Lipton, A. SiMascio, and K. F., Killam, eds.), pp. 1131–1137, Raven Press, New York.

Simpson, G. M., and Pi, E. H., 1981, Pharmacokinetics of antipsychotic agents, in: *Recent Ad-*

vances in Neuropsychopharmacology, Vol. 31 (B. Angrist, ed.,), pp. 365–372, Pergamon Press, Oxford, New York.

Simpson, G. M., Krakov, L., Mattke, D., and St. Phard, G., 1970, A controlled comparison of the treatment of schizophrenic patients when treated according to the neuroleptic threshold or by clinical judgement, *Acta Psychiatr. Scand.* **212**(Suppl.):38–43.

Simpson, G. M., Voitashevsky, A., Young, M. A., and Lee, J. H., 1977, Deanol in the treatment of tardive dyskinesia, *Psychopharmacology* **52**:257–261.

Simpson, G. M., Varga, E., Lee, J. H., and Zoubok, B., 1978, Tardive dyskinesia and psychotropic drug history, *Psychopharmacology* **58**:117–124.

Simpson, G. M., Lee, J. H., Zoubok, B., and Gardos, G., 1979, A rating scale for tardive dyskinesia, *Psychopharmacology* **64**:171–179.

Simpson, G. M., Pi, E. H., and Sramek, J. J., 1981, Adverse effects of antipsychotic agents, *Drugs* **21**:138–151.

Simpson, G. M., Pi, E. H., and Sramek, J. J., 1982, Management of tardive dyskinesia: Current update, *Drugs* **23**:381–393.

Simpson, G. M., Pi, E. H., and Sramek, J. J., 1983, The current status of tardive dyskinesia, *J. Psychiatr. Eval. Treatment* **4**:127–133.

Singer, J. M., 1983, Clozapine and tardive dyskinesia, *Arch. Gen. Psychiatry* **40**:347–348.

Smith, R. C., Tamminga, C. A., Haraszti, J., Pandey, G. N., and Davis, J. M., 1977, Effects of dopamine agonists in tardive dyskinesia, *Am. J. Psychiatry* **134**:753–768.

Smith, R. C., Strizich, M., and Klass, D., 1978, Drug history and tardive dyskinesia, *Am. J. Psychiatry*. **135**:1402–1403.

Tarsy, D., and Baldessarini, R. J., 1977, The pathophysiologic basis of tardive dyskinesia, *Biol. Psychiatry* **12**:431–450.

Turek, I. S., 1975, Drug-induced dyskinesia: Reality or myth? *Dis. Nerv. Syst.* **36**:397–399.

Varga, E., Sugerman, A. A., Varga, V., Zomorodi, A., Zomorodi, W., and Menken, M., 1982, Prevalence of spontaneous oral dyskinesia in the elderly, *Am. J. Psychiatry* **139**:329–331.

Wegner, J. T., and Kane, J. M., 1982, Follow-up study on the reversibility of tardive dyskinesia, *Am. J. Psychiatry* **39**:368–369.

Yassa, R., and Nair, V., 1984, Incidence of tardive dyskinesia in an outpatient population, *Psychosomatics* **75**:479–481.

Yassa, R., Nair, V., and Schwartz, G., 1984. Tardive dyskinesia and the primary psychiatric diagnosis, *Psychosomatics* **25**:135–138.

CHAPTER 9

Clinical Aspects of Tardive Dyskinesia

Epidemiology, Natural History, and Management

GEORGE GARDOS and JONATHAN O. COLE

1. INTRODUCTION

Public awareness often seems to lag behind the appearance of a phenomenon by several years. The history of tardive dyskinesia (TD) is a good example. Although the first reports of the syndrome started to appear in the early 1960s, it was not until the late 1960s and early 1970s that the psychiatric profession became aware of the existence of the condition. As the literature began to swell with ever-increasing numbers of reports on TD during the 1970s, there was gradually increasing concern over this apparently iatrogenic condition, which has created alarm over the use of neuroleptics. The pendulum may be starting to swing back in response to recent reports which suggest that TD is not always irreversible, and that the prognosis is usually better than was at first feared. This chapter will provide an overview of the prevalence, course, prognosis, and management of TD. Because of the enormous amount of relevant research data, the literature review will be selective.

2. PREVALENCE

The point prevalence rate reflects the proportion of patients with TD at a given time and place. The reported prevalence rates of neuroleptic-treated patients with TD tend to vary widely and range from 0.5% to as high as 65%, while the average prevalence from thousands of patients is about 20% (Baldessarini, 1980). Problems that partly account for the large variance and make interpretation of prevalence rates uncertain stem from the following: (1) the criteria for diagnosing TD are variable; (2) reliability of measurement is less

GEORGE GARDOS • West-Ros-Park Mental Health Center, Boston, Massachusetts 02136.
JONATHAN O. COLE • Psychopharmacology Services, McLean Hospital, Belmont, Massachusetts 02178.

than optimal; (3) current antipsychotic drugs and dosage levels vary; (4) there are marked population differences; and (5) spontaneous orofacial dyskinesias are present which are unrelated to neuroleptics and which tend to show a prevalence of about 5% (Casey, 1985). A different approach seeks to estimate the incidence or morbid risk of TD by calculating the rate of new cases of TD in a well-defined population. Incidence reports of TD are conspicuous by their absence in view of the inherent difficulties involved in conducting large-scale prospective epidemiological studies. The authors arrived at a rough estimate of TD incidence at a private psychiatric hospital: 4–5% of drug-treated schizophrenics developed new TD in a year (Gardos and Cole, 1980). In an ongoing prospective study, Kane *et al.* (1982) found a TD incidence of 12% after 4 years of cumulative neuroleptic exposure.

Obviously, neuroleptic-related dyskinesia is widespread and is probably going to remain a problem as long as neuroleptics are used.

3. RISK FACTORS

The pathogenesis of TD in its simplest form involves neuroleptic drugs given to vulnerable persons. Accordingly, risk factors may be grouped into treatment variables and patient variables, as shown in Table 1.

The literature on the contributions of drug variables to the pathogenes is of TD is reviewed by Kane and Smith (1982). Despite the plethora of research data, the relative importance of the different drug variables, such as maximum dose, duration of exposure, and total amount of neuroleptics, remains unclear. The type of neuroleptic drug combinations pejoratively referred to as "polypharmacy" and neuroleptic interruptions have also been proposed as risk factors, but an association between these variables and TD is yet to be clearly

TABLE 1
Tardive Dyskinesia (TD)—Predisposing Factors[a]

Treatment variables		Patient variables	
Neuroleptics		Age	+ +
Maximum dose	+	Sex	+
Duration of exposure	+	Brain damage	+
Total amount	+	Affective disorder	+
Polypharmacy	(?)	Acute EPS	(?)
Type of drug	(?)		
Interruptions	(?)		
Antiparkinson drugs	(?)		
ECT	(?)		
Antidepressants	(?)		

[a] +, Likely association with TD; + +, definite association with TD; (?), possible association with TD; EPS, extrapyramidal side effects; ECT, electroconvulsive therapy.

demonstrated. Antiparkinsonian drugs deserve special mention. While these drugs may uncover or aggravate dyskinetic movements, discontinuation of antiparkinsonian drugs tends to result in a return of dyskinesia to baseline levels. There is very little evidence to connect the administration of central antimuscarinic compounds to *persistent* TD (Gardos and Cole, 1983). Tricyclic antidepressants (Fann *et al.*, 1976) and antihistamines (Sovner, 1976) have been reported to induce TD in a few cases. The role of electroconvulsive therapy in the etiology of TD remains dubious (Chouinard *et al.*, 1979; Kane and Smith, 1982).

Among patient variables, age is by far the most important and best established etiological factor. Smith and Baldessarini (1980) examined the association between age and TD. Up to age 70, there is a linear relationship between older age and the frequency and severity of TD. After age 70, the data suggest an age-by-sex interaction insofar as in women the relationship continues to be linear, while in men the frequency and severity of TD tend to decline. Female sex is often correlated with TD prevalence, but it is unclear whether this association is a consequence of biological differences (Kane and Smith, 1982). Structural brain damage—even in the absence of neuroleptic exposure—has been associated with the emergence of orofacial dyskinesia (Waddington *et al.*, 1985). Crane (1970) and Yagi *et al.* (1976) suggested an association between acute extrapyramidal symptoms and the later emergence of persistent TD, but this important issue has not been resolved because of the lack of appropriate studies.

4. "NATURAL" HISTORY OF TD

In attempting to determine the long-term outlook of TD, it is important to assess the usual course under two conditions: in patients who remain on neuroleptics despite TD and in those who discontinue neuroleptics because of TD. Either of these situations may be regarded as the "natural" history of TD. Table 2 summarizes the literature of long-term studies (at least 12 months) of TD on continued neuroleptic treatment.

The data suggest an overall tendency for the level of dyskinesia to increase over time, but in most instances the magnitude of change is neither large nor clinically significant. Studies in which neuroleptics were continued but attempts were made to reduce dosage tended to show that TD tended to remain the same or improve (Casey and Toenniessen, 1983b). The long-term results of TD following neuroleptic withdrawal were comprehensively reviewed by Jeste and Wyatt (1979). These studies show an even clearer overall trend toward improvement than the neuroleptic dose reduction data. There was considerable variability, both between and within studies; some patients got much better and others became worse. However, many more patients showed improvement, and in 7 of the 20 studies examined by Jeste and Wyatt, more than 50% of patients improved. A more recent review of relevant data (Casey and Toen-

TABLE 2
Course of Tardive Dyskinesia with Continued Neuroleptics

n	Population	Length of observation (months)	Outcome	Reference
100	30 male and 70 female inpatients with TD	15	About half the sample improved; half did not change	Heinrich *et al.*, 1968
91	Chronic inpatients on high- and low-dose trifluoperazine	12	Trend toward increased TD; more in high-than in low-dose group	Crane, 1970
158	Chronic inpatients	12	Slight, nonsignificant increase in TD	Crane, 1972
24	Chronic inpatients with TD	6	No significant changes	Chien and Cole, 1973
79	Chronic inpatients	30	Increase in both scores and number of patients with TD	Crane, 1975
70	Chronic inpatients	24	Trend toward increased dyskinesia	Degkwitz *et al.*, 1976
11	Chronic schhizophrenics with TD	100	Slight increase in TD scores	Cole and Gardos, 1981 (unpublished data)
67	Chronic inpatients	36	Significantly increased AIMS scores	Smith *et al.*, 1981
17	Patients with TD	60	65% reduction of AIMS scores on lowered dosage; 36% reduction on higher dosage	Casey and Toenniessen, 1983
99	58 outpatients, 41 inpatients	36	Prevalence of orofacial dyskinesia increased from 39% to 47%	Barnes *et al.*, 1983
47	25 male and 22 female outpatients	36	Increased dyskinesia scores, more statistically than clinically significant	Gardos *et al.*, 1983

niessen, 1983b) confirms the improved prognosis in TD patients withdrawn from neuroleptics and points out that younger patients improve most when dose is decreased or drugs discontinued. The major caveat is the potential for psychotic decompensation, especially in schizophrenic patients, when neuroleptics are withdrawn. Psychiatric hospitalization and treatment with large doses of neuroleptics is often the unintended result.

5. NEW DYSKINESIA

The previous section focused on patients with long-established dyskinesia. The outlook in those patients who are diagnosed shortly after developing the condition may be considerably better, as suggested first by Quitkin *et al.* (1977). The issue is reversibility of TD or, more precisely, the effectiveness of secondary prevention. If early recognition and timely intervention materially improve prognosis, then secondary prevention would become a worthy clinical goal. Relevant data have been obtained by the authors from an ongoing study

TABLE 3
Characteristics of Patients with New Dyskinesia

Age 36.2 ± 13.5 (range 18–73)			
Sex M = 26			
F = 27			
		n	
Diagnosis			
Schizophrenia		23	
Schizoaffective disorder		15	
Bipolar disorder		7	
Major depression		8	
Length of psychiatric hospitalization (months)			23.2 ± 38.9
Duration of neuroleptic therapy (months)			69.3 ± 57.1
	Baseline	1 year	*t*
AIMS total	8.53	7.47	NS
AIMS global	2.13	1.74	2.81 $p = 0.007$

at McLean Hospital (Gardos *et al.*, 1982). Patients who developed dyskinesia within 12 months and who scored at least one "mild" item on the abnormal involuntary movements scale (AIMS) (Guy, 1976) have been followed prospectively with repeated ratings as well as videotapes. To date, 53 patients have completed at least 1 year on the study (Table 3).

The cohort was predominantly young and evenly divided between men and women. The typical patient was extensively treated with neuroleptics and was hospitalized for about 2 years. Affective disorder was present in a significant proportion of patients. During the year of observation, patients continued to receive regular psychiatric care but also had the additional benefit of psychopharmacological consultations to their treating physicians. Neuroleptic dose reduction was frequently considered, and alternative drug treatments (e.g., lithium) were explored. At the end of the 1 year of observation, the global severity of TD declined significantly, while the total AIMS score was also lower but did not reach statistical significance. These results, while preliminary and subject to change as more data are forthcoming, are encouraging and tend to endorse the value of secondary prevention, in that early case finding and appropriate intervention in an unselected cohort of patients with new TD led to a significant reduction in the severity of TD a year later.

Overall, the research data on course and prognosis indicate a much less gloomy picture than was feared. Long-standing TD rarely progresses to increasingly severe forms even in the presence of continued neuroleptic treatment. Dose reduction and neuroleptic withdrawal, when appropriate, can be expected to produce improvement. The prognosis of recently developed TD, especially in younger patients, appears favorable in the presence of judicious pharmacotherapy.

6. MANAGEMENT OF TD

The clinical management of cases of TD may be rationally approached with a preventive model in focus.

Primary prevention aims at the elimination of risk factors thereby preventing the emergence of new cases. Applied to TD, it is a truism that an ounce of prevention is better than a pound of cure. Since neuroleptic exposure is unquestionably a major etiological factor, the avoidance of unnecessary use of antipsychotic drugs is of the utmost importance. Most experts agree that neuroleptics are rarely, if ever, indicated in anxiety states and are not the drug of first choice in affective disorders.

Secondary prevention involves early case finding and appropriate intervention. Neuroleptic-treated patients need to be examined every 3–6 months for early TD, which often presents as fine vermicular movements or restlessness of the tongue, facial tics, or mild jerky movements of the fingers and toes. The management of such early cases includes reassessment of the need for continued neuroleptics. If patients have to be continued on drugs, then an attempt to try to find the minimal effective dose by means of gradual dose reduction is worth undertaking. The potential benefits of linthium, antidepressants, carbamazepine, or other psychotropic compounds in place of or in combination with neuroleptics needs to be carefully assessed in each case. This is especially important in the presence of affective disorder or when the diagnosis of schizophrenia is in doubt.

Tertiary prevention aims at amelioration of disability resulting from an established disorder. Patients often do not complain of their condition and, in fact, show a rather striking neglect or denial of what is obvious dyskinesia to everybody else. Relatives and treating personnel are often more concerned about the social and psychological implications of dyskinesia.

The general principles underlying the management of the typical nonsevere cases are similar to those of secondary prevention. Reducing further neuroleptic drug exposure through slow and gradual dose reduction until or unless signs of psychotic decompensation appear might reduce the severity of dyskinesia. Continued observation without changing drugs or dosage is also justifiable as long as the dyskinesia remains fairly stable. Changing neuroleptics may on occasion lead to improvement in dyskinesia, but this approach is much more empirical than scientific.

The use of other classes of psychotropic drugs (e.g., lithium) as supplemental medication or as replacement for neuroleptics needs to be considered. The point to remember is that in long-standing mild cases of dyskinesia, the movement disorder is rarely troublesome enough to require direct intervention to reduce its intensity. Furthermore, as cited previously, even with combined neuroleptic treatment the prognosis of TD is not necessarily unfavorable (Casey and Toenniessen, 1983a).

Severe dyskinesia, on the other hand, is a more serious problem. Fortunately, it is rare, although prevalence estimates are not available. The relevant

TABLE 4
Rare But Serious Complications of Tardive Dyskinesia

Dental problems
Impaired chewing or swallowing
Dysarthria
Disordered respiration
Ataxia
Weight loss
Despondency over movements

literature consists mostly of case reports. A series of 19 patients were personally treated by the authors (Gardos *et al.*, 1984). The grotesque, distressing quality of the movements and functional impairment, rather than any specific score on the AIMS scale, were the criteria by which these patients were regarded as having severe TD. The types of functional disturbances that may occur are shown in Table 4.

With regard to symptomatology, severe choreoathetoid dyskinesia is sometimes seen. In addition, marked dystonic symptoms may also contribute to the severe movement disorder, particularly in young males. Less often, tardive akathisia or blepharospasm is responsible for the impairment of functioning (Gardos *et al.*, 1984). Severe cases of dyskinesia need a more radical therapeutic approach and warrant the consideration of drugs given for the express purpose of reducing the severity of movements. The neuropharmacological approach is based on the hypothesized cholinergic–dopaminergic imbalance and usually consists of administration of drugs that reduce dopaminergic activity and increase cholinergic activity in the extrapyramidal system. Table 5 summarizes currently favored and available drugs for the treatment of tardive dyskinesia.

This list is by no means comprehensive. Scholarly reviews of the therapeutic approaches to TD are available (e.g., Baldessarini, 1980; Jeste and Wyatt, 1979). Reading the literature on the treatment of TD leaves the reader rather depressed since there is no effective therapeutic agent for the condition. However, suppression of dyskinesia by heavy doses of dopamine blockers is feasible. It is gratifying to see in clinical practice how severe dyskinetic movements sometimes literally melt away within days of administration of a dopamine blocker. However, as Table 5 emphasizes, adverse effects are nearly always present and are a major limiting factor in the pharmacotherapeutic management of dyskinesia. For instance, dopamine blockers, while tending to ameliorate dyskinesia, often produce marked parkinsonian side effects. Particularly in the elderly, in whom the condition is more frequent, the administration of dopamine blockers often backfires, and the disability from the resultant parkinsonian signs and symptoms may exceed the benefit resulting from improvement in dyskinesia.

TABLE 5
Drugs in the Treatment of Tardive Dyskinesia

Type of drug	Adverse effects
Dopamine antagonists: reserpine, haloperidol, oxypertine, thiopropazate, etc.	Sedation tremor, rigidity, akinesia
Cholinomimetics: lecithin, deanol, choline	Sedation, depression, gastrointestinal complaints, parkinsonism
GABA agonists (putative): diazapam, clonazepam, valproate, γ-acetylenic GABA	Sedation, confusion, ataxia, etc.
Miscellaneous: antiparkinson drugs, phenobarbital, fusaric acid, propranolol, baclofen	

The treatment of dyskinesia is empirical and remains unsatisfactory. Each drug or class of drugs is likely to benefit a minority of patients. Hence, sequential trial of different drugs, at times in combinations, is not an unreasonable approach and may well result in at least partial relief of symptoms (Gardos *et al.*, 1984).

7. SUMMARY

This chapter is an overview of the clinical aspects of TD. The prevalence rate of neuroleptic-induced persistent dyskinesia varies substantially in different studies but is probably around 20%. Longitudinal studies on the course of TD suggest that in most patients it is not a progressive disease. Even on continued neuroleptic treatment, the severity of TD tends to plateau and remain stationary for years. The prognosis is better in younger patients or when neuroleptic dosage is reduced or after drug discontinuation. Early diagnosis and judicious psychopharmacological management leads to significant improvement after 1 year of observation.

Severe dyskinesia with functional impairment is rare. The treatment of such patients is difficult and unsatisfactory. Dopamine antagonists can reliably suppress choreoathetoid dyskinetic manifestations but often at the expense of parkinsonian manifestations. Cholinergic drugs and a variety of other antidyskinetic compounds produce variable benefits and frequent side effects. Prevention (primary, secondary, and tertiary) is at present the more fruitful means of coping with TD.

8. REFERENCES

Baldessarini, R. J. (chairperson), 1980 *ADA Task Force on Tardive Dyskinesia*, American Psychiatric Association, Washington, DC.

Barnes, T. R. E., Kidger, T., and Gore, S. M., 1983, Tardive dyskinesia: A 3-year follow-up study, *Psychol. Med.* **13**:71.

Casey, D. E., 1985, Spontaneous and tardive dyskinesia: Clinical and laboratory studies, *J. Clin. Psychiatry* **46**(4, Sec. 2):42–47.

Casey, D. E., and Toenniessen, L. M., 1983a, Neuroleptic treatment in tardive dyskinesia: Can it be developed into a clinical strategy for long-term treatment? in: *Modern Problems in Pharmacopsychiatry* (J. Bannet and R. H. Belmaker, eds), pp. 65–79, Karger, Basel.

Casey, D. E., and Toenniessen, L. M., 1983b, Tardive dyskinesia: What is the natural history? *Int. Drug. Ther. Newsl.* **18**(No. 4).

Chien, C-P., and Cole, J. O., 1973, Eighteen-months follow-up of tardive dyskinesia treated with various catecholamine-related agents, *Psychopharmacol. Bull.* **9**:38.

Chouinard, G., Annable, L., Ross-Chouinard, A., and Nestoros, J. N., 1979, Factors related to tardive dyskinesia, *Am. J. Psychiatry* **136**:79.

Crane, G. E., 1970, High doses of trifluoperazine and tardive dyskinesia, *Arch. Neurol.* **22**:176.

Crane, G. E., 1972, Pseudoparkinsonism and tardive dyskinesia, *Arch. Neurol.* **27**:426.

Crane, G. E., 1975, Tardive dyskinesia, A review, in: *Neuropsychopharmacology* (J. R. Boissier, H. Hippius, and P. Pichot, eds.), pp. 346–354, Excerpta Medica, Elsevier, Amsterdam.

Degkwitz, R., Consbruch, U., Haddenbrock, S., Neusch, B., Oehlert, W., and Unsold, R., 1976, Therapeutische Risiken bei der Langzeitbehandlung mit Neuroleptika und Lithium: Klinische, histologische and biochemische Befunde, *Nervenarzt.* **47**:81.

Fann, W. E., Sullivan, J. L., and Richman, B. W., 1976, Dyskinesias associated with tricyclic antidepressants, *Br. J. Psychiatry* **128**:490.

Gardos, G., and Cole, J. O., 1980, Overview: Public health issues in tardive dyskinesia, *Am. J. Psychiatry* **137**:776.

Gardos, G., and Cole, J. O., 1983, Tardive dyskinesia and anticholinergic drugs, *Am. J. Psychiatry* **140**:200.

Gardos, G., Cole, J. O., Haskell, D. S., Moore, P., Bartell, Y., Salomon, M., and Schniebolk, S., 1982, A longitudinal study of early dyskinesia: Preliminary report. Proceedings of the 135th meeting of the American Psychiatric Association, Toronto **68D**:171.

Gardos, G., Perenyi, A., Cole, J. O., Samu, I., and Kallos, M., 1983, Tardive dyskinesia: Changes after three years, *J. Clin. Psychopharmacol.* **3**:315.

Gardos, G., Cole, J. O., Salomon, M., and Schniebolk, S., 1984, Severe tardive dyskinesia. Proceedings of the 137th meeting of the American Psychiatric Association, Los Angeles **59D**:140.

Guy, W., 1976, Abnormal involuntary movements scale, in: *ECDEU Assessment Manual for Psychopharmacoloy*, pp. 534–537, US Department of HEW, Publication No. 76–338, US Government Printing Office, Washington, DC.

Heinrich, K., Wegener, I., and Bender, H. J., 1968, Spate extrapyramidale Hyperkinesen bei neuroleptischer Langzeittherapie, *Pharmacopsychiat. Neuropsychopharmacol.* **1**:169.

Jeste, D. V., and Wyatt, R. J., 1979, In search of treatment for tardive dyskinesia: Review of the literature, *Schizophrenia Bull.* **5**:252.

Kane, J., Woerner, M., Weinhold, P., Wegner, J., and Kinon, B., 1982, A prospective study of tardive dyskinesia development: Preliminary results, *J. Clin. Psychopharmacol.* **5**:345.

Kane, J. M., and Smith, J. M., 1982, Tardive dyskinesia: Prevalence and risk factors 1959 to 1979, *Arch. Gen. Psychiatry* **39**:473.

Quitkin, F., Rifkin, A., Gochfeld, L., and Klein, D. F., 1977, Tardive dyskinesia: Are first signs reversible? *Am. J. Psychiatry* **134**:84.

Smith, J. M., and Baldessarini, R. J., 1980, Changes in prevalence, severity, and recovery in tardive dyskinesia with age, *Arch. Gen. Psychiatry* **37**:1368.

Smith, J. M., Burke, M. P., and Moon, C. O., 1981, Long-term changes in AIMS ratings and their relation to medication history, *Psychopharmacol. Bull.* **17**:120.

Sovner, R. D., 1976, Dyskinesia associated with chronic antihistamine use, *N. Engl. J. Med.* **294**:113.

Waddington, J. L., Youssef, H. A., Molloy, A. G., O'Boyle, K. M., and Pugh, M. T., 1985, Association of intellectual impairment, negative symptoms and aging with abnormal involuntary movements ("tardive" dyskinesia) in schizophrenia: Clinical and animal studies, *J. Clin. Psychiatry* 46(4, Sec. 2):24–33.

Yagi, G., Ogita, K., Ohtsuka, N., Itoh, H., and Miura, S., 1976, Persistent dyskinesia after long-term treatment with neuroleptics in Japan, *Keio J. Med.* 25:27.

CHAPTER 10

Treatments for Tardive Dyskinesia

An Overview of Noncatecholaminergic/ Noncholinergic Treatments

LARRY D. ALPHS and JOHN M. DAVIS

1. INTRODUCTION

Since the late 1960s, tardive dyskinesia (TD) has been recognized as a potential sequela of prolonged therapy with neuroleptics. Besides the disfiguring movements that characterize this syndrome, TD is frequently complicated by temperomandibular joint dysfunction, mouth ulcers, accidental trauma secondary to position instability, respiratory problems, and speech pathology. The movements themselves are often a source of social embarrassment and tend to further isolate individuals with concomitant disabling illnesses like schizophrenia. Because of its medical, social, and legal implications, TD is one of the most significant complications of pharmacological treatment for psychiatric illnesses. As a result, considerable effort has been devoted to finding a treatment for its various symptoms; however, no universally successful therapy has yet been demonstrated.

Over the years, dopamine receptor blockers or dopamine-depleting agents have probably been the most widely used of the potential treatments for TD. It has been proposed that these drugs are effective because TD represents the clinical expression of chemical denervation supersensitivity which follows chronic dopamine–receptor blockade by neuroleptics (Carlsson, 1970; Klawans, 1973). Depleting neuronal stores of dopamine available for release or blocking supersensitive receptors with dopaminergic antagonists are proposed mechanisms to diminish dyskinetic symptoms. Despite their effectiveness, use of these drugs is controversial because they are presumably the offending agent as well as a temporary cure.

Cholinergic agonists have been the next most commonly applied treatment

LARRY D. ALPHS • Maryland Psychiatric Research Center, Department of Psychiatry, University of Maryland, Baltimore, Maryland 21228. JOHN M. DAVIS • Illinois State Psychiatric Institute, Chicago, Illinois 60612.

for TD. These drugs have been used with the rationale that symptoms of TD represent an imbalance of central dopaminergic and cholinergic neuronal activity with a relative excess of dopaminergic activity. Enhancing cholinergic activity theoretically reciprocally diminishes dopaminergic activity and so treats TD (Gerlach *et al.*, 1974; Klawans and Rubovits, 1974). In practice, these drugs are only inconsistently effective. A summary of results observed after their administration has recently been provided elsewhere (Alphs and Davis, 1983).

In view of the general lack of success with these drugs, many others have been tried with the hope of finding some that are more efficacious. In this chapter, we will review some of the less commonly applied treatments for TD, with respect to both their theoretical rationale and their therapeutic success.

2. GABAERGIC DRUGS

There is extensive evidence that GABAergic drugs may be useful therapeutically for treating TD. For instance, animal models have shown that GABA agonists decrease dopamine turnover after pretreatment with α-methyl-para-tyrosine (Palfreyman *et al.*, 1978; Lloyd and Worms, 1981). Similarly, they decrease DOPA accumulation (Scatton *et al.*, 1980, 1982), tyrosine hydroxylase activity (Lloyd and Worms, 1981), and the release of dopamine into the synaptic cleft (Lahti and Losey, 1974; Fuxe *et al.*, 1975; Anden and Wachtel, 1977). Neuroanatomical studies have defined a GABA-mediated neuronal pathway within the basal ganglia, the area of the brain related to neuroleptic-induced dyskinesias (Walters and Chase, 1977; Curtis, 1978; Walaas and Fonnum, 1978). Information processed within the striatum is transmitted through GABA-containing striatal efferent neuronal tracts projecting to the substantia nigra and to the ventral medial lateral nucleus of the thalamus. When tested in rats, GABA agonists have been reported to exhibit marked antidyskinetic effects on L-dopa- and piribedil-induced stereotopies (Lloyd and Worms, 1981). Furthermore, the dyskinetic movements of Huntington's disease resemble those of TD, and on postmortem examination, Huntington's disease patients have been found to have low levels of *l*-glutumate decarboxylase, the pyridoxal-dependent enzyme responsible for the synthesis of GABA from glutamic acid (McGeer and McGeer, 1976). Two classes of GABAergic drugs have been utilized in the treatment of TD. These are (1) the GABA transaminase inhibitors: sodium valproate (Linnoila *et al.*, 1976; Linnoila and Viukari, 1979; Chien *et al.*, 1978; Friis *et al.*, 1983; Nair *et al.*, 1980; Gibson, 1978; Lautin *et al.*, 1979; Crowe, 1983; Casey and Hammerstad, 1979; Nagao *et al.*, 1979; Ames and Webber, 1984) acetylenic GABA (Casey *et al.*, 1980a,b), and α-vinyl GABA (Tell *et al.*, 1981; Tamminga *et al.*, 1983; Thaker *et al.*, 1983); and (2) the putative GABA agonists: muscimol (Tamminga *et al.*, 1979), THIP, (Korsgaard *et al.*, 1982b; Thaker *et al.*, submitted for publication), baclofen (Nair *et al.*, 1978, 1980; Gerlach, 1977; Gerlach *et al.*, 1978; Korsgaard, 1976; Stewart *et al.*, 1982; Simpson *et al.*, 1978; Amsterdam and Mendels, 1979; Feder and Moore, 1980; Wolf *et al.*, 1983), and progabide (Sevestre *et al.*, 1982).

Double- and single-blind studies using the GABA transaminase inhibitors have produced mixed, but generally favorable, results. Three double-blind studies of sodium valproate (Linnoila *et al.*, 1976; and Viukari, 1979; Chien *et al.*, 1978; Friis *et al.*, 1983) demonstrated significant improvement in dyskinetic symptoms, whereas two single-blind reports (Gibson, 1978; Nair *et al.*, 1980) showed no significant response. Nair *et al.* (1980) combined sodium valproate treatment with haloperidol at doses which, individually, were clinically ineffective and noted a marked improvement in dyskinetic symptoms. Since patients are inconsistently removed from neuroleptics during treatment for TD, these results suggest that disparate responses to sodium valproate may be explained on the basis of whether or not the patients have been maintained on neuroleptics. Besides the problems with study design, interpretation of results has been complicated because sodium valproate is a nonspecific weak inhibitor of GABA transaminase, and preclinical studies have demonstrated that enzyme inhibition is not homogeneous in all brain areas (Ferkany *et al.*, 1979; Perry and Hansen, 1978). Thus, at the doses used clinically, it may not sufficiently elevate brain levels of GABA to produce its potential therapeutic response.

Studies using increasing doses of γ-acetylenic GABA (75–225 mg/day) (Casey *et al.*, 1980a,b) demonstrated moderate, but significant, improvement in baseline hyperkinesia with maximum improvement occurring in patients on both high doses of γ-acetylenic GABA and neuroleptics. A double-blind study of the GABA agonist γ-vinyl GABA reported a clear antidyskinetic response in all patients treated with the drug (Tamminga *et al.*, 1983). Similarly, a single-blind, placebo-controlled study with γ-vinyl GABA (Tell *et al.*, 1981) demonstrated a 30–80% reduction in hyperkinesia in seven of nine patients. The two patients who did not improve had progressive senile dementia and actually experienced an aggravation of dyskinesia after treatment with γ-vinyl GABA. In both studies of γ-vinyl GABA most patients relapsed when they were returned to placebo.

Generally, these drugs have not been associated with either deterioration or improvement in the patients' mental status, but mild to moderate sedation was commonly observed with both γ-acteylenic GABA and γ-vinyl GABA. An increase in presenting extrapyramidal symptoms was observed with sodium valproate. In one report (Lautin *et al.*, 1979), extrapyramidal symptoms observed after sodium valproate administration to a neuroleptic-free subject were clinically identical to those observed in the same patient when treated with neuroleptics; however, the sodium-valproate-induced symptoms did not respond to anticholinergic therapy, increased with increasing drug dose, and resolved after discontinuance of drug treatment. Parkinsonian symptoms have not been observed after treatment with γ-vinyl GABA alone, but such treatment markedly exacerbates those secondary to neuroleptics.

Results with GABA agonists have been mixed and frequently limited by their side effects. Muscimol attenuated involuntary dyskinetic movements, but its therapeutic efficacy was counterbalanced by an aggravation of psychotic symptoms (Tamminga *et al.*, 1979). A study using tetrahydroisoxazolopyrindol (THIP) (Korsgaard *et al.*, 1982b), a GABA agonist that crosses the blood–brain barrier more readily than muscimol, demonstrated little effect on dyskinesia, and doses could not be increased because of prominent effects of sedation, dizziness, confusion, and vomiting. In a subsequent study, Thaker *et al.* (submitted for publication) observed about a 25% reduction in dyskinetic symptoms in two patients treated with THIP. In both patients, however, therapeutic efficacy was compromised by severe side effects. Neither patient had a previous history of seizures, but both developed grand mal seizures after withdrawal from THIP.

Preliminary studies with progabide, a structural analogue of GABA that is metabolized to GABA in the brain, are promising and have produced complete amelioration of symptoms in some patients (Sevestre *et al.*, 1982). Side effects include liver toxicity, nausea, vomiting, and insomnia in some patients.

Baclofen produced clear improvement in hyperkinesia and oral dyskinesia in two double-blind crossover studies (Korsgaard, 1976; Gerlach, 1977; Gerlach *et al.*, 1978) and moderate improvement over placebo in a third (Stewart *et al.*, 1982). One open study reported relatively specific efficacy for trunkal dyskinesia (Wolf *et al.*, 1983). Other studies with baclofen were less promising. Nair *et al.* (1978, 1980) observed significant improvement only when patients were treated concurrently with haloperidol, and yet another study found no improvement in dyskinesia after treatment of neuroleptic-free patients with baclofen for 7 weeks (Simpson *et al.*, 1978). As with other GABA-mimetic drugs, baclofen tended to aggravate parkinsonism in patients who remained on neuroleptics. Several groups have reported significant deterioration in mental status including increased irritability and auditory hallucinations following its use (Simpson *et al.*, 1978; Wolf *et al.*, 1983). Results from studies with baclofen are heuristically problematic because its mechanism of action remains unclear. Although it is frequently considered a GABAergic agonist, some investigators have suggested that it is actually an antagonist of Substance P and that any effect on the GABAergic system is indirect (Saito *et al.*, 1975). Still other investigators propose that its effects are best explained on the basis of its phenylethylaminelike properties (Wolf *et al.*, 1983).

3. BENZODIAZEPINES

Benzodiazepines were used early in the therapeutic management of TD with mixed results. Thereafter, interest in their use temporarily waned, but with an increased understanding of the central mechanism of action of benzodiazepines, attention to their clinical application has revived. Theoretically, the palliative effects of benzodiazepines on TD are closely linked to the GABAergic system. Recent work has suggested that they interact with the GABA receptor so as to enhance its affinity for GABA (Schmidt *et al.*, 1967; Massotti *et al.*, 1981). Presumably, then, the effect of benzodiazepines in TD would be to enhance the responsiveness of the GABAergic pathways in the brain so as to facilitate GABAergic inhibitory effects on dopaminergic systems. It should be recognized, however, that, as yet, there has not been any firm clinical evidence showing that benzodiazepines potentiate the response to GABAergic drugs. Another mechanism for clinical improvement is suggested by the observation that anxiety increases symptoms of TD. Thus, the antianxiety effects of these drugs may be helpful in reducing hyperkinesia in these patients.

Clinical studies with benzodiazepines in the treatment of TD are disappointing because most of this work has been uncontrolled (Kruse, 1960; Druckman *et al.*, 1962; Itil *et al.*, 1974; O'Flannagan, 1975; Jus *et al.*, 1974; Mehta *et al.*, 1976; Sedman, 1976; Singh, 1976; Sovner and Loadman, 1978; Rosenbaum and de la Fuente, 1979; Cole *et al.*, 1980). Those which have been controlled (Godwin-Austen and Clark, 1971; Bobruff *et al.*, 1981; Cutler, 1981; Singh *et al.*, 1982; Weber *et al.*, 1983) are encouraging, but equivocal, and suggest that more work with these drugs is merited. Singh *et al.* (1982) reported a greater response of limb dyskinesias to benzodiazepines, whereas Bobruff *et al.* (1981) reported a greater response of oral dyskinesias. One study (Bobruff *et al.*, 1981) suggests that the best response to benzodiazepines is observed in patients who are off or receiving low doses of neuroleptics. In most studies, sedation was a confounding side effect which nonspecifically influenced dyskinetic movements and was seldom controlled. In addition, confusion, ataxia, slurred speech, and behavioral changes are particularly prominent in elderly patients on these drugs and limit the usefulness of benzodiazepines as chronic therapeutic agents.

4. LITHIUM CARBONATE

A number of investigators have examined the therapeutic response of TD to lithium carbonate. The rationale for its use rests on its ability to diminish central catecholaminergic activity through several different mechanisms. Among these are (1) interference with the release of monoamines (Colburn *et al.*, 1967); (2) facilitation of the reuptake of catecholamines (Schildkraut *et al.*, 1966); and (3) desensitization of dopaminergic receptors (Ashcroft, 1972). Preclinical studies have further demonstrated that combining haloperidol and lithium carbonate treatment prevents the development of dopaminergic supersensitivity observed after treatment with haloperidol alone (Pert *et al.*, 1978). Thus, combining neuroleptics with lithium treatment may prevent the development of TD.

About half of the clinical studies we reviewed (Prange *et al.*, 1973; Dalen, 1973; Ehrensing, 1974; Reda *et al.*, 1975; Gerlach *et al.*, 1975; Simpson *et al.*, 1976; Crews and Carpenter, 1977; Jus *et al.*, 1978; Pickar and Davies, 1978; Beitman, 1978; Rosenbaum *et al.*, 1980; Mackay *et al.*, 1980; Yassa *et al.*, 1984) have been controlled. Among the controlled studies response to lithium is extremely heterogeneous both within and between studies, and no consistent change is apparent. As indicated previously, animal models suggest that lithium carbonate may be useful prophylactically in the prevention of TD. However, we are not aware of any published studies where this tissue has been seriously addressed.

5. OPIATES

Based on observations that the highest concentrations of enkephalins and enkephalin receptors are found in the caudate nucleus and globus pallidus (Kuhar *et al.*, 1973; Diamond and Borison, 1979), a single report (Bjorndal *et al.*, 1980) has examined the effects of this group of opiate agonists on the hyperkinetic symptoms of TD. Sixty minutes after 3-mg injections of metenkephalin (FK 33-824), a slight but significant decrease in symptoms was consistently observed. This effect lasted for less than 120 min and was most pronounced in those patients who were concurrently treated with neuroleptics. Morphine had a similar, but less pronounced, effect. Overall, the optimum response to both drugs was modest and limited by side effects which included increased parkinsonism, dizziness, ataxia, and sedation. On the other hand, little or no response has been observed in patients treated with both high and low doses of the opiate antagonist naloxone (Bjorndal *et al.*, 1980; Recker *et al.*, 1982).

Des-tyrosine-γ-endorphin, an endorphin opiate agonist, produces neurolepticlike behavioral responses in animal models (de Wied *et al.*, 1978) and is reported to produce antipsychotic effects when given to schizophrenic patients (Verhoeven *et al.*, 1978, 1979). In addition, destyrosine-γ-endorphin diminishes dopamine release in brain slices (Schoemaker and Nickolson, 1980) and interferes with tritiated spiperone binding *in vivo* (Pedigo *et al.*, 1979). Thus, both clinical and preclinical results suggest that endorphins may influence dopaminergic pathways. Three blind clinical studies with this drug (Casey *et al.*, 1981; Tamminga *et al.*, 1981; Korsgaard *et al.*, 1982d) have failed to demonstrate any consistent clinical improvement in TD, parkinsonism, eye blinking, or mental status after single daily doses ranging from 0.5 to 120 mg i.m.

In summary, results with both endorphins and enkephalins have been disappointing but in view of the limited amount of work that has been completed, larger studies with modifications in design are indicated. In particular, issues of whether clinically significant doses of medication are being administered and whether therapeutic efficacy is enhanced by prolonged administration of these drugs need to be addressed.

6. SEROTONERGIC DRUGS

Early clinical evaluations of drugs responsible for elevating brain serotonin are reported to have had a palliative effect on neuroleptic-induced dyskinesias

(Goldman, 1976). In addition, parkinsonian patients treated with tryptophan and pyridoxine have been reported to experience rapid clinical deterioration (Hall *et al.*, 1972; Prange *et al.*, 1973). Using the rationale that the pathophysiology of Parkinson's disease is opposite that of TD, a few more systematic studies using serotonergic precursors have been conducted. A preliminary study with *l*-tryptophan and pyridoxine (Prange *et al.*, 1973) demonstrated some initial improvement but later recurrence of symptoms during treatment. Two subsequent studies (Prange *et al.*, 1973; Jus *et al.*, 1974) failed to demonstrate any improvement after *l*-tryptophan. Similarly, a double-blind study investigating therapy with 5-hydroxytryptophan and carbidopa (Nasrallah *et al.*, 1982) found no improvement in dyskinetic symptoms and an exacerbation of psychotic symptoms in schizophrenic patients with TD.

Four uncontrolled studies have examined the response of TD to cyproheptadine, a drug with both antiserotonergic and antihistaminic effects. An initial report (Goldman, 1976) suggested significant clinical improvement in patients treated with cyproheptadine; a second report (Kurata *et al.*, 1977) found a mixed, but generally favorable, response, and subsequent studies (Gardos and Cole, 1978; Nagao *et al.*, 1979; Cole *et al.*, 1980) failed to find any significant antidyskinetic activity. One patient in these studies (Cole *et al.*, 1980) demonstrated a dramatic improvement of a "rabbit syndrome," a late-occurring extrapyramidal side effect of neuroleptics which is speculated to involve mechanisms physiologically opposite to those of TD.

7. PYRIDOXINE

The biochemical rationale for using pyridoxine in the treatment of TD seems paradoxical since it is a cofactor for the synthesis of dopamine from dopa. Consequently, its administration would be expected to increase synthesis of dopamine, secondarily activate central dopaminergic systems, and, by present theories, worsen the hyperkinetic symptoms of TD. However, a report that pyridoxine countered the effectiveness of L-dopa in parkinsonian patients and dramatically reversed an L-dopa-induced movement disorder (Yahr *et al.*, 1969) led to its clinical application in patients with TD. Although these studies (Crane *et al.*, 1970; Dynes, 1970; Prange *et al.*, 1973) have not produced consistent responses, another controlled study is encouraging. DeVeaugh-Geiss and Manion (1978) used high doses of pyridoxine (100–1400 mg/day) to treat TD and found that the frequency and severity of involuntary dyskinetic movements were reduced and that these symptoms remitted when the medication was discontinued.

8. ESTROGEN

Three open studies from a single group have reported on the efficacy of estrogens in the treatment of TD (Bedard *et al.*, 1977; Villeneuve *et al.*, 1978,

1980). These were stimulated in part by preclinical observations that hormones like estrogen influence synaptic transmission of catecholamines (Labrie *et al.*, 1978; Raymond *et al.*, 1978). Each of these studies found significant clinical improvement in patients with either TD or L-dopa-induced dyskinesias without aggravating either the patients' mental status or neuroleptic-induced parkinsonism. Recently, Glazer *et al.* (1983) examined the relationship between plasma estradiol and TD in men and postmenopausal women. Although they found no correlation between plasma concentrations of estradiol and severity of dyskinetic movements, a multiple regression analysis showed statistically significant correlations between AIMS scores and estradiol levels in women.

9. MISCELLANEOUS TREATMENTS

In the search for more effective therapies for TD, a number of other treatments have been evaluated. Often the theoretical rationale for these treatments has been weak, but these drugs have usually been of some benefit in the treatment of other movement disorders. Three groups have reported on results with phenobarbital (Lipsius, 1977; Sovner and Loadman, 1978; Bobruff *et al.*, 1981). In the largest and best controlled of these studies (Bobruff *et al.*, 1981), phenobarbital was used as an active placebo to control for the sedative effects of clonazepam. In this study, phenobarbital produced a moderate, but significant, reduction in dyskinetic movements which was particularly pronounced for limb and axial movements. As expected, sedation was a prominent side effect of phenobarbital treatment, but it did not appear to account for all the improvement observed.

Opposite results have been reported after the use of diphenylhydantoin either intravenously or orally in patients with TD. Jus *et al.* (1974) observed improvement in over 50% of patients receiving intravenous doses of diphenylhydantoin which produced neither clinical nor polygraphic evidence of sleep. On the other hand DeVeaugh-Geiss (1978) reported that long-term oral administration of diphenylhydantoin exacerbated dyskinetic symptoms in a patient with TD.

Spectrographic analysis of samples of hair from patients with TD has suggested that they may have inadequate stores of manganese. When dyskinetic patients were treated with the mineral supplement manganese chelate, most showed improvement (Kunin, 1976), but it was unclear from this report whether the dyskinesias examined represented true TD or withdrawal dyskinesias. A second open study (Norris and Sams, 1977) also reported favorable results after manganese chelate treatment.

Symptoms of TD and L-dopa-induced dyskinesia have been reported to respond to treatment with prolyl-leucyl-glycinamide (melanocyte-stimulating hormone, release-inhibiting factor I) (Ehrensing, 1974). However, in a followup study of 13 patients with TD, initial improvement was followed by relapse during continued treatment (Ehrensing *et al.*, 1977).

TABLE 1
Summary of Clinical Studies of Tardive Dyskinesia[a]

Reference	No. subjects	Experimental design	Daily dose	Length of treatment	Duration of symptoms	Concurrent medicines	Comments
GABAergic drugs							
Sodium valproate							
Linnoila *et al.* (1976, 1979)	32	Double-blind crossover	900 mg	2 wk	Greater than 1 yr	Neuroleptics	Decreased oral dyskinesia in 17 patients; dystonias of extremities relieved in 7 of 9 patients; no correlation between serum concentration and response
Chien *et al.* (1978)	7	Double-blind crossover with deanol and oxypertine	1200–1600 mg	3 wk	NG	NG	Symptoms in 3 patients minimally improved; more effective than deanol; less effective than oxypertine
Ames and Webber (1984)	1	Open	400–800 mg	5 wk	1.5 yr	None	Symptoms markedly improved
Friis *et al.* (1983)	8	Double-blind crossover with placebo and biperiden	900–2700 mg	4 wk	NG	Neuroleptics	Symptoms significantly improved as compared to improvement after placebo and biperiden
Nair *et al.* (1980)	10	Single-blind	Up to 1400 mg	3 wk	NG	Neuroleptic-free	No improvement with sodium valproate alone; marked improvement with combination of sodium valproate and haloperidol (significantly better than with haloperidol alone)
	10	Single-blind	1400 mg	3 wk	NG	Haloperidol	
Gibson (1978)	25	Blind raters	600 mg	4 wk	1–6 yr	Neuroleptics	Improvement in symptoms of 8 patients; 17 patients no change
Lautin *et al.* (1979)	1	Open	1000–2000 mg	NG	NG	None	Dyskinetic symptoms emerged during treatment; not controlled by benzotropine or trihexyphenidyl; tremor cleared with discontinuation of sodium valproate
Crowe (1983)	8	Open	400–1600 mg	2 wk or longer	NG	Neuroleptics	Symptoms completely alleviated in all but 1 patient
Casey and Hammerstad (1979)	1	Open	900–3000 mg	9 wk with 6–wk washout	Greater than 2 yr	None	Symptoms improved on 3 scales including polygraph; no correlation of improvement with blood levels; partial relapse off medicines; most improved at doses greater than 2100 mg
Nagao *et al.* (1979)	7	Open	400–600 mg	3 wk	Greater than 1 yr	Neuroleptics	3 patients improved overall, but not statistically significant.
γ-Acetylenic GABA							
Casey *et al.* (1980a,b)	10	Single-blind	75, 105, 150, 225 mg	2 wk at each dose	Greater than 0.5 yr	Neuroleptics continued in 9 patients	9 patients improved (30–70%); greatest response occurred in patients maintained on neuroleptics; symptoms recurred off medications; limited by side effects

(cont.)

TABLE 1 (*continued*)

Reference	No. subjects	Experimental design	Daily dose	Length of treatment	Duration of symptoms	Concurrent medicines	Comments
γ-Vinyl GABA							
Tell *et al.* (1981)	9	Single-blind	2000–6000 mg	More than 2 wk	NG	Neuroleptics, constant	Hyperkinesia was reduced 30–80% in 7 patients; in 2 "organic" patients it was increased; 5 patients relapsed on placebo
Tamminga *et al.* (1983)	7	Double-blind crossover	3000 mg	3 wk	NG	Neuroleptic-free 4 wk	Symptoms improved in all patients; average 44% improvement; 148% increase in CSF GABA; effects lasted 1–2 wk after stopping GVG
Thaker *et al.* (1983)	2	Open	3000 mg	3 wk	NG	Neuroleptic-free 4 wk	Moderate to severe symptoms improved
Baclofen							
Nair *et al.* (1978, 1980)	10	Double-blind crossover	30–90 mg	3 wk	NG	Drug-free greater than 3 mo	Significant symptom improvement only when treatment combined with haloperidol (5 mg)
	10	Single-blind	90 mg	2 wk	NG	Haloperidol	
Gerlach *et al.* (1978) and Gerlach (1977)	18	Double-blind crossover	20–120 mg	3 wk	0.5–9 yr	Neuroleptics 16 patients	Symptoms improved in 13 patients; increased pseudoparkinsonism in 13 patients; symptoms recurred 1–2 days after drug withdrawal
Korsgaard (1976)	20	Double-blind crossover	15–60 mg	2 wk with 1-wk washout	1.5–5 yr	Neuroleptics	Symptoms improved in 15 patients; no patients on placebo improved; relapsed off baclofen
Stewart *et al.* (1982)	33	Double-blind placebo controlled		6 wk		Neuroleptics, constant	Greater than 25% symptom improvement in 67% of baclofen-treated patients; greater than 25% improvement in 47% of placebo-treated patients
Simpson *et al.* (1978)	12	Single-blind	20–120 mg	7 wk with 1-wk washout	NG	Neuroleptic-free 4 wk	Symptoms worsened in 10 patients; symptoms moderately improved in 2 patients
Amsterdam and Mendels (1979)	1	Open	60 mg	8 wk	3 yr	Deanol	Oral dyskinesia improved with combination of deanol and baclofen; most improvement after addition of baclofen
Feder and Moore (1980)	1	Open	60 mg	4 wk	3 mos	NG	Symptoms improved with medication and relapsed when medication discontinued; remitted again with medication, then did not recur off medication
Wolf *et al.* (1983)	8	Open	15–90 mg	4–8 wk	NG	Lithium, valium, dilantin, chlorpromazine, propranolol, phenobarbitol	Trunkal dyskinesia improved in 3 patients; drug exacerbated psychiatric condition and parkinsonian features

Muscimol							
Tamminga *et al.* (1979)	7	Double-blind	5–10 mg	Up to 5 days	NG	Neuroleptic-free 5 days–3 mo pretrial	Symptoms transiently improved in all patients (maximum improvement in 2 hr with 48% mean reduction); 2 patients had increased parkinsonism; higher doses accompanied by increased psychosis score
THIP							
Korsgaard *et al.* (1982)	5	Open	10–25 mg single dose	Once	0.5–5 yr	Neuroleptics and anticholinergics, constant	Symptoms unchanged by treatment; modest but significant improvement in parkinsonism; eye blinking decreased. Side effects included: sedation, confusion, dizziness, vomiting, and myoclonic jerks. No significant change in mental status
	13	Blind raters	20–120 mg	4 wk	0.5–11 yr	Neuroleptics and anticholinergics, constant	
Thaker *et al.* (submitted for publication)	2	Double-blind placebo controlled	Up to 120 mg	4 wk	Greater than 2 yr	None	Symptoms 25% reduced in both patients as compared with placebo treatment; withdrawal accompanied by seizures in both patients
Progabide							
Sevestre *et al.* (1982)	6	Open	14–35 mg/kg	6 wk	NG	Neuroleptics	Symptoms of all patients improved; 4 experienced complete remission of symptoms
	10	Double-blind placebo controlled	900–1200 mg	6 wk	NG	Neuroleptics	5 patients demonstrated greater improvement with progabide than placebo
Benzodiazepines							
Chlordiazepoxide							
Kruse (1960)	2	Open	40–75 mg	NG	NG	Neuroleptic-free	Symptoms improved
Druckman *et al.* (1962)	1	Open	NG	NG	4 mo	Neuroleptic-free	Symptoms not improved
Chlorazepate							
Itil *et al.* (1974)	12	Open	15–45 mg	6 wk	NG	Neuroleptic-free, 2 wk	7 patients improved as compared to drug-free period; psychiatric symptoms controlled
Mehta *et al.* (1976)	1	Open	15 mg	6 mo	4 yr	Neuroleptic-free, 3 yr	Symptoms improved; relapsed off medicine
Sovner and Loadman (1978)	1	Open	45 mg	2 wk	NG	Neuroleptic-free	No symptom improvement with chlorazepate or phenobarbital
Cole *et al.* (1980)	1	Open	NG	NG	NG	NG	No improvement
Clonazepam							
Bobruff *et al.* (1981)	21	Double-blind with phenobarbital; parallel groups	2–10 mg	2 wk	NG	Some patients on neuroleptics	Oral dyskinesia more improved with clonazepam than phenobarbital; especially effective in drug-free patients
	7	Open	2–10 mg	2 wk	NG	Some patients on neuroleptics	More effective than phenobarbital in drug-free patients

(cont.)

TABLE 1 (continued)

Reference	No. subjects	Experimental design	Daily dose	Length of treatment	Duration of symptoms	Concurrent medicines	Comments
Cutler (1981)	1	Blind raters	1.5–10 mg	4 mo	NG	Neuroleptics, nortriptylene	Initially greater than 50% improvement; but relapsed after 4 mo
O'Flanagan (1975)	42	Open	1–3 mg	NG	NG	NG	All patients' symptoms improved; diagnosis of TD questionable in some cases; without significant side effects
Sedman (1976)	18	Open	1–4 mg	Greater than 1 wk	NG	Neuroleptics	Symptoms markedly improved in 2 patients; 9 slightly improved; many side effects including drowsiness
Cole *et al.* (1980)	6	Open	1.5 mg	NG	NG	NG	Symptoms moderately improved in all patients
Diazepam							
Godwin-Austen (1971)	6	Double-blind with tetrabenazine and placebo	4 mg	1 wk	6–10 yr	NG	Symptoms improved in 5 patients; better than placebo, but not different from tetrabenazine
Singh *et al.* (1982)	21	Blind raters	5–40 mg	1–24 wk	Greater than 6 mo	Neuroleptics	Significant symptomatic improvement; most pronounced for limb movements
Weber *et al.* (1983)	13	Double-blind crossover	6–25 mg	10 wk	2–6 yr	Neuroleptics, 10 patients; anticholinergics, 8 patients; amantadine, 1 patient	Symptoms improved equally after both placebo and diazepam; no deterioration of mental status
Jus *et al.* (1974)	14	NG	10 mg i.v.	Once i.v.	3 yr (mean)	NG	Symptoms immediately improved in 13 patients, eventually to complete abolition of symptoms; improvement possibly related to somnulence
Singh (1976)	6	Open	10 mg i.v.	Once i.v.	3 yr	NG	Symptoms improved in 5 patients
	3	Open	4–30 mg	3–26 wk	Greater than 8 mo	Neuroleptics	Symptoms improved in patients; not related to sedation; behavior problems in 1 patient
Rosenbaum and de la Fuente (1979)	6	Open	Varied	Several wk	Brief	Neuroleptics, 5 patients; antidepressants, 1 patient	TD-like symptoms associated with benzodiazepine treatment; 2 patients improved off benzodiazepine
Lithium carbonate							
Mackey *et al.* (1980)	11	Double-blind crossover	(Blood level 0.8–1.4 meq/liter)	5 wk with 6-wk washout	Greater than 1 yr	Neuroleptics, 7 patients	No significant symptomatic improvement; 5 patients developed pseudoparkinsonian features
Jus *et al.* (1978)	29	Double-blind crossover with deanol and placebo	200–500 mg (blood level 0.7–0.9 meq/liter)	8 wk each with 1-wk washout	3–17 yr	Neuroleptics; antiparkinsonians	Symptoms improved in 5 patients, worsened in 2 patients; 6 patients dropped out (3 patients improved on deanol; 7 patients improved on placebo)

Prange *et al.* (1973)	2	Double-blind	NG	NG	NG	NG	Symptoms improved in both patients; relapsed off lithium; improvement dose related
Gerlach *et al.* (1975)	5	Double-blind cross-over	12–24 meq/day (blood level 0.8–1.2 meq/liter)	3 wk	0.25–5 yr	Neuroleptics	30–100% symptomatic improvement in 5 patients
Simpson *et al.* (1976)	10	Single-blind	300–1800 mg (blood level 0.6–1.2 meq/liter)	12 wk	NG	Neuroleptics	No significant symptomatic improvement in any patient
	10	Double-blind cross-over	300–1200 mg (blood level 0.6–1.0 meq/liter)	6 wk, no washout	Chronic	Neuroleptic-free 4 wk	No significant symptomatic improvement in any patient
Rosenbaum *et al.* (1980)	19	Blind raters	225–1200 mg (blood level 0.4–1.0 meq/liter)	4–8 wk	0.2–2 yr	Drug-free 2–52 wk	Moderate to marked symptomatic improvement in 11 patients; symptoms worsened in 1 patient; no improvement in 7 patients; depression also improved
Crews and Carpenter (1977)	1	Open	1200 mg (blood level 2.02 meq/liter)	12 days	5 mo	Neuroleptic-free 1 wk	Symptoms aggravated at toxic doses of lithium
Pickar and Davies (1978)	1	Open	(Blood level 0.4–1.0 meq/liter)	NG	NG	NG	Little symptomatic improvement and, then, only at low blood levels of lithium
Beitman (1978)	1	Open	900 mg (blood level 0.9 meq/liter)	9 mo	6 mo	Neuroleptic-free 2 yr	Oral dyskinesia appeared 6 mo after treatment with lithium; prior history of 4-yr treatment with thioridazine; symptoms improved with trihexaphenidyl
Yassa *et al.* (1984)	6	Open	900–1800 mg	16 mo	At least 1 yr	Haloperidol, 1 patient; neuroleptic-free, 5 patients	Symptoms not improved in any patient; worsened in 1 patient
Ehrensing (1974)	1	Open	900–1200 mg (blood level 1–1.1 meq/liter)	52 wk	1 yr	Doxepin	Symptoms improved; symptoms previously improved with prolyl-leucyl-glycinamide
Reda *et al.* (1974, 1975)	6	Open	900 mg (blood level 0.7–1.0 meq/liter)	2 wk (twice)	NG	Neuroleptics, 5 patients	Slight symptomatic improvement
Dalen (1973)	1	Open	(Blood level 0.7–1.5 meq/liter)	NG	NG	Amytriptyline	Symptoms improved
Opiates							
Metenkephalin (FK-33-824)							
Bjorndal *et al.* (1980)	8	Blind ratters	1, 2, 3 mg i.m.	1 time each drug with 7-day wash-out	3–18 yr	Neuroleptics, 6 patients; antiparkinsonians, 4 patients	Slight symptomatic improvement with concomitant increase in bradykinesia in 6 patients continued on neuroleptics; slight increase in parkinsonism in 4 patients; side effects include dizziness, slurred speech, dry mouth, and flushing of face

(cont.)

TABLE 1 (*continued*)

Reference	No. subjects	Experimental design	Daily dose	Length of treatment	Duration of symptoms	Concurrent medicines	Comments
Morphine							
Bjorndal *et al.* (1980)	8	Blind raters	10 mg s.c.	1 time with 7-day washout	3–18 yr	Neuroleptics, 6 patients; antiparkinsonians, 4 patients	Symptoms improved in 4 patients; effect less pronounced that that of metenkephalin; parkinsonism increased in 3 patients with improvement in hyperkinesia
Naloxone							
Bjorndal *et al.* (1980)	8	Blind raters	0.8 mg i.m.	1 time with 7-day washout	3–18 yr	Neuroleptics, 6 patients; antiparkinsonians, 4 patients	No consistent symptomatic response
Recker *et al.* (1982) and Volavka *et al.* (1982)	13	Double-blind placebo controlled	10 mg i.v.	2 blind trials	NG	NG	No symptomatic improvement with placebo or naloxone within 6 hr of administration; plasma β-endorphins significantly elevated by naloxone
Des-tyrosine-γ-endorphin							
Casey *et al.* (1981)	10	Double-blind placebo controlled	0.5, 1.0, 2.0, 4.0, 8.0, 12.0, 16.0 mg, i.m.	1 time each dose with 1–5 days between doses	0.5–10 yr	Neuroleptics, 7 patients	No significant symptomatic response at any dose; no effects on mental status
Tamminga *et al.* (1981)	2	Double-blind placebo controlled	0.5–2.0 mg i.m. (8.5 mg total)	Daily for 8 days	NG	Neuroleptic-free 8–30 days	No significant symptomatic response at any dose; no effect on mental status
Korsgaard *et al.* (1982)	4	Blind raters	20–120 mg i.m.	1 time each dose, with 7 days between doses	1–11 yr, stable 6 mo	Neuroleptics	No consistent symptomatic improvement in TD, parkinsonism, eye blinking, or mental status from 3 to 120 min
Serotonergic drugs							
L-Tryptophan							
Prange *et al.* (1973)	4	Single-blind	Up to 6000 mg	4–8 wk	NG	Pyridoxine	Symptoms improved in all patients on first day; symptoms recurred over time
	3	Double-blind	6000 mg	4 wk	NG	No pyridoxine	Symptoms no more improved by tryptophan than placebo; slight symptomatic improvement with time
Jus *et al.* (1974)	8	Open	120 mg/kg, i.v. (D,L-tryptophan)	1 time	3 yr	NG	No symptomatic improvement in any patients
5HTP + Carbidopa							
Nasrallah *et al.* (1982)	7	Double-blind crossover	200–400 mg 5-HTP; 50–100 mg carbidopa	4 wk	Greater than 2 yr	Neuroleptics, stable for 1 mo	No significant symptomatic improvement; psychosis worsened
Cyproheptadine							
Goldman (1976)	3	Open	4–8 mg	6–12 mo	NG	Neuroleptics	Symptoms improved in all patients
Gardos and Cole (1978) and Cole *et al.* (1980)	5	Open	8 mg	6 wk, 4 patients; 3 wk, 1 patient	NG	Neuroleptics, 3 patients	No consistent symptomatic improvement; drug had antiparkinsonian effects

Nagao *et al.* (1979)	5	Open	8–16 mg	3 wk	At least 1 yr	Neuroleptics	No significant symptomatic improvement; symptoms improved in 1 patient; worsened in 1 patient
Kurata *et al.* (1977)	11	Open	8–24 mg	8 wk	Several yr	Neuroleptics, anticholinergics	Complete symptomatic improvement in 4 patients; partial symptomatic improvement in 4 patients; no symptomatic improvement in 3 patients; no side effects
Pyridoxine							
DeVeaugh-Geiss and Manion (1978)	5	Blind raters	1000–1400 mg	12 wk	2–14 mo	Drug-free, 3 mo	Some symptomatic improvement in 4 patients; relapsed off medications
Prange *et al.* (1973)	4	Single-blind	5 mg	4–8 wk	NG	L-Tryptophan (6000 mg)	Marked symptomatic improvement in 1 patient; moderate improvement in 3 patients; no improvement on tryptophan alone
Dynes (1970)	22	Open	50–200 mg	2 wk	NG	Neuroleptics, 12 patients	No symptomatic improvement in any patient; all patients had brain damage
Crane *et al.* (1970)	9	Open	300 mg	2 wk	NG	Stopped during 2nd wk	Improved in 1 patient
	2	Open	300 mg	10 days	1 yr	Neuroleptics, 1 patient	No symptomatic improvement in either patient
Estrogens							
Villeneuve *et al.* (1978)	20	Open	1.25–2.50 mg premarin	6 wk	NG	Neuroleptics, 19 patients	Oral dyskinesia improved in 16 patients; symptoms stable for completion of study
Bedard *et al.* (1977)	1	Open	0.625 mg premarin	3 wk	NG	L-Dopa, ethopropazine	Symptoms improved markedly and repeatedly when on estrogens, worsened off estrogens; estrogens aggravated parkinsonism
Villeneuve *et al.* (1980)	1	Open	0.624 mg premarin	NG	2 yr	Reserpine; deanol; norethindrone	Symptoms aggravated by progesterone and by return of menstruation; estrogens improved dyskinesia
Miscellaneous treatments							
Phenobarbitol							
Bobruff *et al.* (1981)	21	Double-blind with clonazepam	30–150 mg (11 patients)	2 wk	NG	Neuroleptics, some patients	More effective than clonazepam for limb and axial dyskinesias
Lipsius (197)	1	Open	120 mg	4 mo	NG	Neuroleptic-free 1 wk	Symptoms improved after 1 wk
Sovner and Loadman (1978)	1	Open	200–360 mg	10 wk	NG	Dilantin	No symptomatic improvement
Diphenylhydantoin							
Jus *et al.* (1974)	14	NG	100 mg i.v.	1 time	NG	NG	Symptoms improved in 8 patients by clinical and polygraph assessment; lasted 15–20 min
DeVeaugh-Geiss (1978)	1	Open	300 mg	3.5 yr, then stopped	2 yr	None	Symptoms improved after discontinuing diphenylhydantoin; resumption of diphenylhydantoin treatment increased symptoms

(cont.)

TABLE 1 (*continued*)

Reference	No. subjects	Experimental design	Daily dose	Length of treatment	Duration of symptoms	Concurrent medicines	Comments
Manganese chelate							
Kunin (1976)	15	Open	15–60 mg	NG	NG	NG	Some symptoms improved in 14 patients; symptoms alleviated within first 24 hr in 7 patients
Norris and Sams (1977)	6	Open	20 mg	NG	NG	NG	Symptoms improved in all patients within 1 day; symptoms relapsed in 2 patients when therapy was discontinued and improved when it was reinstated
Propyl-leucyl-glycinamide							
Ehrensing (1974)	1	Open	150 mg	6 days	1 yr	NG	Symptoms improved and relapsed on discontinuing treatment
Ehrensing *et al.* (1977)	13	Open	50–250 mg	7 wk	Greater than 1 yr	Neuroleptics, some patients	Symptoms improved during first 2 wk, but not significant at 7 wk
Calcitonin							
Nicoletti *et al.* (1983)	8	Controlled	200 U.I.	15 days	NG	Neuroleptics	Symptoms improved significantly during calcitonin treatment; not during placebo
Electroconvulsive therapy							
Asnis and Leopold (1978)	4	Blind raters	—	6–13-wk course ECT; 3 times/wk	Less than 6 mo	Drug-free 3 wks	Negligible changes in symptoms
Uhrbrand and Faurbye (1960)	4	Open	—	20–192 courses of ECT	NG	Neuroleptics, reserpine	Onset of dyskinesias associated with ECT in all patients
Price and Levin (1978)	1	Open	—	7 courses ECT over 2 wk	More than 4 mo	Neuroleptics	Symptoms of TD and depression improved for more than 3 mo after ECT
Hemodialysis							
Callen *et al.* (1982)	1	Double-blind		2 sham hemodialyses and 2 hemodialyses over a period of 1.5 wk	2–3 yr	Neuroleptic-free	Symptoms minimally improved for less than 72 hr after sham procedure; symptom-free for 2–3 wk after actual dialysis

[a] NG, not given.

Preliminary studies with calcitonin suggest that its usefulness as an antidyskinetic drug merits further investigation (Nicoletti *et al.*, 1983).

Several groups have examined the effect of electoconvulsive therapy (ECT) on TD. An early report (Uhrbrand and Faurbye, 1960) suggested that the onset of TD was associated with ECT treatment. Later, a single case report (Price and Levin, 1978) indicated that symptoms of TD improved after ECT, but a larger study using blind raters (Asnis and Leopold, 1978) failed to find any antidyskinetic efficacy. Finally, in a double-blind study of the effects of hemodialysis on TD Callen *et al.* (1982) reported that symptoms of TD were markedly improved after actual hemodialysis but not after sham hemodialysis (Table 1).

10. SUMMARY

Of the treatments for TD evaluated in this overview, moderate responses have been most consistently observed after using GABAergic drugs and benzodiazepines. However, many of the studies are limited by the fact that they were poorly controlled, involved small populations, and employed inadequate statistical analysis. In addition, seldom is any attention given to keeping patients drug-free or to carefully differentiating withdrawal dyskinesia from tardive dyskinesia. Since withdrawal dyskinesias would be expected to improve in the presence or absence of treatment, this distinction is critical to the evaluation of any potential therapeutic regimen.

Finally, none of these studies address the important conceptual distinction between treatments which modulate the intensity of symptoms of TD and those which actually alter the underlying disease process. Although difficult to evaluate without expensive and time-consuming prospective studies, this is a central issue in the evaluation of any treatment for TD.

11. REFERENCES

Alphs, L. D., and Davis, J. M., 1983, Cholinergic treatments for tardive dyskinesia, in: *Modern Problems of Pharmacopsychiatry* (J. Bannet and R. H. Belmaker, eds.), Vol. 21, pp. 168–186, S. Karger AG, Basel.

Ames, D., and Webber, J., 1984, Sodium valproate and tardive dyskinesia, *Med. J. Aust.* **140:**350.

Amsterdam, J., and Mendels, J., 1979, Treatment resistant tardive dyskinesia, a new therapeutic approach, *Am. J. Psychiatry* **136:**1197–1198.

Anden, N. E., and Wachtel, H., 1977, Biochemical effects of baclofen: (Beta-paraclorophenyl-GABA) on the dopamine and the noradrenaline in the rat brain, *Acta. Pharmacol. Toxicol.* **40:**310–320.

Ashcroft, G. W., 1972, Modified amine hypothesis for the etiology of affective illness, *Lancet* **1:**573–577.

Asnis, G. M., and Leopold, M. A., 1978, A single-blind study of ECT in patients with tardive dyskinesia, *Am. J. Psychiatry* **135:**1235–1237.

Bedard, P., Langelier, P., and Villeneuve, A., 1977, Oestrogen and extrapyramidal system, *Lancet* **2:**1367–1368.

Beitman, B. D., 1978, Tardive dyskinesia reinduced by lithium carbonate, *Am. J. Psychiatry* **135:**1229–1230.

Bjorndal, N., Casey, D. E., and Gerlach, J., 1980, Enkephalin, morphine and naloxone in tardive dyskinesia, *Psychopharmacology* **69:**133–136.

Bobruff, A., Gardos, G., Tarsy, D., Rapkin, R. M., Cole, J. O., and Moore, P., 1981, Clonazepam and phenobarbital in tardive dyskinesia, *Am. J. Psychiatry* **138:**189–193.

Callen, K. E., Malek-Ahmadi, P., Davis, D., Davis, L. G., and Sorkin, M. I., 1982, The effect of hemodialysis on tardive dyskinesia, *Psychosomatics* **23:**869–870.

Carlsson, A., 1970, Biochemical implications of dopa-induced actions on the central nervous system, with particular relevance to abnormal movements, in: *L-Dopa and Parkinsonism* (A. Barbeau, and F. H. McDowell, eds.), pp. 205–213, Davis, Philadelphia.

Casey, D. E., and Hammerstad, J. P., 1979, Sodium valproate in tardive dyskinesia, *J. Clin. Psychiatry* **40:**483–485.

Casey, D. E., Gerlach, J., Magelund, G., and Christensen, T. R., 1980a, Gamma-acetylenic GABA in tardive dyskinesia, *Arch. Gen. Psychiatry* **37:**1376–1379.

Casey, D. E., Gerlach, J., Magelund, G., and Christensen, T. R., 1980b, Gamma-acetylenic GABA in tardive dyskinesia, *Adv. Biochem. Psychopharmacol.* **24:**577–580.

Casey, D. E., Korsgaard, S., Gerlach, J., Jorgensen, A., and Simmelsgaard, H., 1981, Effect of des-tyrosine-gamma-endorphin in tardive dyskinesia, *Arch. Gen. Psychiatry* **38:**158–160.

Chien, C., Jung, K., and Ross-Townsend, A., 1978, Efficacies of agents related to GABA, dopamine and acetylcholine in the treatment of tardive dyskinesia, *Psychopharmacol Bull.* **14:**20–22.

Colburn, R. W., Goodwin, E. R., Bunney, W. E., and David, J. M., 1967, Effect of lithium on the uptake of noradrenaline by synaptosomes, *Nature* **215:**1394–1397.

Cole, J. O., Gardos, G., Tarsy, D., Granacher, R. P., Sniffin, C., Vanderkolk, B., and Trenholm, I., 1980, Drug trials in persistent dyskinesia, in: *Tardive Dyskinesia—Clinical and Basic Research* (R. Smith, W. Frann, J. Davis, and E. Domino, eds.), pp. 419–428, Spectrum Books, New York.

Crane, G. E., Turek, I. S., and Kurland, A. A., 1970, Failure of pyridoxine to reduce drug-induced dyskinesia, *J. Neurol. Neurosurg. Psychiatry* **33:**511–512.

Crews, E. L., and Carpenter, A. E., 1977, Lithium-induced aggravation of tardive dyskinesia, *Am. J. Psychiatry* **134:**933.

Crowe, B. M., 1983, Symptom control in tardive dyskinesia, *Br. J. Psychiatry* **143:**419–420.

Curtis, D. R., 1978, GABAergic transmission in the mammalian central nervous system, in: *GABA-Neurotransmitters* (P. Krogsgaard-Larsen, J. Scheel-Kruger and H. Kofod, eds.), pp. 17–27, Academic Press, New York.

Cutler, N. R., 1981, Clonazepam and tardive dyskinesa, *Am. J. Psychiatry* **138:**1127–1128.

Dalen, P., 1973, Lithium therapy in Huntington's chorea and tardive dyskinesia, *Lancet* **1:**107–108.

DeVeaugh-Geiss, J., 1978, Aggravation of tardive dyskinesia by phenytoin, *N. Engl. J. Med.* **298:**457–458.

DeVeaugh-Geiss, J., and Manion, L., 1978, High dose pyridoxine in tardive dyskinesia, *J. Clin. Psychiatry* **39:**573–575.

de Wied, D., Kovacs, G. L., Bohus, B., van Ree, J. M., and Greven, H. M., 1978, Neuroleptic activity of the neuropeptide beta-LPH_{62-77} ((Des-tyr^1)-gamma-endorphin; DγTE), *Eur. J. Pharmacol.* **49:**427–438.

Diamond, B. I., and Borison, R. L., 1979, Enkephalins and nigro-striatal function, *Neurology* **28:**1085–1088.

Druckman, R., Seelinger, D., and Thulin, B., 1962, Chronic involuntary movements induced by phenothiazines, *J. Nerv. Ment. Dis.* **135:**69–75.

Dynes, J. B., 1970, Oral dyskinesia—Occurrence and treatment, *Dis. Nerv. Syst.* **31:**854–859.

Ehrensing, R. H., 1974, Lithium and M. R. I. H. in tardive dyskinesia, *Lancet* **2:**1459–1460.

Ehrensing, R. H., Kastin, A. J., Larsons, P. F., and Bishop, G. A., 1977, Melanocyte-stimulating-hormone release-inhibiting factor-I and tardive dyskinesia, *Dis. Nerv. Syst.* **38**:303–306.

Feder, R., and Moore, D. C., 1980, Baclofen and tardive dyskinesia, *Am. J. Psychiatry* **137**:633–634.

Ferkany, J., Butler, I., and Enna, S., 1979, Effects of drugs on rat brain cerebrospinal fluid and blood GABA content, *J. Neurochem.* **33**:29–33.

Friis, T., Rosted Christensen, T., and Gerlach, J., 1983, Sodium valproate and biperiden in neuroleptic-induced akathesia, parkinsonism, and hyperkinesia, *Acta Psychiat. Scand.* **67**:178–187.

Fuxe, K., Hokfelt, T., Ljungdahl, A., Agnati, L., Johansson, O., and de la Mora, M., 1975, Evidence for an inhibitory GABAergic control of the mesolimbic dopamine neurons: Possibility of improving treatment of schizophrenia by combined treatment with neuroleptics and GABAergic drugs, *Med. Biol.* **53**:177–183.

Gardos, G., and Cole, J. O., 1978, Pilot study of cyproheptadine (Periactin) in tardive dyskinesia, *Psychopharmacol. Bull.* **14**:18–20.

Gerlach, J., 1977, The relationship between parkinsonism and tardive dyskinesia, *Am. J. Psychiatry* **134**:781.

Gerlach, J., Reisby, N., and Randrup, A., 1974, Dopaminergic hypersensitivity and cholinergic hypofunction in the pathophysiology of tardive dyskinesia, *Psychopharmacologia* **34**:21–35.

Gerlach, J., Thorsen, K., and Munkvad, I., 1975, Effect of lithium on neuroleptic-induced tardive dyskinesia compared with placebo in a double-blind crossover trial, *Pharmakopsychiatr. Neuropsychopharmakol.* **8**:51–56.

Gerlach, J., Rye, T., and Kristjansen, P., 1978, Effect of baclofen on tardive dyskinesia, *Psychopharmacology* **56**:145–151.

Gibson, A. C., 1978, Sodium valproate and tardive dyskinesia, *Br. J. Psychiatry* **133**:82.

Glazer, W. M., Naftolin, F., Moore, D. C., Bowers, M. B., and MacLusky, N. J., 1983, The relationship of circulating estradiol to tardive dyskinesia in men and post menopausal women, *Psychoneuroendocrinology* **8**:429–434.

Godwin-Austen, R. B., and Clark, T., 1971, Persistent phenothiazine dyskinesia treated with tetrabenazine, *Br. Med. J.* **4**:25–26.

Goldman, D., 1976, Treatment of phenothiazine-induced dyskinesia, *Psychopharmacology* **47**:271–272.

Hall, C. D., Weiss, E. A., Morris, C. E., and Prange, A. J., 1972, Rapid deterioration of patients with parkinsonism following tryptophan-pyridoxine administration, *Neurology* **22**:232.

Itil, T. M., Unverdi, C., and Mehta, D., 1974, Chlorazepate dipotassium in tardive dyskinesia, *Am. J. Psychiatry* **131**:1291.

Jus, A., Jus, K., Gaulter, J., Villenueve, A., Pires, P., Pineau, R., and Villeneuve, R., 1974, Studies on the action of certain pharmacological agents on tardive dyskinesia and on the rabbit syndrome, *Int. J. Clin. Pharmacol. Biopharm.* **9**:138–145.

Jus, A., Villeneuve, A., Gautier, J., Jus, K., Villeneuve, C., Pires, P., and Villeneuve, R., 1978, Deanol, lithium and placebo in the treatment of tardive dyskinesia, *Neuropsychobiology* **4**:140–149.

Klawans, H. L., Jr., 1973, The pharmacology of tardive dyskinesia, *Am. J. Psychiatry* **130**:83–86.

Klawans, H. L., and Rubovits, R., 1974, Effect of cholinergic and anticholinergic agents on tardive dyskinesia, *J. Neur. Neurosurg. Psychiatry* **27**:941–947.

Korsgaard, S., 1976, Baclofen (Lioresal) in the treatment of neuroleptic-induced tardive dyskinesia, *Acta Psychiatry. Scand.* **54**:17–24.

Korsgaard, S., Casey, D. E., and Gerlach, J., 1982a, High-dose des-tyrosine-gamma-endorphin in tardive dyskinesia, *Psychopharmacology* **78**:285–286.

Korsgaard, S., Casey, D. E., and Gerlach, J., Hetmar, O., Kaldan, B., and Mikkelsen, L. B., 1982b, The effect of Tetrahydroisoxazolopyridinol (THIP) in tardive dyskinesia, *Arch. Gen. Psychiatry* **39**:1017–1021.

Kruse, W., 1960, Persistent muscular restlessness after phenothiazine treatment: Report of 3 cases, *Am. J. Psychiatry* **115**:152–153.

Kuhar, M. F., Pert, C. B., and Snyder, S. H., 1973, Regional distribution of opiate receptor binding in monkey and human brain, *Nature* **245**:447–450.

Kunin, R. A., 1976, Manganese in dyskinesias, *Am. J. Psychiatry* **133**:105.

Kurata, K., Hosokawa, K., and Koshino, Y., 1977, Treatment of neuroleptic-induced tardive dyskinesia with cyproheptadine, *J. Neurol.* **215**:295–298.

Labrie, F., Beaulieu, M., Carom, M., and Raymond, V., 1978, The adenohypophyseal dopamine receptors, specificity and modulation of its activity by estradiol, in: *Proceedings of the International Symposium on Prolactin* (A. Robyn and C. Harter, eds.), pp. 121–136, Elsevier/North-Holland Biomedical Press, Amsterdam.

Lahti, R. A., and Losey, E. G., 1974, Antagonism of the effects of chlorpromazine and morphine on dopamine metabolism by GABA, *Res. Commun. Chem. Pathol. Pharmacol.* **7**:31–40.

Lautin, A., Stanley, M., Angrist, B., and Gershon, S., 1979, Extrapyramidal syndrome with sodium valproate, *Br. Med. J.* **2**:1035–1036.

Linnoila, M., and Viukari, M., 1979, Sodium valproate and tardive dyskinesia, *Br. J. Psychiatry* **134**:223–224.

Linnoila, M., Viukari, M., and Hietala, O., 1976, Effect of sodium valproate on tardive dyskinesia, *Br. J. Psychiatry* **129**:114–119.

Lipsius, L. H., 1977, Barbiturates and tardive dyskinesia, *Am. J. Psychiatry* **134**:1162–1163.

Lloyd, K. G., and Worms, P., 1981, Neuropharmacological actions of GABA agonists: Predictability for their clinical usefulness, in: *Amino Acid Neurotransmitters* (F. V. DeFeudis and P. Mendel, eds.), pp. 59–67, Raven Press, New York.

Mackay, A. V., Sheppard, G. P., Saha, B. K., Motley, B., Johnson, A. L., and Marsden, C. D., 1980, Failure of lithium treatment in established tardive dyskinesia, *Psychol. Med.* **10**:583–587.

Massotti, M., Guidotti, A., and Costa, E., 1981, Characterization of benzodiazepine and gamma-aminobutyric recognition sites and their endogenous modulators, *J. Neuroscience* **1**:409–418.

McGeer, P. L., and McGeer, E. G., 1976, Enzymes associated with the metabolism of catecholamines, acetylcholine, and GABA in human controls and patients with Parkinson's disease and Huntington's chorea, *J. Neurochem.* **26**:65–76.

Mehta, D., Mehta, S., and Mathew, P., 1976, Failure of deanol in treating tardive dyskinesia, *Am. J. Psychiatry* **133**:1467.

Nagao, T., Ohshimo, T., Mitsunobu, K., Sato, M., and Otsuki, S., 1979, Cerebrospinal fluid monoamine metabolites and cyclic nucleotides in chronic schizophrenic patients with tardive dyskinesia or drug-induced tremor, *Biol. Psychiatry* **14**:509–523.

Nair, N. P. V., Yassa, R., Ruiz-Navarro, J., and Schwartz, G., 1978, Baclofen in the treatment of tardive dyskinesia, *Am. J. Psychiatry* **135**:1562–1563.

Nair, N. P. V., Lal, S., Schwartz, G., and Thavundayil, J. K., 1980, Effect of sodium valproate and baclofen in tardive dyskinesia: Clinical and neuroendocrine studies, *Adv. Biochem. Psychopharmacol.* **24**:437–441.

Nasrallah, H. A., Smith, R. E., Dunner, F. J., and McCalley-Whitters, M., 1982, Serotonin precursor effects in tardive dyskinesia, *Psychopharmacology* **77**:234–235.

Nicoletti, F., Patti, F., Marano, P., Condorelli, D. F., Scarzella, L., Bergamasco, B., Marino, B., Reggio, A., and Scapagnini, U., 1983, Effects of salmon calcitonin administration on spontaneous or neuroleptically-induced dyskinesias, *Acta Neurol. (Napoli)* **5**:380–384.

Norris, J. P., and Sams, R. E., 1977, More on the use of manganese in dyskinesia, *Am. J. Psychiatry* **134**:1448.

O'Flanagan, P. M., 1975, Clonazepam in the treatment of drug-induced dyskinesia, *Br. Med. J.* **1**:269–270.

Palfreyman, M. G., Huot, S., Lippert, B., and Schechter, P. J., 1978, The effect of gamma-acetylenic GABA, an enzyme-activated irreversible inhibitor of GABA-transaminase, on dopamine pathways of the extrapyramidal and limbic system, *Eur. J. Pharmacol.* **50**:325–336.

Pedigo, N. W., Schallert, T., and Overstreet, D. M., 1979, Inhibition of *in vivo* ^{3}H-spiperone binding by the proposed antipsychotic des-tyr-gamma-endorphin, *Eur. J. Pharmacol.* **60**:359–364.

Perry, T., and Hansen, S., 1978, Biochemical effects in man and rat of three drugs which can increase brain GABA content, *J. Neurochem.* **30**:679–684.

Pert, A., Rosenblatt, J. E., Sivit, C., Pert, C. B., and Bunney, W. E., 1978, Long term treatment with lithium prevents the development dopamine receptor supersensitivity, *Science* **201**:171–173.

Pickar, D., and Davies, R. K., 1978, Tardive dyskinesia in younger patients, *Am. J. Psychiatry* **135**:385–386.

Prange, A. J., Wilson, I. C., Morris, C. E., and Hall, C. D., 1973, Preliminary experience with tryptophan and lithium in the treatment of tardive dyskinesia, *Psychopharmacol. Bull.* **9**:36–37.

Price, T. R. P., and Levin, R., 1978, The effects of electroconvulsive therapy on tardive dyskinesia, *Am. J. Psychiatry* **135**:991–993.

Raymond, V., Beaulieu, M., Labrie, R., and Boissier, J. R., 1978, Potent antidopaminergic activity of estradiol at the pituitary level of prolactin release, *Science* **200**:1173–1175.

Reda, F. A., Scanlan, J. M., Kemp, K., and Escobar, J. L., 1974, Treatment of tardive dyskinesia with lithium carbonate, *N. Engl. J. Med.* **291**:850.

Reda, F. A., Escobar, J. I., and Scanlan, J. M., 1975, Lithium carbonate in the treatment of tardive dyskinesia, *Am. J. Psychiatry* **132**:560–562.

Recker, D., Anderson, B., Yackulic, C., Cooper, T. B., Banay-Schwartz, M., Leon, C., and Dolavka, J., 1982, Naloxone, tardive dyskinesia and endogenous beta-endorphin, *Psychiatry Res.* **7**:321–324.

Rosenbaum, A. H., and de la Fuente, J. R., 1979, Benzodiazepines and tardive dyskinesia, *Lancet* **2**:900.

Rosenbaum, A. H., Maruta, T., Duane, D. D., Auger, R. G., Martin, D. K., and Brenengen, E. E., 1980, Tardive dyskinesia in depressed patients: Successful therapy with antidepressants and lithium, *Psychosomatics* **20**:715–719.

Saito, K., Konishi, S., and Otsuka, M., 1975, Antagonism between Lioresal and substance P in rat spinal cord, *Brain Res.* **97**:177–180.

Scatton, B., Zivkovic, B., Dedek, J., Lloyd, K. G., Constantinidis, J., Tissot, R., and Bartholini, G., 1982, γ-Aminobutyric acid (GABA) receptor stimulation III: Effect of progabide (SL 76002) on norepinephrine, dopamine, and 5-hydroxytryptamine turnover in rat brain areas, *J. Pharmacol. Exp. Ther.* **220**:678–688.

Scatton, B., Zivkovic, B., and Nartholini, G., 1980, Differential influence of GABAergic agents on dopamine metabolism in extrapyramidal and limbic systems of the rat, *Brain Res. Bull.* **5**(Suppl. 2):421–425.

Schildkraut, J. J., Schanberg, S. M., and Kopin, I. J., 1966, The effects of lithium ion on ^{3}H-norepinephrine metabolism in brain, *Life Sci.* **5**:1479–1481.

Schmidt, R. F., Vogel, M. E., and Zimmerman, M., 1967, Die Wirkung von Diazepam auf die prasynaptische Hemmung und andere Ruckenmarkreflexe, *Arch. Pharmak. Exp. Path.* **258**:69–82.

Schoemaker, H., and Nickolson, V. J., 1980, Effects of des-tyr^{1}-gamma-endorphin on dopamine release from various rat brain regions *in vitro*, *Life Sci.* **27**:1371–1376.

Sedman, G., 1976, Clonazepam in the treatment of tardive oral dyskinesia, *Br. Med. J.* **2**:583.

Sevestre, P., Rondot, P., Bathien, N., Morselli, R. L., and VanLandeghem, V. H., 1982, The effect of progabide, a specific GABAergic agonist on neuroleptic-induced tardive dyskinesia—Results of a pilot study, *13th Collegium Internationale Neuro-Pharmacologicum C. I. N. P.*, abstracts. Vol. II, Jerusalem.

Simpson, G. M., Branchey, M. H., Lee, J. H., Voitashevsky, A., and Zoubok, B., 1976, Lithium in tardive dyskinesia, *Pharmakopsychiat. Neuropsychopharmakol.* **9**:76–80.

Simpson, G. M., Lee, J. H., Shrivastava, R. K., and Branchey, M. H., 1978, Baclofen in the treatment of tardive dyskinesia and schizophrenia, *Psychopharmacol. Bull.* **14**:16–18.

Singh, M. M., 1976, Diazepam in the treatment of tardive dyskinesia, *Int. J. Pharmacopsychiatry* **11**:232–234.

Singh, M. M., Nasrallah, H. A., Lal, H., *et al.*, 1980, Treatment of tardive dyskinesia with diazepam: Indirect evidence for the involvement of limbic, possibly GABA-ergic mechanisms, *Brain Res. Bull.* **5**:673–680.

Singh, M. M., Becker, R. E., Pitman, R. K., Nasrallah, H. A., Lal, H., Dufresne, R. L., Weber, S. S., McCalley-Whitters, M., 1982, Drug-induced changes in tardive dyskinesia: Suggestions for a new conceptual model, *Biol. Psychiatry* **17**:729–742.

Sovner, R., and Loadman, A., 1978, More on barbiturates and tardive dyskinesia, *Am. J. Psychiatry* **13**:382.

Stewart, R. M., Rollins, J., Beckham, B., and Roffman, M., 1982, Baclofen in tardive dyskinesia patients maintained on neuroleptics, *Clin. Neuropharmacol.* **5**:365–373.

Tamminga, C. A., Crayton, J. W., and Chase, T. N., 1979, Improvement in tardive dyskinesia after muscimol therapy, *Arch. Gen. Psychiatry* **36**:595–598.

Tamminga, C. A., Tighe, P. J., Chase, T. N., DeFraites, G., and Schaffer, M. H., 1981, Des-tyrosine-gamma-endorphin administration in chronic schizophrenics, *Arch. Gen. Psychiatry* **38**:167–168.

Tamminga, C. A., Thaker, G. K., Ferraro, T. N., and Hare, T. A., 1983, GABA agonist treatment improved tardive dyskinesia, *Lancet* **2**:97–98.

Tell, G. P., Schechter, P. J., Koch-Weser, J., Cantiniaux, P., Chabannes, J. P., and Lambert, P. A., 1981, Effects of gamma-vinyl GABA, *N. Engl. J. Med.* **305**:581–582.

Thaker, G. K., Hare, T. A., and Tamminga, C. A., 1983, GABA system: Clinical research and treatment of tardive dyskinesia, *Mod. Probl. Pharmacopsychiatry*, **21**:155–167.

Thaker, G. K., Tamminga, C. A., Lafferman, J., Alphs, L. A., and Hare, T. A., GABA-mediated neural systems in tardive dyskinesia: Pathophysiology and treatment (submitted for publication).

Uhrband, L., and Faurbye, A., 1960, Reversible and irreversible dyskinesia after treatment with perphenazine, chlorpromazine, reserpine and electroconvulsive therapy, *Psychopharmacologia* **1**:408–418.

Verhoeven, W., van Praag, H. M., and Botter, P. A., 1978, (Des-tyr[1])-gamma-endorphin in schizophrenia, *Lancet* **1**:1046.

Verhoeven, W. M. A., van Praag, H. M., and van Ree, J. M., 1979, Improvement of schizophrenia patients treated with des-tyr[1]-gamma-endorphin (DγTE), *Arch. Gen. Psychiatry* **36**:294–298.

Villeneuve, A., Langlier, P., and Bedard, P., 1978, Estrogens, dopamine and dyskinesias, *Can. Psychiatr. Assoc. J.* **23**:68–70.

Villeneuve, A., Cazejust, T., and Côte, M., 1980, Estrogens in tardive dyskinesia in male psychiatric patients, *Neuropsychobiology* **6**:145–151.

Volavka, J., Anderson, B., and Koz, G., 1982, Naloxone and naltrexone in mental illness and tardive dyskinesia, *Ann. NY Acad. Sci.* **398**:97–102.

Walaas, I., and Fonnum, F., 1978, The distribution of putative momoamine, GABA, acetylcholine and glutamate fibers in the mesolimbic system, in: *GABA-Neurotransmitters* (P. Korsgaard-Larsen, J. Scheel-Kruger, and H. Kofod, eds.), pp. 60–73, Academic Press, New York.

Walters, J. R., and Chase, T. N., 1977, GABA systems and extrapyramidal function, in: *Neurotransmitter Function: Basic and Clinical Aspects* (W. S. Fields, ed.), pp. 193–211, Atratton Intercontinental Medical Book Corporation, New York.

Weber, S. S., Dufresne, R. L., Becker, R. E., and Mastrati, P., 1983, Diazepam in tardive dyskinesia, *Drug Intell. Clin. Pharm.* **17**:523–527.

Wolf, M. E., Keener, S., Mathis, P., and Mosnaim, A. D., 1983, Phenylethylamine-like properties of baclofen, *Neuropsychobiology* **9**:219–222.

Yahr, M. D., Duvoisin, R. C., and Cote, L. D., 1969, Reversal of L-dopa effects in parkinsonism, *Trans. Am. Neurol. Assoc.* **94**:81–84.

Yassa, R., Archer, J., and Cordozo, S., 1984, The long-term effect of lithium carbonate on tardive dyskinesia, *Can. J. Psychiatry* **29**:36–37.

PART III
Tourette Syndrome and Tic

CHAPTER 11

Tourette Syndrome

ELAINE SHAPIRO and ARTHUR K. SHAPIRO

1. INTRODUCTION

In the last 15 years we have seen a remarkable increase in interest and knowledge about Tourette syndrome (TS). In the past TS was considered to be a rare illness, with a poor prognosis and intellectual and psychological deterioration. Recent findings, primarily from pharmacological treatment and biochemical, genetic, neurological, encephalographic, psychological studies, have advanced our understanding of TS.

This chapter summarizes current findings.

2. CLASSIFICATION OF TIC SYNDROMES

The classification of the tic disorders—transient tic of childhood, chronic multiple motor tic, and TS—is based on age of onset, symptomatology, and clinical course (Shapiro *et al.*, 1978). These divisions are primarily heuristic and are not substantiated by pathophysiological evidence. In fact, it is now commonly believed that the three disorders represent a continuum along a severity axis with transient tic of childhood representing the least severe tic condition and TS the most severe. Although the tic conditions are referred to as disorders in DSM-III (American Psychiatric Association, 1980), we will use the more familiar term "Tourette syndrome" in this chapter.

2.1. Transient Tic of Childhood

Transient tic of childhood is virtually indistinguishable from TS except that its course is usually 1–2 years. The symptoms begin during childhood or early adolescence usually with an eye blink or facial tic. However, the whole head or torso or limbs may be involved. Tics can be single or occur sequentially.

ELAINE SHAPIRO and ARTHUR K. SHAPIRO • Tourette and Tic Laboratory and Clinic, Mount Sinai School of Medicine, New York, New York 10029.

Vocal tics are infrequent. Transient tics were referred to as "habit" tics in the past.

2.2. Chronic Multiple Motor TIC

Chronic multiple motor tic has its onset in childhood. It differs from TS in that the symptoms are only motor. It is considered to be a milder form of TS.

2.3. Chronic Multiple TIC or TS

Chronic multiple tic, or TS has its onset between 2 and 17 years. The motor tics usually involve the head and frequently other parts of the body, torso, and upper and lower limbs. Multiple vocal tics are alays present and include various complicated sounds, words, or coprolalia (involuntary utterance of vulgar or obscene words), the last occurring in approximately 30% of the cases (Shapiro and Shapiro, 1982a,b). Other possible associated features of the disorder are echolalia (repetition of another's last words or phrases), palilalia (repetition of one's own last words or phrases), echokinesis (imitation of the movements of others), and mental coprolalia (thinking about obscene words). The movements and the vocalizations can be voluntarily suppressed for minutes to hours.

Other associated problems include hyperactivity, attention deficits (ADD), and academic difficulties, which are reported in a range of 24–55% (Shapiro *et al.*, 1978; Shapiro and Shapiro, 1982a). The relationship between ADD with or without hyperactivity and TS is unclear, but the data suggest a strong association. Behavioral and emotional problems are not intrinsically associated with TS, and many patients function quite normally. However, behavioral problems may occur independently, as they might in any sample of individuals, or as a secondary effect or concomitant of ADD. Children with ADD with or without hyperactivity are often characterized as inattentive, impulsive, and having a low frustration tolerance. These behavioral characteristics may also be present in Tourette children with ADD.

In approximately half the patients, the first symptom is a single motor tic, most frequently an eye blink. However, the initial symptoms may be multiple or may be a vocal tic. The number, type, and intensity of symptoms vary over weeks or months, and the illness is characterized by a fluctuating, waxing and waning clinical course. One symptom is replaced or is added to a preexisting symptom, or an old symptom may reappear. Occasional periods of remission may occur, sometimes as short as a week, sometimes years. These periods of remission are most frequent in the early phases of the illness. However, TS is usually chronic and lifelong. The percent of patients having a spontaneous remission in three studies of 152 patients was 16% (range 7–19%) (Shapiro and Shapiro, 1982a,b).

The symptoms are characteristically sudden, abrupt, and explosive in nature. Movements may be simple motor acts, such as an eye blink, or complex ones, such as jumping, stomping, and kicking. Coprolalia does not have to be present for a diagnosis to be made, and the diagnosis should not be deferred until it appears. Coprolalia occurs in approximately 30% of patients, may not be an early symptom, and can occur 35 years after the initial onset of symptoms (Shapiro *et al.*, 1978; Shapiro and Shapiro, 1982a,b).

2.4. Chronic Motor Tic

Chronic motor tic is another possible subcategory. It has its onset in childhood or after 40. Vocal tics rarely occur, are of lower intensity, and usually are caused by contractions of thoracic, abdominal, or diaphragmatic muscles. Severity and type of symptoms tend not to change over time. Onset of symptoms in adulthood often is limited to a single tic. Tics can be voluntarily suppressed for minutes to hours. Duration of illness is usually lifelong.

2.5. Tic Characteristics

Tics are characteristically rapid, clonic, brief, repetitive, involuntary movements. The involuntary nature of tics was demonstrated in an electromyographic study by Obeso *et al.* (1981). They found that the normal premovement potential was absent prior to spontaneous simple tics in six TS patients, but was present when patients voluntarily demonstrated a simple tic.

Five percent of Tourette patients have slower, more dystonic movements which can be characterized as sustained movements such as squeezing, tightening, or stretching. These tonic movements, lasting from 0.25 to 1.0 second, are often associated with a prodromal sensory feeling. About 3% of our patients report that all of their symptoms are associated with an initial sensory prodromal feeling to which they respond by voluntarily executing a movement in order to relieve or satisfy the sensory feeling (Shapiro and Shapiro, 1982a). These movements, which are voluntarily executed, differ from the involuntary fast tic of TS. A similar sensory experience was described by Bliss *et al.* (1980). We have classified these patients as having sensory tics since the symptoms are slower, have a longer duration than tics, and are voluntary movements in response to a sensory stimulus.

3. NEUROPATHOLOGY

3.1. Pathological Studies

No definitive findings were reported in four autopsy reports. Bing (1925) reported meningitic thickening. Dewulf and Van Bogaert (1940) found no abnormalities. A third report described an arrested development of the caudate nucleus and putamen (Balthazar, 1956). The fourth autopsy by Borak (1969) found extensive pathological changes, but extensive preexisting neuropathological disease in this patient makes it difficult to evaluate the neuropathological findings. Neuropathological study of brain tissue is a promising area and awaits the availability of brains for autopsies. Mount Sinai Medical Center is one of the centers prepared to thoroughly examine specimens.

3.2. Neurological Studies

Soft neurological signs were found in 59% (range, 55–62%) in three samples of 107 patients compared with 20% expected in children (Shapiro *et al.*, 1978). The most frequent findings in an early sample (Shapiro *et al.*, 1978) and in 379 patients of a more recent sample of 666 (unpublished data) included unilateral impairment of rapid alternating movements, pronation or drift of outstretched extremity, increase in tone or reflexes, decrease in associated movements on one side, and decrease in tone and check on one side.

3.3. Electroencephalographic Studies

Electroencephalographic recordings (EEG) were independently rated by two experienced electroencephalographers in 79 Tourette patients (reliability of 0.83) (Shapiro *et al.*, 1978). The percent of children with abnormal EEGs was 71% compared with 25% in adult patients. The abnormalities were nonspecific bilateral sharp waves, diffuse background disorganization, and slowing and infrequent unilateral, temporooccipital slowing and sharp waves. These EEG abnormalities were significantly more frequent in TS patients with ADD than those without ADD (unpublished data).

3.4. Computed Tomography Scans and Myography

Computed tomography scans and myography are reported normal in most studies except for two reports. Caparulo *et al.* (1981) reported abnormal scans in 38% of 16 Tourette patients. Yeragani *et al.* (1983) reported a slight increase in the density of the caudate nuclei bilaterally in a single patient. Normal myographic studies are consistently reported in the literature.

3.5. Neuropsychological Studies

Neuropsychological studies have attempted to assess central nervous system (CNS) impairment in patients with TS and to delineate the site or locus of the dysfunction.

We used a battery of tests consisting of the Wechsler adult intelligence scale (WAIS), Weschsler intelligence scale for children (WISC), Bender–Gestalt, and Rorschach to compare the performance of 50 TS patients to a matched group of 50 psychiatric outpatients (Shapiro *et al.*, 1978). The records were blindly evaluated by two psychologists. The groups differed significantly in the degree of organic impairment, with 68% of the TS patients rated as impaired compared to 28% of the controls. A marked discrepancy between verbal and performance IQ scores and a pattern of impairment on the Bender–Gestalt test indicated that Tourette patients had significant impairment in visual–motor abilities. The data were suggestive of impairment in the right hemisphere.

The results of Shapiro *et al.* were partially confirmed in a study by Incagnoli and Kane (1981). They tested 13 male Tourette subjects between the ages of 10 and 13 using the WISC-R, Bender–Gestalt test, Wide-Range Achievement Test (WRAT), and Halstead–Reitan neuropsychological battery for children. Scores on the WRAT indicated significant impairment on the arithmetic subtest. Visual–motor discrepancy scores showed significant impairment on visual–motor tasks involving the copying of designs. On the Halstead–Reitan battery there was no evidence of generalized cerebral dysfunction. They concluded that the overall pattern represented a dysfunction of nonconstructional visuopractic abilities.

The Halstead–Reitan neuropsychological test battery for children was also used by Joschko and Rourke (1982). They tested three children with TS using a test–retest paradigm. None of the children demonstrated a specific pattern of deficits on the battery. There was evidence of information-processing deficiencies in one subject who demonstrated problems in sustained attention, immediate auditory–verbal memory, verbal fluency, and phenome–grapheme matching. Another child had a 24-point discrepancy between verbal and performance IQ scores, but the neuropsychological tests did not indicate brain impairment. Arithmetic scores on the WRAT were also depressed. The number of subjects in this study was too small for the results to be generalizable. However, the pattern of deficits in at least one subject was consistent with previous studies (Shapiro *et al.*, 1978; Incagnoli and Kane, 1981).

Sutherland *et al.* (1982) tested 32 Tourette patients who were medication-free using a special comprehensive neuropsychological test battery developed for use with neurological subjects at Montreal Neurological Institute. Test results were compared to those for 47 learning-disabled subjects and 30 schizophrenic subjects. The data supported the conclusion of a neuropsychological impairment in children and adults with TS. On the basis of some of the test results, the authors suggested that the most likely locus of abnormality was in the right hemisphere. Deficits were found in memory and copying of visually presented nonverbal material, reduced verbal fluency, and short-term memory for stories. However, other deficits suggested other possible loci. This study also confirmed the finding of Shapiro *et al.* (1978) of impairment in performance IQ scores compared to verbal IQ scores in Tourette subjects over 9 years of age although the magnitude of the difference was not as great as that reported by Shapiro *et al.* (1978).

Although the results of these studies are not directly comparable since different test batteries were used, the impairment in visual–motor abilities found in three of the studies is suggestive of possible right-hemisphere involvement. The findings of Sutherland *et al.* (1982), however, suggest that more than one brain area may be involved. Evidence of CNS abnormality was also demonstrated in some patients.

Neurometric testing (John *et al.*, 1977) of three of our Tourette patients who were unmedicated revealed significant extreme power asymmetries and waveshape incoherences between homologous recording locations over the left and right hemispheres in evoked responses.

An important factor, not adequately considered in these studies, is the effect of ADD on the reported results. Further studies should separate TS patients with and without ADD.

4. PSYCHOPATHOLOGY

Although the prevailing theory for almost 70 years was that TS was caused primarily by psychological conflicts, the evidence in support of this theory was based on clinical case studies of one or more patients.

We conducted a controlled study of the psychodynamic formulation that tics and other symptoms of TS were the symbolic expression of a massive, unconscious conflict about the expression of hostility and aggression which resulted in reaction formation, obsessions, compulsions, symptom substitution, and ultimately an underlying psychotic process. We found that psychopathological states such as hysteria, inhibition of aggression, obsessive–compulsive traits, schizophrenia, or other psychoses did not characterize patients with TS more than a matched outpatient psychiatric sample (Shapiro *et al.*, 1978).

A recent study by Wilson *et al.* (1982), using the behavior problem checklist, reported considerably more disturbance in 21 children with TS compared to an unselected public-school population. The pattern of scores, however, was nonspecific but was comparable to that seen in children in special-education classes. They found a relationship between number of symptoms, low verbal IQ, and number of errors on the Wisconsin Sorting Test. No information was provided about whether the Tourette subjects had an ADD, hyperactivity, or learning disabilities. It is our impression based on considerable clinical experience that the problems characteristic of these Tourette children may be a behavioral concomitant of having ADD and not intrinsically related to TS. The relationship of low verbal IQ to greater behavioral disturbance found in this study offers partial support of this association.

5. ETIOLOGY

The evidence that a higher-than-expected number of Tourette patients have soft signs of neurological abnormality, abnormal EEG's, organic findings on psychological and neuropsychological tests, ADD, hyperactivity, learning disabilities, lefthandedness, and family history of tics and TS, and that the tics of TS are involuntary rather than voluntary, suggests that the etiology of TS is an organic disturbance of the CNS (Shapiro *et al.*, 1978). However, these abnormalities do not establish the anatomical site or the specific mechanism of any putative organic abnormality.

Circumstantial evidence from study of various movement disorders suggests that the site of the disturbance may be in the basal ganglia and involves the synaptic transmission system. Dopaminergic mechanisms have been implicated in the pathophysiology of TS by the ability of haloperidol and other dopamine (DA) blocker medications to suppress tic symptoms and the exacerbation of symptoms by DA agonists. The blocking of DA at postsynaptic receptors, together with the effect of L-dopa on patients with parkinsonism, contributes to the hypothesis that TS is caused by hyperactivity of dopaminergic neurons in the striatum or supersensitivity of postsynaptic receptors. The effectiveness of pimozide, penfluridol, fluphenazine, and other antipsychotic drugs known to block DA receptors provides the most convincing evidence for the DA hypothesis in TS.

5.1. Biochemical Studies

A number of neurotransmitters have been implicated in TS including the DA, serotonergic, noradrenergic, and cholinergic systems.

The accumulated evidence strongly implicates the DA system. The evidence for dopaminergic hyperactivity arises primarily from measurement of the derivatives of DA in the cerebrospinal fluid (CSF). Studies measuring homovanillic acid (HVA) in CSF, a major metabolite of DA, have been inconsistent. Postprobenicid HVA levels were reported as high in two studies of 5 and 12 TS patients (Van Woert *et al.*, 1976; Han-bai and Han-Quin, 1983), as normal in seven patients before and after probenicid (Shapiro *et al.*, 1978), and other studies of 1, 9, and 10 TS patients reported lower levels (Johansson and Roos, 1974; Butler *et al.*, 1979; Cohen *et al.*, 1979; Singer *et al.*, 1982). Singer *et al.* (1982) demonstrated a significant increase in CSF HVA from baseline measured following treatment with haloperidol. They were unable to confirm a previous report (Cohen *et al.*, 1978) relating the degree of reduction of 5-hydroxyindoleacetic acid (5-HIAA) relative to probenicid and severity of symptomatology.

The inconsistency in the data on HVA levels may be largely the result of procedural differences. Obviously, there is a need for a uniform method to measure HVA. Interpretation of the data would also be enhanced if controls were matched for age, sex, and other variables. Nevertheless, despite the small sample size used in these studies, the majority report that HVA levels are decreased in TS patients, suggesting that the increased dopaminergic activity may be supersensitivity of postsynaptic DA receptors.

Friedhoff (1982) has attempted to test the hypothesis of an overactive dopaminergic system in the pathogenesis of TS. Using a technique of pharmacological intervention in fetal rats, Friedhoff produced a state of enduring hyperdopaminergia secondary to an increase in the number of DA receptors.

L-Dopa was administered to TS patients to produce an increase in central DA and decreased number of postsynaptic DA receptors. The termination of L-dopa resulted in temporary improvement in six patients with TS although other clinicians report having treated patients in whom this technique has been unsuccessful.

CSF 5-HIAA concentrations were reported as normal in several studies (Cohen *et al.*, 1974, 1979; Van Woert *et al.*, 1976; Shapiro *et al.*, 1978; Butler *et al.*, 1979). Lower 5-HIAA levels were reported in one study without probenicid administration (Butler *et al.*, 1979) and in two studies following probenicid administration (Cohen *et al.*, 1978; Han-bai and Han-Quin, 1983).

Normal values have been reported for vanillylmandelic acid in CSF and urine (Han-bai and Han-Quin, 1983).

Urinary 3-methoxy-4-hydroxyphenethyleneglycol (MHPG) was normal in one patient, but levels were reduced in 24-hr urine excretion during exacerbation of symptoms (Sweeney *et al.*, 1978). Low levels of urinary MHPG were reported in 21 medication-free TS outpatients compared to controls (Ang *et al.*, 1982), but elevated values have been reported in occasional patients (Cohen *et al.*, 1979; Yeragani *et al.*, 1983). Normal MHPG concentrations in CSF were reported in five TS patients (Singer *et al.*, 1982) and elevated levels in two patients (Cohen *et al.*, 1980).

Normal values have been reported for dopamine-B-hydroxylase (Lake *et al.*, 1977; Shapiro *et al.*, 1978; Cohen *et al.*, 1979), plasma norepinephrine (Lake *et al.*, 1977), plasma oxidase activity, platelet monoamine oxidase (MAO) activity (Lake *et al.*, 1977; Cohen *et al.*, 1979), and erythrocyte catechol-O-methyltransferase (Lake *et al.*, 1977; Cohen *et al.*, 1979). However, in a carefully controlled study of potential genetic markers in TS, platelet MAO was significantly ($p > 0.001$) higher in 24 untreated TS patients compared to a matched control group (Shapiro *et al.*, 1984). The evidence from clinical pharmacotherapy does not suggest abnormalities for norepinephrine, 5-hydroxytriptamine, and acetylcholine (Shapiro *et al.*, 1981).

The data on the role of the cholinergic system in TS are also inconsistent. Hanin *et al.* (1979) found significantly higher red blood cell choline and a higher ratio of red blood cell choline to plasma choline in 20 patients. The effect of physostigmine on tics has been variable; two patients were mildly improved, and three had no change in tics (Shapiro *et al.*, 1978), six patients had a significant decrease in motor and vocal tics (Stahl and Berger, 1980, 1981), and 10 patients had an increase in tics (Tanner *et al.*, 1982). The initial positive results of treatment with medication such as choline and lecithin (Barbeau, 1978, 1980), known to have an effect on acetylcholine, have suggested a relative imbalance between the cholinergic and DA systems. However, choline has not been reported to be effective, and the effectiveness of lecithin in TS has not been confirmed in two studies (Polinsky *et al.*, 1980; Moldofsky and Sandor, 1983).

Initial reports of improvement of the symptoms of TS with clonidine implicated the noradrenergic system (Cohen *et al.*, 1980; Borison *et al.*, 1982).

Low doses of clonidine appear to preferentially stimulate α-adrenergic receptors presynaptically, decreasing the amount of norepinephrine released into the synapse. Shapiro *et al.* (1984) have reported less successful results with clonidine to suppress tic symptoms, suggesting that the noradrenergic system's role may be interactive rather than primary.

Decreased levels of HVA and the appearance of tardive dyskinesia (TD) after long-term treatment with high dosages of neuroleptics are consistent with the hypothesis of a supersensitivity of postsynaptic dopaminergic receptors in the pathogenesis of TS. However, the responsiveness of Tourette symptoms to different types of medication and the complexity and inconsistency of the biochemical data suggest that more than one neurotransmitter system is involved with complicated compensatory feedback effects.

Interpretation of the data is complicated by the variability of the biochemical, genetic, and treatment results reported in the literature and suggests the possibility of pathophysiological heterogeneity among Tourette patients.

5.2. Genetic Factors

There is considerable interest in heredity factors in the transmission of TS. We and other investigators have shown that many Tourette patients have a positive family history for tics and TS, that females, although less frequently affected than males, are more likely to transmit the illness, and that multiple tics may be a milder form of TS. The mode of transmission is as yet unknown (Eldridge *et al.*, 1977; Wassman *et al.*, 1978; Shapiro *et al.*, 1978; Kidd *et al.*, 1980; Baron *et al.*, 1981).

We have previously reported that 40.5% of patients with TS have a family history of tics and 7.8% have a family history of TS. Eighteen of nineteen pairs of identical twins are concordant for TS, whereas only one of eight pairs of fraternal twins is concordant for TS.

Two studies attempted to identify potential genetic markers for TS. In a careful study of five families with two or more members with TS and 30 other patients, HLA antigen and C3, Clg, and total complement levels failed to show either linkage or association (Shapiro *et al.*, 1978). As previously indicated, several studies have reported normal values for dopamine-B-hydroxylase, erythrocyte COMT, and platelet MAO, but significantly higher ($p < 0.001$) values were obtained in a recent carefully controlled study by us in 24 untreated TS patients (Shapiro *et al.*, 1984d).

6. TREATMENT

6.1. Haloperidol

The efficacy of treatment with haloperidol has been reviewed extensively by us and others (Shapiro and Shapiro, 1968, 1981a, 1982a; Abuzzahab and

Anderson, 1973; Shapiro *et al.*, 1973, 1978; Wassman *et al.*, 1978; Nomura and Segawa, 1979). Improvement rates in these reports vary from 61 to 97% of patients, and reduction of symptoms averaged about 80% for all patients. Occasional clinicians report less favorable results (Cohen *et al.*, 1980; Tibbitts, 1981). Almost all studies have been open clinical trials and have not been confirmed by well-controlled, comparative, double-blind studies.

The major limitation of haloperidol is the appearance of adverse effects in about 20–25% of patients. For the majority of patients, improvement in response is possible but requires judicious titration of dosage and concomitant use in some patients of antiparkinson, stimulant, antianxiety, and other drugs (Shapiro *et al.*, 1978, 1981; Shapiro and Shapiro, 1980, 1981a,b).

6.2. Treatment Regimen

Our standard drug regimen is to titrate the dosage in very small increments. We begin treatment with 0.25 mg/day, increasing the dosage by this amount every fifth or seventh day until symptomatic improvement of 50–70% is achieved without adverse effects, until adverse effects occur without symptomatic benefit, or until symptoms decrease and adverse effects occur at the same time. Average dosage is usually 5 mg/day with a range of 2–10 mg/day.

6.3. Adverse Effects

Adverse effects are similar to those reported for all phenothiazines and butryophenomes and primarily include akathisia, akinesia, and extrapyramidal side effects. Some children and adults become irritable and fearful, develop phobias, and often seem unmotivated academically and vocationally. Cognitive impairment is frequently reported and manifests itself in depressed grades and possible school failure. Dysphoric symptoms include crying, brooding, excessive rumination about unpleasant events, and depression.

Although TD is known to occur with haloperidol, none of the 600 patients treated by us with haloperidol have developed TD (Shapiro, E., and Shapiro, A. K., 1982). Withdawal from high doses of haloperidol, or sudden withdrawal, may be accompanied by withdrawal dyskinesia for several weeks.

6.4. Management of Adverse Effects

The dosage at which adverse effects occur is extemely variable since patients metabolize drugs differently. Adverse effects usually are most severe during the first 3 months of treatment. If they occur, dosage is reduced 0.25 mg every 5 days until adverse effects are reduced or disappear, or the dosage can be reduced by half immediately. Benztropine mesylate, 0.5 mg/day used initially to obviate acute dystonia and akinesia, can be increased every day up to 6 mg/day to help reduce extrapyramidal side effects.

7. STIMULANTS

Several reports in the literature conclude that stimulants cause or provoke TS (Golden, 1977; Lowe *et al.*, 1982). However, all of the reports are retrospective, and the response to stimulants was so varied that we concluded that the evidence is inadequate that stimulants cause tics or TS (Shapiro and Shapiro, 1981b). Moreover, these drugs are considered crucial agents in the treatment and management of hyperactive children and in the management of adverse effects of haloperidol (Shapiro and Shapiro, 1982b). Differences of opinion about the use of stimulants in TS require carefully designed prospective studies.

7.1. Blood Levels

We used a specific radioimmunoassay (Shostak and Perel, 1976) to determine haloperidol blood levels in 24 patients. The minimum effective blood level was 2.0 ng/ml (Shapiro *et al.*, 1981). Blood levels were highly correlated with dosage but not with therapeutic response, adverse effects, age, sex, weight, chronicity of illness, or length of treatment.

Using a gas chromatographic procedure, haloperidol blood levels in nine patients with tics and five with TS were significantly associated with age and adverse effects, but not with dosage for the group as a whole (Morselli *et al.*, 1979). However, there was a relationship of blood level and dosage in some patients. Although adverse effects appeared at blood levels exceeding 6 ng/ml and therapeutic responses tended to occur at blood levels between 1 and 4 ng/ml, the authors noted a 15-fold variability for blood level at a constant daily dosage and no evident relationship among measures of dosage, blood level, and therapeutic effects.

Serum levels of haloperidol were measured using a radioreceptor assay in another study of five patients with TS (Singer *et al.*, 1982). Effective daily dosage was low, and no patient exceeded 5 mg/day. No correlation was apparent between doses and serum level. Effective control of symptoms occurred at a serum level between 1 and 4 ng/ml, but this level was associated with significant adverse effects.

7.2. Pimozide

Pimozide, a diphenylbutylpiperidine, is a specific dopamine blocker. In an open clinical trial we compared pimozide and haloperidol in 31 patients who had been on both medications. Twenty-eight patients (90.3%) evaluated themselves as improved on pimozide, one as unchanged, and two as worse. In comparison to haloperidol, 74% of patients treated with pimozide improved over 70% compared to 45% treated with haloperidol ($p < 0.02$). The overall decrease of tics was higher for pimozide (71%) compared to haloperidol (62%). Adverse effects were significantly less severe on pimozide compared to haloperidol ($p < 0.02$) (Shapiro *et al.*, 1983b). We compared pimozide to placebo in a double-blind crossover study in 20 Tourette patients (Shapiro and Shapiro, 1984). Multiple measures of improvement from the doctor, patient, and another collateral were obtained. Patients were videotaped at each revisit, and two independent raters rated the tapes for estimates of severity and improvement under three stimulus conditions. On all measures pimozide was significantly more effective than placebo ($p < 0.0001$). Adverse effects were evaluated as mild by most patients. These results confirm an acute double-blind study (Ross and Moldofsky, 1978) and a clinical description comparing haloperidol with pimozide (Nomura and Segawa, 1979).

7.3. Penfluridol

Penfluridol, a long-lasting diphenylbutylpiperidine derivative with duration of 5–7 days, has been reported to have more specific DA-blocking effects without blocking norepinephrine or serotonin. We were granted permission to use this investigational drug in patients with TS who could not tolerate or did not benefit from treatment with haloperidol.

In an open clinical study of eight patients who had previously failed on haloperidol, six of whom had also failed on pimozide, the average decrease of tic symptoms was significantly ($p < 0.001$) higher for penfluridol (74.4%) than for haloperidol (61.0%) and more effective than pimozide (67.7%). Penfluridol was chosen as the most effective drug by seven patients ($p < 0.0001$). Severity of adverse effects, especially sedation, was significantly less for penfluridol compared to the other drugs ($p < 0.0001$) (Shapiro *et al.*, 1983a).

7.4. Clonidine

Clonidine, an α-2-adrenergic agonist that reduces noradrenergic activity preferentially in the locus ceruleus when used at low dosages, has been clinically described as improving compulsive behavior, behavioral blocking, attention problems, and tics in 50–70% of 25 patients and as ameliorating these behavioral problems more than tics (Cohen *et al.*, 1980).

We compared the amount of improvement on clonidine compared to other medications such as haloperidol, pimozide, penfluridol, and fluphenazine (Shapiro *et al.*, 1983c). We studied treatment effects in two groups of patients, a total of 68 patients. The first group had been treated initially with haloperidol and subsequently with clonidine, and the second group started treatment with clonidine and then went on to treatment with a neuroleptic medication such as haloperidol, pimozide, or penfluridol.

The results for the two samples were similar. On the average, patients showed little improvement on clonidine, an average of 13.5% decrease in tic symptoms. Patients treated with the neuroleptic medication improved significantly more ($p < 0.0001$), an average of 68.8%. Fifty of the sixty-eight patients treated with clonidine had no benefit at all from treatment. In addition, significantly ($p < 0.0001$) more patients rated the neuroleptic medication as better than clonidine. Clonidine did not have a more positive effect on behavioral or psychological symptoms.

7.5. Other Medications

Clonazepam, a benzodiazepine, was reported effective by some investigators and ineffective by others (Gonce and Barbeau, 1977; Shapiro *et al.*, 1981). More than 50 different medications have been used to treat TS (Shapiro *et al.*, 1981). Although most other antipsychotic drugs are considered to be less effective than haloperidol, comparative clinical studies have not been done except for clozapine (Caine *et al.*, 1979), fluphenazine, and trifluoperazine (Borison *et al.*, 1982). Clozapine was ineffective in six patients, and fluphenazine and trifluoperazine were reported to be as effective as haloperidol in 10 patients.

There is no support for the efficacy of the following drugs: antianxiety–sedative–hypnotic drugs, including chlordiazepoxide and diazepam; antidepressant drugs, such as tricyclics, MAO inhibitors, and lithium; anticonvulsants, including carbamazepine; antiparkinson drugs; as well as corticosteroids, baclofen, piribedil, choline, lecithin, 5-hydroxytryptophan, L-dopa in combination with carbidopa, methysergide, β-chlorphenyl GABA, tetrabenazine, disulfiram, benztropine, deanol, and chlorimipramine (Shapiro and Shapiro, 1981a, 1982a; Shapiro *et al.*, 1981).

8. CONCLUSION

It is generally agreed that the tic conditions represent a continuum, with TS having the most complex and severe symptoms. Multiple tics are considered to be a milder form of TS. The finding that some patients have a combination of fast involuntary symptoms and slow voluntary dystonic movements indicates a possible linkage between the dystonias and tics.

Neuropsychological testing with the Halstead–Reitan neuropsychological test battery has not demonstrated a specific pattern of deficits. Other test data, however, consistently demonstrate deficits in visual–motor ability, including arithmetic tasks with a visual–motor component, indicating neuropsychological impairment. The data, although strongly suggesting right-hemispheric involvement, are inconclusive.

We do not as yet know the etiology of TS. The differential response to treatment, the variation in course and symptomatology, and genetic studies all suggest biological heterogeneity. The evidence from biochemical studies is inconclusive although dopaminergic hypersensitivity is implicated by both biochemical and pharmacological data. The effectiveness of haloperidol to alleviate the symptoms of TS particularly suggests abnormal DA effects, the most likely hypothesis being postsynaptic hypersensitivity. However, hypothesized relationships between clinical response to drugs and underlying neurochemical etiology for tics and TS are indirect and inferential. The data at present suggest that the drugs may potentially affect more than one neurotransmitter system with compensatory feedback effects. Iverson and Alpert (1982) have suggested that it may be wise to think in terms of widespread circuits that are functionally related to the DA neurons.

ACKNOWLEDGMENT. This work was supported by the Tourette, Tic and Movement Disorders Foundation, Gatesposts Foundation, Mandell and Elaine Shimberg, Herman and Fay Sarkowsky, M. Gould, H. Stein, and other grantors.

9. REFERENCES

Abuzzahab, F. S., and Anderson, F. O., 1973, Gilles de la Tourette's syndrome: International registry. *Minn. Med.* **56**:492–496.

American Psychiatric Association, 1980, *Diagnostic and Statistical Manual of Mental Disorders*, 3rd ed., American Psychiatric Association Press, Washington, DC.

Ang, L., Borison, R., Dysken, M., and Davis, J. M., 1982, Reduced excretion of MHPG in TS, in: *Advances in Neurology, Gilles de la Tourette Syndrome*, Vol. 35 (A. J. Friedhoff and T. N. Chase, eds.), pp. 171–175, Raven Press, New York.

Balthazar, K., 1956, Über das anatomische Substrat der generalisierten Tic-krankheit (maladie des tics, Gilles de la Tourette): Entwicklungshemmung des Corpus Striatum, *Arch. Psychiatr. Nervenkr.* **195**:531–549.

Barbeau, A., 1978, Emerging treatments: Replacement therapy with choline or lecithin in neurological diseases, *J. Can. Sci. Neurol.* **5**:157–160.

Barbeau, A., 1980, Cholinergic treatment in the Tourette syndrome, *N. Engl. J. Med.* **302**:1310–1311.

Baron, M., Shapiro, E., Shapiro, A., and Rainer, J. D., 1981, Genetic analysis of Tourette syndrome suggesting major gene effect, *Am. J. Hum. Genet.* **33**:767–775.

Bing, R., 1925, Ueber lokale muskelspasmen und Tics, *Schweiz. Med. Wochenschr.* **6**:993–997.

Bliss, J., edited by Cohen, D. J., and Freedman, D. X., 1980, Sensory experiences of Gilles de la Tourette syndrome, *Arch. Gen. Psychiatry* **37**:1343–1347.

Borak, W., 1969, Przypadek encefalopatii Z tikami Tourette'a zespolem anankastycznym i impulsywymi tendencjami do Samovszkodzen, *Psychiatr. Pol.* **3**:111–114.

Borison, R. L., Ang, L., Chang, S., Dysken, M., Comaty, J. E., and Davis, J. M., 1982, New pharmacological approaches in the treatment of Tourette syndrome, in: *Advances in Neurology, Gilles de la Tourette Syndrome*, Vol. 35 (A. J. Friedhoff and T. N. Chase, eds.), pp. 377–382, Raven Press, New York.

Butler, I. J., Koslow, S. H., Seifert, W. E., Jr., Caprioli, R. M., and Singer, H. S., 1979, Biogenic amine metabolism in Tourette syndrome, *Ann. Neurol.* **6**:37–39.

Caine, E. D., Polinsky, R. J., Kartzinel, R., and Ebert, M. H., 1979, The trial use of clozapine for abnormal involuntary movement disorders, *Am. J. Psychiatry* **136**:317–320.

Caparulo, B. K., Cohen, D. J., Rothman, S. L., Young, J. G., Katz, J. D., Shaywitz, S. E., and Shaywitz, B. A., 1981, Computed tomographic brain scanning in children with developmental neuropsychiatric disorders, *J. Am. Acad. Child Psychiatry* **20**:338–357.

Cohen, D. J., Shaywitz, B. A., Johnson, W. T., and Bowers, M., Jr., 1974, Biogenic amines in autistic and atypical children, *Arch. Gen. Psychiatry* **31**:845–853.

Cohen, D. J., Shaywitz, B. A., Caparulo, B., Young, J. G., and Bowers, M. B., Jr., 1978, Chronic, multiple tics of Gilles de la Tourette's disease, *Arch. Gen. Psychiatry* **35**:245–250.

Cohen, D. J., Shaywitz, B. A., Young, J. G., Carbonari, C. M., Nathanson, J. A., Lieberman, D., Bowers, M. B., Jr., and Maas, J. W., 1979, Central biogenic amine metabolism in children with the syndrome of chronic multiple tics of Gilles de la Tourette, *J. Am. Acad. Child Psychiatry* **18**:320–341.

Cohen, D. J., Detlor, J., Young, J. G., and Shaywitz, B. A., 1980, Clonidine ameliorates Gilles de la Tourette syndrome, *Arch. Gen. Psychiatry* **37**:1350–1357.

Dewulf, A., and Van Bogaert, L., 1940, Études anatomo-cliniques de syndromes hypercinetiques complexes, *Mschr. Psychiatr. Neurol.* **104**:53–61.

Eldridge, R., Sweet, R., Lake, C. R., Ziegler, M., and Shapiro, A. K., 1977, Gilles de la Tourette's syndrome: Clinical, genetic, psychologic and biochemical aspects in 21 selected families, *Neurology (Minneap.)* **27**:115–124.

Friedhoff, A. J., 1982, Receptor maturation in pathogenesis and treatment of Tourette syndrome, in: *Advances in Neurology, Gilles de la Tourette Syndrome*, Vol. 35 (A. J. Friedhoff and T. N. Chase, eds.), pp. 133–140, Raven Press, New York.

Golden, G. S., 1977, The effect of central nervous system stimulants on Tourette syndrome, *Ann. Neurol.* **2**:69–70.

Gonce, M., and Barbeau, A., 1977, Seven cases of Gilles de la Tourette's syndrome: Partial relief with clozepam: A pilot study, *Can. J. Neurol. Sci.* **4**:279–283.

Han-bai, C., and Han-quin, L. F., 1983, Tourette syndrome report of 19 cases, *Chinese Med. J.* **96**(1):45–50.

Hanin, I., Merikangas, J. R., Merikangas, K. R., and Kopp, U., 1979, Red-cell choline and Gilles de la Tourette syndrome, *N. Engl. J. Med.* **301**:661–662.

Icagnoli, T., and Kane, R., 1981, Neuropsychological functioning in Gilles de la Tourette's syndrome, *J. Clin. Neuropsychol.* **3**:165–169.

Iverson, S. D., and Alpert, J. E., 1982, Functional organization of the dopamine system in normal and abnormal behavior, in: *Advances in Neurology, Gilles de la Tourette Syndrome*, Vol. 35 (A. J. Friedhoff and T. N. Chase, eds.), pp. 69–76, Raven Press, New York.

Johansson, B., and Roos, B., 1974, 5-Hydroxyindoleacetic acid and homovanillic acid in cerebrospinal fluid of patients with neurological diseases, *Eur. Neurol.* **11**:37–45.

John, E. R., Karmel, B. Z., Corning, W. C., Easton, P., Brown, D., Ahn, H., John, M., Harmony, T., Prichep, L., Toro, A., Gerson, I., Bartlett, F., Thatcher, R., Kaye, H., Valdes, P., and Schwartz, E., 1977, Neurometrics, *Science* **196**:1393–1410.

Joschko, M., and Rourke, B., 1982, Neuropsychological dimensions of Tourette syndrome: Test-retest stability and implications for intervention, in: *Advances in Neurology, Gilles de la Tourette Syndrome*, Vol. 35 (A. J. Friedhoff and T. N. Chase, eds.), pp. 297–304, Raven Press, New York.

Kidd, K. K., Prusoff, B. A., and Cohen, D. J., 1980, The familial pattern of Tourette syndrome, *Arch. Gen. Psychiatry* **37**:1336–1339.

Lake, C. R., Ziegler, M. G., Eldridge, R., and Murphy, D. L., 1977, Catecholamine metabolism in Gilles de la Tourette's syndrome, *Am. J. Psychiatry* **134**:257–260.

Lowe, T. L. Cohen, D. J., Detlor, J., Kremenitzer, M. W., and Shaywitz, B. A., 1982, Stimulant medications precipitate Tourette's syndrome, *JAMA* **247**:1168–1169.

Moldofsky, H., and Sandor, P., 1983, Lecithin in the treatment of Gilles de la Tourette's syndrome, *Am. J. Psychiatry* **140**:1627–1629.

Morselli, P. L., Bianchetti, G., Durand, G., Le Heuzey, M. D., Zarifian, E. and Dugas, M., 1979, Haloperidol plasma level monitoring in pediatric patients, *Ther. Drug Monit.* **1**:35–46.

Nomura, Y., and Segawa, M., 1979, Gilles de la Tourette syndrome in Oriental children, *Brain Dev.* **1**:1103–111.

Obeso, J. A., Rothwell, J. C., and Marsden, C., 1981, Simple tics in Gilles de la Tourette syndrome are not prefaced by a normal premovement EEG potential, *J. Neurol. Neurosurg. Psychiatry* **44**:735–738.

Polinsky, R. J., Ebert, M. H., Caine, E. D., Ludlow, C., and Bassich, C. J., 1980, Cholinergic treatment in the Tourette syndrome, *N. Engl. J. Med.* **302**:1310.

Ross, M. S., and Moldofsky, H., 1978, A comparison of pimozide with haloperidol in Gilles de la Tourette syndrome, *Am. J. Psychiatry* **135**:585.

Shapiro, A. K., and Shapiro, E., 1968, Treatment of Gilles de la Tourette's syndrome with haloperidol, *Br. J. Psychiatry* **114**:345–350.

Shapiro, A. K., and Shapiro, E. S., 1980, *Tics, Tourette Syndrome and Other Movement Disorders: A Pediatrician's Guide*, The Tourette Syndrome Association, Inc., Bayside, NY.

Shapiro, A. K., and Shapiro, E., 1981a, The treatment and etiology of tics and Tourette syndrome, *Compr. Psychiatry* **22**(2):193–205.

Shapiro, A. K., and Shapiro, E., 1981b, Do stimulants provoke, cause or exacerbate tics and Tourette syndrome? *Compr. Psychiatry* **22**(3):265–273.

Shapiro, A. K., and Shapiro, E., 1982a, Tourette syndrome: Clinical aspects, treatment and etiology, *Semin. Neurol.* **2**(4):373–385.

Shapiro, A. K., and Shapiro, E., 1982b, An update on Tourette syndrome, *Am. J. Psychother.* **36**:379–390.

Shapiro, A. K., and Shapiro, E., 1984, Controlled study of pimozide vs placebo in Tourette syndrome, *J. Am. Acad. Child Psychiatry* **23**:161–173.

Shapiro, E., and Shapiro, A. K., 1982, Tardive dyskinesia and chronic neuroleptic treatment of Tourette patients, in: *Advances in Neurology, Gilles de la Tourette Syndrome*, Vol. 35, (A. J. Friedhoff and T. N. Chase, eds.), pp. 413, Raven Press, New York.

Shapiro, A. K., Shapiro, E., and Wayne, H. L., 1973, Treatment of Gilles de la Tourette syndrome with haloperidol: Review of 34 cases, *Arch. Gen. Psychiatry* **28**:92–96.

Shapiro, A. K., Shapiro, E., Bruun, R. D., and Sweet, R. D., 1978, *Gilles de la Tourette Syndrome*, Raven Press, New York.

Shapiro, A. K., Shapiro, E., and Sweet, R. D., 1981, Treatment of tics and Tourette syndrome, in: *Disorders of Movement* (A. Barbeau, ed.), pp. 105–132, MTP Press, Ltd., Lancaster, England.

Shapiro, A. K., Shapiro, E., and Eisenkraft, G. J., 1983a, Treatment of Tourette disorder with penfluridol, *Compr. Psychiatry* **24**(4):327–331.

Shapiro, A. K., Shapiro, E., and Eisenkraft, G. J., 1983b, Treatment of Tourette disorder with pimozide, *Am. J. Psychiatry* **140**(9):1183–1186.

Shapiro, A. K., Baron, M., Shapiro, E., and Levitt, M., 1984, Enzyme activity in Tourette syndrome, *Arch. Neurol.* **41**:282–285.

Shapiro, A. K., Shapiro, E., and Eisenkraft, G. J., 1983c. Treatment of Tourette syndrome with clonidine and neuroleptics, *Arch. Gen. Psychiatry* **40**:1235–1240.

Shostak, M., and Perel, J., 1976, Radioimmunoassay for haloperidol, *Fed. Proc.* **35**:531, 1976.

Singer, H. S., Tune, L. E., Butler, I. J., Zaczek, R., and Coyle, J. T., 1982, Clinical symptomatology, CSF neurotransmitter metabolites, and serum haloperidol levels in Tourette Syndrome, in: *Advances in Neurology, Gilles de la Tourette Syndrome*, Vol. 35 (A. J. Friedhoff and T. N. Chase, eds.), pp. 177–183, Raven Press, New York.

Stahl, S. M., and Berger, P. A., 1980, Cholinergic treatment in the Tourette syndrome, *N. Engl. J. Med.* **302**:1311.

Stahl, S. M., and Berger, P. A., 1981, Physostigmine in Tourette syndrome: Evidence for cholinergic underactivity, *Am. J. Psychiatry* **138**:240–242.

Sutherland, R. J., Kolb, B., Schoel, W. M., Whishaw, I. Q., and Davies, D., 1982, Neuropsychological assessment of children and adults with Tourette syndrome: A comparison with learning disabilities and schizophrenia, in: *Advances in Neurology, Gilles de la Tourette Syndrome*, Vol. 35 (A. J. Friedhoff and T. N. Chase, eds.), pp. 311–322, Raven Press, New York.

Sweeney, D., Pickar, D., Redmond, D. E., Jr., and Maas, J., 1978, Noradrenergic and dopaminergic mechanisms in Gilles de la Tourette syndrome, *Lancet* **1**:872.

Tanner, C. M., Goetz, C. G., and Klawans, H. L., 1982, Cholinergic mechanisms in Tourette syndrome, *Neurology (NY)* **32**(11):1315–1317.

Tibbitts, R. W., 1981, Neuropsychiatric aspects of tics and spasms, *Br. J. Hosp. Med.* **25**:454, 456–457, 459.

Van Woert, M. H., Jutkowitz, R., Rosenbaum, D., and Bowers, M. B., Jr., 1976, Gilles de la Tourette syndrome: Biochemical approaches, in: *The Basal Ganglia* (M. D. Yahr, ed.), pp. 459–465, Raven Press, New York.

Wassman, E. R., Eldridge, R., Abuzzahab, S. S. R., and Nee, L., 1978, Gilles de la Tourette syndrome: Clinical and genetic studies in a Midwestern city, *Neurology (Minneap.)* **28**(3):304–307.

Wilson, R. S., Garron, D. C., Tanner, C. M., and Klawans, H. L., 1982, Behavior disturbance in children with Tourette syndrome, in: *Advances in Neurology, Gilles de la Tourette Syndrome*, Vol. 35 (A. J. Friedhoff and T. N. Chase, eds.), pp. 329–333, Raven Press, New York.

Yeragani, V. K., Blackman, M., and Baker, G. H., 1983, Biological and psychological aspects of a case of Gilles de la Tourette syndrome, *J. Clin. Psychiatry* **44**:27–29.

CHAPTER 12

Clinical Management of Tourette Syndrome with Clonidine

RICHARD DORSEY

1. BACKGROUND

The demonstration of clonidine's effectiveness in Tourette syndrome (TS) (Ashanuddin *et al.*, 1980; Cohen *et al.*, 1979, 1980; Dorsey, 1981) substantially expanded the pharmacological options in the management of this condition. For the preceding decade, the only clearly effective medication was haloperidol, and before that, TS sufferers received a host of treatments ranging from psychotherapy to behavior modification to electroconvulsive treatment.

This chapter will focus on the clinical approach to assessment and treatment of TS patients, using clonidine. In many instances, patients will present to the specialist with the diagnosis already made by themselves or other physicians. Clonidine therapy will frequently be sought if haloperidol has proven ineffective or has been poorly tolerated.

2. DIAGNOSTIC EVALUATION

As in most areas of medicine, particularly neuropsychiatry, the single most important diagnostic tool in TS is the history. The typical patient is a male who began experiencing facial tics between ages 5 and 10, without a known precipitant. Eye blinking is probably the most common initial symptom, followed by facial twitches and head and neck movements. Tics characteristically vary in intensity and pattern over time, with some tendency toward spread downward to the neck, upper extremities, and lower extermities. Sniffing, grunting, and other nonspecific vocalizations are common. Only a minority of patients have coprolalia, which generally appears after several years, if at all. In many

RICHARD DORSEY • Department of Psychiatry, Doctors Hospital West, Columbus, Ohio 43228.

instances, patients can mask obscene words so that the resulting vocalizations sound like a cough or grunt.

Patients typically can suppress most or all of their tics temporarily with considerable voluntary effort, but eventually must "catch up" by release of multiple tics, often (though not invariably) when the patient has managed to get into a situation that affords some privacy. Tics may be worse when the patient is inattenttive to them or when he is anxious, stressed, or using caffeinated beverages. Since patients are usually on their best behavior in the doctor's office, observing them unobtrusively in the waiting room or parking lot can give a more accurate picture of the degree of movement disorder. For pediatric patients, reports by parents, teachers, and other adults are essential in identifying the pattern and severity of the tics.

In examining a patient with suspected TS, the physician should try to answer three questions. First, does the patient have TS? Second, does he also have any associated neuropsychiatric or medical disorders? Third, what course of treatment should be chosen?

The history of the present illness should cover onset, pattern, progression, and severity of symptoms. Inquiry should be made for any precipitating or aggravating factors. For example, some reports have suggested that the use of psychostimulants in the treatment of attention deficit disorder may lead to development of TS (Cohen *et al.*, 1982).

Current therapy, if any, should be noted, along with its effectiveness and side effects.

Mental and behavioral disturbances should be carefully assessed. Although motor tics and vocalizations are characteristic of TS, other behavioral disturbances are reported in a considerable proportion of this patient population. Cohen, in particular, has reported "severe compulsions, irritability, intolerance of frustration, antagonistic behavior and obsessive thoughts" in approximately 40% of his patients, and he tends to view TS as a "disorder of psychomotor inhibition" (Cohen *et al.*, 1980). Other authors have also reported a high incidence of such symptoms as sleep disturbance, learning disability, destructive behavior, inappropriate sexual activity, and antisocial behavior among TS patients (Nee *et al.*, 1980). No one is sure at present whether these symptoms reflect variants of TS, coexisting disorders to which the patient is also genetically predisposed, disturbances developed in response to the existence of TS, or simply coincidental conditions. In some instances, mental disturbance, particularly impaired alertness, concentration, and academic performance, may be due to adverse effects of medication prescribed for TS, particularly haloperidol.

The patient's own familiarity with TS and expectations in seeing a particular physician should be noted. Some patients are quite sophisticated on the basis of previous medical encounters, reading, and contacts with a TS association. Others know only that their family physician has suspected TS, or that they themselves have reached the same conclusion after seeing a television program about this condition.

The past history should review details of development, especially in pediatric patients, noting any delays in reaching usual milestones and any disturbances of normal mental or neurological function. Prenatal disturbances, birth difficulties, and childhood disease should be noted.

Previous diagnoses and the bases for these should be determined, with old medical records being reviewed if available. A diagnosis of TS made by a physician known to have experience with this condition, and to have conducted a thorough evaluation within the preceding 3–6 months, obviates the need for evaluation in as much detail as is advisable for a patient without such assessment. In addition, previous diagnoses may have shaped the patient's thinking with respect to the condition. For example, a severely affected woman who came to us in her 40s was told in childhood that she had "Saint Vitus' dance." When, as an adolescent, she developed barking among her other tics, she was advised that she must have been frightened by a dog. However, at no time was any specific treatment given, and she had grown resigned to her abnormal movements until seeing a television film abut TS, at which point she made the preliminary diagnosis herself. Other middle-aged patients may have received various psychiatric diagnoses and treatments over the years before awareness of TS reached their physicians or the general public.

A careful review of previous treatment offers an invaluable guide to future therapy. Most patients who are correctly diagnosed as having TS and started on usual doses of haloperidol show appreciable improvement with acceptable side effects. Those not satisfied, either with therapeutic results or because of adverse effects, should be checked as to prescribed dosage, duration of therapy, and compliance. For some, total daily dosage may have produced more side effects than would occur with the same dose given once at bedtime. For still others, mild akathisia may have gone unrecognized or been attributed to anxiety or TS itself.

Any associated neurological disease, particularly epilepsy, should be noted, with details of onset, pattern, treatment, and control. Similarly, behavioral disturbances such as those noted previously and any mental illness requiring psychiatric hospitalization or outpatient treatment should be inquired after. Response to any previously prescribed psychotropics, including antipsychotics, antidepressants, and antianxiety medications, should be recorded.

Family history should also indicate the presence of tics, which may be identified in about a quarter of the first-degree relatives of TS patients. Particular note should be taken of any other genetic relatives diagnosed as having TS, with particular reference to any treatment given and the response to such treatment. The social and occupational history of TS patients is particularly important in determining the need for, nature of, and response to treatment. Mild and moderate tics often are not subjectively distressing to patients, but cause problems for them mainly on the basis of the response of others. Children with TS may be subject to mockery or ostracism by their classmates and disciplined by school personnel who mistake their tics for unruliness or attention-seeking behavior. The resultant pressure from peers and teachers may, of course, lead to adjustment reactions with secondary psychiatric symptoms and behavioral problems.

For adults, the social history should focus on the degree of occupational or social impairment resulting fom the tics. Even mild tics may be disabling for such occupations as sales or modeling, while more severe tics may have limited impact on people whose jobs involve minimal public contact.

Proper evaluation of the TS patient also requires a careful physical examination, which may be done by the referring physician, the treating physician, or a consultant. Particular attention should be placed on the neurological examination, looking for other neurological disorders as well as for the "soft" neurological signs which may be seen in patients who have both TS and such conditions as attention deficit disorder. The psychiatric examination should establish the presence or absence of significant psychopathology, including anxiety, depression, insomnia, feelings of unreality, hallucinations, delusions, and impairment of memory, concentration, or orientation. Intelligence should be estimated clinically, with particular reference to the possibility of coexisting mental retardation.

Neurodiagnostic studies, particularly electroencephalography and computerized tomography, should generally be done to exclude seizure disorders and structural central nervous system pathology. Psychological testing may be helpful in some cases, particularly for subtle organic brain syndromes or other psychopathology or when mental retardation is suspected.

Routine laboratory screening, including a complete blood count, urinalysis, and chemistry profile, should be considered, along with thyroid function studies in some cases. Measurement of plasma levels of haloperidol in patients taking this medication can give some indication as to adequacy of dosage by identifying very low or very high levels. A therapeutic range of 3–15 ng has been reported for haloperidol in schizophrenia and possibly in TS patients, though the latter should probably be titrated toward the low end of that range. Those with less than 1 ng are usually underdosed and/or noncompliant, while those with plasma levels of 10–15 ng or higher may be exposed to unnecessary side effects.

3. PRELIMINARY CLASSIFICATION

Once evaluation has been completed, patients should be classified according to the key questions noted in Section 2. First, with respect to TS, decisions should be made as to whether the patient does or does not have this condition. If it is present, the severity should be estimated on both clinical and psychosocial grounds and an analysis made of whether the patient has primarily motor tics, behavioral disturbance, or a combination of the two.

In addition, previous treatment response should be assessed carefully. Also, when close genetic relatives have been treated for TS, their response to medication may be useful in predicting the response of the current patient.

Second, the presence or absence of any other neuropsychiatric disorder should be established, with proper diagnosis of any coexisting conditions evident. The severity and course of these disorders should be noted, along with any previous treatment and response.

Finally, an estimate should be made of the interaction between TS and any coexisting neuropsychiatric disorder. Many patients have adjustment reactions due to the social consequences of their tics, which tend to resolve when tic control has been achieved. Conversely, some patients with mental retardation or psychoses may have the unrealistic expectation that treatment of the TS will solve all their mental or behavioral problems. An accurate appraisal of the case by the physician, shared with the patient or responsible relative initially, can greatly reduce misunderstanding and dissatisfaction.

4. CHOICE OF PHARMACOLOGICAL THERAPY

With TS, as with many other neuropsychiatric disorders, the wise physician will consider "no treatment" as one option in good patient care. Some TS patients come only for confirmation of the diagnosis and do not really want treatment. Others seek advice as to whether treatment is needed or not and are inclined to follow the physician's recommendation. Major considerations include objective severity of the tics, as observed by the physician and reported by the patient or other responsible individuals; the subjective distress the patient experiences as a result of the tics, or the reactions of others to them; and the degree of impairment of social or occupational functioning produced by the patient's condition. In addition, some patients have a philosophical aversion to taking medication unless no practical alternative exists, while others report unusual drug sensitivities or intolerance of medication side effects.

A course of nontreatment has the advantages of avoiding adverse effects, minimizing the cost and inconvenience associated with regular medication and visits to the physician's office, and eliminating concern about unnecessary medication use. To present a balanced picture of risks and benefits, patients should be informed that neither haloperidol nor clonidine has any addiction potential and that there is no reason to suppose that deferring treatment will contribute to an increased severity of illness in the future. In general, when in doubt, omission of treatment and reassessment after 60–90 days constitutes a sensible course of action.

If pharmacological treatment is indicated, haloperidol is generally accepted as the drug of first choice. Abuzzahab and Anderson (1973) reported that 89% of TS patients were improved after 6 months of treatment with haloperidol. Shapiro and Shapiro (1982) reported similar findings. The latter authors found that haloperidol was more effective than clonidine in 61% of TS patients, with the reverse being true in 28% of their cases (Shapiro and Shapiro, 1982).

Haloperidol is generally prescribed on the basis of one or more of the following indications: subjective distress, objective impairment, social/occupational impairment, and patient desire for pharmacological treatment. The best candidates for haloperidol therapy are those who have not previously received any pharmacological treatment and those who have appreciable anxiety associated with their TS. For anxious TS patients, in the rare cases with TS plus schizophrenia, haloperidol's tranquilizing and antipsychotic actions provide what amounts to a therapeutic bonus. In addition, patients who have responded well to haloperidol in the past, but discontinued medication for various reasons (vacation, temporary remission of symptoms), should be started back on this medication. The final indication for haloperidol is unsatisfactory response to clonidine (inadequate therapeutic response and/or undesirable side effects). In patients treated first with clonidine, failure of this medication will generally lead to a trial of haloperidol as the drug of next choice. For patients who have taken both medications, patient preference, guided by physician observation, will usually determine which drug the patient should continue.

The most common haloperidol side effects leading TS patients to seek treatment with clonidine are those frequently experienced by normal people given antipsychotic drugs. Among these are drowsiness, dysphoria, impaired alertness, and difficulty with memory and concentration. Marked weight gain, especially in children, may occur occasionally with haloperidol. A few TS patients, usually older children or young adults, may experience dystonic reactions with this medication. Tardive dyskinesia is extremely rare, but physicians should nonetheless be alert for its possible development. Dry mouth is rarely a significant problem with haloperidol in TS patients, and blurred vision almost never occurs; in both instances, the probable explanation is the relatively low dosage used for TS patients as compared with individuals treated for psychosis.

The latest pharmacological option for the treatment of TS is pimizide, which became available in the United States in 1984. This diphenylbutylpiperidine compound has a pharmacological profile similar to that of haloperidol, with comparable efficacy but perhaps fewer side effects (Shapiro and Shapiro, 1982, 1984).

5. USE OF CLONIDINE

The greatest advantage of clonidine over haloperidol in treating TS appears to be its substantially lower incidence and frequency of side effects (Cohen *et al.*, 1980; Dorsey, 1981; Shapiro and Shapiro, 1982; Brun, 1982; Leckman *et al.*, 1982). Estimates of efficacy by the same authors range from 40 to 70%, with somewhat better improvement in behavioral disturbance than in motor and phonic tics.

Principal indications for use of clonidine are similar to those of haloperidol, as cited previously. Factors favoring the use of clonidine over haloperidol include absence of any associated anxiety or other major psychopathology, unsatisfactory prior experience with haloperidol, and an informed patient preference for this medication. Cardiology consultation should be considered for

TS patients with significant cardiovascular disease before initiation of clonidine therapy, though these cases are rare.

Before prescribing clonidine, the physician should inform the patient of expected therapeutic and adverse effects. Literature available from the Tourette Syndrome Association and personal discussion with other patients taking this medication can be extremely valuable adjuncts, and patients should be encouraged to obtain additional information from these sources. An initial dosage of 0.05–0.10 mg once or twice daily is generally reasonable for older children or young adults. Because of the spontaneous fluctuation in intensity of symptoms with TS and the time necessary to obtain full effect with the given dose of clonidine, dosage changes should be made at intervals of no less than 1 week. Thus, to reach the usually effective dosage of 0.3–0.6 mg/24 hr, a period of at least 4–6 weeks is required for titration. Some experienced physicians may prefer an even more gradual approach.

Unlike haloperidol, clonidine should usually be given on a divided-dose basis for TS. Twice-daily regimens work for some patients, but at higher doses, t.i.d. or q.i.d. administration is appropriate.

The optimum dose for a given patient is that which provides the most satisfactory balance between symptoms and side effects. In many cases, some tradeoff in this regard is necessary, and the maintenance dose may prove to be somewhat lower than the peak dose. In other instances, patients obtain satisfactory symptom control with minimal side effects, and in these instances, no further increase in dose is warranted.

Therapeutic results with clonidine usually are manifested as a substantial reduction in tics, but complete resolution is rare. Occasionally, dramatic improvement may occur within a few weeks, with patients going from severe disability to almost normal functioning. In general, a target of 70–80% improvement should be set initially and agreed to by the patient or family.

Clonidine's most common side effects are sedation and postural hypotension, either of which may be dose-limiting. The former is usually experienced as pure drowsiness, which most patients can distinguish from the adverse side effects of haloperidol noted previously. Postural hypotension generally occurs at doses of 0.4–0.6 mg daily, with tolerance usually developing over time. Patients should be advised to sit before they stand and to stand before they start walking, in order to minimize postural dizziness during the early stages of clonidine therapy.

During the initial phase of therapy, patients should be seen by the physician every 1 or 2 weeks, or more frequently if problems arise. Telephone availability can reduce the need for office visits and provide valuable guidance and reassurance. On each office visit, information should be obtained about the overall frequency and severity of tics and of side effects. The patient should be observed in the office and his condition compared with that on previous visits; vital signs, particularly pulse and blood pressure, should be checked. Once the patient has been stabilized, visits can be decreased to every 2–3 months, with provision for greater frequency of visits if the need arises. In some instances, patients coming considerable distances may prefer to return to physicians in their own communities after stabilization of their regimen, seeing the subspecialty consultant only if new difficulties develop.

Clonidine may be used in combination with other psychotropic medications for specific indications. First, a small number of patients who have not responded satisfactorily to either haloperidol or clonidine alone may show greater benefit when both are used. In this instance, one drug should be titrated to the point of optimal results, weighing therapeutic benefits against side effects. After a reasonable period of time at this dose, the second drug should be added and titrated independently in the same manner.

Very anxious TS patients may benefit from coadministration of a short-acting benzodiazepine (e.g., lorazepam) with clonidine. This approach avoids pharmacological interactions, reduces anxiety and/or insomnia, and may also reduce stress-related exacerbation of TS. Most TS patients, particularly those with a coexisting anxiety component, should be advised not to use caffeine-containing beverages. Alcoholic beverages should also be avoided since their effects may complicate the assessment of the results of pharmacological therapy for TS.

Psychosocial aspects of TS treatment with clonidine must also receive attention. Patients or their parernts need a realistic understanding of the probable therapeutic benefits and adverse effects of the medication and of the untreated course of TS. They also may need to be reminded that not every symptom that develops while a person is taking a particular medication is an actual adverse side effect of that drug, nor is every episode of disturbed behavior a result of TS. The information functions of a good TS-association can provide extremely valuable assistance in this regard. At some point toward the end of the first year of therapy, many patients express a desire for a trial off medication. When clonidine is tapered under these circumstances, symptoms usually recur, satisfying the patient that the medication is still beneficial and reducing recurrent speculation on this point.

No significant late adverse side effects have occurred in our patients treated with clonidine continuously for up to 3 years. Nor does experience with the use of this drug for hypertension suggest the likelihood of any substantial long-term risks.

6. CONCLUSION

Clonidine is a safe and effective pharmacological treatment for TS. Currently accepted as the drug of second choice (after haloperidol) for this condition, on the basis of efficacy, its limited side effects may justify its use as the primary pharmacological agent in patients with milder motor tics, typical TS-related behavioral disturbance, or need for maximal alertness. Careful pretreatment evaluation, with particular emphasis on establishing a firm diagnosis of TS, identifying any coexisting neuropsychiatric disorders, and accurately documenting previous pharmacological treatment and response, will greatly improve the probability of successful results when clonidine is prescribed.

7. REFERENCES

Abuzzahab, F. S., and Anderson, F. O., 1973, Gilles de la Tourette syndrome, *International Registry, Minn. Med.* **56**:492–496.

Ashanuddin, K., Hall, K., Kowall, M., and Prosen, H., 1980, Tourette's syndrome-treatment with clonidine, A centrally acting alpha-adrenergic agonist, *Drug. Intell., Clin. Pharmacol.* **14**(32):288–289.

Brun, R. D., 1982, Clonidine treatment of Tourette's syndrome, *Adv. Neurol.* **35**:403–405.

Cohen, D. J., Young, J. G., Nathanson, J. A., and Shaywitz, B. A., 1979, Clonidine in Tourette's *syndrome, Lancet* **II**(8142):551–553.

Cohen, D. J., Detlor, J., Lowe, T., 1980, Clonidine ameliorates Gilles de la Tourette syndrome, *Arch. Gen. Psychiatry* **37**(12):1350–1357.

Cohen, D. J., *et al.*, 1982, Stimulant medications in Tourette's syndrome, *JAMA* **248**:1062–1063.

Dorsey, R., 1981, Clonidine and Gilles de la Tourette syndrome, *Arch. Gen. Psychiatry* **38**(10):1185.

Leckman, J. F., Cohen, D. J., Detlor, J., Young, J. G., Harchurik, D., and Shawitz, B. A., 1982, Clonidine in the treatment of Tourette syndrome: A review of data, in: *Gilles de la Tourette Syndrome, Advances in Neurology*, Vol. 35, pp. 391–402, Raven Press, New York.

Nee, L. E., Polinsky, R. J., and Ebert, M. H., 1980, Gilles de la Tourette syndrome: Clinical and family study of fifty cases, *Ann. Neurol.* **7**(1):41–49.

Shapiro, A. K., and Shapiro, E., 1984, Controlled study of Pimizide vs. placebo in Tourette syndrome, *J. Am. Acad. Child Psychiatry* **23**(2):161–173.

Shapiro, A. K., and Shapiro, E., 1982, *Advances in Neurology*, Vol. 35, pp. 283–286, Raven Press, New York.

CHAPTER 13

The Pathogenesis of Tourette Syndrome

A Review of Data and Hypotheses

James F. Leckman, Donald J. Cohen, R. Arlen Price, Mark A. Riddle, Ruud B. Minderaa, George M. Anderson, and David L. Pauls

1. TOURETTE SYNDROME

Tourette syndrome (TS) is a rare, chronic neuropsychiatric disorder of unknown etiology that is characterized by waxing and waning multiform motor and phonic tics and a broad range of behavioral symptoms (Shapiro *et al.*, 1978; Cohen *et al.*, 1982; Leckman *et al.*, 1983a).

1.1. Natural History and Phenomenology

The natural history of TS is a complex and partially age-dependent set of phenomena. Attentional problems and difficulties with hyperactivity and impulse control frequently precede the emergence of frank motor and phonic tics (Jagger *et al.*, 1982). Motor tics typically appear well before phonic tics (Table 1). The motor tics usually show a rostral–caudal progression, so that tics involving the face and head usually precede those involving the trunk or the

JAMES F. LECKMAN and MARK A. RIDDLE • Child Study Center, Children's Clinical Research Center, and Departments of Psychiatry and Pediatrics, Yale University, New Haven, Connecticut 06510. DONALD J. COHEN • Child Study Center, Children's Clinical Research Center, and Departments of Psychiatry, Pediatrics, and Psychology, Yale University, New Haven, Connecticut 06510. R. ARLEN PRICE • Department of Psychiatry, University of Pennsylvania, Philadelphia, Pennsylvania 19104. RUUD B. MINDERAA • Eramus Universiteit Rotterdam, Rotterdam, The Netherlands. GEORGE M. ANDERSON • Child Study Center and Department of Laboratory Medicine, Yale University, New Haven, Connecticut 06510. DAVID L. PAULS • Child Study Center and Department of Human Genetics, Yale University, New Haven, Connecticut 06510.

TABLE 1
Rostral–Caudal Progression of Motor Symptoms; Sequence of Phonic and Behavioral Symptoms in Tourette Syndrome[a]

Mean age (years)	Motor symptoms		Phonic symptoms		Behavioral symptoms	
	Symptom	%	Symptom	%	Symptom	%
6					Impulsive, hyperactivity, attentional problems	70
7	Eyes, face, and head tics	95				
8	Shoulder and neck tics	92				
9	Arm and head tics	83				
10	Trunk and leg tics	61				
11			Low noises	84	Compulsive actions: touching objects	55
			Loud noises	67	Touching body	36
			Coprolalia	37	Tapping, self-destructive acts	20
					Biting, hitting	17
12			Echolalia	37	Mimicking others	21

[a] Adapted from Jagger *et al.*, 1982.

extremities (Shapiro *et al.*, 1978; Jagger *et al.*, 1982). Descriptively, these symptoms vary in their complexity, frequency, and the degree of social role disruption they produce (Leckman *et al.*, 1983a). Simple motor tics are usually rapid and meaningless muscular events, such as eye blinking, grimacing, nose twitching, and lip pouting. The complex motor tics are slower and more "purposeful" in appearance. Often they involve repetitive grooming behaviors, such as a movement brushing hair back away from the face, but they can involve any type of movement the body can produce, including touching, clapping, throwing, hopping, or kicking movements, as well as bending, writhing, or the sudden appearance of "dystonic" postures. A small percentage of patients develop self-abusive, complex motor "tics" of biting or hitting (Table 1). Simple phonic tics typically involve linguistically meaningless noises and sounds, such as sniffing, throat clearing, hissing, coughing, or sucking. Complex phonic tics involve the sudden ejaculation of inappropriate words or phrases. Patients may also suddenly alter the pitch and/or volume of their speech or mimic the speech of others.

The frequency of both the motor and phonic symptoms can be quite variable. However, the rapid repetition of a motor or phonic tic or a set combination of tics is not unusual, and the occurrence of "paroxysms" of tics is well described.

1.2. Genetic Factors

Interest in a possible hereditary component to the transmission and expression of TS is long-standing, dating back to the earliest descriptions of the syn-

TABLE 2
Concordance of Tourette Syndrome in Monozygotic and Dizygotic Twins[a]

Zygosity	Number of twin pairs	Status of cotwin (%) Fully concordant	Partially concordant	Discordant
MZ	30	53	23	23
DZ	14	7	14	79

[a] Adapted from Price *et al.*, 1983. MZ, monozygotic; DZ, dizygotic.

drome (Gilles de la Tourette, 1885). Although a great deal of work has been performed in this area, no specific genetic mechanism has been identified (Eldridge *et al.*, 1977; Kidd *et al.*, 1980; Nee *et al.*, 1980; Pauls *et al.*, 1981; Wilson *et al.*, 1978; Shapiro and Shapiro, 1981). The familial concentration of TS has been documented, and the pattern is consistent with some form of vertical transmission (Kidd and Pauls, 1982). Several other important findings have also emerged, including the close relationship between TS and the syndrome of chronic multiple tics (usually chronic motor tics without phonic symptoms), and a sex threshold effect (although boys are more commonly affected with TS than girls, the relatives of girls with TS are at greater risk for TS or chronic multiple tics than are the relatives of boys with TS).

Previous anecdotal case reports of twins have indicated the presence of a very high concordance rate for monozygotic twins for TS in the range of 75–95% (Shapiro and Shapiro, 1981; Ellison, 1964; Wasserman *et al.*, 1978; Bachman, 1981; Jenkins and Ashby, 1983). In contrast, the initial results from a questionnaire study of 30 monozygotic twin pairs in which at least one twin has TS indicate only a 53% concordance for TS and a 76% concordance for TS or chronic multiple tics (Table 2) (Price *et al.*, 1985). Concordant twin pairs were also frequently found to show differing levels in the severity of symptom expression. These more recent data provide support for the importance of environmental (nongenetic) factors in the expression of TS.

1.3. Environmental Factors

The onset of TS symptoms and the transitional periods that occur in its waxing and waning course often seem unrelated to environmental factors. However, clinical experience and pilot epidemiological studies of TS indicate that periods of increased anxiety and emotional stress regularly produce an exacerbation of TS symptoms (Jagger *et al.*, 1982).

A second environmental (nongenetic) factor known to exacerbate TS symptoms is exposure to stimulant medications. Besides leading to a worsening of symptoms in TS patients, chronic exposure to stimulants has also been linked to the emergence of TS symptoms in children who have no prior history of TS (Golden, 1974, 1977; Lowe *et al.*, 1982). Typically, this occurs in children who display inattention and poor impulse control in the classroom and who are diagnosed as having attention deficit disorder with hyperactivity. Frequently, these children have a positive family history of TS or chronic multiple tics and may themselves have a history of transient tics (Lowe *et al.*, 1982).

The sequence of events that lead to the full expression of TS in children treated with stimulants is variable. In most cases, the attentional problems improve over the first weeks of treatment. Usually only after weeks to months do the tics appear, although occasionally the onset is within a day or 2 of starting the stimulant. Usually motor tics precede phonic tics, although the full syndrome can suddenly appear over a few days. With cessation of medication, the symptoms can persist indefinitely, although in some cases the symptoms abate with time, over weeks to months (Lowe *et al.*, 1982). The broad range of response argues against there being a simple one-to-one relationship between exposure to stimulants and the appearance or exacerbation of tics.

1.4. Neurochemical Systems

Although a number of neurochemical systems have been implicated in TS, including the cholinergic, serotonergic, noradrenergic, GABAergic, and neuropeptide systems, the strongest evidence supports the role of dopaminergic systems in the pathophysiology of TS. This evidence includes lowered levels of cerebrospinal fluid homovanillic acid (HVA), a major metabolite of central dopamine (DA), in many TS patients(Cohen *et al.*, 1979a; Butler *et al.*, 1979). Pharmacologically, the beneficial suppressant effects of DA-blocking agents, haloperidol, pimozide, as well as other neuroleptics in 60–80% of TS patients requiring treatment (Shapiro *et al.*, 1973, 1978; Shapiro and Shapiro, 1981; Connell *et al.*, 1967; Messiha *et al.*, 1971; Moldofsky *et al.*, 1974; Ross and Moldofsky, 1977; Eldridge *et al.*, 1977; Borison *et al.*, 1982) and the detrimental effects of stimulant medications, direct DA agonists, and DA precursors on TS symptoms (Golden, 1974, 1977; Lowe *et al.*, 1982; Klempel, 1974) strongly support the importance of DA mechanisms in TS. Other suggestive data include the emergence of TS-like symptoms following withdrawal from chronic neuroleptics in psychiatric patients (Fog and Pakkenberg, 1980; Klawans *et al.*, 1978; Stahl, 1980).

Evidence for noradrenergic involvement in TS comes primarily from pharmacological studies of the beneficial effects of clonidine, an imidazoline derivative (Cohen *et al.*, 1979b; Leckman *et al.*, 1982, 1983b, 1985). A variety of mechanisms of action have been proposed for clonidine's effect on the central nervous system (CNS). Central noradrenergic events have been the most extensively studied in this regard. At low doses, clonidine preferentially stimulates α_2-adrenergic receptors probably located presynaptically (Schmitt *et al.*, 1971; Sharma *et al.*, 1978). Stimulation of these receptors decreases the amount of norepinephrine released into the synapse. Studies using a microiontophoretic technique have shown that clonidine acutely inhibits the spontaneous firing of the locus ceruleus and reduces brain norepinephrine (NE) turnover (Anden *et al.*, 1970; Cedarbaum and Aghajanian, 1976, 1978; Svensson *et al.*, 1975). It also acutely lowers the concentration of the major brain metabolite of NE, 3-methoxy-4-hydroxyphenethylene glycol (MHPG), in rats (Braestrup and Nielsen, 1976), reduces brain MHPG production in monkeys (Maas *et al.*, 1977), and decreases plasma free MHPG concentrations in man (Leckman *et al.*, 1980b, 1981).

Interestingly, the mode of action of clonidine in TS may involve indirect effects on the DA system as evidenced by altered levels of plasma HVA following chronic clonidine treatment (Leckman *et al.*, 1983b). The report that a positive clinical response to haloperidol during a double-blind crossover trial predicted a positive response to clonidine is also consistent with this hypothesis (Borison *et al.*, 1982). Serotonergic mechanisms have been suggested as the link between central adrenergic and dopamine systems (Bunney and DeRiemer, 1982).

2. AMPHETAMINE–STRESS SENSITIZATION

Amphetamine sensitization is a well-described phenomenon in which the repetitive administration of amphetamine leads over a period of days to weeks to the emergence of hyperactivity, stereotypies, and dyskinetic reactions in a variety of animal species (Robbins and Sahakian, 1981). The range of behaviors elicited appears to be under some genetic control (Robbins and Sahakian, 1981; Szechtman *et al.*, 1982; Leith and Kuczenski, 1982). A listing of the observed behaviors contrasted to TS symptoms is presented in Table 3.

Antelman and co-workers demonstrated the interchangeability of repetitive "stress" using a "tail pressure paradigm" and amphetamine administration in inducing stereotyped sniffing in rats (Antelman *et al.*, 1980). Other stressors, including foot shock and food deprivation, are also known to produce stereotyped behaviors in animals (Kokkinidis *et al.*, 1979). Other investigators have documented the synergistic effects of stress and amphetamine administration in producing stereotypies in animals (Sahakian *et al.*, 1975; Eichler and Antelman, 1979; Guisado *et al.*, 1980). More recently, MacLennan and Maier (1983) have shown that stress-induced stereotypies, using foot shock as a stimulus, develop only under conditions in which the animal has no "control" over the stressful stimulus as animals that could "control" the duration of the stimulus did not develop stereotypies despite receiving the same amount of stimulus over the same time course.

Neurochemically, central DA mechanisms have been most widely implicated in the development of amphetamine or stress sensitization in animals (Kobayashi *et al.*, 1976; Anisman, 1978; Robbins and Sahakian, 1981). The evidence linking this sensitization to DA systems includes the antagonism of sensitization by haloperidol and the ability of other agents able to release DA

TABLE 3
Comparison of Stress-Amphetamine-Induced Stereotypies in Animal Studies and Tourette Syndrome in Man[a]

Dimension	Stress–amphetamine-induced stereotypies	Tourette syndrome
Behaviors	Hyperactivity, hyperexcitability, intermittent stereotypies: grooming, eating, sniffing; body twitches, rotational behavior, dyskinesias	Hyperactivity, attentional problems, simple and complex motor and phonic symptoms, "obsessive–compulsive" behaviors, dyskinetic postures, rotational behavior (rare)
Sexual differences	Evidence for sexual dimorphism; estrogens may inhibit stereotypies	Boys more commonly affected than girls
Genetic factors	Some evidence for strain-specific expression of stereotypic behavior	Frequent positive family history of TS and chronic multiple tics
Pathogenesis	Intermittent repetitive stimuli over several days	Unknown; stress and stimulants are likely contributing factors; usually develops over months
Onset	In 21-day-old rats, marked increases in locomotor activity and sniffing in response to stimulants	Mean age of onset of motor tics 7 years; hyperactivity often present from 3–4 years of age
Course	Animals continue to show increased sensitivity to stress and stimulants	Symptoms wax and wane; most show attenuation of symptoms in adulthood
Pharmacological	Suppressed acutely by haloperidol, clonidine? fenfluramine? opiates and estrogen treatment; increased by DA precursors and releasers of DA and by withdrawal from estrogens	Suppressed acutely by haloperidol—70–80% of cases; clonidine is effective in suppressing symptoms after 6–12 weeks; increased by DA precursors and releasers of DA
Stress effects	Similar to the effects of stimulant agents in producing and intensifying behaviors	Possible pathogenetic role in TS; once syndrome is established, stress produces a time-limited increase in symptoms

[a] TS, Tourette syndrome; DA, dopamine.

(phenmetrazine, methylphenidate) to induce stereotypies (Randrup and Munkvrad, 1974). Interestingly, prior treatment with haloperidol appears to facilitate this sensitization process (Gianutsos *et al.*, 1974). Injection of a direct DA agonist, such as apomorphine, into the striatum has also been reported to produce stereotypic gnawing (Ernst and Smelik, 1966). Increased release of DA and increased levels of DA metabolites from the caudate and other brain regions has been reported concurrently with stereotypic behavior in stressed animals (Curzon *et al.*, 1979; Herman *et al.*, 1982). Depletion of DA in the striatum by bilateral injections of 6-hydroxydopamine in the substantia nigra blocks the development of amphetamine-induced stereotypy in adult rats (Creese and Iversen, 1972). These lesions, however, did not block the hyperactivity associated with amphetamine administration. There exists a large and expanding literature concerning the role of dopaminergic mechanisms in amphetamine/stress-induced stereotypies, and basic insights concerning these mechanisms should emerge in the near future. A growing body of literature also suggests that cholinergic, serotonergic, and endogenous opiate mechanisms may be intimately involved.

The role of central NE systems in mediating amphetamine or stress-induced stereotypies remains unclear (Robbins and Sahakian, 1981). Although amphetamine administration (acute and long-term) and stress (acute and chronic) clearly affect central NE systems (Cassens *et al.*, 1980; Segal, 1975), until now there has been relatively little evidence to suggest that these effects on central NE systems have a major impact on the development of amphetamine or stress-induced stereotypies (Hollister *et al.*, 1974; Mogilnicka and Braestrup, 1976). Additional work is needed, however, before any definite conclusions can be reached concerning the modulatory role of central NE systems.

Endocrine factors may also play a role in modulating stimulant-induced stereotypies. As noted in Table 3, evidence has been presented that indicates that central DA systems are sexually dimorphic and that estrogens may exert modulatory effects on striatal function in animals (Gordon, 1980; Robinson *et al.*, 1980; Chiodo *et al.*, 1981). These reports have emphasized the importance of timing and dosage in determining the effects of gonadal steroid hormones on DA systems (Gordon, 1980). Although controversial, it appears that endogenous estrogens can suppress DA activity and that a similar suppression is seen immediately after the administration of exogenous estrogens to ovariectomized rats (Gordon, 1980). However, the relevance of these data to physiological events in man is uncertain because these studies employed pharmacological doses of estrogen.

Other investigators have emphasized the similarities between amphetamine sensitization and the process of electrical kindling, where, following repetitive intermittent electrical stimulation of subcortical limbic areas, animals consistently experience generalized seizures to what previously had been subthreshold levels of electrical stimulation (Post and Kopanda, 1976; Adamec and Stark-Adamec, 1983). Apparently, the intermittent pattern of stimulation is critical as chronic nonintermittent stimulation actually leads to elevation of the amount of electrical stimulation required to reach a seizure threshold (Goddard *et al.*, 1969; Morrell, 1973; Racine *et al.*, 1973). Two other properties of kindling that are relevant are: (1) that in "kindled" animals there is a long-lasting (more than 1 year), interictally maintained potentiation of response, and (2) that when carried out long enough, kindling stimulation leads to spontaneous and recurrent seizures (Adamec and Stark-Adamec, 1983). Although no direct causal link has been established, a number of neurochemical systems have been implicated in the development of kindling, including central DA and NE systems, as well as cholinergic, serotonergic, and GABAergic systems (Corcoran, 1981; Deakin and Green, 1978; Deakin *et al.*, 1981; Delgado and Armand, 1953).

3. ONTOGENY OF CENTRAL DA SYSTEMS

Central DA systems appear to undergo age-specific alterations during the course of development. Although in rats these developmental changes appear to be complete by the third or fourth postnatal week (Coyle and Campochiaro, 1976); Kellogg and Lundborg, 1973; Loizou and Salt, 1970), alterations in man are evident through adolescence (Leckman *et al.*, 1980a; Anderson *et al.*, 1983; Shaywitz *et al.*, 1982). Some evidence also points to possible sex differences in the maturational course of central DA systems in man with boys showing a greater fall in cerebrospinal fluid HVA levels during childhood and adolescence (Shaywitz *et al.*, 1982).

The possible importance of the maturational course of brain DA systems in TS has been pointed out previously (Leckman *et al.*, 1983a; Shaywitz *et al.*, 1982). However, no direct evidence has been presented linking changes in the maturation of central DA systems and TS. In contrast, the DA-mediated stereotypies and electrical kindling seen in animals both appear to be partially age-dependent phenomena. Both newborn rats under 21 days of age and "older" rats over 43 weeks of age show a relative resistance to amphetamine-induced hyperactivity and stereotypic sniffing (Friedhoff, 1982; Shalaby and Spear, 1980; Watanabe *et al.*, 1982). Similarly, rats over 40 months of age also show a relative resistance to the development of kindling.

4. PATHOGENESIS OF TS: A HYPOTHESIS

The clinical expression of TS depends on both an inherited vulnerability and the action of environmental factors during a critical period of brain development. Intermittent, repeated, and "uncontrolled" stresses may activate developing endogenous brain systems which have age-dependent sensitizing effects similar to chronic stimulant medication in vulnerable individuals. Although the nature of the relationship between the underlying genetic vulnerability, the course of brain development, and the basic neurochemical mechanisms involved in the sensitization process is unclear, it is likely that all these factors are closely linked (Table 4).

4.1. Implications for Animal Models of TS

Although some elements of this hypothesis have been proposed previously, particularly the link between amphetamine-induced stereotypies and TS symptoms (Knott and Hutson, 1982; Diamond *et al.*, 1982), a number of testable corollaries of our more general hypothesis have yet to be fully explored in these

TABLE 4
Stress-Diathesis Model for the Pathogenesis of Tourette Syndrome

Genetic endowment[a]	Exposure to "uncontrolled"[b] stress or stimulants during critical period of development (3–15 years)	TS[c] outcome
−	−	No symptoms
−	+	No symptoms
+	−	No symptoms
+	+	CMT[d]
+	+ +	TS

[a] No mode of transmission has been identified in Tourette syndrome. Family study data are compatible with single major locus models. The "genetic" basis, if any, of the associated syndromes of attention deficit disorder and obsessive–compulsive disorder has also not been established and may involve additional genetic diatheses.
[b] The timing and duration of exposure are likely important characteristics, as well as the individual's and family's ability to cope with the stress. Increasing amounts of stress may lead to more extreme clinical manifestations.
[c] TS, Tourette syndrome.
[d] CMT, chronic multiple tics.

animal models. These include: (1) that different genetic strains of animals will show a differential vulnerability to the repetitive administration of stimulants, as well as discrete ranges of stereotypic behaviors, and (2) that the development of sensitization is closely tied to the ontogeny of central DA systems. More specifically, we predict that differences in the ontogeny of central DA systems will be linked to differential vulnerability among different animal strains. We also predict that central NE systems play a critical modulatory role in potentiating the DA-mediated events associated with sensitization. This modulatory role may be responsible in part for the differences observed by MacLennan and Maier (1983) in the development of stress-induced stereotypies. Another issue that could be addressed more fully in the context of such studies would be the sex threshold difference between males and females observed in TS. The possible effects of gonadal steroid hormones on early CNS organization, as well as the known effects of estrogens in suppressing amphetamine-induced stereotypies, need to be examined more fully using physiological doses of estrogens. However, the miniscule amounts of circulating estrogens in prepubertal animals may make it difficult to study this question. The reported sexual dimorphism in stress-induced stereotypies also needs to be documented more fully.

Limitations of this animal model for TS also need to be emphasized. They include the fundamental uncertainty concerning the correspondence of the "stereotypies" observed in animals and the "tics" observed in TS and the failure of the observed animal stereotypies to correspond to the typical presentation of TS in man, i.e., the more or less continuous manifestation of TS symptoms that change in character, e.g., head jerks replaced by a shoulder tic, and that wax and wane in severity over time. A second limitation concerns the correspondence between animal models of "stress" and those complex human experiences described as "stressful." Beyond these limitations, it is also important to note that several of the most critical elements in this model depend on animal studies which have not been replicated yet (Antelman *et al.*, 1980; MacLennan and Maier, 1983).

Another difficulty is that the animal model may not offer an adequate account of the relationship observed clinically between TS and the hyperactivity associated with attentional problems. At a superficial level, motoric hyperactivity is seen both as one of the first manifestations of the increasing sensitization to stimulants in the animal model, as well as one of the earliest clinical manifestations of TS (Table 3). Animal studies have also suggested that the motoric hyperactivity and stereotypies are mediated by two anatomically and functionally dissociable dopaminergic mechanisms (nigrostriatal: stereotypies; mesolimbic: hyperactivity) (Kelly *et al.*, 1975). The difficulty comes in attempting to account for the clinical observation that stimulants are *useful* in treating the hyperactivity, and attentional problems associated with TS (the animal model would predict that the hyperactivity would become worse on stimulants). Typically, those patients who go on to manifest TS symptoms while on stimulants initially respond well to the stimulants. It is only somewhat later that the treatment is discontinued because of the onset of tics. In this regard, recent work indicating that chronic exposure to stimulants leads to a *decrease* in the specific [^{3}H]spiperone binding sites in the mesolimbic area by Akiyama and co-workers (1982) is particularly exciting as it suggests that differential responses to subsequent stimulant exposure might be expected. The differential effects of lesions of the superior colliculus also point in this direction (Dawbarn and Pycock, 1982).

Further efforts to characterize separately the ontogeny of the neuronal mechanisms involved in the hyperactivity versus the stereotypies seen in the amphetamine- or stress-induced model of TS may resolve some of these issues.

4.2. Implications for the Prevention and Treatment of TS

If this hypothesis is correct, then the management of stressful situations and control of exposure to stimulant agents in vulnerable individuals during a particular period of development would be critical to prevent the development of TS symptoms. This hypothesis would suggest also that, at least in some instances, the natural history of TS is the result of detrimental positive feedback systems. Such a system would initially be activated in vulnerable children by exposure to intermittent stress, perceived as uncontrollable by the children, which, in turn, would lead to hyperactivity and other prodromal symptoms of TS which then could lead to increasing stress in the family and the development of the full syndrome with motor and phonic tics. Whether some increased neurochemical responsivity to stress may characterize part of the underlying vulnerability to TS is unclear.

Several implications follow directly from this hypothesis: (1) emotionally supportive and educational efforts with the individual, family, and school are an important clinical enterprise which contribute in a major way to an interruption of such detrimental positive feedback systems, both by reducing the amount of stress and by enhancing the family's ability to cope more effectively with stress; (2) clinical trials with pharmacological agents should rigorously control for the effects of ancillary treatments offered to the family, as interventions that diminished the degree of stress may have beneficial effects in their own right; (3) studies of the natural history of "stimulant induction" or marked exacerbation of TS would be critically dependent on other factors besides the degree of the child's exposure to stimulants (age, timing of exposure vis-à-vis CNS development, prior exposure to haloperidol, history of prior intermittently stressful periods, and some appraisal of how well the individual and the family were able to cope with the stress); (4) in family studies of TS (after controlling for stimulant usage) the families that have been subjected to the greatest amount of uncontrolled stress will show an increased rate of TS and chronic multiple tics compared to TS families with lower levels of stress; and (5) the combined use of haloperidol and stimulants, advocated by some investigators (Shapiro and Shapiro, 1981), may not result in an exacerbation of tics as the haloperidol would be expected to block the sensitizing effects of the stimulants.

5. COMMENT

Amphetamine-induced paranoid psychosis in man has led to the development of the amphetamine–stress sensitization model as a possible animal model of the psychoses (Connell, 1958; Angrist and Gershon, 1970). In our view, while similar mechanisms of sensitization may be involved in the pathogenesis of schizophrenia and bipolar illness, the animal models of amphetamine–stress sensitization developed thus far are likely to be better models of the pathogenesis of TS than of the psychoses.

In addition to the development of better animal models of TS, this hypothesis should lead to more detailed and better controlled clinical studies that can address the relationship between an individual's genetic vulnerability and the timing of pathogenetic stresses and stimulant usage. Detailed studies of the natural histories of male and female pairs of monozygotic twins partially or completely discordant for TS should be particularly useful. Biological studies of their neurochemical responsivity to modestly stressful stimuli under experimental conditions may also be useful in clarifying whether or not an exaggerated, neurochemical response to stressful situations may comprise part of the underlying vulnerability of TS.

ACKNOWLEDGMENT. This research was supported in part by NIMH grant numbers 1 P50 MH30929 and HD-03008; NICNDS grant number NS16648; the Children's Clinical Research Center, NIH grant RR00125; the William T. Grant Foundation; Mr. Leonard Berger; the Baker Foundation; the Gateposts Foundation, Inc.; the Tourette Syndrome Association (Drs. Leckman and Price); and the John Merck Fund (Dr. Leckman).

The authors wish to thank Drs. B. A. Shaywitz, J. A. Gertner, K. K. Kidd, and B. S. Bunney, Ms. S. Ort, Ms. D. Harcherik, and Mr. K. Caruso for their comments and criticisms of an earlier draft of this article.

6. REFERENCES

Adamec, R. E., and Stark-Adamec, C., 1983, Limbic kindling and animal behavior—Implications for human psychopathology associated with complex partial seizures, *Biol. Psychiatr.* **18**:269–293.

Akiyama, K., Sato, M., and Otsuki, S., 1982, Increased ^{3}H-spiperone binding sites in mesolimbic area related to methamphetamine-induced behavioral hypersensitivity, *Biol. Psychiatr.* **17**(2):223–231.

Anden, N. E. Corrodi, H., Fuxe, K., Hokfelt, B., Hokfelt, T., Rydin, C., and Svensson, T., 1970, Evidence for a central noradrenaline receptor stimulation by clonidine, *Life Sci.* **9**:513–523.

Anderson, G. M., Shaywitz, B. A., Riddle, M. A., Hoder, E. L., Lu, X., and Cohen, D. J., 1983, CSF monoamine metabolite and precursor measurements in the rat pup and children, *Pediatr. Res.* **17**:126A.

Angrist, B. M., and Gershon, S., 1970, The phenomenology of experimentally induced amaphetamine psychosis: Preliminary observations, *Biol. Psychiatr.* **2**:95–107.

Anisman, H., 1978, Neurochemical changes elicited by stress, in: *Psychopharmacology of Aversively Motivated Behaviors* (H. Anisman and G. Bignami, eds.), pp. 119–161, Raven Press, New York.

Antelman, S. M., Eichler, A. J., Black, C. A., and Kocan, D., 1980, Interchangeability of stress and amphetamine in sensitization, *Science* **207**:329–331.

Bachman, D. S,. 1981, Pemoline-induced Tourette's disorder: A case report, *Am. J. Psychiatry* **138**:1116–1117.

Borison, R. L., Ang, L., Chang, S., Dysken, M., Comaty, J. E., and Davis, J. M , 1982, New pharmacological approaches in the treatment of Tourette syndrome, in: *Gilles de la Tourette Syndrome, Advances in Neurology*, Vol. 35 (A. J. Friedhoff and T. N. Chase, eds.), pp. 377–382, Raven Press, New York.

Braestrup, C., and Nielsen, M., 1976, Regulation in the use of central norepinephrine neurotransmission induced in vivo by alpha adrenoceptor active drugs, *J. Pharmacol. Exp. Ther.* **198**:596–608.

Bunney, B. S., and DeRiemer, S., 1982, Effects of clonidine on dopamine neuron activity in the substantia nigra: Possible indirect mediation by noradrenergic regulation of the serotonergic raphe system, in: *Gilles de la Tourette Syndrome*, *Advances in Neurology*, Vol. 35 (A. J. Friedhoff and T. N. Chase, eds.), pp. 99–104, Raven Press, New York.

Butler, I. J., Koslow, S. H., Seifert, W. E., Jr., Caprioli, R. M., and Singer, H. S., 1979, Biogenic amine metabolism in Tourette syndrome, *Ann. Neurol.* **6**:37–39.

Cassens, G., Roffman, M., Kuruc, A., Orsulak, P. J., and Schildkraut, J. J., 1980, Alterations in brain norepinephrine metabolism induced by environmental stimuli previously paired with inescapable shock, *Science* **209**:1138–1139.

Cedarbaum, J. M., and Aghajanian, G. K., 1976, Noradrenergic neurons of the locus coeruleus: Inhibition by epinephrine and activation by the alpha-antagonist piperoxane, *Brain Res.* **112**:413–419.

Chiodo, L. A., Caggiula, A. R., and Saller, C. F., 1981, Estrogen potentiates the stereotypy induced by dopamine agonists in the rat, *Life Sci.* **28**:827–835.

Cohen, D. J., Shaywitz, B. A., Young, J. G., Carbonari, C. M., Nathanson, J. A. Lieberman, D., Bowers, M. B., Jr., and Maas, J. W., 1979a, Central biogenic amine metabolism in children with the syndorme of chronic multiple tics of Gilles de la Tourette syndrome: Norepinephrine, serotonin, and dopamine, *J. Am. Acad. Child. Psychiatry* **18**:320–341.

Cohen, D. J., Young, J. G., Nathanson, J. A., and Shaywitz, B. A., 1979b, Clonidine in Tourette's syndrome, *Lancet* **2**:551–553.

Cohen, D. J., Detlor, J., Shaywitz, B. A., and Leckman, J. F., 1982, Interaction of biological and psychological factors in the natural history of Tourette syndrome: A paradigm for childhood neuropsychiatric disorders, in: *Gilles de la Tourette Syndrome, Advances in Neurology*, Vol. 35 (A. J. Friedhoff and T. N. Chase, eds.), pp. 31–40, Raven Press, New York.

Connell, P. H., 1958, *Amphetamine Psychosis*, Oxford University Press, London.

Connell, Ph.H., Corbett, J. A. Horne, D. J., and Matthews, A. M., 1967, Drug treatment of adolescent tiqueurs: A double-blind trial of diazepam and haloperidol, *Psychiatry* **113**:375–381.

Corcoran, M. E., 1981, Catecholamines and kindling, in: *Kindling*, Vol. 2 (J. A. Wada, ed.), pp. 87–104, Raven Press, New York.

Coyle, J. T., and Campochiaro, P., 1976, Ontogenesis of dopaminergic–cholinergic interactions in rat striatum. A neurochemical study, *J. Neurochem.* **27**:673–678.

Creese, I., and Iversen, S. D, 1972, Amphetamine response in rat after dopamine neurone destruction, *Nature* (*New Biol.*) **238**:247–248.

Curzon, G., Hutson, P. H., and Knott, P. J., 1979, Voltammetry *in vivo*: Effect of stressful manipulations and drugs on the caudate nucleus of the rat, *Br. J. Pharmacol.* **66**:127P–128P.

Dawbarn, D., and Pycock, C. J., 1982, Lesions of the superior colliculus in the rat differentiate between nigrostriatal and mesolimbic dopamine systems, *Brain Res.* **135**:148.

Deakin, J. F., and Green, A. R., 1978, The effects of putative 5-hydroxytryptamine antagonists on the behavior produced by administration of tranylcypromine and L-tryptophan or tranylcypromine and L-dopa to rats, *Br. J. Pharmacol.* **64**:201–209.

Deakin, J. F., Owen, F., Cross, A. J., and Dashwood, M. J., 1981, Studies on possible mechanisms of action of electroconvulsive therapy; Effects of repeated electrically induced seizures on rat brain receptors for monoamines and other neurotransmitters, *Psychopharmacology* **73**:345–349.

Delgado, J. M. R., and Armand, B. K., 1953, Increase in food intake induced by electrical stimulation of the lateral hypothalamus, *Am. J. Physiol.* **172**:162–168.

Diamond, B. I., Reyes, M. G., and Borison, R., 1982, A new animal model for Tourette Syndrome, in: *Gilles de la Tourette Syndrome, Advances in Neurology*, Vol. 35 (A. J. Friedhoff and T. N. Chase, eds.), pp. 221–226, Raven Press, New York.

Eichler, A. J., and Antelman, S. M., 1979, Sensitization to amphetamine and stress may involve nucleus accumbens and medial frontal cortex, *Brain Res.* **176**:412–416.

Eldridge, R., Sweet, R., Lake, C. R., Ziegler, M., and Shapiro, A. K., 1977, Gilles de la Tourette's syndrome: Clinical, genetic, psychologic, and biochemical aspects in 21 selected families, *Neurology* **27**:115–155.

Ellison, R. M., 1964, Gilles de la Tourette's syndrome, *Med. J. Aust.* **1**:153–155.

Ernst, A. M., and Smelik, P. G., 1966, Site of action of dopamine and apomorphkne on compulsive gnawing behavior in rats, *Experientia* **22**:837–838.

Fog, R., and Pakkenberg, H., 1980, Theoretical and clinical aspects of the Tourette syndrome (chronic multiple tic). *J. Neural Transm.* **16**(Suppl.):211–215.

Friedhoff, A. J., 1982, Receptor maturation in pathogenesis and treatment of Tourette syndrome, in: *Gilles de la Tourette Syndrome, Advances in Neurology*, Vol. 35 (A. J. Friedhoff and T. N. Chase, eds.), pp. 133–140, Raven Press, New York.

Gianutsos, G., Drawbaugh, R. B., Hynes, M. D., and Lal, H., 1974, Behavioral evidence for dopaminergic supersensitivity after chronic haloperidol, *Life Sci.* **14**:887–899.

Gilles de la Tourette, G., 1885, Etude sur une affection nerveuse, caracterisee par de l'incoordination motrice, acompagnee d'echolalie et de coprolalia, *Arch. Neurol. (Paris)* **9:**19–42.

Goddard, G. V., McIntyre, D. C., and Leech, C. K., 1969, A permanent change in brain function resulting from daily electrical stimulation, *Exp. Neurol.* **25:**295–330.

Golden, G. S., 1974, Gilles de la Tourette's syndrome following methylphenidate administration, *Dev. Med. Child Neurol.* **16:**76–78.

Golden, G. S., 1977, The effect of central nervous system stimulants on Tourette syndrome, *Ann. Neurol.* **2:**69–70.

Gordon, J. H., 1980, Modulation of apomorphine-induced stereotypy by estrogen: Time course and dose response, *Brain Res. Bull.* **5:**679–682.

Guisado, E., Fernadez-Tome, P., Garzon, J., and Del Rio, J., 1980, Increased dopamine receptor binding in the striatum of rats after long-term isolation, *Eur. J. Pharmacol.* **65:**463–464.

Herman, J. P., Guillonneau, D., Dantzer, R., Scatton, B., Semerdjian-Rouquier, L., and Le Moal, M., 1982, Differential effects of inescapable footshocks and of stimuli previously paired with inescapable footshocks of dopamine turnover in cortical and limbic areas of the rat, *Life Sci.* **30:**2207–2214.

Hollister, A. S., Breese, G. R., and Cooper, B. R., 1974, Comparison of tyrosine and dopamine-beta-hydroxylase inhibition with the effects of various 6-hydroxydopamine treatments on d-amphetamine induced motor activity, *Psychopharmacologia* **36:**1–16.

Jagger, J., Prusoff, B. A., Cohen, D. J., Kidd, K. K., Carbonari, C. M., and John, K., 1982, The epidemiology of Tourette's syndrome: A pilot study, *Schizophrenia Bull.* **8:**267–278.

Jenkins, R. L., and Ashby, H. B., 1983, Tourette's disorder in identical twins: Report of a case, *Arch. Neurol.* **40:**249–251.

Kellogg, C., and Lundborg, P., 1973, Inhibition of catecholamine synthesis during ontogenic development, *Brain Res.* **61:**321–329.

Kelly, P. H., Seviour, P. W., and Iversen, S. D., 1975, Amphetamine and apomorphine responses in the rat following 6-OHDA lesions of the nucleus accumbens septi and corpus striatum, *Brain Res.* **94:**507–522.

Kidd, K. K., and Pauls, D. L., 1982, Genetic hypotheses for Tourette syndrome, in: *Gilles de la Tourette Syndrome, Advances in Neurology*, Vol. 35 (A. J. Friedhoff and T. N. Chase, eds.) pp. 243–249, Raven Press, New York.

Kidd, K. K., Prusoff, B. A., and Cohen, D. J., 1980, Familial pattern of Gilles de la Tourette syndrome, *Arch. Gen. Psychiatry.* **37:**1336–1339.

Klawans, H. L., Falk, D. K., Nausieda, P. A., and Weiner, W. J., 1978, Gilles de la Tourette syndrome after long term chlorpromazine therapy, *Neurology (Minneap.)* **28:**1064–1066.

Klempel, K., 1974, Gilles de la Tourette's symptoms induced by L-dopa, *S. Afr. Med. J.* **48:**1379–1380.

Knott, P. J., and Hutson, P. H., 1982, Stress-induced stereotypy in the rat: Neuropharmacological similarities to Tourette Syndrome, in *Gilles de la Tourette Syndrome, Advances in Neurology*, Vol. 35 (A. J. Friedhoff and T. N. Chase, eds.), pp. 233–238, Raven Press, New York.

Kobayashi, R. M., Palkovits, M., Kizev, J. S., Jacobowitz, D. M., and Kopin, I. J., 1976, Selective alterations of catecholamines and tyrosine hydroxylase activity in the hypothalamus following acute and chronic stress, in: *Catecholamines and Stress* (E. Usdin, R. Kuetnansky, and I. J. Kopin, eds.), pp. 28–38, Pergamon Press, Oxford.

Kokkinidis, L., Irwin, J., and Anisman, H., 1979, Shock induced locomotor excitation following acute and chronic amphetamine treatment, *Neuropharmacology* **18:**13.

Leckman, J. F., Cohen, D. J., Shaywitz, B. A., Caparulo, B. K., Heninger, G. R., and Bowers, M. B., Jr., 1980a, CSF monoamine metabolites in child and adult psychiatric patients: A developmental perspective, *Arch. Gen. Psychiatry.* **37:**677–681.

Leckman, J. F., Maas, J. W., Redmond, D. E., Jr., and Heninger, G. R., 1980b, Effects of oral clonidine on plasma 3-methoxy-4-hydroxyphenethyleneglycol (MHPG) in man: Preliminary report. *Life Sci.* **26:**2179–2185.

Leckman, J. F., Maas, J. W., and Heninger, G. R., 1981, Covariance of plasma free 3-methoxy-4-hydroxyphenethyleneglycol and diastolic blood pressure, *Eur. J. Pharmacol.* **70:**111–120.

Leckman, J. F., Cohen, D. J., Detlor, J., Young, J. G., Harcherik, D., and Shaywitz, B. A., 1982, Clonidine in the treatment of Tourette syndrome: A review of data, in: *Gilles de la Tourette Syndrome, Advances in Neurology*, Vol. 35 (A. J. Friedhoff and T. N. Chase, eds.), pp. 391–402, Raven Press, New York.

Leckman, J. F., Detlor, J., and Cohen, D. J., 1983a, Gilles de la Tourette syndrome: Emerging areas of clinical research, in: *Childhood Psychopathology and Development* (S. B. Guze, F. J. Earls, and J. E. Barrett, eds.), pp. 211–229, Raven Press, New York.

Leckman, J. F., Detlor, J., Harcherik, D. F., Young, J. G., Anderson, G. M., Shaywitz, B. A., and Cohen, D. J., 1983b, Acute and chronic clonidine in Tourette's syndrome: A preliminary report on clinical response and effect on plasma and urinary catecholamine metabolites, growth hormone, and blood pressure, *J. Am. Acad. Child Psychiatry* **22**:433–440.

Leckman, J. K., Detlor, J., Harcherik, D. F., Ort, S., Shaywitz, B. A., and Cohen, D. J., 1985, Short- and long-term treatment of Tourette's syndrome with colnidine: A clinical perspective, *Neurology* **35**:343–351.

Leith, N. J., and Kuczenski, R., 1982, Two dissociable components of behavior sensitization following repeated amphetamine administration, *Psychopharmacology* **76**:310–315.

Loizou, L. A., and Salt, P., 1970, Regional changes in monoamines of the rat brain during postnatal development, *Brain Res.* **20**:467–470.

Lowe, T. L., Cohen, D. J., Detlor, J., Kremenitzer, M. W., and Shaywitz, B. A., 1982, Stimulant medications precipitate Tourette's syndrome, *JAMA* **247**:2739–1731.

Maas, J. W., Hattox, S. E., Landis, D. H.,. and Roth, R. H., 1977, A direct method for studying 3-methoxy-4-hydroxyphenethyleneglycol (MHPG) production by brain in awake animals, *Eur. J. Pharmacol.* **46**:221–118.

MacLennan, H. J., and Maier, S. F., 1983, Coping and the stress-induced potentiation of stimulant stereotypy in the rat, *Science* **219**:1091–1092.

Messiha, F. S., Knopp, W., Vanecko, Ss., O'Brien, V., and Corson, S. A., 1971, Haloperidol therapy in Tourette's syndrome: Neurophysiological, biomedical and behavioral correlates, *Life Sci.* **10**:449–457.

Mogilnicka, E., and Braestrup, C., 1976, Noradrenergic influence on the stereotyped behaviour induced by amphetamine, phenethylamine, and apomorphine, *J. Pharm. Pharmacol.* **28**:253–255.

Moldofsky, H., Tullis, C., and Lamon, R., 1974, Multiple tic syndrome (Gilles de la Tourette's syndrome): Clinical, biological, and psychosocial variables and their influence with haloperidol, *J. Nerv. Ment. Dis.* **159**:282–292.

Morrell, F., 1973, Goddard's kindling phenomenon: A new model of "mirror focus," in: *Chemical Modulation of Brain Function* (H. C. Sabelli, ed.), pp. 207-223, Raven Press, New York.

Nee, L. E., Caine, E. D., Polinsky, R. J., Eldridge, R., and Ebert, M. H., 1980, Gilles de la Tourette syndrome: Clinical and family study of 50 cases, *Ann. Neurol.* **7**:41–49.

Pauls, D. L. Cohen, D. J., Heimbuch, R., Detlor, J., and Kidd, K. K., 1981, Familial pattern and transmission of Gilles de la Tourette syndrome and multiple tics, *Arch. Gen. Psychiatry* **38**:1091–1093.

Post, R. M., and Kopanda, R. T., 1976, Cocaine, kindling, and psychosis, *Am. J. Psychiatry.* **133**:627–634.

Price, R. A., Kidd, K. K., Cohen, D. J., Pauls, D. L., and Leckman, J. F., 1985, A twin study of Tourette syndrome, *Arch. Gen Psychiatry* **42**:815–820.

Racine, R. J., Burnham, W. M., Gartner, J. G., and Levitan, D., 1973, Rates of motor seizure development in rats subjected to electrical brain stimulation: Strain and interstimulation interval effects, *Electroencephalo. Clin. Neurophysiol.* **35**:553–556.

Randrup, A., and Munkvard, L., 1974, Pharmacology and physiology of stereotyped behavior, *J. Psychiatr. Res.* **2**:1.

Robbins, T. W., and Sahakian, B. J., 1981, Behavioral and neurochemical determinants of drug-induced stereotypy, in: *Metabolic Disorders of the Nervous System*, pp. 244–491, Pitman Books, Ltd., London.

Robinson, T. E., Becker, J. B,. and Ramirez, V. D., 1980, Sex differences in amphetamine-elicited rotational behavior and the lateralization of striatal dopamine in rats, *Brain Res. Bull.* **5**:539–545.

Ross, M. S., and Moldofsky, H., 1977, Comparison of pimozide with haloperidol in Gilles de la Tourette syndrome, *Lancet* **1**:103.

Sahakian, B. J., Robbins, T. W., Morgan, M. J., and Iversen, S. J., 1975, The effects of psychomotor stimulants on stereotypy and locomotor activity in socially-deprived and control rats, *Brain Res.* **84**:195–205.

Schmitt, H., Schmitt, H., and Fenard, S., 1971, Evidence for an -sympathomimetic component in the effects of catapresan on vasomotor centres: Antagonism by piperoxane, *Eur. J. Pharmacol.* **14**:98–100.

Segal, D. S., 1975, Behavioral and neurochemical correlates of repeated d-amphetamine administration, in: *Neurobiological Mechanisms of Adaptation anad Behavior* (A. J. Mandell, ed.), pp. 247–262, Raven Press, New York.

Shalaby, I. A., and Spear, L. P., 1980, Psychopharmacological effects of low and high doses of apomorphine during ontogeny, *Eur. J. Pharmacol.* **67**:451–459.

Shapiro, A. K., and Shapiro, E., 1981, The treatment and etiology of tics and Tourette syndrome, *Comp. Psychiatry.* **22**:193–205.

Shapiro, A. K., Shapiro, E., and Wayne, H., 1973, Treatment of Tourette's syndrome with haloperidol, review of 34 cases, *Arch. Gen Psychiatry.* **28**:92–97.

Shapiro, A. K., Shapiro, E. S., Bruun, R. D., and Sweet, R. D., 1978, *Gilles de la Tourette Syndrome*, Raven Press, New York.

Sharma, J. N., Sandrew, B. B., and Wang, S. C., 1978, CNS sites of clonidine induced hypotension: A microiontophoric study of bulbar cardiovascular neurons, *Brain Res.* **151**:127–133.

Shaywitz, B. A., Anderson, G. M., Young, J. G., and Cohen, D. J., 1982, comparison of CSF monoamine metabolites and precursor measurements in the rat pup and in children, *Soc. Neurosci. Abstr.* **8**:177.

Stahl, S. M., 1980, Tardive Tourette syndrome in autistic patient after long-term neuroleptic administration, *Am. J. Psychiatry* **137**:1267–1969.

Svensson, T. H., Bunney, B. S., and Aghajanian, G. K., 1975, Inhibition of both noradrenergic and serotonergic neurons in brain by the -adrenergic agonist clonidine, *Brain Res.* **92**:291–306.

Szechtman, H., Ornstein, K., Teitelbaum, P., and Golani, I., 1982, Snout contact fixation, climbing and gnawing during apomorphine stereotypy in rats from two substrains, *Eur. J. Pharmacol.* **80**:385–392.

Wasserman, E. R., Eldridge, R., Abuzzahab, S., and Nee, L., 1978, Gilles de la Tourette syndrome: Clinical and genetic studies in a midwestern city, *Neurology* (*Minneap.*) **28**:304–307.

Watanabe, H., Nakano, S., and Ogawa, N., 1982, Age difference in apomorphine-induced stereotypy in rats: Relationship to plasma and brain concentrations, *Psychopharmacology* **76**:57–61.

Wilson, R. S., Garron, D. C., and Klawans, H. L., 1978, Significance of genetic factors in Gilles de la Tourette syndrome: A review, *Behav. Genet.* **8**:503–510.

PART IV
Huntington's Disease

CHAPTER 14

Biochemical and Pharmacological Aspects of Movement Disorders in Huntington's Disease

RICHARD L. BORISON, ANA HITRI, and BRUCE I. DIAMOND

1. INTRODUCTION

The neurochemistry of Huntington's disease (HD) presents an unusual problem. Since this disorder is progressive and deteriorative, the relative neurochemical balance may be quite different depending on what stage of illness was obtained prior to death and autopsy analysis of the brain. Furthermore, with the significant loss of tissue that occurs with degeneration, the relative concentrations of transmitters, enzymes, and receptors may appear falsely high when expressed per milligram of tissue. Other issues that obscure the neurochemical analysis of HD include the variability of genetic expression (e.g., differences in times of onset of choreas, dementia), which make it unlikely that any two HD brains will necessarily show concordance upon neurochemical analysis. Finally, it is quite likely that brain plasticity may account for homeostatic neurochemical changes in some neurotransmitter systems to compensate for the loss, due to degeneration, of other neurotransmitters and neuroanatomical pathways. Thus, our review of HD neurochemistry may at times seem contradictory, most likely because of one or more of these factors. It is important, however, to remember that HD is not a static illness, and neurochemical analysis at any particular stage of illness may show a wide variability.

2. GABA

The anatomical pathology of HD is in part characterized by a marked loss of small neurons in the putamen and caudate nucleus, which in turn would

RICHARD L. BORISON, ANA HITRI, and BRUCE I. DIAMOND • Department of Psychiatry, Downtown Veterans Administration Medical Center, and Medical College of Georgia, Augusta, Georgia 30910.

communicate with the globus pallidus, which then sends the output of the striatum to other brain nuclei. The nature of the neurotransmitter mediating neostriatal–pallidal communication was largely unknown until the original work of Perry *et al.* (1973) demonstrated a significant decrease of GABA in the autopsied brains of HD patients. The greatest loss of GABA was measured in the substantia nigra, putamen, and globus pallidus (Urquhart *et al.*, 1975). These findings were later confirmed by other investigators (Bird and Iverson, 1974). One group has been able to analyze GABA in samples of frontal cortex taken at biopsy. Their data show that only approximately one-half of the 13 HD patients studied showed a decrease in GABA concentration (Kremzner *et al.*, 1979). Although the loss of GABA proved to be the first biochemical lesion found in HD, measurement of brain GABA proved an untrustworthy marker, as it was known that the GABA concentration of animal brain rapidly increased during the postmortem period. For this reason, two alternative means were sought for quantifying GABA loss in HD, namely, *in vivo* measurement of GABA in cerebrospinal fluid (CSF) and measurement of the GABA biosynthetic enzyme glutamic acid decarboxylase (GAD) in postmortem tissue.

The measurement of GABA in human CSF has been beset by methodological difficulties. It was originally demonstrated (Glaeser *et al.*, 1975) that GABA concentration was fivefold lower in the CSF of HD patients as compared to patients with other neurological illness. This finding could not be confirmed by other groups; there were actually several reports indicating that GABA levels were undetectable even in the CSF of control subjects (Perry *et al.*, 1975; Welch *et al.*, 1975). Later reports even suggested that GABA levels were increased in the CSF of HD patients compared to controls (Achar *et al.*, 1976). An attempt to improve the measurement of GABA in the CSF was made using the radioreceptor assay (Enna *et al.*, 1977a). This technique involves incubation of CSF in the presence of tritiated GABA bound to rat brain synaptosomes. This technique revealed that the GABA concentration in the CSF of 19 HD patients was one-half that of 26 control patients. Since it is believed that GABA loss may be the earliest detectable biochemical lesion in HD, measurement of GABA in the CSF may identify patients at risk for developing HD. Along these lines, Manyam *et al.* (1978) measured GABA concentrations in six HD patients, five normal controls, and 22 subjects at risk for HD. Using an ion-exchange–fluorimetric technique they measured 142 ± 27 pmole/ml of GABA in the CSF of HD patients, whereas 297 ± 87 pmole/ml was measured in control subjects. In subjects at risk, there was a clear bimodal distribution of GABA concentrations, with one group measured at 209 ± 79 pmole/ml and the other at 159 ± 27 pmole/ml. These data cannot be interpreted except on a prospective basis as to the clinical outcome of the at-risk subjects; however, this clearly suggests that some individuals at risk for HD do have abnormalities in central GABA transmission. In attempting to repeat and refine the latter's biochemical procedure, Perry *et al.* (1982) found that the GABA concentration in the CSF of normal subjects should be revised downward to 87 pmole/ml; however, in 22 patients with HD they measured 82 pmole/ml. The authors suggest that the measurement of conjugated GABA, rather than free GABA, may prove more useful in biochemically differentiating patients or subjects at risk for HD. Despite these discrepancies in the measurement of GABA in CSF, they apparently truly reflect brain GABA activity, as pharmacological interventions that increase brain GABA, such as γ-acetylinic GABA, γ-vinyl GABA, and isoniazid, all increase GABA concentration in CSF (Grove *et al.*, 1981; Manyam *et al.*, 1981; Tell *et al.*, 1981).

Perhaps the best marker for GABA in the brain, GAD activity, has consistently been shown to be decreased in HD brains. This was first demonstrated by Bird *et al.* (1973), who observed a 75% decrease in GAD activity in the globus pallidus, caudate nucleus, and putamen. These results have been consistently replicated by others (McGeer *et al.*, 1973; Stahl and Swanson, 1974; Enna *et al.*, 1976a). To differentiate whether the decrease in GAD activity was due to an inhibition of enzyme activity or due to an absolute lessening of enzyme molecules owing to neuronal degeneration, Wu *et al.* (1978) conducted microcomplement fixation tests which revealed that these enzymes resemble those found in control brain, thus making an absolute decrease in GAD molecules responsible for decreased enzymatic activity. This decrease in GAD activity remains the single most consistent neurochemical abnormality present in HD.

Another area of neurochemical controversy concerns possible changes in GABA receptors in HD. Originally, it was observed, using radioligand binding studies, that there are no changes in GABA receptors in HD brains (Enna *et al.*, 1977b). This is quite significant, as it implies that the neurons with which GABA neurons synapse are left intact and suggests that mechanisms aimed at GABA replacement therapy should prove successful in the treatment of HD. Later work has demonstrated that in the caudate nucleus and putamen there may be a significant decrease in GABA receptors in HD brains (Lloyd *et al.*, 1977; Iversen *et al.*, 1979; Olsen *et al.*, 1979; Reisine *et al.*, 1978; Van Ness *et al.*, 1982). In contrast, an apparent denervation hypersensitivity takes place in the substantia nigra, as increased numbers of GABA receptors have been reported in this area (Enna *et al.*, 1976b; Waddington and Cross, 1980). Although the general finding of decreased GABA receptors is most easily explained by neuronal degeneration, it has been suggested that changes in phospholipid metabolism may potentially change the affinity of the GABA receptor in HD (Lloyd and Davidson, 1979). In addition to possible changes in GABA receptors, it has also been observed that benzodiazepine receptors may also be altered in the brains of HD patients (Shoemaker *et al.*, 1982). Although receptor numbers appear normal in the caudate nucleus and globus pallidus, there is a 51% increase in benzodiazepine receptor concentration in the putamen, which is probably associated with gliosis (Shoemaker *et al.*, 1982). An alternative approach to measuring receptors is via autoradiography. Using this technique, Penny and Young (1982) demonstrated a decrease of GABA, benzodiazepine, and muscarinic cholinergic receptors in the caudate nucleus and putamen; however, benzodiazepine receptors were increased in the lateral and medial pallidum. These results suggest that a loss of striatal afferents to the pallidum results in a supersensitivity of GABA and benzodiazepine receptors.

With changes in GABA being the most universal abnormal finding in HD, it has been hoped that GABA replacement, as paralleled by dopamine replacement in Parkinson's disease, may prove of benefit. The most direct replacement therapy would be with GABA itself, and when this was attempted in the presence of dipropylacetic acid, a GABA transaminase (GABA-T) inhibitor, there was no improvement (Shoulson *et al.*, 1976). Similar studies, using GABA-T inhibitors alone, dipropylacetic acid (Lenman *et al.*, 1976; Symington *et al.*, 1978), or γ-acetylinic GABA (Tell *et al.*, 1981), even when plasma or CSF levels showed increased GABA, failed to prove of therapeutic benefit. Another putative GABA-T inhibitor, isoniazid, is of equivocal therapeutic value (Perry *et al.*, 1979; Manyam *et al.*, 1981). The use of GABA-mimetic drugs, imidazole-4-acetic acid or muscimol, has been disappointing (Shoulson *et al.*, 1975, 1978). This suggests that either the GABA system is so disrupted in HD that replacement therapy is impractical, or, more likely, unlike Parkinson's disease, multiple neurotransmitters and their receptors are profoundly altered, making multiple neurotransmitter intervention necessary.

3. ACETYLCHOLINE

The cholinergic system has been best studied by the activity of choline acetyltransferase (ChAc), the biosynthetic enzyme for Ach which is localized only to cholinergic neurons. It was originally observed that in the brains of choreic patients there was a mean decrease by 50% in ChAc activity in the basal ganglia (Bird *et al.*, 1973). This finding was corroborated by others (McGeer *et al.*, 1973; Stahl and Swanson, 1974; Wastek *et al.*, 1976; Wastek and Yamamura, 1978). What was most intriguing was that not all brains showed a decrease in ChAc activity, indicating that the cholinergic system was intact in some choreic patients. Correlating GAD and ChAc activities has presented a picture indicating that GABA loss is the first biochemical lesion to become manifest in HD, and only with the progression of the disease do Ach neurons also degenerate.

An alternative technique for measuring cholinergic system activity is via the measurement of muscarinic cholinergic receptors using tritiated quinuclidinyl benzilate (QNB). Using a similar ligand, Hiley and Bird (1974) demonstrated a 50% decrease in muscarinic binding in the putamen and caudate nucleus in choreic brains. This was confirmed using QNB by several studies (Wastek *et al.*, 1976; Enna *et al.*, 1977c; Wastek and Yamamura, 1978). It is interesting to note that loss of cholinergic receptors can occur even in the presence of normal ChAc activity, suggesting that the GABA (or other) neurons which degenerate early in the course of HD are the sites where cholinergic receptors are located.

The deficit found in brain Ach transmission in HD has suggested that increasing central cholinergic tone may be of therapeutic benefit. This has been attempted using the centrally active cholinesterase inhibitor, physostigmine, with both success (Aquilonius and Sjostrom, 1971; Klawans and Rubovits, 1972) and failure (Tarsy *et al.*, 1974). Similarly, dimethylaminoethanol, which purportedly is metabolized to Ach, has had mixed success in treatment of HD (Walker *et al.*, 1973; Reibling *et al.*, 1975). The use of choline, and the direct receptor agonist arecoline, either fails to affect or actually exacerbates choreic movements (Growdon *et al.*, 1977; Nutt *et al.*, 1978). The partial therapeutic success achieved with cholinergic agents may be explained by the variability in cholinergic loss in HD or, more likely, by the fact that GABA and other neurochemicals play such strategic roles in HD that treatment with cholinomimetics alone is insufficient.

4. NEUROPEPTIDES

If we turn our attention to the role of peptides in HD, it is found that the undecapeptide substance P, the pentapeptide methionine-enkephalin (met-ENK), the octapeptide cholecystokinin (CCK), and angiotensin-converting enzyme (ACE) are the peptides most studied since all of them are concentrated in modest amounts in the basal ganglia, the neuroanatomical region most closely linked to HD.

In the case of ACE, this enzyme converts angiotensin I into angiotensin II and can also deactivate bradykinin (Soffer, 1976). Angiotensin I and II can produce central effects such as eliciting drinking behavior, releasing vasopressin, and producing increases of blood pressure (Severs and Daniel-Severs, 1973). ACE is most highly localized to the basal ganglia of rat (Yang and Neff, 1972) and human brain (Arregui *et al.*, 1977), followed by the substantia nigra (Arregui *et al.*, 1977). It has been demonstrated by Arregui *et al.*, (1977) that ACE activity is markedly reduced in the caudate nucleus, putamen, and globus pallidus of autopsied HD brains. Moreover, these authors (Arregui *et al.*, 1978) have demonstrated that ACE activity is also reduced in the pars reticulata of the substantia nigra and that the kainic acid animal model for HD also shows diminished ACE activity in these brain regions. The behavioral actions of ACE and angiotensin on drinking, and its interaction with dopamine, with vasopressin, and its putative role in memory, still await further elucidation, as does the role of ACE itself within the basal ganglia.

CCK is the most recent peptide to be studied in relation to HD. CCK is a gastrointestinal peptide (33 amino acids) found in the brain (Beinfeld *et al.*, 1983) and now proposed to be a neurotransmitter or neuromodulator (Innis *et al.*, 1979; Pinget *et al.*, 1979). The carboxyterminal octapeptide of CCK-33 represents 50% of brain CCK so far studied (Rehfeld, 1978). CCK is concentrated in the cerebral cortex, basal ganglia, and hippocampus of both rat and human brain (Emson *et al.*, 1980a; Beinfeld *et al.*, 1983). It is an excitatory neuropeptide (Dodd and Kelly, 1981). Moreover, CCK binding sites are also localized to the cerebral cortex and basal ganglia (Saito *et al.*, 1980; Hays *et al.*, 1981). It has been demonstrated by Emson *et al.* (1980b) that CCK is reduced by 50–60% in the brains of HD patients. This reduction was found in the globus pallidus and substantia nigra, and not in the caudate nucleus, putamen, or cerebral cortex. Hays *et al.* (1981) have found that CCK binding sites were significantly reduced in the caudate nucleus and frontal cortex of HD brains. The reduction in receptor activity that they measured was due to an absolute reduction in the number of receptor sites, as receptor affinity was found to be unchanged. In a further study by Hays and Paul (1982), CCK receptors were also found to be reduced in the putamen of HD brains. Animal studies with CCK have demonstrated that this peptide is capable of producing sedation and catalepsy, as well as antagonizing apomorphine-induced stereotypy, a dopaminergic behavior mediated by the striatum. In contrast, direct CCK injections into the caudate nucleus produce behaviors identical to dopamine, and the nature of CCK's pharmacology suggests an interaction with striatal dopamine (Diamond *et al.*, 1983). Clinical studies with CCK are being performed in schizophrenia and Parkinson's disease (T. Chase and C. Tamminga, personal communication), but as yet, there are no published studies of CCK's clinical application in HD.

Substance P is an undecapeptide which has been shown to excite nigral neurons (Walker *et al.*, 1976). Its highest concentration in human brain is found in the substantia nigra (both pars compacta and reticulata), globus pallidus, and hypothalamus (Gale *et al.*, 1978). In HD there is a significant decrease in substance P concentration in the globus pallidus and substantia nigra, whereas no decrease was observed in the caudate nucleus, putamen, or cerebral cortex (Gale *et al.*, 1978). Emson *et al* (1980a) have also shown that substance P is concentrated in the substantia nigra and globus pallidus, but they noted that it was the medial, or internal, segment of the globus pallidus that contained substance P. They explained this concentration to the medial part of the globus pallidus as representing the presence of substance P in the terminals of collaterals of striatonigral axons descending to the nigra. These authors also demonstrated in HD brains that there is a loss of substance P content in the nigra and globus pallidus. Also, in the kainic acid animal model for HD there is a similar reduction in substance P content in the striatum (Jessell *et al.*, 1978). Studies by Buck *et al.* (1981) have found substance P levels to the highest in the substantia nigra, with the pars reticulata containing threefold more substance P than the pars compacta. The globus pallidus and putamen also contained high levels, with the hypothalamus containing twice the concentration of the putamen. In brains from HD patients, these authors observed a 48% decrease in the substance P content of the caudate nucleus, with over a 90% loss in the pars reticulata. No decreases were found in the cerebral cortex, thalamus, or hypothalamus. Animal studies where substance P was chronically injected into the caudata nucleus mimicked the pharmacology of HD patients (Diamond *et al.*, 1979).

Evidence for a neurotransmitter role for met-ENK has come from many studies (Emson, 1979). In animal models for neuropsychiatric diseases, including HD, it has been proposed that met-ENK has significant behavioral and pharmacological significance (Diamond and Borison, 1982). High concentrations of met-ENK in human brain are found in the globus pallidus, substantia nigra, nucleus accumbens, putamen, caudate nucleus, hypothalamus, and central gray (Emson *et al.*, 1980a). The rich met-ENK content of the globus pallidus occurs in the lateral, or external, segment although there are substnatial amounts in the median, or internal, segment (Emson *et al.*, 1980a). These authors suggest that there is a striatopallidal and striatonigral enkephalin pathway in human brain, which is lost in HD brains. They found, as with substance P, that met-ENK content in HD brains is decreased markedly in the globus pallidus and substantia nigra, reflecting the loss of neurons in the caudate nucleus and putamen. The kainic acid model of HD also demonstrates the reduction in striatal met-ENK content (Hong *et al.*, 1977).

Other neuropeptides may also play a role in HD. It has been reported that neurotensin levels are significantly increased in the caudate nucleus and nucleus accumbens of HD brains, and it has been questioned as to whether increased neurotensin and dopamine in the caudate nucleus may mediate choreatic movement (Manberg *et al.*, 1982). Elevated neurotensin levels in HD brain were also measured by Nemeroff *et al.* (1983), who also found elevated levels of somatostatin and thyrotropin-releasing factor in HD caudate nucleus and nucleus accumbens; however, in the amygdala only levels of somatostatin and thyrotropin-releasing factor were increased. By comparison, in living patients plasma growth hormone and prolactin were measured after the administration of dopamine, GABA, and acetylcholine agonists, with the conclusion that there is a loss of somatostatin activity in the hypothalamic–pituitary axis of HD patients (Durso *et al.*, 1983).

5. AMINO ACIDS

The role of amino acids in HD has been studied with much attention paid to glutamate as a putative causative agent in this disorder. Perry *et al.* (1969) found that proline, alanine, valine, isoleucine, leucine, and tyrosine plasma levels were reduced in HD patients, a finding that they later confirmed (Perry *et al.*, 1972). In other studies of amino acid blood levels, Yates *et al.* (1973) found a reduction in plasma tryptophan, but not tyrosine, whereas Phillipson and Bird (1977) found decreases in leucine, isoleucine, valine, and tryptophan in HD patients. Watt and Cunningham (1978) found reduced levels of plasma threonine, lysine, histidine, alanine, isoleucine, and leucine in HD patients and suggested that amino acid differences in HD patients were related to nonspecific factors such as age, weight, and dementia.

Much work has been focused on glutamate's role in HD. Excessive stimulation by glutamate or its analogues causes death of some neurons (Olney *et al.*, 1972), such as neostriatal neurons (Schwarcz *et al.*, 1977). The neocortical efferents to the striatum use glutamate as a neurotransmitter (Divac *et al.*, 1977), and its has been suggested that pathological changes in the brains of HD patients may result from excessive stimulation of glutamate-sensitive neurons (Dom *et al.*, 1976). Causes for glutamate abnormalities may include its overproduction by cortical neurons, synthesis of a neurotoxic glutamate analogue, improper glutamate inactivation, or hypersensitivity of the postsynaptic neuron (Divac *et al.*, 1977). However, Perry *et al.* (1973) found normal concentrations of glutamate aspartate in HD brains. As of yet, no known glutamate antagonist has been clinically tried in HD; however, it has been suggested that folate be reduced in the diet of HD patients (Divac *et al.*, 1977), as it inhibits glutamate uptake into glia (Roberts, 1974).

Glutamate and its analogues produce striatal destruction in animals that mimics the neuropathology in HD (Coyle and Schwarcz, 1976; McGeer and McGeer, 1976). Kainate, a structural conformationally restricted analogue of glutamate, activates postsynaptic glutamate receptors and is currently used as an animal model for HD because it mimics the pathology, biochemistry, and pharmacology of this disease (Coyle and Schwarcz, 1976).

As mentioned previously, no antagonists of glutamate have been tried in HD patients yet; however, baclofen has been shown *in vitro* to inhibit glutamate release from cerebral cortex (Potashner, 1978) and may be of value in the prevention of the progression of HD (Sanberg and Johnston, 1981). It was proposed by Olney and deGubareff (1978) that adult-onset disturbances in glutamate uptake processes underlie the neurodegenerative changes in HD and that recognition of these abnormalities may enable an early diagnosis to be made for this disease. Glutamate has been shown to stimulate the release of dopamine in rat striatum, an effect that was abolished after kainate lesions (Roberts and Anderson, 1979). It has been proposed that in early stages of HD with chorea present, the excessive glutamate stimulation increases dopamine activity. In later stages, where glutamate is ineffective in releasing dopamine because of striatal degeneration, the abnormal neurotransmitter balance plays a role in favoring hypokinesia (Sanberg and Johnston, 1981). It has been shown that intracerebral injection of glutamate into animals produces learning, memory, and perceptual disorders (Freed and Michaelis, 1976).

It has been demonstrated that glutamate is decreased in the CSF of HD patients (Kim *et al.*, 1980) and its uptake affinity is higher in the platelets of these patients (Mangano and Schwarcz, 1980). Glutamate synthetase, the enzyme that converts glutamate to glutamine, is lost in HD brains (Carter, 1981). This enzyme is localized to glia; hence its loss is not due to neuronal loss. There are endogenous inhibitors of kainate binding sites that increase with age (Skerritt and Johnston, 1981), which could explain the results of decreased kainate receptor binding in HD postmortem striatal tissue (Henke, 1979). As of yet, no endogenous ligand of kainate has been found in brain, but methyltetrahydrofolate interacts specifically with kainate receptors on neuronal membranes (Ruck *et al.*, 1980). It is not known whether this folate can cause degeneration similar to that produced by kainate, or whether excessive concentrations of this compound may be found in HD brains.

6. DOPAMINE

The main pathological alterations in adult-onset HD are loss of small neurons in the striatum, proliferation of astrocytes, and diffuse cerebral cortical atrophy. The extrapyramidal manifestations of HD have been related to pathological alterations in corpus striatum (Klawans and Weiner, 1974), whereas the mental symptoms are more likely to be related to changes in the cerebral cortex where atrophy of the third, fifth, and sixth layers occurs (Bruyn *et al.*, 1979; Toglia *et al.*, 1978). The most frequent extrapyramidal symptom in adult-onset HD is chorea.

Numerous clinical studies indicate that chorea in HD can be ameliorated by agents that block dopaminergic transmission in the corpus striatum, whereas agents that enhance striatal dopaminergic activity will exacerbate chorea (Toglia *et al.*, 1978; Klawans and Weiner, 1974; Siegmund *et al.*, 1982; Caraceni *et al.*, 1980).

Dopamine in the striatum orginates in the cell bodies of the substantia nigra. It has been demonstrated that there is no pathological alteration in the dopaminergic cell bodies of the substratia nigra, but the neurons of the striatum which normally receive dopamine input are markedly altered (Klawans and Weiner, 1976). In earlier studies it was found that dopamine content of the striatum is normal in patients dying of HD (Ehringer and Hornykiewicz, 1960). In more recent studies in which different regions of the corpus striatum were analyzed for their dopamine content, it was found that dopamine content of the caudate nucleus was normal, but the levels of dopamine in the putamen were significantly higher than controls (Melamed *et al.*, 1982; Spokes, 1980). There are also reports in the literature on decreased amounts of striatal dopamine in HD (Bernheimer *et al.*, 1973; Hornykiewicz, 1976). These differences in reports regarding striatal dopamine concentrations may be attributed to different types of HD, as well as to the development of more precise methods of dopamine measurements which allow the utilization of small tissue amounts in the assays. Thus, the analysis of dopamine in different regions of the striatum may provide more relevant data than whole striatal assays in earlier studies.

In regard to the type of HD, none of the patients that had the rigid form of the disease had elevated DA levels in the putamen (Mann *et al.*, 1980). Akinesia and rigidity are seen with juvenile onset of HD, as well as in the late course of adult-onset HD following the initial choreatic phase. Based on the experiences with Parkinson's disease, it is generally felt that rigidity and akinesia are related to loss of dopamine influence on striatal neurons. Since dopamine levels are normal (Ehringer and Hornykiewicz, 1960; Mann *et al.*, 1980) or near normal in this disease, the decreased dopamine influence would have to be related to decreased response of damaged striatal neurons to normal levels of dopamine. Thus, it has been proposed that rigidity is related to an overall decreased response to dopamine by certain striatal neurons while chorea is related to an increased response by other neurons (Klawans and Weiner, 1976). In fact, it was proposed a long time ago that the disease of large striatal neurons results in rigidity while disease of small striatal neurons results in chorea (Hunt, 1917).

The fact that enhancement of striatal dopaminergic activity exacerbates chorea but improves the rigid akinetic symptoms is consistent with the notion that dopamine may be acting on different striatal cell populations containing different dopamine receptors. Electrophysiological studies have demonstrated the existence of dopamine excitatory cells and dopamine inhibitory cells in the striatum (McLennan and York, 1969), suggesting the possible existence of two separate dopaminergic mechanisms and possibly two separate types of dopamine receptors. Biochemical studies involving receptor binding studies utilizing [^{3}H]dopamine and [^{3}H]halopridol as a ligand and confirmed the existence of two distinct binding sites for [^{3}H]dopamine in guinea pig striatal membranes and also demonstrated that these two binding sites do not respond identically to either chronic or acute pharmacological manipulations (Weiner *et al.*, 1975). By virtue of selective involvement of distinct subpopulations of dopamine receptors that have different physiological functions, it may be possible to explain the different clinical symptoms seen in hyperkinetic and hypokinetic HD. An attempt to distinguish between different populations of dopamine receptors was reported by Cross and Rossor (1983), who measured D1 and D2 dopamine receptors using tritiated spiroperidol and piflutixol. Both receptor types were decreased by 45–50% in the putamen, but only piflutixol binding was decreased (48%) in the pars reticulata, with the pars compacta showing no change. This loss of D2 receptors would then be consistent with a loss of striatal GABA cell bodies and corticostriatal neurons.

The degeneration of striatal cholinergic neurons bearing dopamine receptors in HD is reflected in decreased levels of acetylcholine and decreased number of dopamine receptor sites (Reisine *et al.*, 1978). This degenerative process results in disruption of the normal physiological equilibrium between dopamine and acetylcholine in the striatum of HD patients. Dopamine and acetylcholine are antagonistic in their action on striatal neurons, and decreased levels of acetylcholine will tip the balance toward dopamine.

Moreover, the degeneration of GABAergic striatal neurons as reflected in decreased levels of GABA in the striatum of HD patients results in the loss of feedback inhibition of dopamine neurons in the substantia nigra, which causes an increase in synthesis and release of dopamine from the terminals of nigrostriatal dopaminergic neurons. However, there are data in the literature suggesting that the nigrostriatal projections gradually adapt to the loss of striatal mechanisms that normally control the rates of dopamine synthesis and release, thus suggesting that the hyperactivity of nigrostriatal neurons is transitory and subsides when the lesion becomes chronic as seen in HD patients with slowly progressive neuronal degeneration in the striatum (Melamed *et al.*, 1982). The mechanism involved in the adaptation of dopaminergic neurons probably occurs through selective involvement of distinct subpopulations of dopamine receptors during the course of the disease.

The functional activity of dopamine-containing neurons is usually estimated by measurements of the brain concentrations of the major dopamine metabolites, 3,4-dihidhroxyphenylacetic acid and homovanillic acid (HVA). Since about 75% of the dopamine in the mammalian brain is present in the nigrostriatal system, the level of HVA measured in lumbar cerebrospinal fluid is believed to reflect the functional activity of the nigrostriatal dopaminergic pathway. There have been several reports of decreased concentrations of HVA in lumbar CSF and HD (Caraceni *et al.*, 1977; Chase, 1972; Johanson and Ross, 1974). There are also reports suggesting that CSF HVA levels were even lower in a subgroup of HD patients characterized by increased tone and slowness of voluntary movements (Cunha *et al.*, 1981). The rigid akinetic cases of HD have particularly marked atrophy of the corpus striatum, and hence the net dopamine content in such patients is likely to be very low, although there are reports that failed to reveal any significant differences between the rigid and nonrigid patients (Spokes, 1979). Absence of correlation of CSF HVA with the severity of abnormal movements suggests that factors other than dopamine metabolism, including alterations in other neurotransmitter systems, may be responsible for the severity of chorea (Cunha *et al.*, 1981).

7. SEROTONIN

The role of serotonin (5-hydroxytryptamine) in HD is not as well elaborated as the role of either acetylcholine or dopamine. There are many synaptic connections in the striatum that utilize serotonin for their signal transmission across the synaptic gap. The main serotonin-containing neurons lie within the raphe nuclei of the midbrain and pons (Dahlstrom and Fuxe, 1964). The rostral projections terminate mostly in the globus pallidus and limbic structures (Anden *et al.*, 1966). Since the corpus striatum is the most severly degenerated area of the brain in HD, it would be expected that the loss of striatal influence on the pallidum may produce chorea, through the loss of antagonism of the serotonergic input to the globus pallidus (Lee *et al.*, 1968). Consequently, administration of the immediate metabolic precursor of serotonin, e.g., 5-hydroxytryptophan, results in exacerbation of chorea secondary to increased serotonergic activity (Lee *et al.*, 1968; Birkmayer and Hornykiewicz, 1962). Since 5-hydroxytryptophan worsens choreiform movements in HD, logically, agents that block serotonergic effects in the central nervous system should decrease these movements. Methysergide is a serotonin antagonist, and unlike 5-hydroxytryptophan, it does not penetrate the blood–brain barrier very well after intravenous injection in rats (Doepfner, 1962). However, in spite of its incomplete penetration into the brain, methysergide has been shown to have significant central effects in animals (Banna and Anderson, 1968; Weiner *et al.,* 1973) and in man (Dewhurst, 1968). When methysergide was used in the treatment of HD, none of the patients showed any improvement in their choreiform movements (Klawans *et al.*, 1972; Ringel *et al.*, 1973). These conflicting results of worsening of chorea with serotonin but failure to improve chorea with a serotonin antagonist are even more complicated by the fact that patients had worsening of chorea with 5-hydroxytrytophan in spite of the fact that they were simultaneously treated with methysergide (Lee *et al.*, 1968). One of the possible explanations for this discrepancy is the possibility that 5-hydroxytryptophan may be producing chorea through a mechanism other than direct serotonergic agonism on the pallidum, and that serotonin plays a limited role in the production of chorea (Ringel *et al.,* 1973).

8. SUMMARY

A review of the neurochemistry of HD reveals that there are many inconsistencies regarding changes in various neurotransmitter systems; however, the most consistent finding is a reduction in brain GABA content. The discrepancies in neurochemical findings may be partially explained by the fact that since HD is a progressive deteriorative disorder, neurochemical analyses performed on brains obtained at different stages of illness may yield differing neurochemical results. Despite our present knowledge of some of the biochemical deficits found in HD, pharmacological interventions based on these various neurotransmitter systems have failed to produce satisfactory therapeutic benefits.

9. REFERENCES

Achar, V. S., Welch, K. M. A., Chabi, E., Bartosh, K., and Meyer, J. S., 1976, Cerebrospinal fluid gama-aminobutyric acid in neurologic disease, *Neurology* **26:**777–780.

Anden, N. E., Dahlstrom, A., Fuxe, K., Larson, K., Olson, L., and Ungerstedt, U., 1966, Ascending monoamine neurons to the telencephalon and diencephalon, *Acta. Physiol. Scand.* **67:**313–326.

Aquilonius, S. M., and Sjostrom, R., 1971, Cholinergic and dopaminergic mechanisms in Huntington's chorea, *Life Sci.* **10:**404–414.

Arregui, A., Bennett, J. P., Jr., Bird, E. D., Yamamura, H. I., Iversen, L. L., and Snyder, S. H., 1977, Huntington's chorea: Selective depletion of activity of angiotensin converting enzyme in the corpus striatum, *Ann. Neurol.* **2:**294–298.

Arregui, A., Emson, P. C., and Spokes, E. G., 1978, Angiotensin-converting enzyme in substantia nigra: Reduction of activity in Huntington's disease and after intrastriatal kainic acid in rats, *Eur. J. Pharmacol.* **52:**121–124.

Banna, N. R., and Anderson, E. G., 1968, The effects of 5-hydroxytryptamine antagonists on spinal neuronal activity, *J. Pharmacol. Exp. Ther.* **162:**319–325.

Beinfeld, M. C., 1983, Cholecystokinin in the central nervous system: A minireview, *Neuropeptides* **3:**411–427.

Bernheimer, H., Birkmayer, W., Hornykiewicz, O., Jellinger, K., and Seitelberger, F., 1973, Brain dopamine and the syndromes of Parkinson and Huntington. Clinical, morphological and neurochemical correlations, *J. Neurol. Sci.* **20**:415–455.

Bird, E. D., and Iverson, L. L., 1974, Huntington's chorea: Postmortem measurement of glutamic acid decarboxylase, choline acetyltransferase and dopamine in basal ganglia, *Brain* **97**:457–472.

Bird, E. D., Mackay, A. V. P., Rayner, C. N., and Iversen, L. L., 1973, Reduced glutamic-acid-decarboxylase activity of postmortem brain in Huntington's chorea, *Lancet* **1**:1090–1092.

Birkmayer, W., and Hornykiewicz, O., 1962, The L-dihydroxyphenylalanine effect in Parkinson's syndrome in man: On the pathogenesis and treatment of Parkinson's akinesis, *Arch. Psychiatrie Nervenkr.* **203**:560–574.

Bruyn, G. W., Bots, G. Th. A. M., and Dom, R., 1979, Huntington's chorea: Current neuropathological status, *Adv. Neurol.* **23**:83–93.

Buck, S. H., Burks, T. F., Brown, M. R., and Yamamura, H. I., 1981, Reduction in basal ganglia and substantia nigra substance P levels in Huntington's disease, *Brain Res.* **209**:464–469.

Caraceni, T. A., Calderini, G., Consolazione, A., Riva, E., Algeri, S., Girott, F., Spreafico, R., Banciforti, A., Dall'olio, A., and Morselli, L., 1977, Biochemical aspects of Huntington's chorea, *J. Neurol. Neurosurg. Psychiatry* **40**:581–587.

Caraceni, T. A., Girotti, F., Giovannini, P., Pederzoli, M., and Parati, E. A., 1980, Effects of dopamine agonist in Huntington disease hyperkinesia, *Ital. J. Neurol. Sci.* **3**:155–161.

Carter, C. J., 1981, Loss of glutamine synthetase activity in the brain in Huntington's disease, *Lancet* **1**:782–783.

Chase, T. N., 1972, Biochemical and pharmacologic studies of monoamines in Huntington's chorea, *Adv. Neurol.* **22**:533–542.

Coyle, J. T., and Schwarcz, R., 1976, Lesion of striatal neurones with kainic acid provides a model for Huntington's chorea, *Nature* **263**:244–246.

Cross, A., and Rossor, M., 1983, Dopamine D-1 and D-2 receptors in Huntington's disease, *Eur. J. Pharmacol.* **88**:223–229.

Cunha, L., Oliviera, C. R., Diniz, M., Amaral, R., Concalves, A. F., and Pio-Abreu, J., 1981, Homovanilic acid in Huntington's disease and Sydenham's chorea, *J. Neurol. Neurosurg. Psychiatry* **44**(3):258–261.

Dahlstrom, A., and Fuxe, K., 1964, Evidence for the existence of monoamine-containing neurons in the central nervous system I, demonstration of monoamines in the cell bodies of brain stem neurons, *Acta. Physiol. Scand.* **62**(Suppl.):231–255.

Dewhurst, W. G., 1968, Methysergide in mania, *Nature* **219**:506–507.

Diamond, B. I., and Borison, R. L., 1982, Regulatory peptides in animal paradigms of neuropsychiatric illness, in: *Regulatory Peptides: From Molecular Biology to Function* (E. Costa and M. Trabucchi, eds.), pp. 541–548, Raven Press, New York.

Diamond, B. I., Comaty, J. S., Sudakoff, G. S., Havdala, H. S., Walter, R., and Borison, R. L., 1979, Role of substance P in the striatum, *Adv. Neurol.* **23**:505–516.

Diamond, B. I., Pasinetti, G., Hitri, A., and Borison, R. L., 1983, CCK and dopamine in the striatum: Implications for parkinsonism, *Adv. Neurol.* **40**:483–488.

Divac, I., Fonnum, F., and Storm-Mathisen, J., 1977, High affinity uptake of glutamate in terminals of corticostriatal axons, *Nature* **266**:377–378.

Dodd, J., and Kelly, J. S., 1981, The actions of cholecystokinin and related peptides on pyramidal neurones of the mammalian hippocampus, *Brain Res.* **205**:337–350.

Doepfner, W., 1962, Biochemical observations on LSD-25 and diseril, *Experientia (Basel)* **18**:256–257.

Dom, R., Malfroid, M., and Baro, R., 1976, Neuropathology of Huntington's chorea: Studies of the ventrobasal complex of the thalamus, *Neurology* **26**:64–68.

Durso, R., Tamminga, C. A., Denaro, A., Ruggeri, S., and Chase, T. N., 1983, Plasma and growth hormone response to dopaminergic, GABA-mimetic and cholinergic stimulation in Huntington's disease, *Neurology* **33**:1229–1232.

Ehringer, H., and Hornykiewicz, O., 1960, Distribution of noradrenaline and dopamine (3-hydroxytyramine) in the human brain and their behavior in diseases of the extrapyramidal system, *Klin. Wochenschr.* **38:**1236–1239.

Emson, P. C., 1979, Peptides as neurotransmitter candidates in the CNS, *Progr. Neurobiol.* **13:**61–116.

Emson, P. C., Arregui, A., Clement-Jones, V., Sandberg, B. E., and Rossor, M., 1980a, Regional distribution of methionine-enkephalin and substance P-like immunoreactivity in normal human brain and in Huntington's disease, *Brain Res.* **199:**147–160.

Emson, P. C., Rehfeld, J. F., Langevin, H., and Rossor, M., 1980b, Reduction in cholecystokinin-like immunoreactivity in the basal ganglia in Huntington's disease, *Brain Res.* **198:**497–500.

Enna, S. J., Bennett, J. P., Jr., Bylund, D. B., Synder, S. H., Bird, E. D., and Iversen, L. L., 1976a, Alterations of brain neurotransmitter receptor binding in Huntington's chorea, *Brain Res.* **116:**531–537.

Enna, S. J., Bird, E. D., Bennett, J. P., Jr., Bylund, D. B., Yamumaura, H. I., and Iversen, L. L., 1976b, Huntington's chorea: Changes in neurotransmitter receptors in the brain, *N. Engl. J. Med.* **294:**1305–1309.

Enna, S. J., Wood, J. H., and Snyder, S. H., 1977a, γ-Aminobutyric acid (GABA) in human cerebrospinal fluid: Radioreceptor assay, *J. Neurochem.* **28:**1121–1124.

Enna, S. J., Stern, L. Z., Wastek, G. J., and Yamamura, H. I., 1977b, Cerebrospinal fluid γ-aminobutyric acid variations in neurological disorders, *Arch. Neurol.* **34:**683–685.

Enna, S. J., Stern, L. Z., Wastek, G. J., and Yamamura, H. I., 1977c, Minireview: Neurobiology and pharmacology of Huntington's disease, *Life Sci.* **20:**205–212.

Freed, W. J., and Michaelis, E. K., 1976, Effects of intraventricular glutamic acid on the acquisition, performance, and extinction of an operant response, and on general activity, *Psychopharmacology* **50:**293–299.

Gale, J. S., Bird, E. D., Spokes, E. G., Iversen, L. L., and Jessell, T., 1978, Human brain substance P: Distribution in controls and Huntington's chorea, *J. Neurochem.* **30:**633–634.

Glaeser, B. S., Hare, T. A., Vogel, W. H., Olewiler, D. B., and Beasley, B. L., 1975, Low GABA levels in CSF in Huntington's chorea, *N. Engl. J. Med.* **292:**1029–1030.

Grove, J., Schecter, P. J., Tell, G., Koch-Weser, J., Sjoerdsma, A., Warter, J. M., Marsecaux, C., and Rumbach, L., 1981, Increased gamma-aminobutyric acid (GABA), homocarnosin and beta-alanine in cerebrospinal fluid of patients treated with gamma-vinyl GABA (4-amino-hex-5-enoic acid), *Life Sci.* **28:**2431–2439.

Growdon, J. H., Cohen, E. L., and Wurtman, R. J., 1977, Huntington's disease: Clinical and chemical effects of choline administration, *Ann. Neurol.* **1:**418–422.

Hays, S. E., and Paul, S. M., 1982, CCK receptors and human neurological disease, *Life Sci.* **31:**319–322.

Hays, S. E., Goodwin, F. K., and Paul, S. M., 1981, Cholecystokinin receptors are decreased in basal ganglia and cerebral cortex of Huntington's disease, *Brain Res.* **225:**452–456.

Henke, J., 1979, Kainic acid binding in human caudate nucleus: Effect of Huntington's disease, *Neurosci. Lett.* **14:**247–251.

Hiley, C. R., and Bird, E. D., 1974, Decreased muscarinic receptor concentration in post-mortem brain in Huntington's chorea, *Brain Res.* **80:**355–358.

Hong, J. S., Yang, H. Y. T., and Costa, E., 1977, On the location of methionine enkephalin neurons in rat striatum, *Neuropharmacology* **16:**451–453.

Hornykiewicz, O., 1976, Neurohumoral interactions and basal ganglia function and dysfunction, in: *The Basal Ganglia: Research Publication, Association for Research in Nervous and Mental Disease,* Vol. 55 (M. D. Yahr, ed.), pp. 269–280, Raven Press, New York.

Hunt, J. R., 1917, Progressive atrophy of the globus pallidus (primary atrophy of the pallidal system), *Brain* **40:**58–76.

Innis, R. B., Correa, F. M., Uhl, G. R., Schneider, B., and Snyder, S. H., 1979, Cholecystokinin octopeptide-like immunoreactivity: Histochemical localization in rat brain, *Proc. Natl. Acod. Sci. U.S.A.* **76:**521–525.

Iversen, L. L., Bird, E., Spokes, E., Nicholson, S. H., and Suckling, C. J., 1979, Agonist specificity of GABA binding sites in human brain and GABA in Huntington's disease and schizophrenia, *Munksgoard* 179–180.

Jessell, T. M., Emson, P. C., Paxinos, G., and Cuello, A. C., 1978, Topographic projections of substance P and GABA pathways in the striato- and pallido-nigral system: A biochemical and immunohistochemical study, *Brain Res.* **152**:487–498.

Johanson, B., and Ross, B. E., 1974, 5-Hydroxyindolacetic acid and homovanillic acid in cerebrospinal fluid of patients with neurological diseases, *Eur. Neurol.* **11**:37–45.

Kim, J. S., Kornhuber, H. H., Holzmuller, B., Schmid-Burgk, W., Mergner, T., and Krzepinski, G., 1980, Reduction of cerebrospinal fluid glutamic acid in Huntington's chorea and in schizophrenic patients, *Arch. Psychiatr. Nervenkr.* **228**:7–10.

Klawans, H. L., and Rubovits, R., 1972, Central cholinergic–anticholinergic antagonism in Huntington's chorea, *Neurology* **22**:107–116.

Klawans, H. L., and Weiner, W. J., 1974, The effect of *d*-amphetamine on choreiform movement disorders, *Neurology* **24**:312–318.

Klawans, H. L., and Weiner, W. J., 1976, The pharmacology of choreatic movement disorders, *Prog. Neurobiol.* **6**:49–80.

Klawans, H. L., Rubovits, R., Ringel, S. P., and Weiner, W. J., 1972, Observations on the use of methysergide in Huntington's chorea, *Neurology* **22**:930–933.

Kremzner, L. T., Berl, S., Stellar, S., and Cote, L. J., 1979, Amino acids, peptides, and polyamines in cortical biopsies and ventricular fluid in patients with Huntington's disease, *Adv. Neurol.* **23**:537–546.

Lee, D. K., Markham, C. H., and Clark, W. G., 1968, Serotonin (5-hydroxytryptamine) metabolism in Huntington's chorea, *Life Sci.* **7**:707–712.

Lenman, J. A., Ferguson, I. T., Fleming, A. M., Herzberg, L., Robb, J. E., and Turnbull, M. J., 1976, Sodium valproate in chorea, *Br. Med. J.* **2**:1107–1108.

Lloyd, K. G., and Davidson, L., 1979, (^{3}H) GABA binding in brains from Huntington's chorea patients: Altered regulation by phospholipids? *Science* **205**:1147–1149.

Lloyd, K. G., Dreksler, S., and Bird, E. D., 1977, Alterations in ^{3}H-GABA binding in Huntington's chorea, *Life Sci.* **21**:747–753.

Manberg, P. J., Nemeroff, C. B., Iverson, L. L., Rossor, M. N., Kizer, J. S., and Prange, A. J., 1982, Human brain distribution of neurotensin in normals, schizophrenics, and Huntington's choreics, *Ann. NY Acad. Sci.* **400**:354–365.

Mangano, R. M., and Schwarcz, R., 1980, Glutamate uptake into human platelets, characterization and observations in Huntington's disease, *Soc. Neurosci. Abstr.* **6**:508.

Mann, J. J., Stanley, M., Gershon, S., and Rossor, M., 1980, Mental symptoms in Huntington's disease and a possible primary aminergic neuron lesion, *Science* **210**:1369–1371.

Manyam, N. V., Hare, T. A., Katz, L., and Glaeser, B. S., 1978, Huntington's disease. Cerebrospinal fluid GABA levels in at-risk individuals, *Arch. Neurol.* **35**:728–730.

Manyam, B. V., Katz, L., Hare, T. A., Kaniefski, K., and Tremblay, R. D., 1981, Isoniazid-induced elevation of CSF GABA levels and effects on chorea in Huntington's disease, *Ann. Neurol.* **10**:35–37.

McGeer, E. G., and McGeer, P. L., 1976, Duplication of biochemical changes of Huntington's chorea by intrastriatal injections of glutamic and kainic acids, *Nature* **263**:517–519.

McGeer, P. L., McGeer, E. G., and Fibiger, H. C., 1973, Choline acetylase and glutamic acid decarboxylase in Huntington's chorea: A preliminary study, *Neurology* **23**:912–917.

McLennan, H., and York, D. H., 1969, The action of dopamine on neurons of caudate nucleus, *J. Physiol.* **189**:393–402.

Melamed, E., Hefti, F., and Bird, E. D., 1982, Huntington chorea is not associated with hyperactivity of nigrostriatal dopaminergic neurons: Studies in postmortem tissues and in rats with kainic acid lesions, *Neurology* **32**:640–644.

Nemeroff, C. B., Youngblood, W. W., Manberg, P. J., Prange, A. J., and Kizer, J. S., 1983, Regional brain concentrations of neuropeptides in Huntington's chorea and schizophrenia, *Science* **221**:972–975.

Nutt, J. G., Rosin, A., and Chase, T. N., 1978, Treatment of Huntington disease with a cholinergic agonist, *Neurology* **28:**1061–1064.
Olney, J. W., and deGubareff, T., 1978, Glutamate neurotoxicity and Huntington's chorea, *Nature* **271:**557–559.
Olney, J. W., Sharpe, L. G., and Feigin, R. D., 1972, Glutamate-induced brain damage in infant primates, *J. Neuropathol. Exp. Neurol.* **31:**464–488.
Olsen, R. W., Van Ness, P. C., and Tourtellotte, W. W., 1979, Gamma-aminobutyric acid receptor binding curves for human brain regions: Comparison of Huntington's disease and normal, *Adv. Neurol.* **23:**697–704.
Penny, J. B., and Young, A. B., 1982, Quantitative autoradiography of neurotransmitter receptors in Huntington's disease, *Neurology* **32:**1391–1395.
Perry, T. L., Hansen, S., Diamond, S., and Stedman, D., 1969, Plasma-aminoacid levels in Huntington's chorea, *Lancet* **1:**806–808.
Perry, T. L., Hansen, S., and Lesk, D., 1972, Plasma amino acid levels in children of patients with Huntington's chorea, *Neurology* **22:**68–70.
Perry, T. L., Hansen, S., and Kloster, M., 1973, Huntington's chorea, deficiency of Y-aminobutyric acid in brain, *N. Engl. J. Med.* **288:**337–342.
Perry, T. L., Hansen, S., and Kennedy, J., 1975, CSF amino acids and plasma-CSF amino acid ratios in adults, *J. Neurochem.* **24:**587–589.
Perry, T. L., Wright, J. M., Hansen, S., and MacLeod, P. M., 1979, Isoniazid therapy of Huntington disease, *Neurology* **29:**370–375.
Perry, T. L., Hansen, S., Wall, R. A., and Gauthier, S. G., 1982, Human CSF GABA concentrations: revised downward for controls, but not decreased in Huntington's chorea, *J. Neurochem.* **38:**766–773.
Phillipson, O. T., and Bird, E. D., 1977, Plasma glucose, nonesterified fatty acids and amino acids in Huntington's chorea, *Clin. Sci. Mol. Med.* **52:**311–318.
Pinget, M., Straus, E., and Yalow, R. S., 1979, Release of cholecystokinin peptides from a synaptosome-enriched fraction of rat cerebral cortex, *Life Sci.* **25:**339–342.
Potashner, S. J., 1978, Baclofen: Effects of amino acid release, *Can. J. Physiol. Pharmacol.* **56:**150–154.
Rehfeld, J. F., 1978, Immunochemical studies on cholecystokinin II. Distribution and molecular heterogeneity in the central nervous system and small intestine of man and hog, *J. Biol. Chem.* **253:**4022–4030.
Reibling, A., Reyes, P., and Jameson, H. D., 1975, Dimethylaminoethanol ineffective in Huntington's disease, *N. Engl. J. Med.* **293:**724.
Reisine, T. D., Fields, J. Z., Bird, E. D., Spokes, E., and Yamamura, H. I., 1978, Characterization of brain dopaminergic receptors in Huntington's disease, *Comm. Psychopharmacol.* **2**(2):79–84.
Ringel, S. P., Weiner, W. J., Rubovits, R., and Klawans, H. L., 1973, Methysergide in Huntington's chorea, in: *Advances in Neurology*, Vol. 1, pp. 769–776, Raven Press, New York.
Roberts, P. J., 1974, Glutamate receptors in the rat central nervous system, *Nature* **252:**399–401.
Roberts, P. J., and Anderson, S. D., 1979, Stimulatory effect of L-glutamate and related amino acids on [^{3}H]dopamine release from rat striatum: An *in vitro* model for glutamate actions, *J. Neurochem.* **32:**1539–1545.
Ruck, A., Kramer, S., Metz, J., and Brennan, M. J., 1980, Methyltetrahydrofolate is a potent and selective agonist for kainic acid receptors, *Nature* **287:**852–853.
Saito, A., Sankaran, H., Goldfine, I. D., and Williams, J. A., 1980, Cholecystokinin receptors in the brain; Characterization and distribution, *Science* **208:**1155–1156.
Sanberg, P. R., and Johnston, G. A., 1981, Glutamate and Huntington's disease, *Med. J. Aust.* **2:**460–465.
Schwarcz, R., Bennett, P. J., Jr., and Coyle, J. T., Jr., 1977, Loss of striatal serotonin synaptic receptor binding induced by kainic acid lesion: Correlations with Huntington's disease, *J. Neurochem.* **28:**867–869.

Severs, W. B., and Daniels-Severs, A. E., 1973, Effects of angiotensin on the central nervous system, *Pharmacol. Rev.* **25**:415–423.

Shoemaker, H., Morelli, M., Deshmukh, P., and Yamamura, H. I., 1982, 3H-RO5-4864 benzodiazepine binding in the kainate lesioned striatum and Huntington's diseased basal ganglia, *Brain Res.* **248**:396–401.

Shoulson, I., Chase, T. N., Roberts, E., and VanBalgooy, J. N. A., 1975, Huntington's disease. Treatment with imidazole-4-acetic acid, *N. Engl. J. Med.* **293**:504–505.

Shoulson, I., Katrizinel, R., and Chase, T. N., 1976, Huntington's disease: Treatment with dipropylacetic acid and gamma-aminobutyric acid, *Neurology* **26**:61–63.

Shoulson, I., Goldblatt, D., Charlton, M., and Joynt, R. J., 1978, Huntington's disease: Treatment with musciomol, a GABA-mimetic drug, *Ann. Neurol.* **4**:279–284.

Siegmund, R., Schmeisser, G., and Heidrich, R., 1982, Therapeutic experiences in treatment of hyperkinesias with the neuroleptic pimozide, *Psychiatr. Neurol. Med. Psychol. (Leipz)* **34**(5):307–308.

Skerritt, J. H., and Johnston, G. A. R., 1981, Postnatal development of GABA and kainate binding sites and their endogenous inhibitors in rat brain, *Aust. Neurosci. Proc.* **1**:65C.

Soffer, R. L., 1976, Angiotensin-converting enzyme and the regulation of vaso-active peptides, *Ann. Rev. Biochem.* **45**:73–94.

Spokes, E. G. S., 1979, Dopamine in Huntington's disease. A study of post-mortem brain tissue, *Adv. Neurol.* **23**:481–493.

Spokes, E. G. S., 1980, Neurochemical alterations in Huntington's chorea: A study of post-mortem brain tissue, *Brain* **103**:179–210.

Stahl, W. L., and Swanson, P. D., 1974, Biochemical abnormalities in Huntington's chorea brains, *Neurology* **24**:813–819.

Symington, G. R., Leonard, D. P., Shannon, P. J., and Vajda, F. J. E., 1978, Sodium valproate in Huntington's disease, *Am. J. Psychiatry* **135**:352–354.

Tarsy, D., Leopold, N., and Sax, D. S., 1974, Physostigmine in choreiform movement disorders, *Neurology* **24**:28–33.

Tell, G., Bohlen, P., Schechter, P. J., Koch-Weser, J., Agid, Y., Bonnet, A. M., Coquillat, G., Chazot, G., and Fischer, C., 1981, Treatment of Huntington disease with γ-acetylenic GABA, an irreversible inhibitor of GABA-transaminase: increased CSF GABA and homocarnosine without clinical amelioration, *Neurology* **31**:207–211.

Toglia, J. U., McGlamery, M., and Sambandham, R. R., 1978, Tetrabenazine in the treatment of Huntington's chorea and other hyperkinetic movement disorders, *J. Clin. Psychiatry* **39**(1):81–87.

Urquhart, N., Perry, T. L., Hansen, S., and Kennedy, J., 1975, GABA content and glutamic acid decarboxylase activity in brain of Huntington's chorea patients and control subjects, *J. Neurochem.* **24**:1071–1075.

Van Ness, P. C., Watkins, A. E., Bergman, M. O., Tourtellotte, W. W., and Olsen, R. W., 1982, γ-Aminobutyric acid receptors in normal human brain and Huntington disease, *Neurology* **32**:63–68.

Waddington, J. L., and Cross, A. J., 1980, Characterization of denervation supersentitivity in the striatonigral GABA pathway of the kainic acid lesioned rat and in Huntington's disease, *Brain Res. Bull.* **5**(Suppl. 2):825–828.

Walker, J. E., Hoehn, M., Sears, E., and Lewis, J., 1973, Dimethyl aminoethanol in Huntington's chorea, *Lancet* **1**:1512–1513.

Walker, R. J., Kemp, J. A. Yajima, H., Kitagawa, K., and Woodruff, G. N., 1976, The action of substance P on mesencephalic reticular and substantia nigral neurones of the rat, *Experientia* **32**:214–215.

Wastek, G. J., and Yamamura, H. I., 1978, Biochemical characterization of the muscarinic cholinergic receptor in human brain: Alterations in Huntington's disease, *Mol. Pharm.* **14**:768–780.

Wastek, G. J., Stern, L. Z., Johnson, P. C., and Yamamura, H. I., 1976, Huntington's disease:

Regional alteration in muscarinic cholinergic receptor binding in human brain, *Life Sci.* **19**:1033–1039.

Watt, J. A., and Cunningham, W. L., 1978, Plasma amino acid levels in Huntington's chorea, *Br. J. Psychiatry* **132**:394–397.

Weiner, W. J., Goetz, C., and Klawans, H. L., 1973, Serotonergic and antiserotonergic influences on apomorphine induced stereotyped behavior, *Acta Pharmacol. Toxicol.* **36**:155–160.

Weiner, W. J., Hitri, A., Caryey, P., Koller, W. C., Nausieda, P., and Klawans, H. L., 1975, ^{3}H dopamine binding studies in guinea pig striatal membrane suggesting two distinct dopamine receptor sites, *Adv. Neurol.* **23**:687–695.

Welch, K. M. A., Chabi, E., Achar, V. S., Bartosh, K., and Meyer, J. S., 1975, GABA in human CSF: The significance of measurement in neurological disease, *5th Annual Meeting of the Society for Neuroscience*, Washington D. C. Abstract No. 502.

Wu, J. Y., Bird, E. D., Chen, M. S., and Huang, W. M., 1978, Studies of neuro-transmitter enzymes in Huntington's chorea, *Adv. Neurol.* **23**:527–536.

Yang, H. Y. T., and Neff, N. H., 1972, Distribution and properties of angiotensin converting enzyme of rat brain, *J. Neurochem.* **19**:2443–2450.

Yates, C. M., Magill, B. E., Davidson, D., Murray, L. G., Wilson, H., and Pullar, I. A., 1973, Lysosomal enzymes amino acids and acid metabolites of amines in Huntington's chorea, *Clin. Chem. Acta* **44**:139–145.

CHAPTER 15

The Effects of Neuroleptics on Longevity in Huntington's Disease

CHARLES N. STILL and THOMAS J. GOLDSCHMIDT

1. INTRODUCTION

Huntington's disease (HD) comes from the observations of George Sumner Huntington, who gained eponymic immortality by his description of chronic progressive hereditary chorea in 1872. Huntington's paper includes the following major features of the disease:

1. Its hereditary nature: "When either or both the parents have shown manifestation of the disease, . . . one or more of the offspring almost invariably suffer from the disease, if they live to adult age. But if by chance these children go through life without it, the thread is broken and the grandchildren and great grandchildren of the original shakers may rest assured that they are free from the disease. . . . Unstable and whimsical as the disease may be in other respects, in this it is firm, it never skips a generation to again manifest itself in another; once having yielded its claims, it never regains them."
2. Its rarity: "hereditary chorea . . . is confined to certain and fortunately a few families."
3. Its abnormal involuntary movements: "It begins as an ordinary chorea might begin, by the irregular and spasmodic action of certain muscles, as of the face, arms, etc. These movements gradually increase, when muscles hitherto unaffected take on the spasmodic action, until every muscle in the body becomes affected (excepting the involuntary ones). . . ."

CHARLES N. STILL and THOMAS J. GOLDSCHMIDT • Department of Neuropsychiatry and Behavioral Science, University of South Carolina School of Medicine, Columbia, South Carolina 29203. *Present address of T. J. G.:* Department of Neurology and Psychiatry, Tulane University School of Medicine, New Orleans, Louisiana 70112.

4. Its progressive nature: ". . . increasing by degrees, and often occupying years in its development, until the hapless sufferer is but a quivering wreck of his former self."
5. Its delayed age of onset: "I do not know of a single case that has shown any marked signs of chorea before the age of thirty of forty years, while those who pass the fortieth year without symptoms of the disease, are seldom attacked."
6. Its behavioral abnormalities, including suicide: "In nearly all the families . . . in which the choreic taint exists, the nervous temperament greatly preponderates, and in my grandfather's and father's experience, which conjointly covered a period of 78 years, nervous excitement in a marked degree almost invariably attends upon every disease these people may suffer from, although they may not when in health be over nervous. . . . The tendency to insanity, and sometimes that form of insanity which leads to suicide is marked."
7. Its progressive dementia: "As the disease progresses the mind becomes more or less impaired, in the many amounting to insanity, while in others mind and body both gradually fail until death relieves them of their sufferings."
8. Its resistance to treatment: "I have never known a recovery or even an amelioration of symptoms in this form of chorea; when once it begins it clings to the bitter end. No treatment seems to be of any avail, and indeed nowadays its end is so well known to the sufferer and his friends, that medical advice is seldom sought."

2. NATURAL HISTORY OF HD

The gene for HD is known to be inherited in a mendelian autosomal dominant mode, with complete penetrance, so that each HD gene carrier will eventually show clinical manifestations of HD if death does not occur first. Since each child of an affected parent has a 50% risk of inheriting the HD gene, the majority of HD cases can be confirmed by a documented family history (Bruyn, 1968). Before the recent discovery of a recombinant DNA marker for the HD gene on the short arm of chromosome 4, there was no fully reliable method of detecting the HD gene before clinical onset, commonly recognized by the appearance of abnormal involuntary choreiform movements, usually subtle intermittent facial twitches or piano-playing finger movements. It is not yet certain that all cases of HD are in fact caused by the HD gene identified on chromosome 4, or by any other single gene at a single locus (Martin, 1984). Nevertheless, there is new hope for preclinical detection of HD gene carriers, identification of the HD gene product, and development of corrective gene therapy (Shoulson, 1984).

Most HD families are found among the lower socioeconomic groups of society, frequently manifesting alcoholism, antisocial behavior, and major mental disorders, often requiring long-term institutional care (Bruyn, 1968). A study of 962 members of HD kindreds delineated manic–depressive versus demented groups, noting that "increasing irritability, insomnia, malaise are especially apt to be immediate precursors of an outbreak"(Davenport and Muncey, 1916). Streletzki (1961) identified 67 cases of paranoid psychosis among 1200 documented cases of HD in Germany. Paranoia marked the onset of HD in 20 patients, half of whom were diagnosed schizophrenic. Paranoia usually merged with dementia by the third year of illness. To summarize, behavioral changes in HD generally conform to the following sequence: (1) early stage—aggression, antisocial behavior, irritability, sexual promiscuity, (2) intermediate stage—affective/mood changes (depression more common), schizophrenia/hallucinations/delusions (less common), (3) late stage—apathy, abulia, self-neglect with obvious dementia. Depression is the most common psychological disorder in early HD (McHugh and Folstein, 1975; Martin, 1984).

Studies of the age of onset in HD have shown a high concordance among monozygotic twins, contrasted with dizygotic twins and full siblings, suggesting that genetic mechanisms control the variability in age of onset, even though the HD gene is present from conception. Moreover, first-degree relatives show closer ages of onset than more distant relatives, suggesting the action of a discrete number of modifier genes on expression of the HD gene (Martin, 1984). Regarding the natural history of HD, Quadfasel (1973) stated, "It is impossible to predict that the same form will occur regularly in the same family, or which manifestation—mental deterioration, social deterioration, involuntary movements or rigidity—will regularly make its appearance first or last, or appear at all, or at what age."

Huntington apparently overlooked the juvenile form of HD, despite its existence among HD families living in Long Island, Connecticut, and Massachusetts during his lifetime. As noted by Davenport and Muncey (1916), "The disease has made its onset in one or more cases in every decade up to and including the eighth. . . . in over a sixth of the cases the age of onset is at 20 years or under, much as in Sydenham's chorea." In juvenile-onset cases, the source of the HD gene is nearly always the father, who may not show choreiform involuntary movements before HD appears in the child as the akinetic–rigid Westphal variant, with akinesia, rigidity, ataxia, dystonia, myoclonus, seizures, dementia, and a rapid downhill course to death (Bruyn, 1968; Martin, 1984). While seizures are common in juvenile-onset HD, epilepsy is no more common in adult-onset HD than in the general population (Oepen, 1963).

Late-onset HD, defined by initial clinical manifestations at age 50 or later, shows slower progression and less obvious dementia than HD of midlife onset. Affected offspring of late-onset females also have late-onset HD, but affected offspring of late-onset males have significantly earlier ages of onset. Late-onset offspring apparently inherit the HD gene from an affected mother, in most cases. These findings that the sex of the affected parent modifies age of onset in HD offspring have been explained by putative maternally transmitted factors, e.g., mitochondrial DNA or other extrachromosomal factors which influence the age of onset. For example, juvenile onset is not likely to be inherited from an HD mother, whose early-onset extrachromosomal factors would bring early self-destruction (Martin, 1984).

Calculations to determine the duration of illness (survival time), require a precisely known age of onset. Longevity (life-span) may be calculated from the dates of birth and death, thereby reflecting the maximum age attained by an individual or by a group. On the other hand, life expectancy refers to average length of life for a specified population, usually calculated from a life table. Life tables assume that each member of a cohort is born within a specified time interval, being reared under essentially identical conditions of diet, environment, and biopsychosocial interaction, without migration into or away from the study area (Rockstein *et al.,* 1977). While such ideal conditions do not exist for HD families, useful data have been reported (Bruyn, 1968).

Reed *et al.* (1958) conducted a massive prevalence survey of HD in the lower peninsula of Michigan as of April 1, 1940. From this project, 229 kindreds having 801 affected members were identified. Michigan residents with HD numbered 363 males and 398 females, with a mean age of onset of 35.3 years, based on the first appearance of choreiform movements. In some cases, onset was based on manifestations of behavioral changes preceding chorea. The mean age for institutionalization of 89 males was 48.11 years, and for 102 females, 35.85 years. The mean age at death was 53.04 years for males and 54.11 years for females. Survival from onset to death was 15.78 years for males and 15.93 years for females. In noninstitutionalized patients, suicide was an important cause of death, accounting for 7.8% of male deaths and for 6.4% of female deaths (Chandler *et al.,* 1960). From the 1940 Michigan prevalence (Reed *et al.,* 1958) of 4.1 HD cases per 100,000 population, Jones (1979) calculated a 1970 U.S. prevalence of 4.8 cases per 100,000 population. Recognizing that his calculations assume that the average age of onset of HD is precisely known, Jones concluded that for both sexes, about 5 years of life are lost to HD for each thousand persons born.

In Michigan, Reed *et al.* (1958) found no correlation between the age of onset and survival time from onset to death. In Germany, Wendt *et al.* (1960) reported a negative correlation, especially in women. Based on 762 German cases of HD (377 males and 385 females), the mean age of onset was found to be 44.25 years for males and 43.69 for females. The mean age at death for 246 HD patients was 56.47 years for males and 56.28 years for females. In the Wendt series, mean duration of illness (survival time from onset to death) was 12.23 years for males and 13.52 years for females. The cause of death for 462 German cases of HD is as follows (Wendt *et al.,* 1960):

1.	Intercurrent (secondary) illness	27%
2.	Heart and circulatory disorders	25%
3.	Cachexia and exhaustion	18%
4.	Apoplexy	7%

5. Infectious diseases	8%
6. Suicide	6%
7. Carcinoma	4%
8. Gastrointestinal disorders	2%
9. Renal disorders	1%
10. Accidents—euthanasia	1%
11. Heart–metabolic–blood disorders	1%
Total	100%

To avoid missing late-onset HD cases, Wendt *et al.* (1959) urged that non-extinct cohorts be excluded in calculating mean ages of onset. However, Wendt derived his admonition from data on cohorts born from 1870 to 1899, thereby excluding an unknown number of early-onset HD cases and at-risk offspring who were of prime military age during the 1914–1918 World War (Still and Young, 1974). Moreover, the mean life-span of German HD patients approximates that of Michigan HD patients, whose mean age of onset was nearly 10 years younger (Chandler *et al.*, 1960).

The differential diagnosis includes a formidable list of choreiform disorders which may be confused with HD, as follows (Bruyn, 1973):

1. Hereditary disorders—senile chorea, familial paroxysmal choreathetosis, hereditary juvenile chorea, ataxia–telangiectasia, acanthocytosis, Down's syndrome, dystonia musculorum deformans, Fabry's disease, Fahr's disease, Hallervorden–Spatz disease, Lesch–Nyhan syndrome, oculocerebrorenal syndrome, porphyria, phenylketonuria, Pick's disease, spinocerebellar ataxia, striatonigral degeneration, Sturge–Weber–Dimitri syndrome, neurolipidosis, Wilson's disease.
2. Metabolic disorders—Addison's disease, beriberi, hyperalaninemia, hypercalcemia, hypernatremia, hypocalcemia, hypoglycemia, hypomagnesemia, hypoparathyroidism, hepatocerebral degeneration, myxedema, thyrotoxicosis, uremia, vitamin B_{12} deficiency, chorea gravidarum.
3. Toxic disorders—alcohol, amphetamines, anticholinergics, azides, bilirubin, carbon monoxide, lead, lithium, mercury, oral contraceptives, phenytoin, ethosuximide, isoniazid, imipramine, levodopa, methylphenidate, α-methyldopa, neuroleptics (tardive dyskinesia).
4. Infectious disorders—diphtheria, encephalitis (acute viral, measles, mumps, pertussis, scarletina, varicella), syphilis, subacute bacterial endocarditis, subacute spongiform encephalopathy (Creutzfeldt–Jakob disease), Sydenham's chorea, Rocky Mountain spotted fever, typhoid fever, subacute sclerosing panencephalitis.
5. Vascular disorders—cerebral palsy, cerebrovascular accident, lupus erythematosus, polycythemia vera, purpura, vasculitis.
6. Neoplastic disorders—primary and metastatic brain tumors.
7. Traumatic disorders—burns, electrical shock, head injury.

8. Degenerative disorders—Alzheimer's disease, Canavan's spongy degeneration, supranuclear palsy, thalamic centromedian atrophy.

Clinical diagnosis of HD is not a sure thing, especially in the absence of an adequate family history of the disease (Bruyn, 1968). Alzheimer's disease is the condition most commonly misdiagnosed as HD (Martin, 1984).

3. PATHOLOGY, PATHOPHYSIOLOGY, AND PHARMACOTHERAPY OF HD

Two decades after Huntington wrote, "I know nothing of its pathology," postmortem brain studies were believed to show evidence of meningoencephalitis, perhaps related to syphilis. Associating severe atrophy of the caudate nucleus with massive astrogliosis, Alzheimer (1911) insisted that neostriatal–cortical neuronal loss represented the true pathology of HD.

Distinctive microscopic features of HD include 70–80% depletion of small Golgi type II spiny neurons, which apparently contain GABA, enkephalins, and substance P. Moreover, a majority of striatal spiny neurons show dendritic abnormalities (Martin, 1984).

Electron micrographs of HD cortical and caudate biopsies reveal disarray of all membrane systems within affected neurons and glia, with severe disruption of mitochondrial inner membranes and large intracellular amounts of lipofuscin pigment (Tellez-Nagel *et al.*, 1974; Roizin *et al.*, 1976). Ultrastructural studies of the nucleus accumbens show nuclear membrane indentations, also found in the caudate nucleus, where astrocyte/neuron ratios are substantially increased. Patients who die at an older age, after an appreciably longer course of HD, show a lower percentage of neurons with nuclear membrane indentations and a lower astrocyte/neuron ratio in the caudate nucleus than patients who die at a younger age after a relatively shorter duration of HD (Roos and Bots, 1983).

While still unknown, the pathogenesis of HD appears to involve a metabolic disorder affecting glial–neuron interaction, associated with a defective gene product which leads to cell death, either by failure to supply an essential precursor or by failure to remove an accumulating toxin (Bruyn *et al.*, 1979). Early in the course of HD, positron emission tomography shows a significant decrease in radiolabeled glucose uptake in the basal ganglia, without detectable caudate atrophy on computed tomography (Martin, 1984).

Fibroblast cultures from HD patients show dose-related loss of viability in the presence of physiological concentrations of glutamate associated with decreased numbers of glutamate receptors and increased glutamate binding, supporting the view that HD develops via a glutamate-dependent neurotoxic process which may be retarded by chronic attenuation of glutamatergic neurotransmission. Clinical trials have been aimed at decreasing the availability of glutamate in the striatum (Shoulson, 1984). However, the pathophysiology of HD involves at least a dozen neurotransmitters. Neurotransmitters that are decreased in HD include GABA, acetylcholine, substance P, enkephalins, and cholecystokinin. In contrast, thyrotropin-releasing hormone, somatostatin, and neurotensin are increased in HD. Dopamine may be increased, but is usually unchanged, as are 5-hydroxytryptamine, vasoactive intestinal peptide (VIP) and norepinephrine (Martin, 1984). With involvement of multiple neurotransmitter systems, pharmacotherapy limited to one specific neurotransmitter seems likely to fall short of controlling HD. Clinically, dopamine-blocking neuroleptic drugs have shown beneficial therapeutic effects on chorea in HD, but at the risk of drug-induced parkinsonism and/or tardive dyskinesia with prolonged use (Still, 1984).

Historically, three major groups of neuroleptic (nerve-seizing) drugs became available for the treatment of mental illness during the 1950s, as follows (Delay and Deniker, 1968): (1) phenothiazines [chlorpromazine (CPZ)]—1952, (2) rauwolfia (reserpine)—1954, and (3) butyrophenones (haloperidol)—1959. These drugs were found to be useful in the pharmacological management of HD, ameliorating acute and chronic psychotic states, as well as controlling agitation and excitement, but at the cost of psychomotor apathy, complicated by diencephalic and extrapyramidal side effects, especially drug-induced parkinsonism. Without satisfactory animal models of psychosis, experimental neuroleptics are evaluated for efficacy by studying their potent extrapyramidal effects in laboratory animals (APA, 1980). Their antipsychotic—antichoreic effects originally made neuroleptics the drugs of choice for treating choreic HD patients (Whittier, 1967). Reserpine was found to deplete dopamine in the striatum, pallidum, and substantia nigra, leading to drug-induced parkinsonism, as with phenothiazine therapy. Subsequently, phenothiazines were found to block dopamine-sensitive receptors in the central nervous system (Hornykiewicz, 1966).

The key pathophysiology of HD is briefly summarized as follows:

1. Dopamine—normal or slightly increased in HD. Drugs which deplete dopamine from presynaptic terminal (reserpine, α-methyltyrosine, tetrabenazine) inhibit choreiform movements, as do neuroleptics which block postsynaptic dopamine receptors, (CPZ, fluphenazine, haloperidol). The dopaminergic dominance associated with chorea in HD apparently reflects relative sparing of presynaptic dopaminergic neurons, with higher concentrations of dopamine in the striatum, substantia nigra, and certain limbic regions, resembling the effects of chronic treatment with haloperidol (Spokes, 1979). Similarly, drugs that enhance central dopaminergic activity (levodopa, bromocriptine) worsen chorea (Goetz *et al.*, this volume, Chapter 16).
2. Acetylcholine—despite a 50% reduction in choline acetyltransferase activity in HD striatum, cholinergic drugs (choline, deanol, lecithin, physostigmine) have not been effective in treating HD patients, whose chorea is usually made worse by anticholinergics (Martin, 1984; Goetz *et al.*, this volume, Chapter 16).

3. GABA is the most conspicuously decreased neurotransmitter in HD, with an associated 85% reduction in striatal glutamic acid decarboxylase activity, probably reflecting losses of GABA-containing neurons. GABA itself fails to cross the blood–brain barrier, and repeated efforts to restore brain GABA levels have produced no therapeutic benefits in HD (Martin, 1984).

Neuroleptics which block dopamine receptors remain the standard drugs for controlling abnormal involuntary movements in HD, with somewhat higher dosage levels needed for effective treatment of associated behavioral disorders (Bruyn, 1968; Hornykiewicz, 1979). However, the apparent benefits of neuroleptics may be offset by delayed reappearance of abnormal involuntary choreiform movements resembling those for which the neuroleptic drugs were orginally administered (Crane, 1973). Analysis of tardive dyskinesia (TD), which develops after prolonged neuroleptic therapy, has disclosed several important distinguishing features from HD. A detailed account of the variety of abnormal involuntary movement disorders associated with TD is found elsewhere in this volume (Goetz *et al.*, Chapter 16).

Shoulson (1981) has reported a functional evaluation study of haloperidol and perphenazine therapy in 22 HD patients (10 males, 12 females) over a period of 13–40 months. This study noted a significant decline in overall functional capacity within 2 years. Even though the severity of chorea was reduced in 12 of 18 patients, adverse side effects in 10 patients (lethargy, immobility, unsteadiness, akathisia, parkinsonism, and/or TD) led to discontinuing neuroleptics in five patients. Shoulson (1984) states that more patients with HD gain greater benefit from avoiding the use of neuroleptic drugs than otherwise. We do not agree with this view, concerning either quality of life or length of life (Still, 1984).

4. COHORT STUDY OF NEUROLEPTIC EFFECTS IN HD

Development of an HD registry from institutional records of the South Carolina Department of Mental Health (SCDMH) led to the identification of more than 1500 individuals at risk, with detailed data for 205 cases of HD, representing more than 70 kindreds. From this larger population, a case-control cohort study was done according to the following criteria:

1. Birth date between January 1, 1900 and December 31, 1930.
2. SCDMH hospital admission at least 1 year before death.
3. Clinical onset of abnormal involuntary movements before 1960, to minimize risk of confusion with TD.
4. At least one affected first-degree relative and/or geographic origin in a community known to harbor HD over a period of at least two generations.
5. Exclusion of suicides and juvenile-onset cases.

Age at onset was estimated by the kindred-manifest criteria of Whittier (1967), including chorea, ataxia, dysarthria, dysphagia, incoordination, dystonia, seizures, dementia, emotional lability, depression, hallucinations, delusions, and/or a persistent pattern of abnormal (aggressive, antisocial, paranoid, violent) behavior.

Duration of illness (survival time) was calculated from age at onset to age at death, recorded in SCDMH hospital records. The exact date of death was ascertained for each of 14 males and 22 females included in this study.

Recorded causes of death for 36 HD patients are as follows:

1.	Cardiovascular disorders	27%
2.	Pneumonia	24%
3.	Cachexia	20%
4.	Septicemia	7%
5.	Asphyxia	4%
6.	Seizures	3%
7.	Renal disorders	2%
8.	HD—unspecified	13%
	Total	100%

With respect to cause of death, there were no remarkable differences between HD patients who received neuroleptics and those who did not.

The control group consisted of seven males and eight females who never received neuroleptic drugs while hospitalized for the management of HD. The neuroleptic therapy group consisted of seven males and 14 females, each of whom received at least the equivalent of 10 g of CPZ, calculated on the basis of conversion ratios derived from a table of equivalent doses of commonly used antipsychotic drugs (APA, 1980), as shown in Table 1.

The mean duration in days of neuroleptic therapy was calculated from

TABLE 1
Huntington's Disease: Neuroleptic Drugs

Generic name	Conversion ratio
Chlorpromazine (CPZ)[a]	1:1
Fluphenazine[a]	1:50
Haloperidol[b]	1:50
Mesoridazine[a]	1:2
Perphenazine[a]	1:10
Thioridazine[b]	1:1
Trifluoperazine[b]	1:20
Trifluopromazine[a]	1:4

[a] Huntington's disease, females only.
[b] Huntington's disease, males and females.

TABLE 2
Huntington's Disease: Neuroleptic Therapy Group
Birth Cohorts 1900–1930

	Males (7)[a]	Females (14)[a]
Duration (days)	593.4 ± 277.9	509.9 ± 190.7
Total dose (grams CPZ)	101.1 ± 28.8	125.5 ± 62.7

[a] Mean ± SEM.

medication administration records for each HD patient, without adjustment for route of administration, i.e., oral versus parenteral. Mean dose-duration data for the HD neuroleptic therapy group are found in Table 2. While neuroleptic drugs are not usually prescribed for persons at risk for HD before the onset of clinical manifestations, no drug data are available for either control or neuroleptic therapy groups except while hospitalized.

If neuroleptic therapy has no adverse effect on the course of HD, there should no be significant difference in survival time from age of onset to death in the HD control group versus the HD neuroleptic therapy group, assuming that the groups are comparable. As shown in Table 3, HD control group males have a significantly later age of onset than HD control group females, yet a shorter survival time, with no significant difference in age at death. The HD control group included two males with late-onset HD, while all HD control group females had age of onset under 50. Similarly, HD neuroleptic therapy group males had a significantly later age of onset than HD females in the neuroleptic group, since three HD males in this group had late-onset HD, but only one female. Unfortunately, neither the HD control group nor the HD neuroleptic therapy group was identified on the basis of which parent was affected by HD, nor by age of onset in the affected parent. However, there were no significant differences between neuroleptic and control groups of the same sex with regard to age of onset, age at first SCDMH admission, age at death, or survival time. Males of the HD neuroleptic therapy group showed greater lon-

TABLE 3
Huntington's Disease: Control Group
Birth Cohorts 1900–1930

	Males (7)[a]	Females (8)[a]	p^b
Onset	43.86 ± 2.79	35.60 ± 2.53	<0.05
Admission 1	48.49 ± 3.26	49.06 ± 3.02	NS
Death	55.15 ± 3.20	53.75 ± 2.78	NS
Survival	11.29 ± 1.21	18.15 ± 1.85	<0.01

[a] Mean ± SEM (statistical significance is determined with Student's *t* test).
[b] NS, not significant.

TABLE 4
Huntington's Diease: Neuroleptic Therapy Group
Birth Cohorts 1900–1930

	Males (7)[a]	Females (14)[a]	p
Onset	48.20 ± 2.93	35.24 ± 2.15	<0.01
Admission 1	54.85 ± 3.24	49.96 ± 2.44	<0.05
Death	60.40 ± 2.91	51.64 ± 2.50	<0.02
Survival	12.20 ± 0.73	16.39 ± 1.96	>0.05

[a] Mean ± SEM (statistical significance is determined with Student's *t* test).

therapy group, with a corresponding later age at first SCDMH admission, reflects a tendency toward longer survival time, as described by Wendt *et al.* (1960). Moreover, the Wendt effect may explain the similar longevity of HD males in the control and treated groups. However, HD control group females (with no late-onset cases) showed a significantly longer survival time than HD control group males (Table 3). Such was not the case in the HD neuroleptic therapy group, where there was no significant difference in survival time.

While it seems clear that neuroleptic therapy did not shorten life-span or have a deleterious effect on survival time of hospitalized HD patients, further studies will be necessary to explore the possibility that neuroleptic drugs may benefit HD patients by exerting protective effects on membrane phospholipids. Foster (1979) has reported that CPZ prevents irreversible membrane degeration during episodes of acute anoxia in experimental animals, and CPZ has antiarrhythmic effects on the heart (Baldessarini, 1980). In addition, CPZ has been found to prevent fetal death in malnourished pregnant rats, with a nicotinamide-sparing effect thought to enhance functioning of the pyridine nucleotide coenzymes, which are essential for biooxidative processes (Fratta *et al.*, 1964).

Whatever the reasons, neuroleptic therapy has not been associated with increased mortality, but appears to have reduced mortality in chronically institutionalized psychiatric patients, as noted by Born (1979): ". . . introduction of CPZ for the control of schizophrenic inpatients from about 1955 onwards accounts for their relative protection against cardiac mortality at a time when it was increasing rapidly in Norway and elsewhere." Craig and Lin (1981) found that the neuroleptic drug era population sample drawn from 15,000 psychiatric inpatients showed a 30% reduction in mortality among patients hospitalized gevity than females with HD in the treated group, with no significant difference in survival time (Table 4). The later age of onset for HD males of the neuroleptic less than 1 year, and a 50% lower mortality among longer-term patients, particularly elderly men.

A generation after the introduction of CPZ, neuroleptic therapy continues to offer a new lease on life for patients with major mental disorders, including HD and other neurobehavioral disorders.

5. SUMMARY

HD is a progressive heredofamilial disorder involving the central nervous system, characterized by autosomal dominant transmission, abnormal involuntary choreiform movements or hypokinetic rigidity, behavioral abnormalities, and dementia. Differential diagnosis of its usual choreiform presentation includes at least 80 distinct disorders of the nervous sytem, making a family history of HD the critical feature. Death in HD patients is usually due to cardiovascular disease, cachexia, or intercurrent illness such as pneumonia. Uncertainty regarding age at onset continues to compromise survival studies, despite general agreement that survival time shows a strong negative correlation with age of onset.

The case-control cohort study of longevity reported here involves 14 HD males and 22 HD females born from 1900 to 1930. All were admitted to SCDMH hospitals at least 1 year before death. Suicides and juvenile-onset cases were excluded from the study. HD males in both control and neuroleptic therapy groups had later age of onset than HD females. HD female controls showed longer survival time than HD male controls. HD males in the neuroleptic therapy group showed greater longevity than HD females in the treated group, whereas HD control males showed no greater longevity than HD females in either group. This unexpected finding suggests beneficial effects for HD males receiving neuroleptic therapy. Phenothiazines have antiarrhythmic effects and protective effects on membrane phospholipids. Since cardiovascular disease is more likely to affect males under age 60, improved longevity may reflect added protection from cardiac death during chronic neuroleptic therapy. Large-scale studies of psychiatric inpatients (other than HD) receiving long-term neuroleptic treatment showed a 30–40% reduction in mortality, especially among elderly men. Further studies will be necessary to clarify interactions between neuroleptic drugs and the molecular pathogenesis of HD, now believed to involve widespread membrane abnormalities leading to cell death in the brain, with less obvious damage to other organ systems.

6. REFERENCES

Alzheimer, A., 1911, Concerning the anatomical basis of Huntington's chorea and choreatic movements in general, *Neurol. Centralbl.* **30**:891–892.

American Psychiatric Association (APA), 1980, *Tardive Dyskinesia*, American Psychiatric Association Task Force, Report 18, p. 4.

Baldessarini, R. J., 1980, Drugs and the treatment of psychiatric disorders, in: *The Pharmacological Basis of Therapeutics*, 6th ed. (A. G. Gilman, L. S. Goodman, and A. Gilman, eds.), pp. 391–447, Macmillan, New York.

Born, G. V. R., 1979, Possible role for chlorpromazine in protection against myocardial infarction, *Lancet* **1**:822.

Bruyn, G. W., 1968, Huntington's chorea. Historical, clinical and laboratory synopsis, in: *Diseases of the Basal Ganglia, Handbook of Clinical Neurology*, Vol. 6 (P. J. Vinken and G. W. Bruyn, eds.), pp. 298–378, North-Holland Publishing Company, Amsterdam.

Bruyn, G. W., 1973, Clinical variants and differential diagnosis, in: *Huntington's Disease, Advances in Neurology*, Vol. 1 (A. Barbeau, T. N. Chase, and G. W. Paulson, eds.), pp. 51–56, Raven Press, New York.

Bruyn, G. W., Bots, G. Th. A. M., and Dom, R., 1979, Huntington's chorea: Current neuropathological status, in: *Huntington's Disease, Advances in Neurology*, Vol. 23 (T. N. Chase, N. S. Wexler, and A. Barbeau, eds.), pp. 83–93, Raven Press, New York.

Chandler, J. H., Reed, T. E., and DeJong, R. N., 1960, Huntington's chorea in Michigan, III. Clinical observations, *Neurology* **10**:148–153.

Craig, T. J., and Lin, S. P., 1981, Mortality among psychiatric inpatients: Age-adjusted comparison of populations before and after psychotropic drug era, *Arch. Gen. Psychiatry* **38**:935–938.

Crane, G. E., 1973, Tardive dyskinesia and Huntington's chorea: Drug-induced and hereditary dyskinesias, in: *Huntington's Disease, Advances in Neurology*, Vol. 1 (A. Barbeau, T. N. Chase, and G. W. Paulson, eds.), pp. 115–122, Raven Press, New York.

Davenport, C. B., and Muncey, E. B., 1916, Huntington's chorea in relation to heredity and eugenics, *Am. J. Insanity* **73**:195–222.

Delay, J., and Deniker, P., 1968, Drug-induced extrapyramidal syndromes, in: *Diseases of the Basal Ganglia, Handbook of Clinical Neurology*, Vol. 6 (P. J. Vinken and G. W. Bruyn, eds.), pp. 248–266, North-Holland Publishing Company, Amsterdam.

Foster, C. S., 1979, Chlorpromazine in prophylaxis of myocardial infarction, *Lancet* **1**:1249.

Fratta, I., Zak, S. B., Greengard, P., and Sigg, E. B., 1964, Fetal death from nicotinamide-deficient diet and its prevention by chlorpromazine and imipramine, *Science* **145**:1429–1430.

Hornykiewicz, O., 1966, Dopamine (3-hydroxytyramine) and brain function, *Pharmacol. Rev.* **18**:925–964.

Hornykiewicz, O., 1979, Pharmacology of Huntington's disease, in: *Huntington's Disease, Advances in Neurology*, Vol. 23 (T. N. Chase, N. S. Wexler, and A. Barbeau, eds.), pp. 679–686, Raven Press, New York.

Huntington, G., 1872, On chorea, *Med. Surg. Reporter* **26**:317–321.

Jones, M. B., 1979, Years of life lost due to Huntington disease, *Am. J. Hum. Genet.* **31**:711–717.

Martin, J. B., 1984, Huntington's disease: New approaches to an old problem, *Neurology* **34**:1059–1072.

McHugh, P. R., and Folstein, M. F., 1975, Psychiatric syndromes of Huntington's chorea: A clinical and phemomenologic study, in: *Psychiatric Aspects of Neurologic Disease* (D. F. Benson and D. Blumer, eds.), pp. 267–286, Grune & Stratton, New York.

Oepen, H., 1963, Paroxysmal disorders in Huntington's chorea, *Arch. Psychiat. Nervenkr.* **204**:245–261.

Quadfasel, F. A., 1973, Why Huntington's disease now instead of Huntington's chorea, *Psychiatr. Forum* **4**:61–62.

Reed, T. E., Chandler, J. H., Hughes, E. M., and Davidson, R. T., 1958, Huntington's chorea in Michigan, I. Demography and genetics, *Am. J. Hum. Genet.* **10**:201–225.

Rockstein, M., Chesky, J. A., and Sussman, M. D., 1977, Comparative biology and evoluation of aging, in: *Handbook of the Biology of Aging* (C. E. Finch and L. Hayflick, eds.), pp. 3–34, Van Nostrand Reinhold Co., New York.

Roizin, L., Kaufman, M. A., Willson, N., Stellar, S., and Liu, J. C., 1976, Neuropathologic observations in Huntington's chorea, in: *Progress in Neuropathology*, Vol. 3 (H. M. Zimmerman, ed.), pp. 447–488, Grune & Stratton, New York.

Roos, R. A. C., and Bots, G. Th. A. M., 1983, Nuclear membrane indentations in Huntington's chorea, *J. Neurol. Sci.*, **61**:37–47.

Shoulson, I., 1981, Huntington disease: Functional capacities in patients treated with neuroleptic and antidepressant drugs, *Neurology* **31**:1333–1335.

Shoulson, I., 1984, Huntington's disease: A decade of progress, *Neurol. Clin.* **2**:515–526.

Spokes, E. G. S., 1979, Dopamine in Huntington's disease: A study of postmortem brain tissue, in: *Huntington's Disease, Advances in Neurology*, Vol. 23 (T. N. Chase, N. S. Wexler, and A. Barbeau, eds.), pp. 481–493, Raven Press, New York.

Still, C. N., 1984, Involuntary movement disorders, *Neurol. Clin.* **2**:71–89.

Still, C. N., and Young, S. R., 1974, Age of onset in Huntington's disease: a reappraisal of the Wendt hypothesis, *Am. J. Hum. Genet.* **26**:84A.

Streletzki, F., 1961, Psychoses in the course of Huntington's chorea with special reference to the formation of delusions, *Arch. Psychiat, Nrvenkr.* **202**:202–214.

Tellez-Nagel, I., Johnson, A. B., and Terry, R. D., 1974, Studies on brain biopsies of patients with Huntington's chorea, *J. Neuropathol. Exp. Neurol.* **33**:308–332.

Wendt, G. G., Landzettel, I., and Solth, L., 1959, The age of onset of illness in Huntington's chorea, *Acta Genet.* (*Basel*) **9**:18–32.

Wendt, G. G., Landzettel, I., and Solth, K., 1960, Length of illness and life expectancy in Huntington's chorea, *Arch. Psychiat. Nervenkr.* **201**:298–312.

Whittier, J. R., 1967, Clinical aspects of Huntington's disease, in: *Progress in Neurogenetics* (A. Barbeau and J. R. Brunette, eds.), pp. 632–644, Excerpta Medica Foundation, Amsterdam.

PART V

Neuroleptic Drugs in the Production of Movement Disorders

CHAPTER 16

Movement Disorders Induced by Neuroleptic Drugs

CHRISTOPHER G. GOETZ, HAROLD L. KLAWANS, and CAROLINE M. TANNER

1. INTRODUCTION

Neurology and psychiatry were once considered to be such closely related disciplines that one practitioner commonly treated patients suffering from illnesses of either type. This close relationship has waned, however, and neurologists and psychiatrists now maintain separate journals and professional organizations. Although this separation has advantages, new problems have been created. Nowhere are the problems more evident than in the field of neuropharmacology, where drugs used to treat neurological disease often affect psychiatric function and drugs used to treat psychiatric illness produce neurological symptoms. Behavioral disorders by neurologists and movement disorders by psychiatrists are increasingly reported in their own respective specialty journals. However, clinicians in one area often use vague or technically imprecise terminology when describing the other field. As a result, a psychiatrist reading a neurologist's report of a behavior or a neurologist reading a psychiatrist's report of an abnormal movement cannot easily understand the description in terms of classification and presumed pathophysiology or pathogenesis. In this chapter, we hope to contribute to a reciprocal educational effort directed toward ameliorating this problem. This chapter will discuss drug-induced abnormal movements associated with the administration of neuroleptic agents and their classification, description, and pharmacology.

All drug-related abnormal involuntary movements have been grouped together under a single heading—extrapyramidal symptoms or syndrome (EPS) (Ayd, 1961a; Detre and Jarecki, 1971; Boston Collaborative Drug Surveillance Program, 1973). Therefore, the term EPS has included acute dystonias, aki-

CHRISTOPHER G. GOETZ, HAROLD L. KLAWANS, and CAROLINE M. TANNER • Department of Neurological Sciences, Rush–Presbyterian–St. Lukes Medical Center, Chicago, Illinois 60612.

nesia, parkinsonism, rabbit syndrome, akathisia, covert dyskinesia, withdrawal dyskinesia, and tardive dyskinesia. Although the above conditions all represent examples of extrapyramidal dysfunction, these disorders differ in clinical description, natural history, prognosis, or management, making the term EPS functionally obsolete. At the same time, however, several of the terms describe disorders which are neurologically and pathophysiologically identical. A reevaluation of this older terminology will provide an accurate foundation for description and diagnosis and a functional basis for therapy. Since the newer classification presented here is based directly on extrapyramidal anatomy and pharmacology, a brief description of the extrapyramidal system will facilitate an understanding of the various clinical disorders.

2. THE EXTRAPYRAMIDAL SYSTEM

The extrapyramidal system involves the nuclei and pathways of the central nervous system that relate to motor activity, but are not directly part of the classical descending pyramidal tracts. Three major characteristics help to contrast this system with the pyramidal system. First, the extrapyramidal system is multisynaptic; second, it does not have its own descending spinal cord tract; and third, lesions in the extrapyramidal system are characterized by poor integration of motor activity but not weakness (Brodal, 1969; Carpenter, 1976).

The pyramidal system most simplistically viewed is a two-cell system. To understand the basic anatomy, it is best to use an example and trace the pathway for volitional movement of the right foot. The first cell body is located in the left motor cortex, and its axon travels through the left internal capsule and left descending pyramidal tracts of the brain stem and crosses in the medullary pyramidal decussation continuing in the right lateral cortical spinal tract to the right anterior horn cell. The second cell body is located in the right anterior horn cell, and its axon travels out of the spinal cord to the muscles of the right foot. Damage to the pyramidal system anywhere along this pathway will result in right-foot weakness. The presence of spasticity or flacidity, increased or decreased reflexes, atrophy, and fasiculations will help in differentiating which cell of the two-part system is damaged.

In contrast, the extrapyramidal system involves inputs from multiple brain stem and deep cortical nuclei that modulate this pyramidal function. Such nuclei include the substantia nigra in the midbrain and the basal ganglia, a collective term for the striatum (caudate nucleus and putamen), globus pallidus, and subthalamic nucleus. The nuclei send ascending information primarily to the motor cortex via the thalamus for modulation of the pyramidal system. Since there is no extrapyramidal spinal cord pathway, information from extrapyramidal nuclei must be first integrated and then descend in the final pathway of the lateral cortical spinal tract. When the extrapyramidal nuclei and integrating axons are damaged, there is abnormal motor function, but since the two-cell pyramidal system is uninvolved, there is no weakness. Instead, poor control or integration of movement, either hypokinetic (pathologically slow) or hyperkinetic (pathologically overactive) movements, are seen. Hence, while the term extrapyramidal indicates that this is a motor system that is nonpyramidal, it should not suggest that extrapyramidal function is independent of or in conflict with pyramidal activity. The two systems are closely integrated both anatomically and functionally so that strength and coordination are effected simultaneously in the normal situation.

Pharmacological studies of extrapyramidal function have focused primarily on the striatum, where there are high concentrations of two major putative neurotransmitters that act antagonistically to one another, dopamine and acetylcholine (Bartholini *et al.*, 1973). Striatal dysfunction in the form of structural disease or biochemical imbalance of dopaminergic–cholinergic activities has been linked to a number of abnormal involuntary movement disorders. These movement disorders, while all extrapyramidal, have distinctly different pharmacological bases, so that considering them all as a single entity (EPS) is confusing and therapeutically hazardous. For example, parkinsonism is felt to relate to a diminished activity of dopamine at selected striatal dopamine receptor sites (Klawans, 1968). In Parkinson's disease, cells whose axons project to the striatum die, leaving the striatum depleted of dopamine. Parkinsonism due to primary striatal degeneration is also seen, and drugs that block striatal dopamine receptors will induce the same parkinsonian effect (Marsden *et al.*, 1975). Although parkinsonism is a disorder of the dopaminergic system, changes in striatal cholinergic function will influence parkinsonian features. Acetylcholine antagonizes striatal dopaminergic function, so that in parkinsonism where there is an absolute underactivity of dopamine, there is a relative overactivity of acetylcholine. Hence, there are two potential arms of therapy for parkinsonism—amplification of dopaminergic function or antagonism of cholinergic activity (Klawans, 1968).

In contrast to parkinsonism, the various forms of chorea are felt to relate to overactivity of dopamine at selected dopamine receptors in the striatum (Klawans, 1973). These receptors are most likely distinct from the dopaminergic receptors related to the pathophysiology of parkinsonism. Similarly to parkinsonism, however, the same antagonism between dopamine and acetylcholine can be demonstrated. While chorea is aggravated by dopaminergic precursors or agonists and ameliorated by dopamine receptor blockade or dopamine depletion, cholinergic alterations will also modify the abnormal movement. Absolute overactivity of dopaminergic function can be viewed as a relative underactivity of cholinergic function, and hence reasonable treatment for chorea would include cholinergic amplification. This treatment has been shown to be effective with the use of short-acting cholinergic agents, and reports of putative cholinergic precursors have in many cases been favorable (Klawans and Rubovits, 1972; Barbeau, 1978; Davis *et al.*, 1976). Other neurotransmitters may also play a role in the pathophysiology of chorea and parkinsonism, but these have not been proven.

Clearly, since the pathophysiology of parkinsonism and chorea are in many ways opposed to one another, the classification of both movements under a heading such as extrapyramidal syndrome is misleading and can often result in erroneous treatment. The common statement that anticholinergic drugs are useful for the treatment of EPS is applicable only to parkinsonism and acute dystonias. Anticholinergic agents aggravate chorea of any etiology.

The movement disorders discussed in the next sections are descriptively distinct clinical entities although all are associated with neuroleptic administration. Each disorder is discussed in terms of its pathogenesis, pathophysiology, and treatment.

3. PARKINSONISM

The four cardinal signs of idiopathic parkinsonism—bradykinesia, resting tremor, rigidity, and loss of postural reflexes—can also be produced by neuroleptic drugs (drug-induced parkinsonism). All four signs may be present simultaneously and to the same degree, or a patient may have only one sign. The pathophysiology and pharmacology of these variants are the same, so that all should be considered under the same heading of parkinsonism (Selby, 1968).

Drug-induced parkinsonism, related to phenothiazine or butyrophenone therapy, is felt to relate to dopaminergic receptor site blockade in the striatum (Klawans *et al.,* 1973). The clinical features of parkinsonism appear to result whenever striatal receptor sites do not receive appropriate dopaminergic stimulation. In the case of idiopathic Parkinson's disease, where presynaptic nigrostriatal dopaminergic cells die, there is not enough dopamine to stimulate the receptor sites: in neuroleptic-induced parkinsonism, the receptors themselves are pharmacologically blocked. The same pharmacologic result—underactivity of striatal dopaminergic function—is seen in both conditions, so that the clinical features of each syndrome are similar. As a result, the identification of parkinsonian features in a patient receiving neuroleptic therapy indicates that there is underactivity of dopaminergic striatal function. From a pharmacological perspective, the presence of one or more parkinsonian features has the same significance.

Parkinsonian features usually develop days to weeks after beginning neuroleptic therapy and are generally self-limited to weeks or months. Bradykinesia or akinesia is the earliest, most common, and frequently the only marked manifestation of drug-induced parkinsonism. Masklike facial expression, loss of arm swing or associated movements while walking, and hesitant initiation and slow performance of volitional movements and speech are characteristic features (Marsden *et al.,* 1975). Although akinesia is often listed by authors as separate from drug-induced parkinsonism, this distinction is misleading and inaccurate. Akinesia is a cardinal manifestation of parkinsonism. Since true akinesia is probably not seen in other neurological conditions, it should always be considered as an indication of striatal dopaminergic underactivity (Klawans,

1968). It should be managed from the same pathophysiological perspective as all parkinsonian features and should not be considered as distinct.

Similarly, resting tremor may occur independently or in conjunction with other parkinsonian signs. The tremor is apparent when the body part is at complete rest and disappears when movement is initiated. The tremor may involve primarily the extremities or may preferentially affect the mouth, chin, and lips. In the latter case, such tremor has been termed "rabbit syndrome" (Villeneuve, 1972). Although the term is descriptive and picturesque, oral tremor has the same character and pharmacological significance as resting tremor in other body regions. From a pathophysiological perspective, it does not deserve a separate diagnostic category. Like other parkinsonian tremors, oral tremor readily reverses when neuroleptic therapy is discontinued and responds promptly to anticholinergic drug therapy. The identification of rabbit syndrome as a parkinsonian feature is especially important because it primarily involves buccal masticatory muscles and may be confused with tardive dyskinesia (Sovner and Dimascio, 1977; Jus *et al.*, 1974), and titubation often seen with essential or familial tremors (see below).

Resting tremor, whether oral–facial or limb, tends to occur coincident with commencement or increase in neuroleptic medication and is often associated with other parkinsonian features. It reflects decreased activity of dopamine at the striatal level and is therefore treated in the same manner as other parkinsonian features.

Although tremor is one of the hallmarks of parkinsonism, all that shakes is not parkinsonian tremor. Benign essential tremor, senile tremor, and familial tremor all describe a form of postural tremor where shaking is absent when the patient is at complete rest, but prominent when the patient has his hands extended, maintaining a posture. This form of tremor is idiopathic in nature, but it is frequently aggravated by lithium and agents with adrenergic activity (Young and Shahani, 1979). We have seen 12 patients referred because of "tricyclic-induced parkinsonism," and all had benign essential tremor aggravated by tricyclic administration. These patients did not show resting tremor or the other manifestations of parkinsonism. They responded in most cases to either withdrawal of the tricyclic drug or administration of propranolol, a β-adrenergic blocking agent.

The other features of parkinsonism, rigidity of the extremities or trunk and abnormal postural reflexes, are especially important to identify. Rigidity is a sign elicited by passively moving the extremity or neck. The hypertonicity or increased resistance is appreciated by the examiner, often with an intermittent resist/release component that is present throughout the range of motion at a joint (cogwheel rigidity). Although the patient does not have specific complaints related to this phenomenon, the patient or relatives may complain of stooped posture, flexed arms, and small-stepped, shuffling gait.

Loss of postural reflexes is the parkinsonian feature with the highest morbidity since patients tend to fall forward (propulsion) or backward (retropulsion). Patients may complain of unsteadiness or propulsive gait ("walking on my tiptoes'' or a need to run) or may fall without losing consciousness and without tripping or other apparent cause. Postural reflexes are assessed by having the patient assume a comfortable standing posture and then providing a backward thrust to the sternum while standing behind the patient. Normal individuals will maintain this position or correct for a strong thrust with one step. Persons with abnormal postural reflexes will take several backward steps involuntarily (retropulsion) or fall *en bloc*, unable to correct at all for the postural threats.

Occasionally, parkinsonism may be confused with the primary psychiatric diagnosis for which the neuroleptic drugs were prescribed. Catatonic schizophrenic patients may show a lack of spontaneity, expressionless facies, and fixed postures that may appear parkinsonian (Jaspers, 1963). In these cases, the absence of cogwheel rigidity and resting tremor can be helpful in differential diagnosis, although, as stated, individual parkinsonian features may rarely occur independently without the other accompanying signs. Similarly, the psychomotor retardation associated with severe depression in manic–depressive psychosis may closely resemble parkinsonian akinesia, making differential diagnosis difficult (Marsden *et al.,* 1975). The qualities of sadness and despair seen in retarded psychotic depression, however, may be helpful distinguishing features.

Drug-induced parkinsonism is frequently a self-limited condition, often resolving spontaneously by 3–6 months (Klawans, 1973). Mild cases do not require treatment, but if the condition is troublesome to the patient, a short course of antiparkinsonian therapy can be instituted. The best treatment is, of course, to remove the causative agent and, hence, decrease the dose of neuroleptic. When the patient's psychiatric condition makes this impossible, a short course of anticholinergic medication is beneficial; trihexyphenidyl hydrochloride, 2–10 mg/day, or benztropine mesylate, 1–8 mg/day, is commonly used. These agents are felt to act pharmacologically at the striatum to effect a new dopaminergic–cholinergic balance. Dopamine and acetylcholine appear to act antagonistically in the striatum. Since dopaminergic receptors are blocked by the neuroleptic drugs, there is a relative cholinergic excess. The anticholinergic drugs, by blocking striatal cholinergic receptors, reestablish better dopaminergic–cholinergic balance, and motor function is improved (Klawans, 1968).

Some clinicians treat drug-induced parkinsonism with the dopaminergic agents levodopa or amantadine. However, levodopa administration has been associated with acute, dramatic exacerbation of psychotic behavior in schizophrenic patients (Yaryura-Tobias *et al.,* 1972). It has been suggested that schizophrenic behavior is related to enhanced activity of limbic dopaminergic systems, and that the ameliorative effect of the neuroleptic drugs relates to their ability to block limbic dopamine receptors (Snyder *et al.,* 1974). Drug-induced parkinsonism would be considered a side effect related to more diffuse, in this case, striatal, dopaminergic receptor site blockade. While levodopa may compete for receptor sites with neuroleptics at the striatum and hence abate the drug-induced parkinsonism, it would be expected also to distribute itself in the limbic system and aggravate ambient psychosis. Although anecdotal cases of amantadine-related aggravation or precipitation of psychosis are available, this milder dopaminergic agent has not been associated with the frequent and dramatic exacerbation seen with levodopa. As a dopaminergic agent, however, amantadine has been shown to have the pharmacological potential to aggravate or induce psychiatric behavior in parkinsonian patients (Weiner and Bergen, 1977).

It is again important to emphasize that whatever treatment is delivered, it should only be short term since drug-induced parkinsonism is self-limited. Frequent decreases in antiparkinsonian medication (monthly) will allow the physican to reevaluate the need for these drugs. As long as the patient shows disabling parkinsonian features, treatment is indicated, but prolonged or indefinite anticholinergic treatment without attempts to reevaluate parkinsonian signs is unjustified. There is theoretical and laboratory evidence to suggest that anticholinergic medication may lower the threshold for late dopaminergic hypersensitivity which manifests as tardive dyskinesia in these patients (Rubovits and Klawans, 1972; Crane, 1974). In order to avoid sudden exacerbation of parkinsonian features, withdrawal of anticholinergic medication should be slow.

Although drug-induced parkinsonism is listed as an early extrapyramidal complication of neuroleptic therapy, it does not only occur when medication is started but may occur whenever a patient's dose of neuroleptic is increased. Hence, a patient on a stable dose of neuroleptic for several years whose dose is suddenly increased may develop drug-induced parkinsonism. Because it occurs after the patient has been on neuroleptics for a prolonged period, this condition may often be misdiagnosed as tardive dyskinesia, especially when the parkinsonism has prominent tremor in the mouth or chin regions. The rhythmicity of the movement at rest and its diminution with movement help to define the movement as a parkinsonian tremor, and the history of recent increase in medication dose confirms the diagnosis. Although it has been suggested that intravenous administration of agents with opposite effects on central cholinergic systems should help to differentiate parkinsonian tremor from the choreiform movements of tardive dyskinesia, it is our experience that these diagnoses can be made clinically, without the use of unpleasant and potentially harmful agents.

4. DRUG-INDUCED DYSTONIAS

A dystonia is an abnormal and involuntary sustained posture affecting any body area. Dystonias are common early side effect of neuroleptic agents (acute dystonias), but they may rarely occur as late-onset adverse effects (tardive dystonias).

The acute dystonias generally occur within the first few days after neuroleptic treatment is initiated. The body areas most commonly affected are the eyes and neck.

Patients and oculogyric crisis often complain of inability to move their eyes in the vertical plane and may have double vision, blurred vision, and, rarely, pain on attempted gaze. Most often the eyes maintain a sustained and involuntary upward gaze. The severe dystonic displacement of the eyes may itself be painful as may other severe, contorting dystonias. The abnormal posture of the head and neck including opisthotonus, where the head and neck are in a retrocollic position, give the patient a bizarre appearance. Torticollis, where the neck is deviated to one side, can be painful and disabling. Other muscles may be involved in acute drug-induced dystonia, but these are much less common (Kartzinel and Chase, 1977).

The incidence of dystonia with different neuroleptics seems to parallel the differential incidence of drug-induced parkinsonism, the piperazine agents being the most hazardous (Ayd, 1961b). Agents with a high incidence of parkinsonism have a high incidence of drug-induced dystonia, while those with a low incidence of parkinsonism have a low incidence of dystonia. The simultaneous administration of anticholinergic agents is felt to decrease the incidence of neuroleptic-induced dystonia, and the acute administration of anticholinergic agents almost invariably reverses them. Physiologically, acute neuroleptic-induced dystonia is felt to represent a sudden disruption of basal ganglia function in some way related to dopamine. This alteration is most probably acute dopaminergic receptor blockade since all offending agents are capable of blocking striatal dopamine receptors. The ability of anticholinergic agents to prevent and ameliorate these dystonias suggests that dopamine–acetylcholine balance is involved in these events. Acute therapy involves intravenous or preferably intramuscular injection of an anticholinergic agent. This treatment will ameliorate the dystonia rapidly, but since the anticholinergic effect is short lived, oral anticholinergic agents should be prescribed for the next 24–48 hr. If the patient's psychosis requires continued neuroleptic therapy, he should be placed on maintenance anticholinergic treatment for several weeks or switched to another neuroleptic (e.g., thioridazine) with a lower propensity to cause dystonia.

The ability of neuroleptics to elicit acute dystonic reactions disappears to a great extent as the duration of therapy is extended (Marsden, *et al.,* 1975). New dystonias are rare after the first few weeks, and dystonias that occur in the acute phase are usually no longer present after months of therapy. As a result, the anticholinergic agents used to treat and/or prevent dystonia can be decreased and withdrawn in most patients after 1–2 months of use. As already discussed, these drugs should be withdrawn slowly. Drug-induced dystonias are most common among younger patients given neuroleptics for vomiting and in young adults (especially between the ages of 20 and 40) being started on neuroleptic or depot neuroleptic therapy (Ayd, 1961b).

Dystonias can be confused with bizarre mannerisms associated with psychotic behavior. Because the former are temporally related to drug administration and respond rapidly to anticholinergic drugs, the two can be differentiated. A patient who presents initially with prominent dystonias and behavioral alterations must be evaluated for Wilson's disease and dystonia musculorum deformans, two genetic neurological disorders associated with prominent abnormal postures and behavioral disorders (McDowell *et al.*, 1978). A family history of a similar condition is helpful in making these diagnoses but not always elicited. These conditions are progressive and show poor response to drug therapy, unlike the usually rapid abatement with anticholinergic drugs seen with neuroleptic-induced dystonias.

Although most dystonic reactions related to neuroleptics are acute and follow the initiation of drug therapy or recent increase in dosage, Burke *et al.* (1982) have described a late-onset dystonia. They reported 42 patients treated with chronic neuroleptic agents who developed a tardive form of dystonia. Five had experienced prior dystonic reactions with introduction of neuroleptic therapy, and in 6 of 16 patients whose past histories were detailed, abnormal birth or developmental delay could be identified. This condition has been treated successfully as a variant of tardive dyskinesia with dopamine-depleting agents (Fahn, 1983). Since in most cases, tardive dystonia is accompanied by mild chorea, this clinical combination of movement disorders should prompt the clinician to obtain a detailed drug history. Some cases of tardive dystonia involving the face and neck closely resemble spontaneously occurring oral–facial–mandibular dystonia or Meige syndrome (Goetz and Klawans, 1984.

5. TARDIVE DYSKINESIA

Tardive dyskinesia (TD) is a term applied to a group of movement disorders which occur only after chronic administration of neuroleptic agents. Characteristically, the movements are irregular and variable, in contrast to a tremor, where the movements are rhythmic and more regular. The early description of this syndrome stressed abnormal motor movements affecting the face, the so-called buccolingual masticatory syndrome. This consisted of involuntary mouthing, chewing, sucking, and licking movements of the tongue. Later, the syndrome description was broadened to include a variety of abnormal muscular manifestations, including choreoathetoid type movements of the fingers, hands, arms, and feet, and ballistic (violent flinging) movements particularly of the arms, as well as axial hyperkinesias, and diaphragmatic movement resulting in grunting and difficult breathing. As the condition was studied more, the mouthing movements were further refined to include puckering, panting, smacking, and tongue movements inside the mouth (Klawans *et al.*, 1980a).

The pathophysiology of TD is felt to relate to striatal dopaminergic hypersensitivity (Klawans, 1973). Chronic blockade of striatal dopamine receptor sites by these drugs may induce a slowly developing hypersensitivity. This hypersensitivity is most commonly manifested as an abnormal choreic behavioral response that is elicited when neuroleptic dose is lowered in these patients (Goetz and Klawans, 1984).

As indicated, TD is by definition an iatrogenic disorder related specifically to chronic neuroleptic administration. Phrases such as spontaneous TD are, hence, self-contradictions and should be discarded. Spontaneous lingual–facial–buccal dyskinesias (i.e., chorea) may be a manifestation of Huntington's disease, senile chorea, and other degenerative neurological syndromes, but are not tardive dyskinesia.

TD usually beings after 1–2 years of chronic neuroleptic treatment and hence is temporally distinct from the acute extrapyramidal syndromes discussed in Section 2 (Crane, 1968; Jus *et al.*, 1976). It is important to distinguish TD from other hyperkinetic syndromes, especially stereotypic mannerisms and agitation related to psychotic behavior and akathisia associated with early administration or recent elevation in dose of neuroleptics.

A wide variety of mannerisms are seen in the psychotic population, and these peculiar behaviors may resemble oral facial dyskinesias or limb chorea (Bleuler, 1951). In the vast majority of cases, however, they are highly ritualistic in character and consistent in anatomical distribution, which distinguishes them from the unpredictable irregularity of choreiform movements. Furthermore, mannerisms are usually conscious and symbolically significant to the patient whereas tardive dyskinesia is often completely unrecognized, or of little concern. The former movements also are seen as part of the early psychotic state and do not usually emerge after prolonged drug treatment (Kraepelin, 1919). A psychotic or mentally deranged patient with early-onset chorea or dystonia should raise the additional important diagnostic considerations of extrapyramidal disorders like Huntington's chorea, dystonia musculorum deformans, and Wilson's disease.

An initial glance at TD patients might suggest that they are agitated, since there is an abundance of movement often with shifting in the chair, irregular tapping of feet, or strange mouthing movements. While these behaviors might suggest an overly nervous person, TD does not resemble the hyperactivity seen in psychotic states like manic–depressive illness or involutional psychosis. In the latter cases, a generalized psychomotor activation is seen characterized by overactivity, but well-coordinated movements and often exaggerated expressions of affect.

Finally, akathisia must be differentiated from TD. Akathisia is a state of motor restlessness that occurs early in the course of or after augmentation of neuroleptic dose (Ayd, 1961b; Marsden *et al.*, 1975; Ayd, 1961a). It is not clear, however, that this effect relates specifically to extrapyramidal dysfunction. Akathisia does not occur naturally, except in very rare occurrences of Parkinson's disease, in any extrapyramidal disorder (Selby, 1968). While it does occur coincident with neuroleptic administration, there are no data to confirm that akathesia relates to neuroleptic effects on dopaminergic activity or the extrapyramidal system. While it can be mildly modified by the administration of anticholinergic agents, the effect of this drug class on akathisia is much less marked than in cases of drug-induced parkinsonism (Marsden *et al.*, 1975). The hallmark of akathisia is that there is a striking complaint of inner tension, a feeling of a need to move about and resist inactivity. Usually, the patient feels forced to shuffle his feet restlessly and may feel unable to sit or lie down. Akathisia and TD are not mutually exclusive, and Fahn has reported that many TD patients can be identified early by a careful interview focusing on the gradual development of akathisia (Fahn, personal communication).

A recent report suggesting new diagnostic categories for late-onset drug-related dyskinesias included "withdrawal dyskinesia" and "covert dyskinesia" (Gardos *et al.*, 1978). The former was defined as self-limited TD, resolving within 12 weeks after drug withdrawal, and covert dyskinesia was defined as dyskinesia seen with discontinuation or reduction of neuroleptic dose without spontaneous abatement. These distinctions, of theoretical interest, are not practical, since in a patient who develops dyskinesia after medication withdrawal, the diagnosis would first be withdrawal dyskinesia for 12 weeks, and if there was no abatement, the diagnosis would be changed to covert dyskinesia. The change in diagnosis gives no insight into pharmacology and pathophysiology, and the time of 12 weeks is at best arbitrary. As the authors of the report admitted, these distinctions may "confuse rather than clarify clinical thinking."

The basic pathogenic event in these proposed dyskinesias is chronic striatal dopamine receptor site blockade, and the resultant pathophysiology is postulated to be dopamine receptor denervation hypersensitivity (Klawans *et al.*, 1980b). Work with animal models, where guinea pigs or rats were treated with chronic neuroleptics, demonstrated behavioral dopaminergic hypersensitivity associated with parallel striatal dopamine receptor site alterations. The behavioral, as well as the biochemical, changes were reversible in many cases when neuroleptic administration was interrupted early (Klawans *et al.*, 1980a). These data suggest that this model may be an accurate one of TD in humans and that a single and unifying pathophysiology appears to underlie the different phases of hypersensitivity development in animals. In humans, TD as a clinical phenomenon appears to have a spectrum of presentations and prognoses. Unless diagnostic distinctions clarify specific differences in pathophysiological mechanisms, biochemistries, or treatments, there is little justification for separate categories. Presently the term TD seems sufficient and accurate to describe the large groups of late-onset choreiform disorders associated with neuroleptic treatment.

Since TD is felt to relate to striatal dopaminergic hypersensitivity, treatment has focused on altering dopaminergic–cholinergic balance. Reserpine, which depletes the brain of dopamine, has been successful for short periods of time (Duvoisin, 1972). Augmentation of cholinergic activity has gained recent interest, and presumed dietary precursors of central acetylcholine, choline chloride and lecithin, have been reported to be of some use (Barbeau, 1978; Davis *et al.*, 1975). No single effective therapy is available (Klawans *et al.*, 1980b). Reinstitution or augmentation of neuroleptic dose will abate the movements by presumed increase of striatal dopaminergic blockade, but this is treatment with the etiological agent and, except in severe, perhaps life-threatening cases, should be avoided. Manipulation of other neurotransmitter systems has been suggested to be effective in the control of TD. Baclofen, a presumed agonist of the γ-aminobutyric acid (GABA) system, has been reported to abate TD in some patient studies (Korsgaard, 1976). GABA is felt to act as a dopaminergic antagonist at the level of the substantia nigra and possibly striatum.

6. SUMMARY

This chapter has focused on the clinical, anatomical, pharmacological, and treatment distinctions between a number of disorders traditionally grouped together as EPS. The differential diagnosis of each and the distinction between movement disorders has been discussed. The emphasis has been directed toward clear definitions, and the elimination of redundant subcategories or diagnoses.

ACKNOWLEDGMENT. Dr. Goetz is the recipient of a NINCDS Teacher-Investigator Award.

7. REFERENCES

Ayd, F. J., Jr., 1961a, A survey of drug-induced extrapyramidal reactions, *JAMA* **175**:1054–1060.

Ayd, F. J., Jr., 1961b, Neuroleptics and extrapyramidal reactions in psychiatric patients, *Rev. Can. Biol.* **20**:451–459.

Barbeau, A., 1978, Phosphatidylcholine (Lecithin) in neuroleptic disorders, *Neurology* **28**:358.

Bartholini, G., Stadler, H., and Lloyd, K. G., 1973, Cholinergic–dopaminergic interactions in the extrapyramidal system, in: *Progress in the Treatment of Parkinsonism* (D. B. Calne, ed.), pp. 233–241, Raven Press, New York.

Bleuler, E. P., 1951, *Textbook of Psychiatry*, Dover, New York.

Boston Collaborative Drug Surveillance Program, 1973, Drug induced extrapyramidal symptoms: A cooperative study, *JAMA* **224**:889–891.

Brodal, A., 1969, Pathways mediating supraspinal influences on the spinal cord, in: *Neurological Anatomy*, pp. 151–243, Oxford University Press, New York.

Burke, R. E., Fahn, S., Jankovic, J., Marsden, C. D., Lang, A. E., Gollomp, S., and Ilson, J., 1982, Tardive dystonia: Late onset and persistent dystonia caused by antipsychotic drugs, *Neurology* **32**:1335–1346.

Carpenter, M. D., 1976, *Human Neuroanatomy*, William & Wilkins, Baltimore.

Crane, G. E., 1968, Tardive dyskinesias in patients treated with major neuroleptics: A review of the literature, *Am. J. Psychiatry.* **123**(Feb. Suppl.):40–48.

Crane, G. E., 1974, Factors pre-disposing to drug-induced neurological side effects, in: *Advances in Biochemical Psychopharmacology*, Vol. 9 (I. S. Forrest, C. J. Carr, and E. Usdin, eds.), pp. 269–279, Raven Press, New York.

Davis, K. L., Berger, P. A., and Hollister, L. E., 1975, Choline for tardive dyskinesia, *N. Engl. J. Med.* **293**:152.

Davis, K. L., Hollister, L. E., Barchas, J. D., and Berger, P. A., 1976, Choline in tardive dyskinesia and Huntington's disease, *Life Sci.* **19**:1507–1515.

Detre, T. P., and Jarecki, H. G., 1971, *Modern Psychiatric Treatment*, Lippincott, Philadelphia.

Duvoisin, R. C., 1972, Reserpine for tardive dyskinesia, *N. Engl. J. Med.* **286**:611.

Fahn, S., 1983, Treatment of tardive dyskinesia: Use of dopamine depleting agents, *Clin. Neuropharmacol.* **6**:151–158.

Gardos, G., Cole, J. O., and Tarsy, D., 1978, Withdrawal syndromes associated with antipsychotic drugs, *Am. J. Psychiatry* **135**:1321–1324.

Goetz, D. G., and Klawans, H. L., 1984, Tardive dyskinesia, in: Movement Disorders (J. Jankovic, ed.), *Neurol. Clin.* **2**:605–614.

Jaspers, K., 1963, *General Psychopathology*, pp. 179–185, University of Chicago Press, Chicago.

Jus, K., Jus, A., and Gautier, J., 1974, Studies on the action of certain pharmacological agents on tardive dyskinesia and on the rabbit syndrome, *Int. J. Clin. Pharmacol.* **9**:138–145.

Jus, A., Pineau, R., Lanchance, R., Pelchat, G., Jus, K., Pires, P., and Villeneuve, R., 1976, Epidemiology of tardive dyskinesia. Part I, *Dis. Nerv. Syst.* **37**:210–214.

Kartzinel, R., and Chase, T. N., 1977, Pharmacology of dystonia, in: *Clinical Neuropharmacology*, Vol. 2 (H. L. Klawans, ed.), pp. 43–54, Raven press, New York.

Klawans, H. L., Jr., 1968, The pharmacology of parkinsonism: A review, *Dis. Nerv. Syst.* **29**:805–816.

Klawans, H. L., 1973, *The Pharmacology of Extrapyramidal Movement Disorders*, Karger, Basel.

Klawans, H. L., and Rubovits, R., 1972, Central cholingeric–anticholinergic antagonism in Huntington's chorea, *Neurology* **22**:107–112.

Klawans, H. L., Bergen, D., and Bruyn, G. W., 1973, Prolonged drug induced parkinsonism, *Confin. Neurol.* **35**:368–377.

Klawans, H. L., Carvey, P., Nausieda, P. A., Goetz, C. G., and Weiner, W. J., 1980a, Effect of dose and type of neuroleptic in an animal model of tardive dyskinesia, *Neurology* **30**:302.

Klawans, H. L., Goetz, C. G., and Perlik, S., 1980b, Tardive dyskinesia: Review and update, *Am. J. Psychiatry* **137**:900–908.

Korsgaard, S., 1976, Baclofen (Lioresal) in the treatment of neuroleptic-induced tardive dyskinesia, *Acta Psychiatr. Scand.* **54**:17–24.

Kraepelin, E., 1919, *Dementia Praecox and Paraphrenia*, Livingstone, Edinburgh.

Marsden, C. D., Tarsy, D., and Baldessarini, R. J., 1975, Spontaneous and drug-induced movement disorders in psychiatric patients, in: *Psychiatric Aspects of Neurologic Disease* (D. F. Benson and D. Blumer, eds.), pp. 14–32, Grune & Stratton, New York.

McDowell, F. H., Lee, J. E., and Sweet, R. D., 1978, Extrapyramidal disease, in: *Clinical Neurology* (A. B. Baker and L. H. Baker, eds.), pp. 1–67, Harper & Row, Hagerstown, Maryland.

Rubovits, R., and Klawans, H. L., 1972, Implication of amphetamine induced stereotyped behavior as a model for tardive dyskinesias, *Arch. Gen. Psychiatry* **27**:502–507.

Selby, G., 1968, Parkinson's disease, in: *Handbook of Clinical Neurology, Diseases of the Basal Ganglia*, Vol. 6 (B. Vinken and G. W. Bruyn, eds.) pp. 173–211, Elsevier North Holland, Amsterdam.

Snyder, S. H., Banerjee, S. P., Yamamura, H. I., and Greenberg, D., 1974, Drugs, neurotransmitters and schizophrenia: Phenothiazines, amphetamines, and enzymes synthesizing psychotomimetic drugs aid schizophrenia research, *Science* **184**:1243–1253.

Sovner, R., and Dimascio, A., 1977, The effect of benztropine mesylate in the rabbit syndrome and tardive dyskinesia, *Am. J. Psychiatry* **134**:1301–1302.

Villeneuve, A., 1972, The rabbit syndrome: A peculiar extrapyramidal reaction, *Can. Psychiatr. Assoc. J.* **17**(Suppl. 2):SS69–72.

Weiner, W. J., and Bergen, D., 1977, Prevention and management of the side effects of levodopa, in: *Clinical Neuropharmacology*, Vol. 2 (H. L. Klawans, ed.), pp. 1–20, Raven Press, New York.

Yaryura-Tobias, J. A., Diamond, B., and Merlis, S., 1972, Psychiatric manifestations of levodopa, *Can. Psychiatr. Assoc. J.* **17**:SS123–128.

Young, R. R., and Shahani, B. T., 1979, Pharmacology of tremor, in: *Clinical Neuropharmacology*, Vol. 4 (H. L. Klawans, ed.), pp. 139–156, Raven Press, New York.

CHAPTER 17

Movement Disorders and Neuroleptic Medication

THOMAS R. E. BARNES and MALCOLM P. I. WELLER

1. PATHOPHYSIOLOGICAL CONSIDERATIONS

Skilled voluntary movements involving fine adjustments are initiated by the large pyramidal-shaped Betz cells in the motor cortex. The decussating pathway receives contributions from the cerebellum and the complex extrapyramidal system. The term "extrapyramidal system" originally referred to an anatomical concept although more recently it has been considered a functional unit. Within this system attention has been focused on the basal ganglia, and, specifically, the dopaminergic nigrostriatal system, as being important in motor planning (Marsden, 1980). Figure 1 illustrates, diagrammatically, the postulated interaction among dopaminergic, cholinergic, and GABAergic neurons in the nigrostriatal system (Kebabian and Calne, 1979; Marsden and Jenner, 1980). For a more elaborate representation of the neurotransmitter systems in this area, see Hornykiewicz (1981).

The most obvious effects of lesions in this area, in rodents, are disturbances of motor control (Ungerstedt, 1971; Glick *et al.*, 1977). In humans, pathological changes in the substantia nigra and locus ceruleus are associated with the akinetic–rigid syndrome of Parkinson's disease. It has long been known that the pigmented cells of the substantia nigra are pale and depleted of neurotransmitter in this condition (Blocq and Marinesco, 1893, and Tretiakoff, 1919, cited by Yahr, 1981; Ehringer and Horynkiewicz, 1960). Degenerative changes in the caudate nucleus and putamen are associated with the abnormal involuntary movements of Huntington's chorea (Adams and Victor, 1977). However, despite these clinicopathological associations, the precise role of the basal ganglia in motor control is unknown.

THOMAS R. E. BARNES • Department of Psychiatry, Charing Cross and Westminster Medical School, London W6 8RF, England. MALCOLM P. I. WELLER • Academic Department of Psychiatry, Royal Free Hospital School of Medicine, The Royal Free Hospital, London NW3 2QG, England.

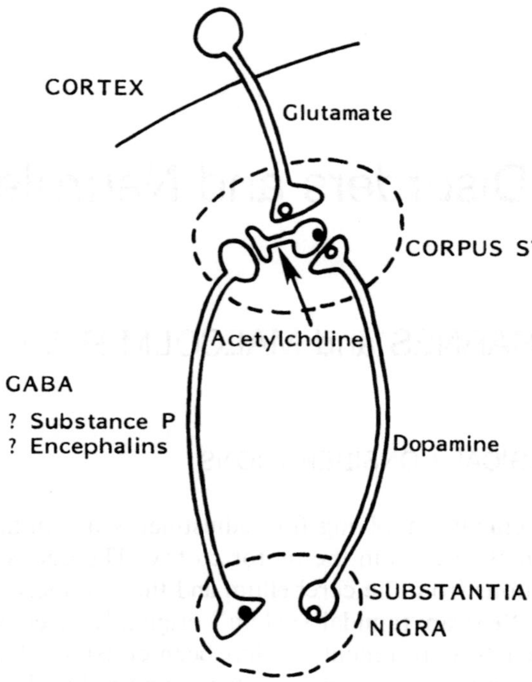

FIGURE 1. Schematic representations of the nigrostriatal system: ●, D1 receptor; ○, D2 receptor (Poirier *et al.*, 1975; Wolfarth, 1976; Hassler *et al.*, 1978; Kebabian and Calne, 1979).

It is generally assumed that the extrapyramidal side effects induced by antipsychotic drugs are related to the dopamine receptor-blocking action of these drugs, principally in the nigrostriatal system. Figure 1 indicates the possible sites for two distinct types of dopamine receptor which have been identified by ligand-binding studies (Schwarcz *et al.*, 1978; Spano *et al.*, 1980). D1 receptors are linked to dopamine-sensitive adenylate cyclase and are labeled preferentially by [3H]dopamine and [3H]apomorphine, while D2 receptors are independent of this enzyme and are labeled preferentially by [3H]butyrophenones (Kebabian and Calne, 1979; Iversen, 1978). Snyder (1981) has pointed out that while phenothiazines are active at both types of receptor, butyrophenones tend to be selective for the D2 group. As butyrophenones are potent at elevating plasma prolactin, inducing motor side effects, and controlling psychotic symptoms, Snyder assumes that all these actions can be medicated via D2 receptors. The absence of a selective or preferential D1 antagonist (Iversen, 1978) limits the extent to which such speculation can be tested. Creese (1982) has recently reviewed the evidence for the existence of three classes of dopamine receptor, D1, D2 (a and b), and D3, and their anatomical locations in the nigrostriatal system. Alternatively, Seeman (1980) and Sokoloff *et al.* (1980) recognize four classes of dopamine receptor, also distinguished by their distribution within the striatal complex. The therapeutic implications of such classifications remain unclear (Calne, 1981).

2. MOVEMENT DISORDERS

Antipsychotic drug-induced movement disorders are considered here within two main categories: early-onset and late-onset. The division is convenient and clinically relevant and may also reflect the pathophysiological mechanisms responsible. The early-onset conditions are thought to be a consequence of inhibition of central dopamine action; a drug effect that seems to be gradually compensated for over time. These conditions are, therefore, relatively transient and tend to diminish or disappear with chronic drug treatment. The late-onset conditions, principally tardive dyskinesia, are thought to be associated with postsynaptic dopamine receptor supersensitivity, a state apparently permanent while drug treatment continues, and possibly persistent for variable periods after drug withdrawal. Thus, late-onset disorders tend to be chronic conditions. For both early- and late-onset problems the likelihood of an individual developing a movement disorder, and the subsequent chronicity and severity of signs and symptoms, may reflect subtle and variable interactions between age-related neurochemical and neuropathological changes and drug effects.

2.1. Early-Onset Movement Disorders

2.1.1. Parkinsonism

As previously discussed, idiopathic Parkinson's disease results from a degeneration of the nigrostriatal dopaminergic system. As antipsychotic drugs produce a functional deficiency of dopamine within this system, it is unremarkable that drug-treated patients manifest a similar condition. While many signs and symptoms are shared by the idiopathic and drug-induced disorders, the latter seem less likely to present with the characteristic, resting, 3–5 Hz, "pill-rolling" tremor and festinant gait (Schwab and England, 1968; Young and Shahani, 1979; Hershey *et al.*, 1982). Clinical examination of patients with drug-induced parkinsonism may reveal a curious, persistent resistance to passive movement of the limbs, called "lead-pipe" rigidity, or a succession of resistances which are rapidly overcome, known as "cogwheel" rigidity.

The incidence of the condition in routine clinical practice has been estimated at 10–15% (Ayd, 1961; Marsden *et al.*, 1975), but a wide variation has been reported in the literature, which may reflect the sensitivity of the clinically scrutiny and the rating instrument used (see Mackay, 1981, for review). As McClelland (1976) has pointed out in this context, ". . . high prevalence rates for extrapyramidal signs do not, of course, reflect morbidity. When patients might be said to be *suffering* from drug-induced parkinsonism is quite a different matter.''

Anticholinergic drugs are generally considered helpful. There is some theoretical rationale to justify their use. A functional balance between dopaminergic and cholinergic activity in the basal ganglia is felt to be important for normal extrapyramidal function (Barbeau, 1962; Anden and Bedard, 1971). Parkinsonism represents a relative dominance of cholinergic over dopaminergic activity, and the administration of anticholinergic medication is thought to restore this balance. An alternative explanation is that anticholinergic agents merely lower the plasma levels of antipsychotic drugs (Loga *et al.*, 1975; Rivera-Calimlin *et al.*, 1976; Jus *et al.*, 1977). The consensus of opinion regarding their clinical use seems to be that these agents can adequately control extrapyramidal reactions, but should be prescribed only if and when these develop, and not routinely (Simpson, 1970; DiMascio and Demirgian, 1970; McClelland *et al.*, 1974; Raleigh, 1977). Although initially helpful, these drugs may be maintained unnecessarily. If they are discontinued after 3 months, about 90% of patients can continue treatment on the same antipsychotic regime without an exacerbation of extrapyramidal symptoms (Orlov *et al.*, 1971; McClelland *et al.*, 1974; Coleman and Hayes, 1975; Johnson, 1978). Nevertheless, bradykinesia, characterized by lessening of spontaneity, paucity of gestures, diminished conversation, and apathy, and which is responsive to anticholinergic medication (Rifkin *et al.*, 1978), may be mistaken for depression (Martin and Townsend, 1974; Knights *et al.*, 1979; Weller, 1981) or residual schizophrenic deficit. Since tricyclic antidepressants are strongly anticholinergic (Janowsky *et al.*, 1972), a beneficial response to these drugs may be falsely attributed to an antidepressant effect and compound diagnostic error.

Different antipsychotic agents vary in their propensity to induce parkinsonism. The more potent piperazine phenothiazines, such as trifluoperazine, and butyrophenones, such as haloperidol (see Fig. 2), are particularly prone to induce this problem. Potency, however, does not seem a sufficient explanation since the drugs are commonly used in roughly equipotent doses. Other explanations have been advanced. Many neuroleptic drugs cause dry mouth, blurred vision, constipation, and other symptoms of antimuscarinic activity. This instrinsic anticholinergic effect differs between the various antipsychotic drugs (Miller and Hiley, 1974; Snyder *et al.*, 1974); those which possess this property strongly may be the least likely to cause parkinsonism since they have an inbuilt antidote to the primary dopaminergic blockade. The relationship between dose and response is generally different for the antidopaminergic effect and the anticholinergic effect, leading to possible improvement in parkinsonism with increasing dosage (Hollister, 1978; Weller, 1981). This apparent paradox can be understood by reference to Fig. 3. Parkinsonism will be most severe when the disparity between the two dose–response curves is maximal (indicated by the arrows). As the dose is increased from this point, the distance between the two conceptual dose–response curves diminishes.

An additional mechanism has been proposed to explain the variation in

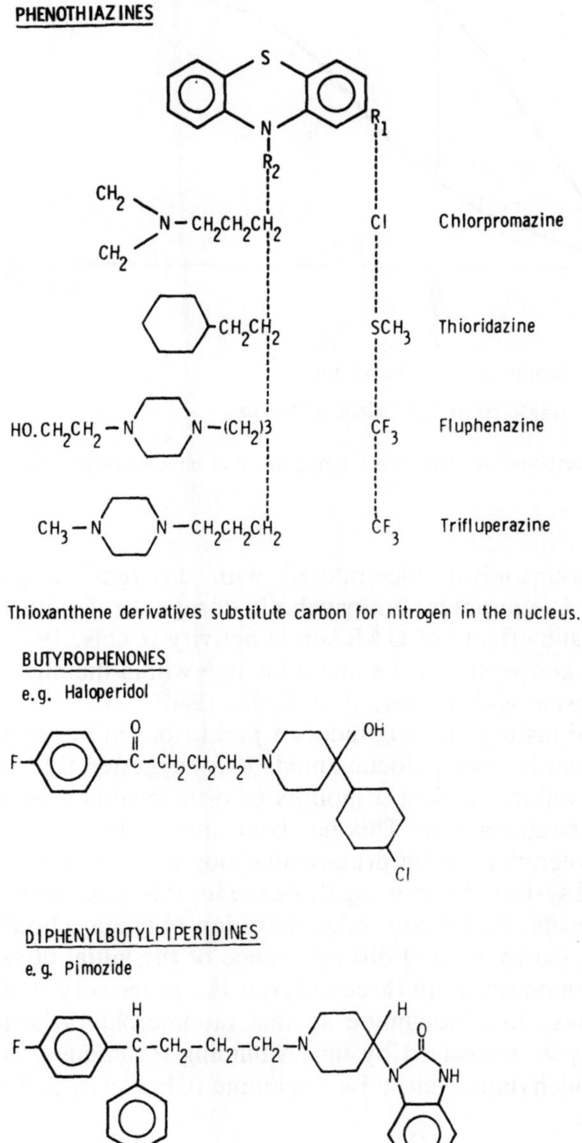

FIGURE 2. Chemical structures of some commonly used antipsychotic drugs.

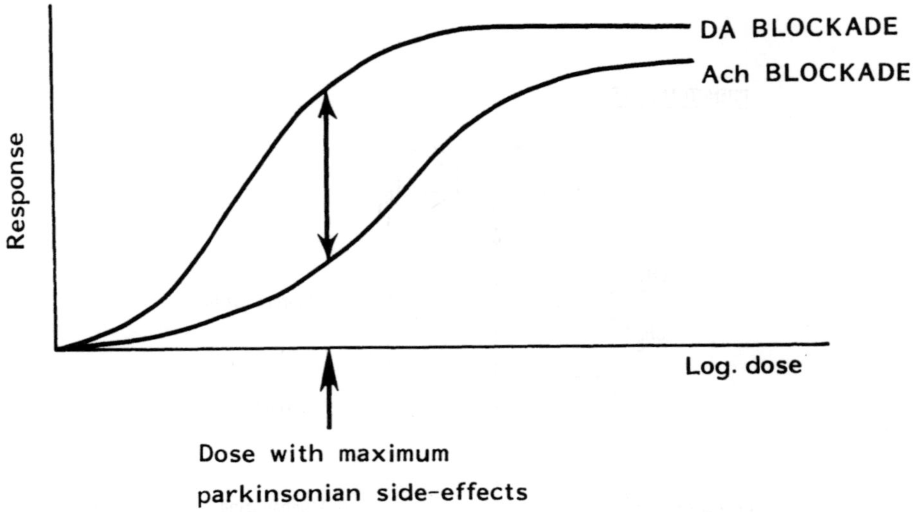

FIGURE 3. Dose–response relationship for hypothetical antipsychotic drug with intrinsic anticholinergic activity.

severity of parkinsonism encountered with different antipsychotic drugs (Weller, 1981). At equipotent antipsychotic dosage the various drugs have different antagonistic effects of GABAergic activity (Cools, 1977; Marco *et al.*, 1976), which is known to exert some influence within the nigrostriatal system (see Fig. 1) (Costall and Naylor, 1978; Gale, 1980).

The natural history of drug-induced parkinsonism in patients continuing on drug treatment is poorly documented. It is suggested that parkinsonism is most common within the first 3 months of drug treatment and subsequently may improve spontaneously. This has been interpreted as indicating the development of tolerance to the drug-related dopamine receptor antagonism in the nigrostriatal system. Supporting evidence for this hypothesis has come from animal experiments. Continuous administration of antipsychotic drugs to rats, over 6 months, led to gradual disappearance of the initial dopamine receptor antagonism, as monitored by three criteria: the propensity to develop stereotypies in response to a dopamine agonist (apomorphine), striatal dopamine receptor affinity as measured by ligand-binding techniques, and the *in vitro* stimulation of adenylate cyclase by dopamine (Clow *et al.*, 1979, 1980a,b).

2.1.2. Acute Dystonias

Acute dystonic reactions are abnormal postures produced by sustained muscle spasms that develop slowly. The muscles of the head and neck may be principally affected, leading to involuntary spasm of the tongue and mouth known as trismus, blepharospasm, oculogyric crises, facial grimacing, torticollis, retrocollis, and opisthotonos. These drug effects usually occur within

24–48 hr of starting neuroleptic medication (Ayd, 1961; American College of Neuropsychopharmacology—Food and Drug Administration Task Force, 1973), and 90% occur by the fourth day of treatment (Swett, 1975). They are probably the least common drug-induced movement disorders with a reported incidence between 2 and 10% (Ayd, 1961; Swett, 1975).

Acute dystonic reactions can be relieved within minutes by intramuscular or intravenous anticholinergic drugs (Coleman and Hayes, 1975), which suggests that dystonias, like parkinsonism, are related to central dopamine antagonism. However, current theories of the pathophysiology reject this simple notion. Following acute antipsychotic drug administration there is an increase in presynaptic dopamine release, perhaps due to preferential presynaptic dopamine receptor blockade. In addition, an acute type of postsynaptic receptor supersensitivity is thought to develop. This has been demonstrated in animals (Christensen, 1973) and, in contrast to the chronic supersensitivity discussed in Section 2.2.2, is manifest only on drug withdrawal. As the antipsychotic drug concentration decreases, either following a single dose or between doses, the supersensitive receptor becomes progressively exposed to an enhanced presynaptic release of neurotransmitter, leading to an exaggerated dopaminergic response (Marsden and Jenner, 1980). Thus, dystonias are provoked by the interaction of a number of dynamic neurochemical drug effects. This explanation seems to account for the early onset of dystonias and their relatively transient nature. If this proposed mechanism is correct, knowledge of the pharmacokinetics of a drug might allow tentative prediction regarding the likelihood of dystonic reactions developing. Garver *et al.* (1976) investigated the pharmacokinetics of a dopamine antagonist in red blood cells, considering that this would be analogous to the situation at "critical brain sites." Thirteen schizophrenic patients were each administered a single oral dose of butaperazine, 40 mg. Dystonic reactions occurred in eight patients during falling plasma drug concentrations: at 23–28 hr postdrug in seven cases and at around 56 hr in the remaining case. Red blood cell kinetics were better than plasma drug levels at differentiating those subjects who developed dystonias from those who did not.

Nevertheless, if dystonias are related to enhanced dopaminergic activity centrally, the therapeutic effect of anticholinergic drugs appears to be paradoxical. However, there is evidence from animal experiments that anticholinergics can antagonize selected central effects of antipsychotics (Wauquier *et al.*, 1975) via pre- and perhaps also postsynaptic action (Ondrusek *et al.*, 1981).

Apart from anticholinergic agents, the only drug treatment with any claims for success is benzodiazepine medication. A number of studies and case reports have suggested that diazepam is an effective treatment for acute dystonias (Donlon, 1973; Schnell, 1972; Gagrat *et al.*, 1978; Director and Muniz, 1982). Facilitation of GABA transmission in the strionigral system is thought to exert an inhibitory effect on the nigrostriatal dopaminergic system (see Fig. 1) and might account for the therapeutic effect. Alternatively, or perhaps in addition, the muscle relaxant properties of the benzodiazepines may be relevant.

2.1.3. Acute Akathisia

Akathisia is the most common and perhaps the most distressing of all the antipsychotic drug-induced movement disorders (Ayd, 1961) with a reported incidence around 20% (Braude, et al., 1983). It is an acute, dose-related side effect usually beginning within the first few days of antipsychotic drug treatment but not within the first 48 hr (Marsden *et al.*, 1975). However, there is a recent report of akathisia symptoms occurring within 90 min of the administration of oral dopamine antagonists metoclopramide and droperidol (Barnes *et al.*, 1982) and within 15–30 min of intravenous metoclopramide (Jungmann and Schöffling, 1982).

Acute akathisia is a syndrome of motor restlessness, and patients with this side effect will commonly refuse medication because of the agitation, fidgetness, and emotional unease they are experiencing (Van Putten, 1974). The term "akathisia" is of Greek derivation and has been translated as "inability to sit." Precise diagnostic criteria have not been established, partly because of the difficulty in differentiating mild akathisia from restlessness related to psychiatric illness, and partly because of the composite nature of the condition, which comprises both a subjective complaint of restlessness and motor signs. Clinicians have differed in their opinion regarding the relative importance of these aspects. Delay and Deniker (1968) considered akathisia an extrapyramidal motor disorder, while Sovner and DiMascio (1978) stated it was basically a sensory phenomenon. However, Crane *et al.* (1971) and Van Putten (1975) have described the condition as principally emotional, and various behavioural presentations have been documented (Van Putten, 1978; Kekich, 1978; Kumar, 1979). It is unclear whether all these reports describe genuine manifestations of the condition or whether some merely reflect the inherent diagnostic difficulties.

Mild akathisia predominantly presents as subjective restlessness, a feeling of being unable to keep the legs still, and possibly abnormal limb sensations. There are usually few observable motor features. With more severe akathisia, characteristic motor signs are present. The movements seen are not abnormal (dyskinetic) but, rather, characteristic patterns of restless, fidgety movements such as rocking from foot to foot or raising or lowering alternate feet as though "walking on the spot" (Braude *et al.*, 1983). Patients often report that the condition is worse on standing: for example, when waiting in line at the supermarket or waiting for meals or medication on the ward.

In more serious cases patients may also be unable to maintain a seated or lying position and feel compelled to stand up and pace about. These severe features have been described as "tasikinesia" (Delay and Deniker, 1968; Raskin, 1972). Some of these signs and symptoms may be misinterpreted as exacerbations of psychotic agitation and excitement. Further antipsychotic medication will, however, tend to aggravate rather than improve the problem.

There is no satisfactory explanation of the pathophysiological mechanism underlying the development of akathisia. Marsden and Jenner (1980) have suggested that the condition may result from postsynaptic dopamine receptor blockade in the mesocortical dopamine system. Animal experiments indicate that this system may exert an inhibitory role on spontaneous locomotor behaviour (Iversen, 1971; Tassin *et al.*, 1978; Koob *et al.*, 1981). Lesions of the prefrontal cortex can lead to increased activity in the subcortical dopamine systems, with an associated increase in locomotor activity (Carter and Pycock, 1980; Pycock *et al.*, 1980a,b). The equivalent result in man, following mesocortical (prefrontal) dopamine receptor blockade by antipsychotic drugs, may be the psychological and motor manifestations of akathisia. Such an explanation of the pathophysiological mechanism of akathisia in terms of central dopamine systems seems plausible. However, recent animal work has identified dopamine receptors in the spinal cord (Dupelj and Geber, 1981) which can modulate motor output (Carp and Anderson, 1982); thus, antipsychotic drug activity at these spinal dopamine receptors might also be involved in the production of some of the observed phenomena.

The only reliable treatment for akathisia is withdrawal of antipsychotic medication, which is often unrealistic with acute psychiatric patients. Various drugs have been suggested as potential remedies, but the condition appears to be relatively resistant to treatment. Anticholinergic agents have an uncertain reputation, some investigators claiming that success with these drugs is virtually diagnostic of akathisia (Van Putten, 1975), while others suggest that the response is unsatisfactory (Marsden *et al.*, 1975; Sovner and DiMascio, 1978). The condition does not appear to get worse with physostigmine, an anticholinesterase (Ambani *et al.*, 1973). The results of a recent study (Braude *et al.*, 1983) suggest that akathisia is an invariable accompaniment of severe parkinsonism, and in such cases both conditions will respond to anticholinergic drugs. However, in the majority of cases, akathisia is not associated with severe parkinsonism, and anticholinergics provide little or not benefit.

Catecholamine agonists have also been tested as potential remedies. There is one report of methylphenidate producing a beneficial effect in akathisia (Carman, 1972). Also, amantadine, a drug that potentiates dopaminergic activity via both pre- and postsynaptic actions (Bailey and Stone, 1975), has been found to be effective in both parkinsonism and akathisia (Schwab *et al.*, 1969; Greenblatt *et al.*, 1977). Theoretically, both these agents could exacerbate psychotic symptoms, but in practice, this has not been found with amantadine (Kelly and Abuzzahab, 1971; DiMascio *et al.*, 1976). There have also been encouraging reports regarding the success of diazepam (Donlon, 1973; Gagrat *et al.*, 1978; Director and Muniz, 1982), and a β-adrenoceptor blocker, propranolol (Lipinski *et al.*, 1984), but, as with all the drug treatment mentioned, a double-blind placebo-controlled trial is required to confirm their therapeutic value.

2.2. Late-Onset Movement Disorders

2.2.1. "Tardive" Akathisia

Unlike parkinsonism and acute dystonias, the early-onset type of akathisia already mentioned does not tend to disappear after the first 3 months of antipsychotic drug treatment (Baldessarini, 1979). This suggests that tolerance does not develop to the drug-related dopamine receptor blockade that causes this condition. This reinforces the idea of the mesocortical system being involved, as it has been demonstrated that tolerance to dopamine receptor blockade due to antipsychotic drugs does not occur in this system, perhaps because of the absence of presynaptic dopamine receptors (Bannon *et al.*, 1982). Thus, early-onset akathisia may persist and become "chronic" with time. However, there appears to be a distinct form of akathisia that appears for the first time only after long-term antipsychotic treatment. This late-onset, or "tardive," akathisia is characterised by a relatively mild subjective sense of restlessness and a lack of response to conventional anticholinergic treatment. The condition is commonly associated with tardive dyskinesia (Simpson, 1977; Goetz *et al.*, 1980), which may account for the regular inclusion of motor restlessness in clinical descriptions of tardive dyskinesia (Delay and Deniker, 1968; Raskin, 1972; Fahn, 1978). Brandon *et al.* (1971) studied antipsychotic drug-induced movement disorders in a psychiatric inpatient population and found akathisia to be significantly in excess in patients with orofacial dyskinesia, compared with the rest of the sample. Barnes *et al.* (1983a), in a 3-year prospective study of psychiatric patients on long-term antipsychotic medication, found that akathisia at follow-up was present in 29% of patients with moderate or severe orofacial dyskinesia compared with only 12% of those with mild or absent dyskinesia (Gardos *et al.*, 1978; Hershon *et al.*, 1972).

Tardive akathisia and tardive dyskinesia have similar pharmacological characteristics, in that both may be precipitated, or worsen, following withdrawal or reduction of antipsychotic drug treatment, and restarting or increasing the dose of drug will alleviate the symptoms (Braude and Barnes, 1983). This suggests that some aspects of pathophysiology may be common to both conditions.

2.2.2. Tardive Dystonia

Dystonia, like akathisia, is often thought of as a disorder associated, almost exclusively with the start of drug treatment. However, dystonic movements have been regularly included in the lists of diverse involuntary movements seen in chronic patients (Barnes *et al.*, 1983b). Burke *et al.* (1982) provided information on patients who had developed dystonic movements on long-term antipsychotic medication. They considered this "tardive dystonia" to be a serious physical disability. This chronic and distressing condition may be differentiated from acute dystonic reactions by its coexistence with "late-onset hyperkinesias" (orofacial and trunk and limb dyskinesia), some persistence after drug withdrawal, and resistance to conventional treatments such as anticholinergic agents.

2.2.3. Tardive Dyskinesia

The term "tardive" dyskinesia implies late onset, and it is commonly assumed that the length of time that a patient has been receiving antipsychotic drugs is an important factor in the development of the syndrome. But there is little convincing evidence to support this idea. What has emerged from the epidemiological studies of this condition is that age is the major variable influencing both prevalence and severity (Smith and Baldessarini, 1980; Kane and Smith, 1982; Barnes *et al.*, 1983a). It seems likely that the development of the condition is dependant on age-related changes in specific dopamine systems within the extrapyramidal system. These degenerative changes may render patients vulnerable to the development of orofacial dyskinesia if antipsychotic or other dopamine antagonist drugs are administered. It is certain that the older a patient is when starting antipsychotic drug treatment, the earlier the condition develops (Jus *et al.* 1976).

The condition is characterized by orofacial movement; the buccolinguo–masticatory triad and more widespread cephalic movements, involving the periorbital muscles, facial muscles, and neck muscles, may be present. Individuals tend to present with a characteristic combination of movements which remains relatively constant. Trunk and limb dyskinesia is sometimes included in the syndrome description. It has been suggested that orofacial dyskinesia and trunk and limb dyskinesia should be considered separate subsyndromes (Barnes and Kidger, 1979; Granacher, 1981); evidence to support this view derives from a number of sources. First, the two subsyndromes have emerged as discrete, phenomenological clusters, or components, following the application of multivariate statistical methods in epidemiological studies (Kennedy *et al.*, 1971; Crane *et al.*, 1971; Kidger *et al.*, 1980). Second, a number of pharmacological studies have found a discrepant response between orofacial dyskinesia and trunk and limb dyskinesia (Fann *et al.*, 1974; Gardos *et al.*, 1976; Casey and Denney, 1977; Simpson *et al.*, 1977; Casey and Hammerstad, 1979; Bobruff *et al.*, 1981). Third, it has been demonstrated that patients with severe tardive dyskinesia are far more aware of abnormal trunk and limb movements than orofacial dyskinesia (Rosen *et al.*, 1982).

In addition to consideration of tardive dyskinesia as two subsyndromes, it may also be important to study the nature of the abnormal movements observed. In a sample of antipsychotic drug-treated psychiatric patients, Barnes, Rossor, and Trauer (1983b) were able to classify all trunk and limb "dyskinesias" into a variety of recognized abnormal movement categories, including choreiform and dystonic movements. Stereotypies, mannerisms, and facial tics were also seen, which are problably related to psychiatric illness rather than antipsychotic drug treatment. Such movements contribute to the problems of assessment of tardive dyskinesia in any sophisticated quantitative or qualitative fashion. At present, there are no generally agreed-upon diagnostic criteria for tardive dyskinesia, and many of the rating instruments available are unsatisfactory (Barnes, 1984). The problems of assessment of tardive dyskinesia have been reviewed by Gardos *et al.* (1977) and Barnes and Trauer (1983).

Jus *et al.* (1977) noted that the reported prevalence of tardive dyskinesia varied widely (0.5–50%). This is partly because of the lack of consistent diagnositc criteria already mentioned and partly because of the differences in patient samples, especially with regard to age (Smith and Baldessarini, 1980). Prevalence figures of around 50% have been reported for inpatient psychiatric populations (Smith *et al.*, 1979; Barnes and Kidger, 1979), while the prevalence in outpatient samples has usually been lower (Chouinard *et al.*, 1979; Barnes *et al.*, 1983a, Smith *et al.*, 1979).

Understanding of the pathophysiology of tardive dyskinesia is based on the hypothesis, derived by analogy with Huntington's chorea and L-dopa-induced dyskinesias, that the condition is related to antipsychotic drug-induced supersensitivity of postsynaptic dopamine receptors in the striatum (Carlsson, 1970; Klawans, 1973). Such a theory is compatible with the clinical observations that the condition may appear for the first time or worsen on antipsychotic drug withdrawal, is exacerbated by anticholinergic drugs, and is relieved in the short term by an increase in antipsychotic dosage although the condition eventually reappears.

Pharmacological strategies for the treatment of tardive dyskinesia have developed, based on this explanation of the pathophysiological mechanism. Agents affecting the dopaminergic, cholinergic and GABAergic systems have been tested. These drug studies have been reviewed by Mackay and Sheppard (1979) and Jeste and Wyatt (1982). These authors conclude that few of the potential remedies provide more than temporary amelioration, and none could be recommended as a treatment of choice. Increasing the dosage of a patient's antipsychotic medication usually reduces the dyskinesia, presumably because of effective blockade of the supersensitive postsynaptic dopamine receptors. However, benefit is invariably temporary, owing to further postsynaptic dopamine receptor proliferation, i.e., increased supersensitivity in response to the reduced dopaminergic transmission (Kazamatsuri *et al.*, 1972). Thus, progressive increase in antipsychotic dosage might effectively control the dyskinesia temporarily, but it would hardly constitute good clinical practice to prescribe increasing doses of a drug in order to suppress the side effect caused by the same drug. The potential for a further increase in postsynaptic dopaminergic sensitivity is unknown in humans, but lesions of the nigrostriatal pathway in rodents can produce a doubling of dopamine receptor binding in the corpus striatum compared to that achieved by chronic neuroleptic blockade (Burt *et al.*, 1976).

In his extensive review of tardive dyskinesia, Gerlach (1979) concludes that the treatment of tardive dyskinesia is "first and foremost the withdrawal of the provoking neuroleptic." Even following complete antipsychotic drug withdrawal the condition persists in a proportion of patients. Marsden *et al.* (1975) estimated that the risk of persistent tardive dyskinesia might be about 30% of those developing the condition. Jeste *et al.* (1979) carried out a follow-up study of 21 patients withdrawn from antipsychotic drugs and reported that the condition persisted for at least 7 months in nine (43%). In all but one of the other 12 cases, the condition disappeared within 3 months. Possible factors related to the persistence or reversibility of tardive dyskinesia on drug withdrawal may include the number of drug interruptions of more than 2 months duration, the total duration of antipsychotic drug exposure, duration of tardive dyskinesia, and age (Degkwitz, 1968; Crane, 1973; Quitkin *et al.*, 1977; Jeste *et al.* 1979; Smith and Baldessarini, 1980).

The manifestation of tardive dyskinesia after many years free of antipsychotic drugs might be explained in some cases as the development of spontaneous orofacial dyskinesia. However, the persistence of the condition despite drug withdrawal raises the possibility of morphological central nervous system changes, related to prolonged antipsychotic treatment. Neuropathological investigation of this question has involved only uncontrolled, postmortem studies on small numbers of patients. These have been reviewed by Jellinger (1977), who concluded that the majority of changes in human brain following chronic antipsychotic drugs were nonspecific or related to the normal aging process and lethal disease.

3. SUMMARY

This chapter briefly reviews the movement disorders associated with antipsychotic drug administration, in terms of phenomenology, current pathophysiological theories, and treatment strategies. The potential for an individual antipsychotic drug (dopamine antagonist) to induce early-onset movement disorders may depend on its spectrum of activity on various brain neurotransmitter systems, while the development of late-onset motor syndromes may depend on an interaction between drug effects and age-related degenerative changes in relevant brain systems. Parkinsonism, dystonic reactions, and acute akathisia are classified here as early-onset conditions, while tardive dystonia, tardive dyskinesia, and tardive akathisia are considered late-onset problems. Evidence for the existence of tardive akathisia is presented. This akathisia variant may coexist with, or may even be part of, the tardive dyskinesia syndrome. The importance of considering orofacial and trunk and limb dyskinesia as distinct and separate subsyndromes of tardive dyskinesia is discussed. While orofacial dyskinesia appears to represent a discrete clinical entity, the trunk and limb movements seen in a chronic psychiatric population may represent a variety of specific dyskinesias associated with drug treatment, such as choreiform movements, tardive akathisia, or tardive dystonia, as well as the stereotypies and mannerisms related to psychotic illness.

ACKNOWLEDGMENT. Dr. M. P. I. Weller was the recipient of a grant from the North East Thames Regional Health Authority to investigate reaction times in schizophrenia.

4. REFERENCES

Adams, R. D., and Victor, M., 1977, *Principles of Neurology*, McGraw-Hill, New York.
Ambani, L. M., Van Woert, M. H., and Bowers, M. B., 1973, Physostigmine effects on phenothiazine-induced extrapyramidal reactions, *Arch. Neurol.* **29:**444–446.

American College of Neuropsychopharmacology—Food and Drug Administration Task Force, 1973, Neurologic syndromes associated with antipsychotic-drug use. *N. Engl. J. Med.* **289**:20–23.

Anden, N. E., and Bedard, P., 1971, Influence of cholinergic mechanisms on the function and turnover of brain dopamine, *J. Pharm. Pharmacol.* **23**:460–462.

Ayd, F., 1961, A survey of drug-induced extrapyramidal reactions, *JAMA* **175**:1054–1060.

Bailey, E. V., and Stone, T. W., 1975, The mechanism of action of amantadine in parkinsonism: A review, *Arch. Int. Pharmacodyna. Thera.* **216**:246–262.

Baldessarini, R. J., 1979, The "neuroleptic" antipsychotic drugs, 2. Neurologic side-effects, *Post-grad. Med.* **65**:123–128.

Bannon, M. J., Reinhardt, J. R., Jr., Bunney, E. B., and Roth, R. H., 1982, Unique response to antipsychotic drugs is due to absence of terminal autoreceptors in mesocortical dopamine neurones, *Nature* **296**:444–446.

Barbeau, A., 1962, The pathogenesis of Parkinson's diseases. A new hypothesis, *Can. Med. J.* **87**:802–807.

Barnes, T. R. E., 1984, Rating tardive dyskinesia, *Br. J. Psychiatry* **145**:338.

Barnes, T. R. E., and Kidger, T., 1979, Tardive dyskinesia and problems of assessment, in: *Current Themes in Psychiatry*, Vol. 2 (R. N. Gaind and B. L. Hudson, eds.), pp. 145–162, Macmillan, London.

Barnes, T. R. E., and Trauer, T., 1982, Reliability and validity of a tardive dyskinesia videotape rating technique, *Br. J. Psychiatry* **140**:508–515.

Barnes, T. R. E., Braude, W. M., and Hill, D. J., 1982. Acute akathisia after oral droperidol and metoclopramide preoperative medication, *Lancet* **2**:48–49.

Barnes, T. R. E., Kidger, T., and Gore, S. M., 1983a, Tardive dyskinesia: A 3-year follow-up, *Psychol. Med.* **13**:71–81.

Barnes, T. R. E., Rossor, M., and Trauer, T., 1983b, A comparison of purposeless movements in psychiatric patients treated with antipsychotic drugs, and normal individuals, *J. Neurol. Neurosurg. Psychiatry* **46**:540–546.

Bobruff, A., Gardos, G., Tarsy, D., Rapkin, R. M., Cole, J. O., and Moore, P., 1981, Clonazepam and phenobarbital in tardive dyskinesia, *Am. J. Psychiatry* **138**:189–192.

Brandon, S., McClelland, H. A., and Protheroe, C., 1971, A study of facial dyskinesia in a mental hospital population, *Br. J. Psychiatry* **118**:171–184.

Braude, W. M., and Barnes, T. R. E., 1983, Late-onset akathisia: An indicant of covert dyskinesia *Am. J. Psychiatry* **140**:611–2.

Braude, W. M., Barnes, T. R. E., and Gore, S. M., 1983, Clinical characteristics of akathisia: A systematic investigation of acute psychiatric inpatient admissions, *Br. J. Psychiatry* **143**:139–150.

Burke, R. E., Fahn, S., Jankovic, J., Marsden, C. D., Lang, A. E., Gollom, P. S., and Ilson, J., 1982, Tardive dustonia and inappropriate use of neurolaptic drugs, *Lancet* **1**:1299.

Burt, D. R., Creese, I., and Snyder, S. H., 1976, Properties of 3H haloperidol and 3H dopamine binding associated with dopamine receptors in calf brain membranes, *Mol. Pharmacol.* **12**:800–812.

Calne, D. B., 1981, Dopamine receptors in movement disorders, in: *Movement Disorders* (C. D. Marsden and S. Fahn, eds.), pp. 41–58, Butterworth Scientific, London.

Carlsson, A., 1970, Biochemical implications of dopa-induced action on the central nervous system with particular reference to abnormal movements, in: L*-Dopa and Parkinsonism* (A. Barbeau and F. H. McDowell, eds.), pp. 205–212, F. A. Davis, Philadelphia.

Carman, J. S., 1972, Methylphenidate in akathisia, *Lancet* **2**:1093.

Carp, J. S., and Anderson, R. J., 1982, Dopamine receptor-mediated depression of spinal monosynaptic transmission, *Brain Res.* **242**:247–254.

Carter, C. J., and Pycock, C. J., 1980, Behavioural and biochemical effects of dopamine and noradrenaline depletion within the medial prefrontal cortex of the rat, *Brain Res.* **192**:163–176.

Casey, D. E., and Denney, 1977, Pharmacological characterization of tardive dyskinesia, *Psychopharmacology* **54**:1–8.

Casey, D. E., and Hammerstad, J. P., 1979, Sodium valproate in tardive dyskinesia, *J. Clin. Psychiatry* **40**:483–485.

Chouinard, G., Annable, L., Ross-Chouinard, A., and Nesteros, J. N., 1979, Factors related to tardive dyskinesia, *Am. J. Psychiatry* **136**:79–83.

Christensen, A. V., 1973, Acute and delayed effects of a single dose of a neuroleptic drug, *Acta Physiol. Scand.* **396**:78.

Clow, A., Jenner, P., Theodorou, A., and Marsden, C. D., 1979, Striatal dopamine receptors become supersensitive while rats are given trifluoperazine for six months, *Nature* **278**:59–60.

Clow, A., Theodorou, A., Jenner, P., and Marsden, C. D., 1980a, Changes in rat striatal dopamine turnover and receptor activity during one year's neuroleptic administration, *Eur. J. Pharmacol.* **63**:135–144.

Clow, A., Theodorou, A., Jenner, P., and Marsden, C. D., 1980b, Cerebral dopamine function in rats following withdrawal from one year of continuous neuroleptic administration, *Eur. J. Pharmacol.* **63**:145–157.

Coleman, J. H., and Hayes, P. E., 1975, Drug-induced extrapyramidal effects—A review, *Dis. Nerv. Syst.* **36**:591–593.

Cools, A. R., 1977, Two functionally and pharmacologically distinct dopamine receptors in the rat brain, in: *Advances in Biochemical Psychopharmacology*, Vol. 16 (E. Costa and G. L. Gessa, eds.), pp. 215–225, Raven Press, New York.

Costall, B., and Naylor, R. J., 1978, The relationship between extrapyramidal and mesolimbic GABA and acetylcholine in the mediation of neuroleptic activity. Paper presented at IInd World Congress of Biological Psychiatry, Barcelona, Abstract no. 452.

Crane, G. E., 1973, Persistent dyskinesia, *Br. J. Psychiatry* **122**:395–405.

Crane, G. E., Naranjo, E. R., and Chase, C., 1971, Motor disorders induced by neuroleptics—A proposed new classification, *Arch. Gen. Psychiatry* **24**:179–184.

Creese, I., 1982, Dopamine receptors explained, *Trends Neurosci.* **5**:40–43.

Degkwitz, R., 1968, Extrapyramidal motor disorders following long-term treatment with neuroleptic drugs, in: *Psychotropic Drugs and Dysfunctions of the Basal Ganglia* G. E. Crane, and R. Gardner, Jr., eds.), pp. 22–32, Washington, US Public Health Service Publications.

Delay, J., and Deniker, P., 1968, Drug-induced extrapyramidal extrapyramidal syndromes, in: *Handbook of Clinical Neurology*, Vol. 6 (P. J. Vinken and G. W. Bruyn, eds.), pp. 248–266, North Holland, Amsterdam.

DiMascio, A., and Demirgian, E., 1970, Antiparkinsonian drug overuse, *Psychosomatics* **11**:596–601.

DiMascio, A., Bernardo, D. L., Greenblatt, D. J., and Marder, J. E., 1976, A controlled trial of amantadine in drug-induced extrapyramidal disorder, *Arch. Gen. Psychiatry* **33**:599–602.

Director, K. L., and Muniz, C. E., 1982, Diazepam in the treatment of extrapyramidal symptoms: A case report, *J. Clin. Psychiatry* **43**:160–161.

Donlon, P. T., 1973, The therapeutic use of diazepam for akathisia, *Psychosomatics* **14**:222–225.

Dupelj, M., and Geber, J., 1981, Dopamine as a possible neurotransmitter in the spinal cord, *Neuropharmacol.* **20**:145–148.

Ehringer, H., and Hornykiewicz, O., 1960, Verteilung von Noradrenalin und Dopamin (3-Hydroxytyramin) im Gehirn des Menschen und ihr Verhalten bei Erkrankungen des Extrapyramidalen Systems, *Klin. Wochenschr.* **38**:1238–1239.

Fahn, S., 1978, Tardive dyskinesia may only be akathisia, *N. Engl. J. Med.* **299**:202–203.

Fann, W. E., Lake, C. R., and McKenzie, G. M., 1974, Adrenergic and cholinergic factors in extrapyramidal disorders, *Psychopharmacol. Bull.* **10**:52–53.

Gagrat, D., Hamilton, J., and Belmaker, R. H., 1978, Intravenous diazepam in the treatment of neuroleptic acute dystonias and akathisia, *Am J. Psychiatry* **135**:1232–1233.

Gale, K., 1980, Chronic blockade of dopamine receptors by antischizophrenic drugs enhances GABA binding in substantia nigra, *Nature* **283**:569–570.

Gardos, G., Cole, J. O., and Sniffin, C., 1976, An evaluation of papaverine in tardive dyskinesia, *J. Clin. Pharmacol.* **16**:304–310.

Gardos, G., Cole, J. O., and La Brie, R., 1977, The assessment of tardive dyskinesia, *Arch. Gen. Psychiatry* **34**:1206–1212.

Gardos, G., Cole, J. O., and Tarsy, D., 1978, Withdrawal syndromes associated with antipsychotic drugs, *Am. J. Psychiatry* **135**:1321–1324.

Garver, D. L., Davis, J. M., Dekirmenjian, H., Jones, F. D., Casper, R., and Haraszti, J., 1976, Pharmacokinetics of red blood cell phenothiazine and clinical effects, *Arch. Gen. Psychiatry* **33**:862–866.

Gerlach, J., 1979, Tardive dyskinesia, *Dan. Med. Bull.* **26**:209–245.

Glick, S. D., Jerussi, T. P., Cox, R. D., and Fleischer, L. N., 1977, Pre- and post-synaptic actions of apormorphine: Differentiation by rotatory effects in normal rats, *Arch. Int. Pharmacodyn. Ther.* **225**:303–307.

Goetz, C. G., Dysken, M. W., and Klawans, H. L., 1980, Assessment and treatment of drug-induced tremor, *J. Clin. Psychiatry* **41**:310–315.

Granacher, R. P., 1981, Differential diagnosis of tardive dyskinesia: An overview, *Am. J. Psychiatry* **138**:1288–1297.

Greenblatt, D. J., DiMascio, A., Harmatz, J. S., Bernardo, D. L., and Marder, J. E., 1977, Pharmacokinetics and clinical effects of amantadine in drug-induced extrapyramidal symptoms, *J. Clin. Pharmacol.* **17**:704–708.

Hassler, R., Ahn, E. T., Wagner, A., and Kim, J. S., 1978, Experimenteller Nachweis von intrastriatelen Synapsentypen und Axon-Kollateralen durch Isoliepung des Fundus striati von allen extrastriatelen Verbingungen, *Anat. Anz.* **143**:413–436.

Hershey, L. A., Gift, T., and Rivera-Calimlin, L., 1982, Not Parkinson's disease, *Lancet* **2**:49 (letter).

Hershon, H. I., Kennedy, P. F., and McGuire, R. J., 1972, Persistence of extra-pyramidal disorders and psychiatric relapse after withdrawal of long-term phenothiazine therapy, *Br. J. Psychiatry* **120**:41–50.

Hollister, L. E., 1978, *Clinical Pharmacology of Psychotherapeutic Drugs*, Churchill Livingstone, New York.

Hornykiewicz, O., 1981, Brain neurotransmitter changes in Parkinson's disease, in: *Movement Disorders* (C. D. Marsden and S. Fahn. eds.), pp. 41–58, Butterworth, London.

Iversen, L. L., 1978, More than one type of dopamine receptor in the brain, *Trends Neurosci.* **1**:5–6.

Iversen, S. D., 1971, The effect of surgical lesions to frontal cortex and substantia nigra on amphetamine responses in rats, *Brain Res.* **31**:295–311.

Janowsky, D. S., Davis, J. M., El-Yousef, M. K., and Sekerke, H. J., 1972, A cholinergic-adrenergic hypothesis of mania and depression, *Lancet* **2**:632–635.

Jellinger, K., 1977, Neuropathologic findings after neuroleptic long-term therapy, in: *Neurotoxicology* (L. Roizin, H. Shiraki, and N. Grčević, eds.), pp. 25–42, Raven Press, New York.

Jeste, D. V., and Wyatt, R. J., 1982, Therapeutic strategies against tardive dyskinesia: Two decades of experience, *Arch. Gen. Psychiatry* **39**:803–816.

Jeste, D. V., Potkin, S. G., Sinha, S., Feder, S., and Wyatt, R. J., 1979, Tardive dyskinesia reversible and persistent, *Arch. Gen. Psychiatry* **36**:585–590.

Johnson, D. A. W., 1978, Prevalence and treatment of drug-induced extrapyramidal symptoms, *Br. J. Psychiatry* **132**:27–30.

Jungmann, E., and Schöffling, K., 1982, Akathisia and metoclopramide, *Lancet* **2**:221 (letter).

Jus, A., Pineau, R., La Chance, R., Pelchat, G., Jus, K., Pires, P., and Villeneuve, R., 1976, Epidemiology of tardive dyskinesia, Parts I and II, *Dis. Nerv. Syst.* **37**:210–214, 257–261.

Jus, A., Gautier, J., Villeneuve, A., Jus, K., Pires, P., and Gagnon-Binette, M., 1977, Chronology of combined neuroleptics and antiparkinsonian administration, *Am. J. Psychiatry* **134**:1157.

Kane, J. M., and Smith, J. M., 1982, Tardive dyskinesia: Prevalence and risk factors 1959 to 1979, *Arch. Gen. Psychiatry* **39**:473–481.

Kazamatsuri, H., Chien, C-P., and Cole, J. O., 1972, Therapeutic approaches to tardive dyskinesia: A review of the literature, *Arch. Gen. Psychiatry* **27**:491–499.

Kebabian, J. W., and Calne, D. B., 1979, Multiple receptors for dopamine, *Nature* **277**:93–96.

Kekich, W. A., 1978, Neuroleptics: Violence as a manifestation of akathisia, *JAMA* **240**:2185.

Kelly, J. T., and Abuzzahab, Sr., F. S., 1971, The antiparkinson properties of amantadine in drug-induced parkinsonism, *J. Clin. Pharmacol.* **11**:211–214.

Kennedy, P. F., Hershon, H. I., and McGuire, R. J., 1971, Extrapyramidal disorders after prolonged phenothiazine therapy, *Br. J. Psychiatry* **118**:509–518.

Kidger, T., Barnes, T. R. E., Trauer, T., and Taylor, P. J., 1980, Sub-syndromes of tardive dyskinesia, *Psychol. Med.* **10**:513–520.

Klawans, H. L., Jr., 1973, The pharmacology of tardive dyskinesia, *Am. J. Psychiatry* **130**:82–86.

Knights, A., Okasha, M. S., Salih, M. A., and Hirsch, S. R., 1979, Depressive and extrapramidal symptoms and clinical effects: A trial of fluphenazine versus flupenthixol in maintenance of schizophrenic out-patients, *Br. J. Psychiatry* **135**:515–523.

Koob, G. F., Stinus, L., and Le Moal, M., 1981, Hyperactivity and hypoactivity produced by lesions to the mesolimbic dopamine system, *Behav. Brain Res.* **3**:341–359.

Kumar, B. B., 1979, An unusual case of akathisia, *Am. J. Psychiatry* **136**:1088.

Lipinski, J. F., Zubenko, G. S., Cohen, B. M., and Barreira, P. J., 1984, Propranolol in the treatment of neuroleptic-induced akathisia, *Am. J. Psychiatry* **141**:412–414.

Loga, S., Curry, S., and Lader, M., 1975, Interactions of orphenadrine and phenobarbitone with chlorpromazine: Plasma concentrations and effects in man, *Br. J. Clin. Pharmacol.* **2**:197–208.

Mackay, A. V., 1981, Assessment of anti-psychotic drugs, *Br. J. Clin. Pharmacol.* **11**:225–236.

Mackay, A. V. P., and Sheppard, G. P., 1979, Pharmacotherapeutic trials in tradive dyskinesia, *Br. J. Psychiatry* **135**:489–499.

Marco, E., Mao, C. C., Cheney, D. L., Revuelta, A., and Costa, E., 1976, The effects of antipsychotics on the turnover rate of GABA and acetylcholine in rat brain nuclei, *Nature* **264**:363–365.

Marsden, C. D., 1980, The enigma of the basal ganglia and movement, *Trends Neurosci.* **3**:284–287.

Marsden, C. D., and Jenner, P., 1980, The pathophysiology of extrapyramidal side-effects of neuroleptic drugs, *Psychol. Med.* **10**:55–72.

Marsden, C. D., Tarsy, D., and Baldessarini, R. J., 1975, Spontaneous and drug-induced movement disorders in psychotic patients, in: *Psychiatric Aspects of Neurological Disease* (D. F. Benson, and D. Blumer, eds.), pp. 219–265, Grune & Stratton, New York.

Martin, I. C. A., and Townsend, R. A., 1974, Implications of sustained released phenothiazines: A study of fluphenazine decanoate, *Br. J. Psychiatry* **124**:173–176.

McClelland, H. A., 1976, Discussion on assessment of drug-induced extrapyramidal reactions, *Br. J. Clin. Pharmacol.* **3**:401–403.

McClelland, H. A., Blessed, G., Bhate, S., Ali, N., and Clarke, P. A., 1974. The abrupt withdrawal of antiparkinsonian drugs in schizophrenic patients, *Br. J. Psychiatry* **124**:151–159.

Miller, R. J., and Hiley, C. R., 1974, Anti-muscarinic properties of neuroleptics and drug-induced parkinsonism, *Nature* **248**:596–597.

Ondrusek, M. G., Kilts, C. D., Frye, G. D., Mailman, R. B., Mueller, R. A., and Breese, G. R., 1981, Behavioral and biochemical studies of the scopolamine-induced reversal of neuroleptic activity, *Psychopharmacology* **73**:17–22.

Orlov, P., Kasparian, G., DiMascio, A., and Cole, J. O., 1971, Withdrawal of antiparkinsonian drugs, *Arch. Gen. Psychiatry* **25**:410–412.

Poirier, L. J., Filion, M., Larochelle, L., and Péchadre, J. C., 1975, Physiopathology of experimental parkinsonism in the monkey, *Can. J. Neurol. Sci.* **2**:255–263.

Pycock, C. J., Kerwin, R. W., and Carter, C. J., 1980a, Effect of lesion of coritcal dopamine terminals on subcortical dopamine receptors in rats, *Nature* **286**:74–76.

Pycock, C. J., Carter, C. J., and Kerwin, R. W., 1980b, Effect of 6-hydroxydopamine lesions of the medial prefrontal cortex on neurotransmitter systems in subcortical sites in the rat, *J. Neurochem.* **34:**91–99.

Quitkin, F., Rifkin, A., Gochfeld, L., and Klein, D. F., 1977, Tardive dyskinesia: Are first signs reversible? *Am. J. Psychiatry* **134:**84–87.

Raleigh, F. R., Jr., 1977, Reducing unnecessary antiparkinsonian medication in antipsychotic therapy, *J. Am. Pharmaceut. Assoc.* **17:**101–106.

Raskin, D. E., 1972, Akathisia: A side-effect to be remembered, *Am. J. Psychiatry* **129:**345–347.

Rifkin, A., Quitkin, F., Kane, J., Struve, F., and Klein, D., 1978, Are prophylactic antiparkinsonian drugs necessary? A controlled study of procyclidine withdrawal, *Arch. Gen. Psychiatry* **35:**483–489.

Rivera-Calimlin, L., Nasrallah, H., Strauss, J., and Lasagna, L., 1976, Clinical response and plasma levels: Effect of dose, dosage schedules, and drug interactions on plasma chlorpromazine levels, *Am. J. Psychiatry* **133:**646–652.

Rosen, A. M., Mukherjee, S., Olarte, S., Varia, V., and Cardenas, C., 1982, Perception of tardive dyskinesia in outpatients receiving maintenance neuroleptics, *Am. J. Psychiatry* **139:**372–373.

Schnell, R. G., 1972, Drug induced dyskinesia treated with intravenous diazepam, *J. Fla. Med. Assoc.* **59:**22–23.

Schwab, R. S., and England, A. C., Jr., 1968, Parkinson syndromes due to various specific causes in: *Handbook of Clinical Neurology, Diseases of the Basal Ganglia* Vol. 6 (P. J. Vinken and G. W. Bruyn, eds.), pp. 227–247, Amsterdam, North Holland.

Schwab, R. S., England, A. C., Jr., Poskanzer, D. G., and Young, R. R., 1969, Amantadine in the treatment of Parkinson's disease, *JAMA* **208:**1168.

Schwarcz, R., Creese, I., Coyle, J. T., and Synder, S. H., 1978, Dopamine receptors localised on cerebral cortical afferents to rat corpus stratum, *Nature* **271:**766–768.

Seeman, P., 1980, Brain dopamine receptors, *Pharmacol. Rev.* **32:**229–313.

Simpson, G. M., 1970, Long-acting antipsychotic agents and extrapyramidal side effects, *Dis. Nerv. Syst.* **31:**12–14.

Simpson, G. M., 1977, Neurotoxicology of major tranquillizers, in: *Neurotoxicology* Vol. 1 (L. Roizin, H. Shiraki, and N. Grcevic, eds.), pp. 1–7, Raven Press, New York.

Simpson, G. M., Voitashevsky, A., Young, M. A., and Hillary Lee, J., 1977, Deanol in the treatment of tardive dyskinesia, *Psychopharmacology* **52:**257–261.

Smith, J. M., and Baldessarini, R. J., 1980, Changes in prevalence, severity, and recovery in tardive dyskinesia with age, *Arch. Gen. Psychiatry* **37:**1368–1370.

Smith, R. C., Crayton, J., Dekirmenjian, H., Klass, D., and Davis, J. M., 1979, Blood levels of neuroleptic drugs in non responding chronic schizophrenic patients, *Arch. Gen Psychiatry* **36:**579–584.

Snyder, S. H., 1981, Dopamine receptors, neuroleptics, and schizophrenia, *Am J. Psychiatry* **138:**460–464.

Snyder, S., Greenberg, D., and Yamamura, H. I., 1974, Antischizophrenic drugs and brain cholinergic receptors: Affinity for muscarinic sites predicts extrapyramidal effects, *Arch. Gen. Psychiatry* **31:**58–61.

Sokoloff, P., Martres, M-P., and Schwartz, J. C., 1980, Three classes of dopamine receptor (D-2, D-3, D-4) identified by binding studies with ^{3}H-apomorphine and ^{3}H-domperidone, *Naunyn-Schmiedeberg's Arch. Pharmacol.* **315:**89–102.

Sovner, R., and DiMascio, A., 1978, Extrapyramidal syndromes and other neurological side effects of psychotropic drugs, in: *Psychopharmacology: A Generation of Progress* (M. A. Lipton, A. DiMascio, and K. F. Killiam, eds.), pp. 1021–1032, Raven Press, New York.

Spano, P. F., Memo, M., Govoni, S., and Trabucchi, M., 1980. Similarities and dissimilarities between dopamine and neuroleptic receptors: Further evidence for type 1 and 2 dopamine receptors in the CNS, in: *Long-Term Effects of Neuroleptics: Advances in Biochemical Psychopharmacology,* Vol. 24 (F. Cattabeni, G. Racagni, P. F. Spano, and E. Costa, eds.), pp. 113–121, Raven Press, New York.

Swett, C., 1975, Drug-induced dystonia, *Am. J. Psychiatry* **132:**532–534.

Tassin, J. P., Stinus, L., Simon, M., Blanc, G., Thierry, A. M., Le Moal, M., Cardo, B., and

Glowinski, J., 1978, Relationship between the locomotor hyperactivity induced by A10 lesions and the destruction of the frontocortical dopaminergic innervation in the rat, *Brain Res.* **141**:267–281.

Ungerstedt, U., 1971, Postsynaptic supersensitivity after 6-hydroxy-dopamine induced degeneration of the nigro-striatal dopamine system, *Acta Physiol. Scand.* **367**:69–93.

Van Putten, T., 1974, Why do schizophrenic patients refuse to take their drugs? *Arch. Gen. Psychiatry* **31**:67–72.

Van Putten, T., 1975, The many faces of akathisia, *Comp. Psychiatry* **16**:43–47.

Van Putten, T., 1978, During refusal in schizophrenia: Causes and prescribing hints, *Hosp. Commun. Psychiatry* **29**:110–112.

Wauquier, A., Niemegeers, C. J. E., and Lal, H., 1975, Differential antagonism by the anticholinergic dexetimide of inhibitory effects of haloperidol and fentanyl on brain self-stimulation, *Psychopharmacologia* **41**:229–235.

Weller, M. P. I., 1981, Schizophrenia, neuroleptics and Parkinson's disease, in: *Research Progress in Parkinson's Disease* (F. Clifford-Rose and R. Capildeo, eds.) pp. 67–71, Pitman Medical, Turnbridge Wells.

Wolfarth, S., 1976, Experimental basis of the therapy of Parkinson's disease and the cholinergic-dopaminergic equilibrium in basal brain nuclei, *Pol. J. Pharmacol. Pharm.* **28**:469–493.

Yahr, M. D., 1981, *Introduction in Research Progress in Parkinson's disease* (F. Clifford-Rose, and R. Capildeo, eds.), pp. 3–8, Pitman Medical, Tunbridge Wells.

Young, R. R., and Shahani, B. T., 1979, Pharmacology of tremor, in: *Clinical Neuropharmacology*, Vol. 4 (H. L. Klawans, ed.), pp. 139–156, Raven Press, New York.

CHAPTER 18

Drug-Induced Movement Disorders

J. M. S. PEARCE and C. C. CLOUGH

1. INTRODUCTION

The introduction of effective antipsychotic drugs in the early 1950s produced a revolution in the care of acute and chronic psychoses (Delay *et al.*, 1952). Many patients with schizophrenia who previously would have been confined to mental hospitals for life were able to lead almost normal lives in the community, their psychotic symptoms being adequately controlled. This pharmacological miracle was not without its drawbacks. It soon became apparent that these drugs were not only attended by the usually recognized side effects such as cholestatic jaundice, but also had remarkable effect on the control of movement, i.e., on the extrapyramidal system. The first problem to be noted in this regard was drug-induced parkinsonism (Steck, 1954). Early on in their use it was felt that a drug's antipsychotic action paralleled its ability to produce parkinsonism. The impassive facies, flexed posture, and other parkinsonian features soon became a commonplace finding in mental institutions. Initially, this did not cause undue concern as it was felt that this was a small price to pay for control of mentally disturbed patients.

In 1957 other types of abnormal movement were first reported (Schonecker, 1957). It became clear that neuroleptic drugs such as phenothiazines could also cause dyskinesia. Many case reports of a variety of abnormal movements following, and soon epidemiological evidence was available to demonstrate irrefutably a causal relationship with these drugs (Crane, 1973). The most common form of dyskinesia followed months or years after starting medication, and thus these were termed tardive dyskinesias. Others had a more explosive occurrence following one or two doses of the drug, suggesting an idiosyncratic response (Marsden and Jenner, 1980: De Silva *et al.*, 1973). This acute reaction was often dystonic in form. Other subacute responses were also noted, such as akathisia, a state of mental and motor restlessness with a desire to keep constantly on the move. One of the initial difficulties in characterizing clinically

---

J. M. S. PEARCE • Department of Neurology, Hull Royal Infirmary, Hull, England.
C. C. CLOUGH • Department of Neurology, Queen Elizabeth Hospital, Birmingham, England.

the different reactions was their confusion with pretreatment psychomotor disturbances in mentally disturbed patients. Since the days of Freud, it had been recognized that many psychiatric patients, particularly those with schizophrenia, had associated movement disorders. These usually took the form of peculiar stereotypies often acquired during long stays in mental institutions. Much of the ritualistic activity was learned and had cultural significance. Different forms of behavior are recognized in psychotic patients in certain countries but do not necessarily occur elsewhere. It remains likely that other movement disorders of untreated schizophrenic patients have an organic basis. Stereotypic movements, by definition, lack purpose. They may range from hyperkinetic repetitive tics to hypokinetic postures typical of catatonic schizophrenia. Against this background of a preexisting disturbance of movement, it was not surprising that drug-related dyskinesias were considered to be peculiar to the schizophrenic or otherwise brain-damaged population. This notion was dismissed as phenothiazines gained other uses, for instance in the treatment of nausea. It became apparent that dyskinetic responses to neuroleptics could also occur in patients considered to have normal brains.

During the 1960s the clinician's pharmacological armamentarium expanded rapidly, and in its wake other drug-induced movement disorders have been recognized. The newer biochemical characterization of idiopathic parkinsonism also added to our understanding of the important effect of various neurotransmitters on the control of movement. This shed some light on the pathophysiology of the drug-induced movement disorders. Parkinson's disease is primarily a state of dopamine deficiency (Cotzias *et al.*, 1967). There is also evidence of a concurrent cholinergic overactivity as cholinomimetic agents such as physostigmine can increase parkinsonian signs (Duvoisin, 1967). These findings established the concept of a dopaminergic–cholinergic balance existing within the basal ganglia (Yahr *et al.*, 1969). The treatment of Parkinsonism with levodopa has had a great impact on this disease, reducing the relative mortality from three times that of a normal population to a near normal life expectancy (Zumstein and Seigfried, 1976). However, within 2–3 years of its introduction, it was realized that levodopa too could cause abnormal involuntary movements, ranging from orofacial dyskinesia to dystonia and choreic states indistinguishable from Huntington's disease (McDowell and Sweet, 1976). This gave further credence to the concept that dyskinetic states were related to relative dopaminergic overactivity within the basal ganglia. Levodopa-induced dyskinesia is probably the most common cause of generalized chorea in the community and remains a major problem in the treatment of Parkinson's disease. If dyskinesia is related to the selective stimulation of a particular group of dopamine receptors within the striatum of DA1 type, and if this receptor could be blocked specifically without involving other receptors, it might be possible to control dyskinesia without aggravating parkinsonism. So far no such drugs have been developed; all drugs that block the action of dopamine can cause parkinsonism. In the treatment of Parkinson's disease such dopamine-blocking drugs will always reverse the beneficial antiparkinsonian effects of administered L-dopa, and similarly the treatment of other dyskinetic states has been bedeviled by the occurrence of parkinsonism in return for control of extra movements.

From a clinician's point of view, it is convenient to classify the drug-induced movement disorders into three groups: acute, subacute, and chronic. In the past, alternative taxonomies have tried to divide the drug-induced effects into reversible and irreversible. This has proved impracticable since it is impossible to predict with any certainty the future course of the chronic dyskinesias. From a practical point of view, the following simple classification merely indicates the time course such drug reactions may pursue and the way they are likely to present to the attention of the clinician. In general, it has little bearing on the presumed pathophysiology.

2. ACUTE DYSKINESIAS

The most frequently encountered acute dyskinesias are acute dystonic reactions seen after the administration of neuroleptic drugs (De Silva *et al.*, 1973; Marsden *et al.*, 1975). This can occur with all phenothiazine derivatives and butyrophenones. The fact that dystonia results from giving chlorpromazine, a drug blocking both dopamine and noradrenaline receptors, and also haloperidol, which blocks dopamine specifically, indicates that the common denominator is the capacity to block the action of dopamine. In general medical practice, acute dystonia is most commonly seen after giving drugs for the treatment of nausea. Metoclopramide (Maxolon) is also prone to cause such reactions, especially in young patients. It was hoped that the development of domperidone would avoid this complication because it was thought to be a purely peripheral dopamine antagonist not involving brain receptors. Recent reports of acute dystonia following the use of domperidone have unfortunately demonstrated otherwise (Debontridder, 1980).

Acute dystonia is likely to be idiosyncratic in that only 2% of patients who received neuroleptics will have such a reaction (Marsden and Jenner, 1980), and it commonly occurs after only one or two doses of the drug. Why some patients rather than others are affected is a mystery. As young people are predominantly affected, genetic factors are indicated rather than degenerative disease related to aging. Could those patients affected be carrying the gene for a genetically determined basal ganglia disorder such as torsion dystonia? If they are heterogeneous, this could cause them to have an increased susceptibility to dystonia. There is little evidence for this speculation, and if it is true, one might expect a greater number of adverse drug reactions among the Jewish population, who have a higher prevalence of torsion dystonia, but this does not appear to be the case.

The onset of symptoms is explosive, within 12–48 hr of taking the first dose. Patients often present as an emergency to the physician or more usually the hospital emergency room. The patient is extremely distressed and frightened by the bizarre symptoms. Spasms may be intermittent and can be provoked by casual observation. This often wrongly suggests to relatives and medical minds a hysterical cause of these movements. The main features can be remembered as the three O's—oculogyric spasm, oromandibular dystonia, and opisthotonos. Any one feature can present in isolation, often with bizarre symptomatology. Not uncommonly patients are dismissed as hysterical, implying psychogenic origins for their symptoms. This is one of the more common causes of trismus, the teeth remaining clenched; alternatively, the mouth may remain fixedly open. Bulbar muscles are often involved, with difficulty swallowing or a strangled quality to speech. The tongue may adopt a fixed, twisted position, obstructing ingestion of fluids and causing problems with speech. The patient may complain of tightness or pulling of the muscles around the mouth and neck. Sometimes it may be difficult to observe any objective changes though the patient may indicate, with a stroking movement of the hand, the pulling or dragging sensation being experienced down the neck.

There may be observable and powerful contraction of the muscles around the mouth and platysma of the neck. This may produce a painful-looking transverse smile (risus sardonicus) or jutting forward of the chin. This orobuccolingual dystonia is indistinguishable from that seen in the idiopathic chronic dystonia of Meige's or Breugel's disease. Indeed, other dystonic syndromes can also be closely mimicked, confirming incidentally their organic basis. They include blepharospasm and spasmodic torticollis, which as chronic idiopathic syndromes have previously been considered to be psychogenic. Another common feature in acute dystonia is the occurrence of oculogyric spasm; this is similar in all respects to that seen in postencephalitic parkinsonism (Duvoisin and Yahr, 1965). These are the only two common organic causes of oculogyric crises. The eyes are deviated upward and sometimes laterally; the patient can override this movement voluntarily for a few seconds, but the eyes inevitably will resume their former fixed position. Doll's-eye reflexes are usually intact, and the pupils, though often dilated owing to accompanying autonomic overactivity, will react normally. Associated with oculogyric spasm, the head will pursue the eye movement adopting positions of retrocollis or occasionally torticollis. Oculogyric spasm can recur if its etiology is not recognized and the offending drug is continued; this particularly applies to those long-acting phenothiazines given by depot injection. In this case the delay between symptoms and administration of the drug can be delayed by several days.

As indicated previously, young people are more likely to be affected, and the full-blown dystonic reaction is usually confined to them. In addition to the oral, facial, cervical, and ocular dystonias, limbs can assume dystonic postures, and the patient is incapacitated and confined to bed. At this stage opisthotonos is common, the back arching away from the bed. When such a dramatic picture occurs acutely in a young person who is conscious, the suspicion of tetanus or hysterical fits is often erroneously provoked (De Silva *et al.*, 1973). Asymmetrical spasms of the face and limbs may be mistaken for focal epilepsy. The drug history must be pursued assiduously.

3. PATHOPHYSIOLOGY OF ACUTE DYSKINESIA

The mechanism underlying this acute phenomenon is unknown, but there are several factors that shed some light on it. It is an idiosyncratic reaction, and there is no evidence to relate it to other drug-induced movement disorders such as parkinsonism. Nor does the occurrence of an acute reaction increase the risk of subsequent subacute or chronic drug-induced problems. The similarity of this acute dystonia to other naturally occurring clinical entities suggests that the problem is one of dopamine overactivity or sensitivity within the basal ganglia. The patterns of acute dystonia described here resemble those seen in L-dopa-treated parkinsonian patients, and these are clearly dopamine induced and will almost invariably disappear on lowering the dose of L-dopa. If one accepts this hypothesis of increased dopamine activity within the striatum, the paradox is evident. How can drugs that act as a dopamine antagonist, simultaneously produce states of dopamine hyperactivity? It is here that recent theories implicate selective activation of dopamine autoreceptors on presynaptic sites (Marsden *et al.*, 1975). It is postulated that neuroleptic drugs act preferentially on the presynaptic dopamine receptor in doses that do not block postsynaptic receptors. There is good evidence that there is feedback control of dopamine release mediated through this presynaptic receptor. Blocking this receptor would prevent feedback control by dopamine, thus leading to increased dopamine release. Thus, blockade of the presynaptic receptor leads to increased dopamine release and could theoretically lead to a state of dopamine hyperactivity. The blockade of the postsynaptic receptors by neuroleptics is due to competitive blockade; this could now be overcome by the increased release of dopamine. With repeated administration of drugs the evidence of increased dopamine turnover eventually disappears. In addition, there is evidence from animal experiments of changes in the postsynaptic receptors; 24 hr after the acute administration of a neuroleptic drug rats showed increased evidence of stereotypy and other movements suggestive of increased dopamine stimulation, which continued for another week (Moller-Nielsen *et al.*, 1976). It is suggested that this may reflect a state of increased dopamine sensitivity of the postsynaptic receptor following a period of dopamine blockade. This may explain why dystonic reactions are delayed in certain patients for up to 4 days after initiation of the drug. A state of dopamine receptor supersensitivity emerging from blockade might thus overreact to any release of presynaptic dopamine.

Treatment of acute dystonia is invariably successful, and patients will respond promptly to parenteral administration of either an anticholinergic drug, such as bentropine, 2 mg, or a benzodiazepine, e.g., diazepam, 5 to 10 mg. Anticholinergic agents are believed to act by their ability to prevent the dopamine release produced by neuroleptics. Diazepam acts predominantly as a sedative but may have some GABAergic activity. If diazepam alone is used, there is a risk of dystonia recurring over the subsequent 48 hr, and it is probably wise to add an oral anticholinergic drug for 72 hr. The patient must be advised in future to avoid that particular drug, but it is not known whether it is necessary to ban the future use of all phenothiazines and butyrophenones if care is taken with their subsequent prescription.

4. SUBACUTE REACTIONS

The most common subacute response to neuroleptics is drug-induced parkinsonism. This is second only to idiopathic disease as a cause of a parkinsonian syndrome. In long-stay mental hospitals, the prevalence is about 10–15% (Marsden and Jenner, 1980). Higher figures are recorded (up to 90%) when potent drugs are used and signs of parkinsonism specifically sought. The drugs most commonly implicated are those used to control psychosis, i.e., the phenothiazines, butyrophenones, diphenyl–butyl piperidines, and thioxanthenes. Most drugs within these groups can produce parkinsonian signs, but their ability to do so varies considerably although naturally to some extent the dosage does influence matters. The ability of antipsychotic drugs to produce sedation, hypotension, and parkinsonian and other side effects is related to their differing capabilities of blocking a variety of neurotransmitters. They all have in common the propensity for dopamine blockade, in addition to blocking in varying degrees, cholinergic, α-adrenergic, histaminergic, and tryptaminergic receptors. It is clear that parkinsonism is produced by dopamine receptor blockade. However, the ability to produce parkinsonism is not related to this alone, but also to the intrinsic antocholinergic properties of the drug used. The greater the degree of anticholinergic activity, the greater the potential for reducing the parkinsonian side effects. In this regard, there are generally recognized three groups of phenothiazines related to differing properties (British Medical Association and the Pharmaceutical Society of Great Britian, 1985):

Group 1. Chlorpromazine and promazine, characterized by pronounced sedation and moderate anticholinergic and extrapyramidal side effects.

Group 2. Thioridazine pericyazine, characterized by moderate sedative effects, marked antocholinergic activity, and less parkinsonian effect than groups 1 or 3.

Group 3. Fluphenazine, perphenazine, and prochlorperazine, which have few sedative effects and less anticholinergic effects than Group 1 and 2; they have a marked propensity for causing parkinsonism.

Butyrophenones (e.g., droperidol and haloperidol), diphenyl–butyl piperidines (e.g., pimozide), and thioxanthenes (e.g., clopenthixol and flupenthixol) resemble Group 3 in their action and so have marked parkinsonian effects.

Other drugs can produce parkinsonism by other modes of action. Reserpine, which initially was used as a hypotensive, is a potent agent by virtue of its ability to deplete brain stores of dopamine and other neurotransmitters by interfering with presynaptic vesicular storage. Tetrabenazine, a synthetic analogue of reserpine, is now used almost exclusively to control chorea, hemiballismus, and dyskinesia because of this action. Parkinsonian signs almost inevitably will result from its use in return for controlling the abnormal movements.

GABA agonists should theoretically produce parkinsonism but rarely do so in practice, although there have been case reports of sodium valproate (Epilim) causing parkinsonism (Lautin *et al.*, 1979). Similarly, GABA agonists have yet to find a place in the management of choreic disorders such as Huntington's disease as their effect on movements is minimal. The main cause of drug-induced parkinsonism is the antipsychotic drugs, which do this through blockade of dopamine receptors. An interesting question is whether all patients placed on these drugs could develop parkinsonism. To a certain degree, parkinsonism is related to the dosage, and it is conceivable that everybody could develop parkinsonism if placed on a large enough dose of antipsychotics. The age spectrum of people adversely affected by these drugs in this manner does closely follow that for idiopathic parkinsonism, so that younger patients are less likely to be affected. This is probably because older patients having a greater likelihood of subclinical nigrostriatal dopamine deficiency, and thus the drug acts as a provocative factor in a striatum already deprived of adequate dopamine. Once the offending drug has been withdrawn, it is unlikely that persistent parkinsonism is due to permanent damage produced by the drug. A more reasonable explanation is that a number of patients actually have latent Parkinson's disease, which has an incidence in the over-60 age group of about 1% (Pearce, 1978). In this group of patients antipsychotic drugs produce further dopamine depletion, which leads to the presentation as parkinsonism which initially is ascribed to the drug. When recovery is not complete, it is wise to review the history as it is likely that a preparkinsonian state existed prior to its administration. For instance, the patient's spouse may have noticed some slowness of movement, impassivity of facies, or mental inflexibility, suggestive of early parkinsonism. Sometimes, however, no adequate historical evidence is available, and the diagnosis of latent parkinsonism brought to light by neuroleptic drugs remains a supposition. It is clear from this that patients with Parkinson's disease should avoid antipsychotic drugs. Certainly catastrophic results can follow, and severe akinetic, mute parkinsonian crises may ensue.

Parkinsonian signs may be delayed following administration of neuroleptics, probably because of their long half-life. Rarely, acute symptoms may develop within hours of taking a neuroleptic, especially the piperidine derivatives. Usually, the onset of symptoms is insidious gradually coming on over weeks or months. Ninety percent of cases present within 3 months of treatment; thus the doctor who sees the patient infrequently in a specialist clinic is the most likely to make the diagnosis as he will be more impressed by the dramatic change. Symptoms and signs can in all ways resemble those of idiopathic parkinsonism with the triad of tremor, bradykinesia, and rigidity (Pearce, 1977). It is said that drug-induced parkinsonism tends to be symmetrical and that tremor is not the predominant component. In practice, this information is of little value in deciding etiology, as, not infrequently, drug-induced parkinsonism has asymmetrical features and tremor in much the same way as idiopathic parkinsonism. Different symptoms and signs may predominate, and some patients may only have one component of the parkinsonian triad. Commonly, many may merely look parkinsonian with the characteristic unblinking gaze that results from facial bradykinesia. This reptilian deadpan facies may be mistaken for evidence of depression, and the error may be compounded by further psychotropic therapy. If the drug is not withdrawn or is increased at this stage, further signs will emerge. A stooped, flexed posture with failure of arm swing is typical, and the characteristic shuffling gait is well known. Tremor, though a less common component, is of the same type as in idiopathic disease, i.e., 4–6 Hz frequency, a rhythmical resting tremor with pill-rolling quality, which generally disappears or is reduced by action. There may be an added fast, irregular component which is similar to an exaggerated physiological tremor and separate from the parkinsonian tremor. Perioral tremor occurs and has been named the rabbit syndrome for obvious reasons.

The course of events is usually for bradykinesia to appear first with facial impassivity and loss of the normal associated stereotypic movements. Rigidity then occurs some weeks later and can be detected in the neck, trunk, and limbs. Both lead-pipe and cogwheel rigidity can occur. Patients may complain of stiffness, pains, and cramps in the limbs, and it is not unusual for antiarthritic treatment to have been wrongly instigated. Postural abnormalities are important and include retropulsion and propulsion, which take place with initiation of walking. A sudden push may reveal these abnormalities and cause the patient to fall *en bloc*, like a tree being felled. This instability can severely affect a patient's confidence in his gait, for if he suddenly attempts to turn around, he may fall dramatically. Similarly, sitting down may become a less than graceful affair: instead of lowering himself gently into the chair, the patient drops heavily.

Catastrophic responses (malignant neuroleptic syndrome) to these drugs may occur within a few days of taking neuroleptics especially in patients with already compromised nigrostriatal pathways. This akinetic crisis resembles that seen sometimes when anticholinergic drugs are withdrawn suddenly from patients with idiopathic disease. A state of "pseudocoma" may rapidly develop. These may be patients who have received massive doses of neuroleptics in an attempt to control severe psychosis, and so it is not unreasonably assumed that this state of apparent consciousness, without will to move, is a hysterical or otherwise psychogenic reaction. The patient at this stage is totally akinetic, unable to move any part of his body. He is mute but is conscious, and his eyes will follow to command. This crisis can lead rapidly to death from bronchopneumonia if appropriate nursing care is not given and fluids are not administered nasogastrically or via a drip. Anticholinergic agents should be given parenterally and usually result in swift improvement in 1–2 days. The clues to the correct diagnosis lie in recognizing the accompanying parkinsonian rigidity and the history of neuroleptic administration.

Drug-induced parkinsonism will generally respond to stopping the offending drug. Improvement usually results within weeks, though complete resolution of parkinsonism may take up to 3–4 months. Rarely, patients have taken up to 2 years to recover fully. Failure to completely resolve suggests preexisting basal ganglia disease. As the mechanism is one of dopamine blockade, theoretically L-dopa drugs and dopamine agonists should not help (Pearce and Pearce, 1971). Larger doses of L-dopa drugs in practice may help, and this indicates a competitive uptake by the dopamine receptors. Of greater benefit are anticholinergic drugs (benztropine, benzhexol), and these are commonly prescribed. They are usually attended by side effects of their own such as dry mouth, visual blurring, and constipation. These drugs should be prescribed if the patient is troubled by his parkinsonian symptoms and is impatient for speedy recovery. A more difficult question is: should we use anticholinergics when neuroleptics need to be continued because of uncontrolled psychosis? There is some evidence that suggests an increased incidence of tardive dyskinesia when anticholinergic drugs are given with neuroleptic drugs (Mindham, 1977). Thus, if possible, this combination should be avoided. Obviously, the best course is to discontinue neuroleptic drugs if parkinsonian symptoms ensue, rather than adding further therapy. Often, however, their antipsychotic action is still required. In this case a compromise is reached by reducing the dose of neuroleptic drug to that required to control psychosis, minimizing the risk of symptomatic parkinsonism. A further option is to change the type of neuroleptic to one with more intrinsic anticholinergic activity and thus less potential for parkinsonism.

It is now no longer considered that a drug's antipsychotic action goes hand in hand with its ability to cause parkinsonism, as presumably preferential blockade of mesolimbic dopamine receptors can take place without necessarily causing complete blockade of nigrostriatal dopamine receptors. When neuroleptics were first introduced in the 1950 and it became apparent that they caused parkinsonism, it was thought that this was necessary to control psychosis. However,it is clear that the majority of psychotic patients can be managed correctly without the development of this disorder.

It must be conceded that in certain intractable noisy, aggressive psychotics, some degree of drug-induced parkinsonism is an inevitable consequence of adequate control of their mental disorder. Equally, much drug-induced disease is avoidable, and constant reappraisal and simplification of drug regimes is necessary in all such patients.

5. AKATHISIA

Akathisia is a common accompaniment of neuroleptic treatment though its occurrence may pass unnoticed. Its incidence varies from 5 to 50% depending on the potency of the drug used (Marsden *et al.*, 1975). A figure of 20% is accepted as that occurring in general psychiatric practice. Akathisia refers to restlessness and the irresistible urge to move which these patients experience. They stand, shuffling their feet, rather like a bull pawing the floor before charging. Increased fidgeting is the first sign, and initially these movements may be difficult to distinguish from those due to mental agitation or psychosis for which the drugs are given. The patient finds it difficult to sit still and is continuously getting up and down, pacing the floor. If he is asked to remain still, tension will eventually build up and express itself in some form of movement such as rocking to and fro or hopping from one foot to the other. Extrapyramidal signs may be present such as cog-wheel rigidity or bradykinesia. The latter is usually mild and evident as an impassivity of facial movement or lack of arm swing when walking. This mixture of signs and symptoms should always alert the clinician to the question of a drug-induced disorder.

Ekbom's disease, or restless legs, is often confused with akathisia but is a different entity. The main difference is the crawling, creeping discomfort experienced in Ekbom's disease within the leg muscles, which leads to the compulsion to move, kick the legs, or walk about in order to relieve the symptoms. These "fidgety feet" classically occur at rest, especially at night, and the patient rapidly learns that the best form of relief is to get up and walk around. On the other hand, akathisia is usually painless; although cramplike sensations can occur, they are not usually significant, and the desire to move is the predominant complaint. Akathisia can develop within days to weeks of starting the drug. Fifty percent of such reactions take place within the first month and 90% within 3 months. Thus, it is not an idiosyncratic reaction like the acute dyskinetic reactions and generally its incidence will increase with the passage of time. This reaction usually disappears after the drug is stopped. Persistent akathisia can occur coming and going over many months. Indeed, even if the drug is continued, akathisia may remit and relapse spontaneously. Thus, it is not an absolute indication to discontinue therapy. In many patients akathisia is mild and does not cause undue concern; control of the mental symptoms may then take priority. Akathisia does not increase the risk of other long-term effects such as tardive dyskinesia. If akathisia becomes a problem, we should review the decision to continue neuroleptics. The first step is to reduce the dose of the drug or switch to one with less potency. If akathisia is severe and disabling, all neuroleptic drugs should be stopped. Adjuvant therapy with benzodiazepines such as clonazepam or diazepam may be useful, possibly because of their sedative effect. These drugs are potential GABA agonists and may also interfere with movement control through this mechanism.

The pathophysiology of this condition remains a mystery. It has been postulated that there is a preferential blockade of the presynaptic receptor. This receptor would normally regulate the release of dopamine into the synaptic cleft. As indicated previously, dopamine release would be followed by activation of the presynaptic receptor thus preventing further dopamine release. In the presence of a drug that preferentially blocked the presynaptic receptor leaving available receptor sites on the postsynaptic receptor, dopamine would continue to be released leading to a superdopaminergic state manifest as akathisia. This explanation is akin to that postulated in acute dystonias, but there is no evidence that the two reactions occur in the same population. As an explanation of akathisia there are many faults with this theory, the main one being the concomitant occurrence of parkinsonian signs which must indicate adequate postsynaptic dopamine blockade within the nigrostriatal system. It would have to be postulated that akathisia is the result of dopamine interference within the mesocortical system and of simultaneous but different reaction within the nigrostriatal system, blocking dopamine and leading to parkinsonian signs. Certainly the involvement of the mesocortical projections could explain the marked psychological flavor of the symptoms of akathisia. Another argument against this explanation is the occurrence of akathisia with reserpine. This drug does not, as far as we know, interfere with receptors but is a dopamine depletor which leads to a hypodopaminergic state; thus, it is difficult to invoke explanations involving overactivity of dopamine, and the basis of akathisia remains unknown.

6. OTHER DRUGS

Chorea is a well-recognized but rare complication of the contraceptive pill. It is believed to be a direct hormonal effect of the estrogen component, and the chorea invariably disappears on stopping the drug. Others have postulated that a vascular event within the basal ganglia secondary to hormonal influences is the cause. A vascular etiology was suggested because of the sudden strokelike onset of symptoms in occasional cases. However, structural changes have never been demonstrated, and the complete resolution of symptoms makes this improbable. Generalized chorea can start within 6 days to 9 months of starting the pill. Initially the movements may be mistaken for excessive fidgeting or a manifestation of anxiety or hysteria. There may be severe generalized movements with dysarthria and an unstable gait. About 60% of cases reported have shown unilateral signs (Riddoch *et al.*, 1971). One-third had a history of rheumatic fever, but fewer than this had a definite history of Sydenham's chorea. However, patients who develop such chorea are more likely than average to have had an episode either of rheumatic chorea or of chorea gravidarum in the past. Several patients have developed chorea gravidarum during a subsequent pregnancy. The chorea seen with the contraceptive pill is in no way different from these other conditions and similarly may be accompanied by anxiety, mental agitation, and depression. Rarely, a psychotic state, "chorea insaniens," has been reported (Sale and Kalucy, 1981). Drug treatment should be avoided wherever possible as the prognosis is always good. Persistent chorea has not been reported following use of contraceptives. Recovery usually takes place within 2–3 months of stopping the drug, although sometimes it may be as short as 3 days or as long as 6 months. When there is severe chorea or associated mental problems, a short course of a dopamine-blocking agent such as haloperidol is indicated, in doses just large enough to control symptoms without untoward depressive effects. The patient should obviously avoid all estrogen-containing drugs. Time will show whether the progesterone-only pill can also produce this side effect. Subsequent pregnancies should also be carefully supervised as there is a greatly increased risk of the reappearance of chorea.

Other drugs rarely induce movement disorders. Phenytoin commonly causes a dose-dependent ataxia, and its chronic use may rarely cause cerebellar degeneration and permanent dysarthria, nystagmus, and ataxia. Less commonly, choreoathetosis may be seen. This may be isolated or may be associated with ataxia. Invariably the level of phenytoin is in the toxic range, and the abnormal movements respond to reduction of the dose. Finally, abuse of narcotic agents such as methadone has resulted in chorea though it is not known whether this is a permanent effect. Amphetamine abuse may similarly cause an agitated motor disturbance with chorea and is usually associated with psychoses. If the drug is stopped, symptoms usually abate.

7. CHRONIC MOVEMENT DISORDERS: TARDIVE DYSKINESIA

Tardive dyskinesia (TD) is the most serious problem encountered with the use of neuroleptic drugs. TD describes those movement disorders that develop after a period of taking these drugs, and it is this delay which is so characteristic of the syndrome. Once it has occurred, it persists for many months or years. It has rightly provoked a public outcry, and its prevention has become one of the major concerns of neuropsychiatry, requiring the action of a task force whose findings were published in the *American Journal of Psychiatry* in October 1980 (APA Task Force, 1980).

The initial problem in description of this disorder was the separation of what was a new phenomenon related to drugs from those movement disorders which only existed in the chronically psychotic population or the normal aging population. We no longer have large populations of "drug-naïve" psychotic patients to study. The sort of hyperkinetic movements seen by Kraepelin and Bleuler, and still seen occasionally in treated psychotic patients today, are not like those seen as a result of neuroleptics. These stereotypies and complex mannerisms are more organized and semipurposeful and often have a special meaning to the patient. Other repetitive movements such as head rocking or destructive behavior may be learned from other patients. These patients are more likely to be long-term psychotics residing in the large poorly attended mental institutions.

The cornerstone of TD is the occurrence of facial dyskinesia which has been renamed the buccolingual–masticatory syndrome (BLM). It is clear that the incidence of this phenomenon is greatly increased in schizophrenic patients treated with neuroleptics. The BLM syndrome is frequently seen in the normal aging population and is often wrongly attributed to poor fitting dentures. In general, it does indicate an underlying basal ganglia disorder. BLM is the main feature of TD and is the most consistent component of this syndrome. It is in no way different from that seen in old people but is more likely to be severe and distressing to the patient. Because BLM may exist alone in a good number of elderly people, it is difficult to assess the prevalence of dyskinesia due to neuroleptic administration. Thus, figures vary enormously, but it is estimated that about 20% of chronically institutionalized patients exhibit TD (Crane, 1973). These patients will, of course, be more susceptible than others owing to the likelihood of the prescription of large doses of neuroleptics over a long time. However, there is no evidence that the type of underlying disease or the age of the patient affects the risk of the development of TD. Phenothiazines are the main culprits, and there does not appear to be any differing risk for differing types. Butyrophenones and other neuroleptics are equally implicated, although these drugs are less likely to have been given in large doses for lengthy periods. It is probable that these two factors, i.e., the size of dose and the length of treatment, are the most important determinants of TD. Cases are rarely seen when the period of administration is less than 3 months, and most patients develop the syndrome after several years of continuous administration. Often, features of dyskinesia will begin when the drug is reduced or stopped, but this is not an indication to immediately restore the previous dose. Commonly though, TD presents when the patient has been on neuroleptic treatment for years, though others may present after the phenothiazine has been stopped.

Usually it is a chronic disorder and may persist indefinitely. However, we are by no means sure of its natural history, and recent studies have suggested that a large proportion of symptoms will remit spontaneously if further neuroleptics are not prescribed. If it develops while on treatment one study has shown that about 20% will get better completely, 20% will improve, but in 50% the dyskinesia will remain unchanged, though will not progress (Degkwitz, 1969). The likelihood of recovery is probably influenced by the length and total dosage of treatment, but much more important, also by the age of the patient. Children can develop TD but will almost invariably recover if the drug is stopped. Young adults similarly are much more likely to recover; indeed, some patients will recover even if the drug is continued; one study showed that about 40% of patients improved anyway after 15 months of unchanged treatment (Marsden *et al.*, 1975). Generally, the older the patient and the longer the duration of the TD, the less the likelihood of recovery.

Predictors of TD are few. The occurrence of acute dystonic syndrome correlates poorly with the subsequent development of TD. The occurrence of parkinsonian signs does show a correlation. If parkinsonism occurs early in treatment, these patients are probably at greater risk of the development of TD, and thus this is a clear indication to stop treatment unless absolutely necessary.

8. CLINICAL FEATURES OF TD

The clinical spectrum of TD is wide. Many patients will have only one feature, most commonly the BLM syndrome. Most patients, especially those in mental institutions, will not notice these abnormal movements and are not unduly concerned by them. Others will experience physical distress and will also be considerably embarrassed by these movements, especially in company. The first symptom is usually a writhing, darting movement of the tongue. This will progress to licking of the lips and involuntary protrusion and retraction of the tongue ("fly-catcher's tongue"). Pouting of the lips is prominent along with lip smacking and blowing out and sucking in of the cheeks. These movements can progress to rotary movements of the whole mouth and chomping and chewing movements of the jaw. Movements may be continuous and increased in company. At their worst they can interfere with speech, and there may be difficulty in eating because of obstruction from the tongue. BLM mainly affects the lower face but can be associated with involuntary blinking and blepharospasm. Odd darting, upward movements of the eyes have also been recorded.

Generalized features commonly develop, including dyskinetic movements of the limbs which can be a combination of choreic and athetoid movements along with the assumption of dystonic postures. Choreic movements are manifest as brief contractions of proximal of distal limb muscles with flexion of the trunk and asymmetrical movements of the shoulder girdle with head bobbing and twisting. The patients may appear restless, continuously rising from a chair to pace the floor; the gait may be interrupted by pelvic muscle contractions giving the characteristic dancing appearance. Athetoid movements affect the musculature with predominantly twisting, turning movements of the arms and legs, with ulnar deviation of the hands and fanning of the fingers. The legs may plantar-flex and turn inward with inversion of the foot. These are the classical dystonic postures, which can be seen in a variety of basal ganglia disorders and are not specific. At its worst, generalized choreoathetosis may ensue, and this picture is indistinguishable from Huntington's disease. Clinically, the presence of BLM favors drug-induced TD, and there may be other extrapyramidal features such as cogwheel rigidity. These features may exist in Huntington's disease, but the combination of dyskinesia and parkinsonism always makes it more likely that drugs are implicated. Obviously, the lack of family history and absence of dementia are most important to distinguish such generalized TD from Huntington's disease, which almost invariably demonstrates the pattern of an autosomal dominant gene. In addition, there must be a history of prolonged neuroleptic administration prior to the development of movements to make the diagnosis of TD. A more difficult problem arises when a patient with known Huntington's disease is treated with phenothiazines, such as thiopropazate or haloperidol, to control the abnormal movements. The question here is whether these neuroleptics are actually contributing to the movement disorder by the production of an added TD. Dopamine-blocking drugs such as haloperidol can be helpful in controlling Huntington's chorea as this is predominantly the result of dopamine overactivity. If used, they will usually result in parkinsonian signs if the chorea is adequately controlled. It is well recognized that occasionally, when these drugs are stopped, the chorea may worsen. It is possible that this rebound effect is simply due to the progression of the underlying disease process, but the development of a further dopamine receptor sensitivity due to neuroleptic administration cannot be ruled out; thus, it is possible that TD may add to the difficulties in controlling the chorea of Huntington's disease.

In summary, the most common clinical picture of TD is of a buccolingual masticatory syndrome associated with a mild choreic movement affecting the limbs. In young people the dystonic elements are more pronounced, and the dystonic postures may be maintained for many months or even years. This condition has recently been coined *tardive dystonia* (Burke *et al.*, 1982), but this is probably an unnecessary addition to an already crowded nosology. Dystonia merely reflects one end of the spectrum of dyskinetic disorders following drug administration, and even though it may predominate, there is usually evidence of BLM, chorea, or athetosis. The natural history of this condition is exactly the same as in other TD.

Drugs other than phenothiazines have been implicated in the production of TD. Antihistamines (Thach *et al.*, 1975), tricyclics (Fann *et al.*, 1976), and anticholinergic (Fahn and David, 1972) drugs, used alone, have each been incriminated. However, this must be a rare complication, and a causal relationship cannot be presumed from single case reports. It does indicate that any drug that does alter neurotransmission within the brain should be used cautiously on a long-term basis and the need for repeat prescriptions should be reviewed regularly. Perhaps of greater importance is the fact that many antiemetic drugs or drugs taken for vertigo are derived from phenothiazines and do present a real risk for the development of TD if used on a prolonged basis. Prochlorperazine (Stemetil) is a particular culprit and is often prescribed for vague symptoms of dizziness, dysequilibrium, and nausea and on a long-term basis without any thought to the possible catastrophic consequences.

9. PATHOPHYSIOLOGY OF TD

The underlying pathophysiology of TD is not known. The likelihood is that the continued administration of dopamine-blocking drugs has resulted in a state of supersensitivity to dopamine at the postsynaptic receptor (Klawans, 1973). It is presumed that this development can become a chronic state. This concept fits well with the appearance of TD as the dosage of neuroleptic is reduced. Thus, presynaptic dopamine is released into the synaptic cleft onto the supersensitive dopamine receptor within the nigrostriatum. Much of the symptomatology of TD can be replicated by the administration of L-dopa in parkinsonism. In the early stages of treatment, abnormal movements are clearly dose-related and thus due to excess dopamine. Other clinical phenomena are less easy to explain. For instance, why does TD emerge when patients are still on neuroleptics, or when there has been a delay between stopping the drug and the appearance of dyskinesia? It is presumed that the former phenomenon is due to the preferential blockade of different dopamine receptor sites, so that mesolimbic dopamine receptors are blocked but relative sensitivity still occurs within the nigrostriatum. The circumstances of the reaction of postsynaptic receptor sites vary within the two systems. The different types of dopamine receptor must also have a role to play. As indicated previously, D1 receptors are thought to be more relevant in the production of dyskinesia and D2 in parkinsonian symptoms (Kebabian and Calne, 1979). Evidence of dyskinetic symptoms after a delay may well be due to persistence of blockade in the nigrostriatal system resulting in parkinsonian signs which may not have been recognized. It is known that such blockade can persist for up to 2 years after the drug is stopped. It is only when parkinsonian signs have subsided that TD emerges because of the emergence of dopamine supersensitivity. The fault with this argument is swiftly seen: parkinsonian signs commonly coexist with dyskinesia. One proffered explanation is that different states, i.e., dopamine sensitivity and dopamine blockade, exist at the same time at different sites within the dopamine systems. Many of these neurochemical explanations are simplistic, and we confess ignorance of the role of other important systems.

Cholinergic, GABAergic, and neuropeptide systems are known to exist within the striatum and striatal–nigral connections; these must at least play some modifying role. The lynch pin remains the dopamine systems, since only through manipulating this has any definite impact been made on the modification of dyskinesia. Physostigmine (a cholinomimetic agent) administered parenterally does, to some extent, control the movement disorder (Klawans and Rubovits, 1974), but other means of raising acetylcholine levels, such as oral administration of choline or lecithin, have been as ineffective in this disorder as in the chorea of Huntington's disease. Modification of GABAergic connections by using a GABAagonist such as sodium valproate similarly has had little effect on choreic diseases, though there must be some striatal–nigral interference as high doses of sodium valproate have produced parkinsonism which promptly disappeared when the drug was stopped (Lautin *et al.*, 1979). Investigation of neuropeptide systems within the basal ganglia is at an early stage. On a hopeful note, it has been noted that high dosage of an opiate antagonist, naloxone, can modify Parkinson's disease, and thus it appears that the opiate systems also play some role in modifying movement disorders (Agnoli *et al.*, 1980). Their clinical value remains uncertain.

10. DIFFERENTIAL DIAGNOSIS

The diagnosis of TD is made when a dyskinetic problem develops after the prolonged administration of a neuroleptic drug. Obviously, it is necessary to consider other diseases that may produce such a clinical picture. It is clearly important to take a drug history and to establish at what point during treatment the abnormal movements occurred. Length of drug exposure and dosage are important and make it easier to establish a causal connection although it is by no means clear in what way these factors influence the development of tardive dyskinesia. Genetic factors must play a role as clearly some patients are more susceptible than others. The differential diagnosis is not large. *Huntington's chorea* and *benign familial chorea* are distinguished by a family history suggestive of an autosomal dominant inheritance. Sporadic cases of Huntington's disease, *benign essential tremor*, and *senile chorea* may occur, and these may indeed be difficult to distinguish if there is not a clear-cut history of drugs. *Hemiballismus* is distinguished by its sudden onset, involvement of half the body, and the ballistic quality of the movements. It usually occurs in old people. A vascular lesion of the subthalamic nucleus of Luys is usually the cause. Choreic diseases, especially hemichorea, may occur as part of an arteritic disease especially *systemic lupus*, and this can usually be ruled out by the absence of DNA antibodies and antinuclear factor. Chorea is a rare manifestation of *polycythaemia* and should respond to venesection or P32. *Rheumatic chorea* is a self-limiting condition of the young. Other *metabolic causes* include *hypocalcemia, uremia, hepatic failure, and thyrotoxicosis*. These are ruled out by clinical state and simple biochemical tests. *Brain neoplasm* and *arteriovenous malformations* present very rarely as movement disorders. Usually there are other clinical pointers with involvement of other tracts. Computerized tomographic scan and angiograms rule these out.

Dystonic diseases of the young may be the result of birth trauma or other intrauterine factors, and the history of problems should date back to infancy with delayed milestones, in particular a delay in the onset of walking, and clumsiness in running from an early age. *Torsion dystonia* is usually an autosomal dominant disorder and thus family history is important; its course is slowly progressive. *Hallevorden–Spatz* disease lacks an antecedent family history, but this disorder too is characterised by relentless progression of symptoms. There is prominent oral and bulbar dystonia with generalized dystonic postures and dementia; death results from immobility and pneumonia. *Wilson's disease* should always be considered in the differential diagnosis of an extrapyramidal syndrome occurring before middle life. Symptoms include coarse tremor, a characteristic dysarthria, and occasionally choreic movements of the limbs, rigidity and fixity of facial muscles giving the appearance of a vacuous smile. There may be an element of ataxia with intention tremor. The diagnosis is made by finding low ceruloplasmin levels, low serum copper, and increased urinary copper excretion. Kayser–Fleischer rings, a rusty brown discoloration in the deepest layer of the limbus of the cornea, are invariably present once neurological signs have occurred. *Lesch–Nyhan syndrome* may present with choreoathetosis, but usually this dates from birth; self-mutilation is a pronounced feature, hyperuricemia is present, and there may be evidence of gouty tophi.

A much more familiar problem is the patient who has been treated for parkinsonism with L-dopa in whom it has not been recognized that the parkinsonism was itself drug induced. Choreic movements ensue and are often attributed to the L-dopa. If treatment is stopped, *L-dopa-induced chorea* invariably disappears; if there is any remnant of dyskinesia with mild parkinsonian signs, the likelihood is that it is drug induced.

TABLE 1

Suggested Guidelines for the Avoidance and Management of Tardive Dyskinesia

1. Consider carefully the indications for prolonged neuroleptic therapy; the illness should be serious (e.g., chronic psychosis) and there should be objective evidence of benefit.
2. Seek alternative therapies with neuroses, mood and character disorders.
3. Use lower doses with elderly patients and children; strive for minimum effective doses; avoid multiple drugs; discontinue antiparkinsonian agents as soon as possible.
4. Advise patients and families of risks and benefits; arrive at a mutual decision when the use of neuroleptic exceeds 1 year. Note discussion and agreement in clinical record.
5. Examine patients regularly for early signs of choreoathetosis and oral–lingual dyskinesia. Consider alternative neurological diagnoses.
6. Reevaluate patient and document indications and response at least every 3 months and attempt to reduce dose.
7. At the earliest sign of dyskinesia, lower the dose, change to a less potent agent, or, ideally, stop treatment and await remission as long as psychiatric status permits.
8. Treat dyskinesia with simple agents first (e.g., diazepam, deanol, choline, or lecithin in high doses, possibly lithium); stay alert to new experimental therapies, if only to bide time and offer hope. Reinstitute neuroleptics only as a last measure for disabling dyskinesias, using the lowest dose possible.

It is hoped in the future to reduce the incidence of TD by preventing the unnecessary use of neuroleptic drugs. The task force set up to investigate this problem has made several helpful proposals (APA Task Force, 1980). (See Table 1.) The most important point is careful consideration of the indications for prolonged neuroleptic therapy. Patients should have evidence of chronic psychosis, and there must be a demonstrable benefit from the use of neuroleptics. From time to time, say every 3 months, an assessment must be made as to whether the correct dosage is being used or whether a smaller dose could be employed or the drug discontinued. The benefits of such continued therapy are balanced against its potential risks. Rather than trying to control the psychosis completely with high dosage, it may be more acceptable to control some symptoms on a much smaller dosage.

Although it is generally regarded that the treatment of psychiatric illness is the main cause of TD, the general practitioner also has a duty to make sure that his patients are not on phenothiazines or other neuroleptic therapy for long periods for other indications. This means the regular assessment of patients treated for vertigo, nausea, or dysequilibrium with drugs such as prochlorperazine, cinnarizine, and metaclopramide. Such therapy has little to offer patients with chronic symptoms of dizziness. Neurotic disorders should not be treated with neuroleptics if at all possible; alternative psychotherapy or courses of benzodiazepines may be acceptable alternatives. Once the decision has been made to use neuroleptics, it is doubtful whether the potency of the drug is important since weaker drugs are often used in larger doses. The choice of neuroleptic is based on whether sedative action, anticholinergic action, or the tendency not to cause hypotension is required. If parkinsonism occurs, the drug should be either reduced until signs disappear or discontinued if possible. The use of multiple drugs should be avoided, and it is probably wise to avoid the addition of anticholinergic agents. If dyskinesias occur, the choice is to lower the dose, change to a less potent agent, or, ideally, stop treatment altogether. The recognition of TD has meant that the treatment of psychosis has become an altogether more difficult task. It requires skill on the part of the clinician to judge when these drugs are to be used and his continued alertness to the possibility of dyskinesia. It requires the clinician to be on guard, scrutinizing the patient for evidence of movement disorder, and to assess whether continued treatment is needed. The patient and relatives should be informed of the possible effects of long-term treatment in much the same way a patient having an operation is told of the risks. The day is coming when consent forms may well be needed before the patient starts on treatment!

The treatment of TD, once it has occurred, is extremely difficult. When the drug is stopped, at least half the patients will show either improvement or disappearance of the movement disorder within the next few months. The reintroduction of dopamine antagonists may temporarily control the movement disorder but should be avoided at all costs as the prognosis is affected adversely by continuing them. In the majority of cases, symptoms will be mild and the patient able to put up with them once they are adequately explained to him. Some patients do require further treatment, and as a first step, a simple remedy such as a benzodiazepine should be tried, e.g., diazepam, 2–5 mg three times a day. It has been claimed that clonazepam has a more specific effect in view of its GABAergic properties, and this could be used as an alternative, 0.5 mg twice a day, increasing if necessary to a dose that does not cause untoward drowsiness. Often benzodiazepines can cause symptoms to abate but usually at the cost of sedation. It is also necessary to make sure that the patient is not on any drugs that might further aggravate TD; these include anticholinergic agents and any drug with intrinsic anticholinergic activity such as tricyclic antidepressants; it is well accepted that levodopa may aggravate dyskinetic disorders. The role of dopa agonists such as bromocriptine and pergolide is less certain, but it is likely that they too will aggravate matters. If TD is severe and has run a prolonged course, it may be necessary to try other methods. Drugs suoh as reserpine and tetrabenazine, which deplete presynaptic dopamine, can be used. These may well control the movement disorders, but often unacceptable parkinsonian or depressive features emerge. Finally, one is left considering reintroducing dopamine-blocking agents. Haloperidol is often employed and is started in small doses, e.g., 0.5 mg. b.d., and built up to adequately control the movement disorder without attempting to completely suppress all extra movements. Often the end result is unsatisfactory; parkinsonian symptoms emerge and cause more nuisance to the patient, even before there is adequate control of the movement disorder. Dystonic symptoms can be especially difficult to control and may show no response whatsoever.

There is animal evidence that lithium carbonate can prevent dopamine receptor supersensitivity (Pert *et al.*, 1978). These are encouraging results but would mean the introduction of lithium at the same time as neuroleptic drugs. Generally this would appear unacceptable, though if further evidence accrues, it might be an indication for earlier introduction of lithium in bipolar depressive illness or hypomania.

The proof that any of the drugs mentioned are efficacious in treating TD is poor. Mackay and Sheppard (1979) have reviewed all the known trials and have found most lacking in one way or another. It is suggested that future trials should be double blind with adequate numbers of fit and cooperative patients. There should be clear-cut diagnostic definition (Klawans and Rubovits, 1974), and an appropriate recognized rating scale should be used, such as the Rockland TD rating scale (Simpson *et al.*, 1979). Comparable groups should be matched for age, duration, and severity of TD. Observation and rating of TD should be carried out under standard conditions with background medication kept constant.

It is obvious from this that carrying out rigorous trials of drugs in TD is extremely difficult and why, so far, reports on any one drug have been conflicting.

11. FUTURE PERSPECTIVES

The last decade has seen no new major groups of neuroleptic drugs; the problems of drug-induced parkinsonism, akathisia, and TD are likely to remain as problems of understanding dopaminergic transmission and blockade, enhanced by further knowledge of presynaptic release and recycling in the synaptic mechanism. Although recent dopamine agonists (pergolide, lisuride) have been developed, so far they show little advantage over conventional dopamine replacement in parkinsonism, though the incidence of dyskinesia is less.

Dopaminergic factors figure in the chorea of Huntington's disease and are likely to be important in drug-induced dyskinesia of levodopa in parkinsonism. We must not ignore the importance of noradrenergic transmitters and serotoninergic transmitters, not only in the nigrostriatum, but in the important efferent paths from the pallidum. The worsening of on/off dyskinesia produced by physostigmine reminds us of the potential of cholinergic factors in drug-induced movement disorders.

The side effects and therapeutic efficacy of drugs acting on the basal ganglia have shed much light on the still partially discovered murky depths of the caverns, deep in the brain, wherein movements—normal and abnormal—originate.

12. SUMMARY

The judicious use of neuroleptic drugs has improved the lives of a great many patients. Their use can be attended by the occurrence of abnormal movements—many of these will disappear with stopping the offending drug. A small number will persist with chronic dyskinesia, which responds poorly to treatment. In the future it is hoped that more careful prescription of these drugs will cut down the incidence of patients afflicted in this way. Further understanding of the pathophysiology of these conditions is required before useful suggestions can be made regarding prophylaxis or treatment.

13. REFERENCES

Agnoli, R., Ruggiere, S., Falaschi, P., *et al.*, 1980, Are the enkephalins involved in Parkinson's disease, clinical and neuroendocrine responses to Naloxone administration, in: *Neural Peptides and Neuronal Communications* (E. Costa and M. Trabucchi, eds.), pp. 511–521, Raven Press, New York.

APA Task Force, 1980, Tardive dyskinesia: Summary of a task force report of the American Psychiatric Association, by the task force on late neurological effects on antipsychotic drugs, *Am. J. Psychiatry* **137**:1163–1172.

Ayd, F. J., Jr., 1961, A survey of drug-induced extrapyramidal reactions, *JAMA* **175**:1054–1060.

British Medical Association and the Pharmaceutical Society of Great Britain, 1985, *British National Formulary*, No. 4, pp. 123–124.

Burke, R. E., Fahn, S., Jankovic, J., Marsden, C. D., Lang, A. E., Gollomps, S., and Illson, J., 1982, Tardive dystonia: Late-onset and persistent dystonia caused by antipsychotic drugs, *Neurology (NY)* **32**:1335–1346.

Cotzias, G. C., Van Woert, M. H., and Schiffer, L. M., 1967, Aromatic amino acids and modification of parkinsonism, *N. Engl. J. Med.* **276**:374–379.

Crane, G. E., 1973, Persistent dyskinesia, *Br. J. Psychiatry* **122**:395–405.

Debontridder, O., 1980, Dystonic reactions after domperidone (letter), *Lancet* **2**:1259.

Degkwitz, R., 1969, Extrapyramidal motor disorders following long-term treatment with neuroleptic drugs, in: *Psychotropic Drugs and Dysfunctions of the Basal Ganglia* (G. E. Craine and R. J. Gardner, eds.), pp. 21–33, Public Health Service Publication No. 1938, Washington, DC.

Delay, J., Deniker, P., and Harl, J. M., 1952, Utilisation en therapeutique psychiatrique d'une phenothiazine d'action centrale elective, *Ann. Med.-Psychol.* **110**:112–117.

De Silva, K. L., Muller, P. J., and Pearce, J., 1973, Acute drug-induced extrapyramidal syndromes, *Practitioner* **211**:316–320.

Duvoisin, R. C., 1967, Cholinergic–anticholinergic antagonism in parkinsonism, *Arch. Neurol.* **17**:124–136.

Duvoisin, R. C., and Yahr, M.D., 1965, Encephalitis and parkinsonism, *Arch. Neurol.* **12**:227–239.

Fahn, S., and David, E., 1972, Oral-facial-lingual dyskinesia due to anticholinergic medication, *Trans. Am. Neurol. Assoc.* **97**:277–279.

Fann, W. E., Sullivan, J. L., and Richman, B. W., 1976, Dyskinesias associated with tricyclic antidepressants, *Br. J. Psychiatry* **128**:490–493.

Kebabian, J. W., and Calne, D. B., 1979, Multiple receptors for dopamine, *Nature* **277**:93–96.

Klawans, H. L., Jr., 1973, The pharmacology of tardive dyskinesias, *Am. J. Psychiatry* **130**:82–86.

Klawans, H. L., and Rubovits, R., 1974, Effect of cholinergic and anticholinergic agents on tardive dyskinesia. *J. Neurol. Neurosurg. Psychiatry* **37**:941–947.

Lautin, A., Stanley, M., Angrist, B., and Gershon, S., 1979, Extrapyramidal syndrome with sodium valproate, *Br. Med. J.* **2**:1035–1036.

Mackay, A. V. P., and Sheppard, G. P., 1979, Pharmacotherapeutic trials in tardive dyskinesia, *Br. J. Psychiatry* **135**:489–499.

Marsden, C. D., and Jenner, P., 1980, The pathophysiology of extrapyramidal side-effects of neuroleptic drugs, *Psychol. Med.* **10**:55–72.

Marsden, C. D., Tarsy, D., and Baldessarini, R. J., 1975, Spontaneous and drug-induced movement disorders in psychotic patients. in: *Psychiatric Aspects of Neurologic Disease* (F. D. Benson and D. Blumer, eds.), pp. 219–266, Grune & Stratton, New York.

McDowell, F. H., and Sweet, R. D., 1976, in: *Advances in Parkinsonism* (W. Birkmayer and O. Hornykiewicz, eds.), pp. 603–612, Roche, Basel.

Mindham, R. H. S., 1978, The major tranquillizers, in: *Side Effects of Drugs Annual* (M. N. G. Dukes, ed.), Elsevier, New York.

Moller-Nielsen, I., Christensen, A. V., and Fjalland, B., 1976, Receptor blockade and receptor super-sensitivity following neuroleptic treatment, in: *Antipsychotic Drugs, Pharmacodynamics and Pharmacokinetics* (G. Sedval, B. Uvras, and B. Zotterman, eds.), pp. 257–260, Pergamon Press, Oxford.

Pearce, J., and Pearce, I., 1971, Current concepts of parkinsonism, *Postgrad. Med. J.* **47**:794–799.

Pearce, J. M. S., 1977, Symptomatic parkinsonism, *Postgrad. Med. J.* **53**:726–728.

Pearce, J. M. S., 1978, Aetiology and natural history of Parkinson's disease, *Br. Med. J.* **2**:1664–1670.

Pert, A., Rosenblatt, J. E., Sivit, C., Pert, C. B., and Bunney, W. E., Jr., 1978, Long-term treatment with lithium prevents the development of dopamine receptor supersensitivity, *Science* **201**:171–173.

Riddoch, D., Jefferson, M., and Bickerstaff, E. R., 1971, Chorea and the oral contraceptives, *Br. Med. J.* **4**:217–218.

Sale, I., and Kalucy, R., 1981, Psychosis associated with oral contraceptive-induced chorea, *Med. J. Aust.* **1**:79–80.

Schonecker, M., 1957, Ein Eigentumliches Syndrom im Oralen Bereich bei Megaphen applikation, *Nervenarzt* **28**:35.

Simpson, G. M., Lee, J. H., Zoubok, B., and Gardos, G., 1979, A rating scale for tardive dyskinesia, *Psychopharmacology* **64**:171–179.

Steck, H., 1954, Le Syndrome extrapyramidal et di-encephalique au cours des traitements au largactil et an serpasil, *Ann. Med.-Psychol.* **112**:737–743.

Thach, B. T., Chase, T. N., and Bosma, J. F., 1975, Oral–facial dyskinesia associated with prolonged use of antihistamine decongestant, *N. Engl. J. Med.* **293**:486–488.

Yahr, M. D., Duvoisin, R. C., Schear, M. J., Barrett, R. E., and Hoehn, M. M., 1969, Treatment of parkinsonism with levodopa, *Arch. Neurol.* **21**:343–354.

CHAPTER 19

Movement Disorders Induced by Psychotherapeutic Agents

Clinical Features, Pathophysiology, and Management

DANIEL TARSY and ROSS J. BALDESSARINI

1. INTRODUCTION

The antipsychotic drugs (APD) are a group of compounds that are effective in a wide variety of psychotic disorders, produce characteristic extrapyramidal syndromes in man, and profoundly affect muscle tone, posture, and movement of laboratory animals. Their effect on motor behavior led to their designation as "neuroleptic" drugs and distinguishes them from the sedative–hypnotics, anxiolytics, and other centrally active agents. Several classes of antipsychotic drugs have been developed, including phenothiazines, butyrophenones, diphenylbutylpiperidines, indolones, and dibenzoxazepines, as well as reserpine and its congeners. All produce similar, distinctive extrapyramidal syndromes, although reserpine has been associated with a more limited variety of neurological effects, and clozapine, a dibenzodiazepine, has a particularly low incidence of such effects. Early extrapyramidal effects (occurring within several hours or days to weeks) include acute dyskinesias (acute dystonias), parkinsonism, and akathisia (motor restlessness) and are quickly reversible with reduction or discontinuation of APD drug treatment. Tardive dyskinesia (TD) refers to involuntary movements which appear later (typically after at least 3 months) and are sometimes irreversible.

DANIEL TARSY • Department of Neurology, Boston University and Harvard Medical School, and Neurology Section, New England Deaconess Hospital, Boston, Massachusetts 02215.
ROSS J. BALDESSARINI • Department of Psychiatry and Neuroscience Program, Harvard Medical School and Massachusetts General Hospital, and Mailman Research Center, McLean Hospital, Belmont, Massachusetts 02178.

2. ACUTE DYSKINESIA (ACUTE DYSTONIA)

2.1. Clinical Features

The acute dyskinesias usually occur soon after beginning treatment with APD and consist primarily of intermittent or substained muscular spasms and abnormal postures of the eyes, face, neck, and throat. Oculogyric crises or other aversive eye movements, blepharospasm, trismus, forced jaw opening, grimacing, protrusion or twisting of the tongue, distortions of the lips, or glossopharyngeal contractions may occur, and dysarthria, dysphagia, jaw dislocation, and respiratory stidor with cyanosis may result. When the neck is affected, there may be spasmodic torticollis or retrocollis. The trunk is more prominently affected in children, who may show opisthotonos, scoliosis, lordosis, trunk flexion, writhing movements, or tortipelvis with a characteristic dystonic gait. Continuous slow writhing movements of the extremities with exaggerated postures of hyperpronation and adduction may also occur. It is important to realize that milder acute dyskinesias, indistinguishable in appearance from some forms of TD, may also be seen. These include tongue protrusion, lip smacking, blinking, athetosis of the fingers and toes, shoulder shrugging, and a variety of myoclonic muscle contractions of the face, neck, and extremities (Sarwer-Foner, 1960; Delay and Deniker, 1968). Subtle forms of these disturbances, such as muscular cramps or tightness of the jaw and tongue with difficulty in chewing or speaking, may occur by themselves or precede the more obvious manifestations.

The acute dyskinesias are often painful and usually frightening. When severe, they are sometimes confused with seizures, tetanus, rabies, encephalitis, meningitis, or subarachnoid hemorrhage. In addition, because signs often spontaneously remit and exacerbate and may temporarily respond to reassurance and suggestion, they have been mistaken for hysterical (conversion) reactions.

Acute dyskinesias are the earliest APD-induced syndromes to appear and may begin within hours of a single dose of a neuroleptic agent. Quite often, however, there is a delay of 12–36 hr between drug administration and the appearance of dyskinesia. They may either remit or fluctuate spontaneously over several hours or days but often reappear on reintroduction of other drugs of equal potency. Long-acting phenothiazines given by injection, such as fluphenazine decanoate, may produce acute dyskinesias within 72 hr of each administration. Persistent dyskinesia does not occur following acute exposure to ordinary doses of antipsychotic drugs. There also appear to be no well-documented cases of acute dyskinesia following the use of reserpine or its analogues. Following routine clinical use of neuroleptic agents, the overall incidence of acute dyskinesia is approximately 2.5–5% (Ayd, 1961), making this the least frequent drug-induced extrapyramidal syndrome. However, the more potent agents, such as the piperazine phenothiazines and the butyrophenones, are more likely to produce acute dyskinesias and in some series have produced dystonic reactions in 50% or more of cases (Marsden *et al.*, 1975).

2.2. Mechanisms

Until recently the pathophysiology of acute dyskinesias was obscure. Although it has usually been assumed that interference with synaptic dopamine (DA) mechanisms is responsible, several clinical observations raise doubts about this hypothesis. The fact that reserpine rarely, if ever, produces acute dyskinesias, despite its ability to produce acute presynaptic depletion of DA, suggests that acute inhibition of DA transmission is insufficient for their production. The similarity between APD-induced acute dyskinesias, dystonias produced by L-dopa in patients with Parkinson's disease, and some dystonias seen in TD suggests the alternative possibility that activation rather than blockade of DA mechanisms may be responsible (Marsden *et al.*, 1975; Marsden and Jenner, 1980).

Acute dystonic reactions have been described in baboons and monkeys (Meldrum *et al.*, 1977; Liebman and Meale, 1980). Like humans, baboons appear to show individual susceptibility as only 3 of 25 animals developed acute dystonias following haloperidol treatment (Meldrum *et al.*, 1977). However, all three also developed dystonias to a number of other APD, including chlorpromazine and pimozide, but not to thioridazine, thereby paralleling the typical clinical experience. Catecholamine depletion produced by pretreatment with reserpine or reserpine plus methyl-*p*-tyrosine greatly reduced or abolished haloperidol-induced dystonia in this study suggesting that acute dystonias are dependant on a presynaptic catecholaminergic mechanism (Meldrum *et al.*, 1977). Since small doses of APD that are relatively specific DA receptor antagonists (pimozide, haloperidol) produce acute dyskinesias, antagonism of DA receptors is presumably more important than antagonism of norepinephrine receptors.

The synthesis and release of DA in the nigrostriatal system increases acutely in response to blockade of DA receptors. This compensatory increase in DA synthesis and metabolic turnover is a complex function mediated by several mechanisms including long-loop striatonigral feedback connections, local intrastriatal feedback connections, and effects on DA autoreceptors (Marsden and Jenner, 1980). It has been suggested that activation of nigrostriatal DA turnover may account for acute dyskinesia (Marsden *et al.*, 1975; Meldrum *et al.*, 1977; Marsden and Jenner, 1980). First, similar to acute dyskinesia, the activation of DA turnover following DA receptor blockade is a relatively transient phenomenon which declines with repeated drug exposure. Second, acute dyskinesias typically occur after some delay following administration of APD. Patients not infrequently manifest dyskinesias as late as 24–36 hr after receiving a single dose. In one study acute dyskinesias occurred 23–56 hr after a single dose of butaperazine, at a time when plasma and red cell butaperazine concentrations were falling (Garver *et al.*, 1976). If acute dyskinesias are due to an acceleration of DA turnover, the question remains how they can be manifested in the presence of a DA receptor antagonist? Striatonigral DA turnover peaks earlier than would readily account for the appearance of acute dyskinesia (Marsden and Jenner, 1980; Matthysse, 1973).

Perhaps the somewhat delayed manifestation of dyskinesia is due to the subsequent fall in striatal concentrations of APD coupled with a continued increase in DA availability. It has also been found that a single dose of APD can produce supersensitivity of DA receptors which is evident 24–48 hr after acute drug administration (Kolbe *et al.*, 1981). It has therefore been suggested that enhanced DA release acting on incompletely blocked and increasingly sensitive DA receptors may be responsible for acute dyskinesias (Marsden and Jenner, 1980; Kolbe *et al.*, 1981)—that is, a hypothesis of dopaminergic *excess*.

The fact that anticholinergic drugs immediately abort acute dyskinesia has traditionally been interpreted as evidence for acute DA deficiency with restoration of dopamine–acetylcholine balance by cholinergic blockade. The observation that apomorphine, a direct DA receptor agonist, also reverses acute dyskinesia (Gessa *et al.*, 1972) is cited as additional evidence that acute dyskinesia results from acute DA deficiency. However, in view of the formulation proposed by Marsden and Jenner (1980), the ability of anticholinergic drugs and apomorphine to inhibit DA turnover may be more relevant mechanisms. Anticholinergic drugs block the compensatory increase in DA turnover produced by APD while apomorphine reduces nigrostriatal DA turnover (Nybäck *et al.*, 1970) and, in small doses, suppresses nigrostriatal firing rates by activation of presynaptic autoregulatory DA receptors (Skirboll *et al.*, 1979).

Acute dyskinesias are far more common in children and young adults than older individuals. In addition, paralleling idiopathic torsion dystonias, children frequently show more severe and generalized involvement of the trunk and extremities while adults usually show more restricted involvement of the neck, face, tongue, and arms (Ayd, 1961). If acute dyskinesia is due to activation of DA mechanisms, then perhaps children and young adults respond to APD with a brisker activation of nigrostriatal DA turnover because of the larger numbers of nigrostriatal neurons and higher concentrations of striatal DA and tyrosine hydroxylase found in their brains (McGeer and McGeer, 1977).

Although acute dyskinesias are regarded as idiosyncratic, their incidence appears to be especially common after treatment with agents of high antidopamine and low anticholinergic potency, and to be somewhat dose-dependent. In one study patients with acute dyskinesia had higher erythrocyte concentrations of phenothiazine than patients without dyskinesia (Garver *et al.*, 1976). If red blood cell concentration reflects membrane-bound brain levels, this may be an additional factor accounting for increased susceptibility in some individuals.

2.3. Treatment

Acute dyskinesia is the most easily recognized and most readily treated APD extrapyramidal syndrome. However, it is important to be aware of minor, but uncomfortable, dystonic manifestations which sometimes escape diagnosis. The frequent delay between administration of a single dose of APD and the appearance of dyskinesia is a particular cause of misdiagnosis. Intravenous or intramuscular benztropine, an anticholinergic agent, is highly effective for acute and severe dyskinesias, while oral anticholinergics may suffice for milder forms. If the APD is discontinued, repeated administration of benztropine or a similar antiparkinsonism agent may remain necessary for a period of 24–48 hr due to the more rapid half-life of the latter agent. Diphenhydramine is similarly effective, presumably because of its anticholinergic and sedative properties, and may be given parenterally or by mouth. Intravenous diazepam has also been recommended in acute dyskinesia (Korczyn and Goldberg, 1972; Gagrat *et al.*, 1978), but although it offers the advantage of ready availability, it must be given with caution because of the risk of respiratory depression, especially in patients being treated with other sedatives.

The value of prophylactic anticholinergic drugs is controversial. Some studies show no reduction in incidence of extrapyramidal syndromes as a group, although severity of symptoms may be reduced (DiMascio and Demirgian, 1970). However, the incidence of acute dyskinesia does appear to be reduced by prophylactic treatment (Keepers *et al.*, 1983). In young patients, in whom the incidence of dystonia is highest and the likelihood of anticholinergic toxicity is relatively low, such treatment may be worthwhile (Keepers *et al.*, 1983).

3. PARKINSONISM

3.1. Clinical Features

Any or all of the cardinal neurological signs of Parkinson's disease, including akinesia, rigidity, postural abnormalities, and tremor, may be produced by antipsychotic drugs. As a result, this syndrome, sometimes referred to as "pseudoparkinsonism," bears close resemblance to spontaneous forms of Parkinson's disease.

Bradykinesia or akinesia is the most common and sometimes the only manifestation of drug-induced parkinsonism. This lack of motility produces the masked facial expression, absent armswing, slow initiation of motor activity, soft and monotonous speech, and flexed posture that are so common among APD-treated psychiatric patients. Reduced facial expression and armswing are particularly useful early signs of akinesia and may be accompanied by subjective symptoms such as muscle fatigue or weakness. Drug-induced parkinsonism usually produces symmetrical akinesia and rigidity from the onset of symptoms, while Parkinson's disease, especially in early stages, is characteristically asymmetrical in distribution. It is obviously important to distinguish drug-induced akinesia from psychomotor retardation commonly associated with underlying psychiatric illness such as depression or catatonia (Rifkin *et al.*, 1975). APD may produce behavioral inhibition ranging in severity from subtle akinesia to catatonic stupor (Gelenberg and Mandel, 1977). If such signs are mistaken for psychiatric decompensation, more rather than less APD is likely to be given, and the problem worsened.

Rigidity of the extremities, neck, or trunk, not necessarily accompanied by cogwheeling, may appear days to weeks after the onset of akinesia. Characteristic parkinsonian postural abnormalities including flexed posture and impairment of righting responses are also common, and a propulsive or retropulsive gait sometimes occurs.

The characteristic "pill-rolling" resting tremor of Parkinson's disease is uncommon in APD-induced parkinsonism. Instead, the tremor associated with drug-induced parkinsonism is usually a moderate- to high-frequency action tremor similar to essential tremor or parkinsonian action tremor. The "rabbit syndrome" refers to a focal, perioral tremor similar to perioral tremors sometimes seen in Parkinson's disease; this syndrome is typically reversible when APD are discontinued, and it responds to antiparkinsonism medication. In some patients atypical tremors appear which do not resemble any of the classical tremors of Parkinson's disease or other extrapyramidal disorders.

Signs of parkinsonism may begin within several days of APD treatment and gradually increase in incidence so that 50–75% of cases appear within 1 month and 90% of cases within 3 months (Ayd, 1961). Following piperazine phenothiazines or other potent APD, the progression of signs is often telescoped so that acute akinesia with mutism may appear within 48 hr followed within days by rigidity and a resultant "akinetic hypertonic" syndrome (Delay and Deniker, 1968).

Severe parkinsonian reactions may also occur following sudden discontinuation of anticholinergic drugs even if APD have been discontinued at the same time. This is probably the result of a cholinergic rebound phenomenon (also seen in Parkinson's disease) together with the continued DA blocking effects of slowly excreted APD in the absence of protection by the more rapidly excreted anticholinergic drug (Simpson and Kunz-Bartholini, 1968). After discontinuation of APD the majority of patients are free of extrapyramidal signs within a few weeks, but in some patients, usually elderly individuals, these signs may persist for several months or occasionally for as long as a year (Marsden and Jenner, 1980; Klawans *et al.*, 1973). Cases of prolonged drug-induced parkinsonism are usually identifiable by the appearance of some detectable improvement over time. Permanent parkinsonism after drug discontinuation usually occurs in elderly individuals, is probably due to underlying Parkinson's disease made clinically manifest by APD, and is usually slowly progressive despite discontinuation of APD treatment.

Although it is often stated that tolerance develops for the parkinsonian effects of antipsychotic drugs, few prospective clinical studies have dealt adequately with this phenomenon. Some observations that withdrawal of anticholinergic drugs after several months of administration leads to the appearance of a relatively small incidence of parkinsonism may be an artifact due to the substantial number of patients placed on anticholinergic drugs prophylactically or to treat acute dyskinesia rather than parkinsonism (Gelenberg, 1982). Other studies of anticholinergic withdrawal have shown a greater than 50% incidence of parkinsonism even after months or years of treatment with APD (Korczyn and Goldberg, 1976; Manos *et al.*, 1981).

Parkinsonism occurs following the use of reserpine, phenothiazines, thioxanthenes, and the butyrophenones. Following phenothiazines or butyrophenones, the incidence has varied between 5 and 60%, although incidence in routine psychiatric practice is about 10–15% (Marsden *et al.*, 1975). The reported incidence of drug-induced parkinsonism appears to be determined by the potency of the APD involved and the sensitivity of clinical examination. Studies using potent APD and utilizing sensitive methods to detect parkinsonism have yielded an incidence of greater than 90% (Marsden *et al.*, 1975).

In spite of this nearly universal susceptibility, individual predisposition plays an important role. Attempts to correlate total drug dosage with incidence of parkinsonism have usually failed to show a clear relationship. In one study, the total dose of trifluoperazine required to produce a similar degree of parkinsonism ranged from 20 to 480 mg (Simpson and Kunz-Bartholini, 1968). Dramatic examples of severe parkinsonism appearing within several days of treatment with small doses of APD are not uncommon.

There are few clues as to the basis of individual susceptibility to parkinsonian effects of APD. An increased prevalence of Parkinson's disease has been reported in families of patients with drug-induced parkinsonism, but the increment is relatively small and has not been confirmed (Marsden *et al.*, 1975). In large studies, the age distribution of APD-induced parkinsonism closely parallels that of Parkinson's disease, with a sharp rise in incidence after age 40 (Ayd, 1961). However, young adults are frequently affected, and occasional cases are also reported in children and neonates (Keepers *et al.*, 1983). The fact that levels of DA and tyrosine hydroxylase and nigral cell counts all decline with advancing age (McGeer and McGeer, 1977; Smith and Baldessarini, 1980) may account for increased susceptibility to the effects of DA antagonists in older individuals. Chase *et al.* (1970) reported that in patients with APD-induced parkinsonism, cerebrospinal fluid (CSF) levels of homovanillic acid (HVA) were lower than in those without parkinsonism, suggesting that the ability to mount a compensatory increase in DA turnover was less efficient in affected patients. Van Praag and Korf (1976) measured responses of HVA levels in CSF to probenecid (believed to be a measure of DA turnover) before and after treatment with APD of a group of patients with acute psychosis. Patients showing more severe parkinsonism had a greater increase in HVA response following APD treatment than patients without motor effects, suggesting more effective DA receptor blockade in those patients. Although this result differs from that of Chase *et al.* (1970), Van Praag and Korf (1976) also found that HVA responses to probenecid prior to APD treatment were significantly lower among patients who later developed parkinsonian signs than those who did not. These observations suggest that development of APD-induced parkinsonism may depend on an individual susceptibility more closely related to basal, pretreatment DA turnover than posttreatment HVA responses to APD treatment.

3.2. Mechanisms

Many anatomical and physiological studies have demonstrated a dopaminergic nigrostriatal pathway, whose functional importance in relation to Parkinson's syndrome has been reviewed extensively (Hornykiewicz, 1973; Marsden *et al.*, 1975; Marsden and Jenner, 1980). Reserpine depletes brain DA, norepinephrine, and serotonin by interfering with presynaptic vesicular storage mechanisms, thereby permitting increased degradation by monoamine oxidase. Reserpine and tetrabenazine, a synthetic analogue of reserpine with a similar mechanism of action, can produce parkinsonism. In animals, reserpine induces a state of catalepsy characterized by profound akinesia and rigidity that is dramatically reversed by L-dopa. The phenothiazines, thioxanthenes, and butyrophenones produce a state of functional deficiency of DA by their antagonistic actions at synaptic receptors that utilize DA. In laboratory animals, these drugs induce catalepsy and antagonize the behavioral effects of the dopamine agonists L-dopa, apomorphine, and amphetamine. Although a significant role for norepinephrine blockade has not been excluded entirely by these observations, the weight of evidence indicates that APD-induced parkinsonism is primarily due to interference with extrapyramidal dopaminergic mechanisms (Marsden and Jenner, 1980).

The effects of drugs used to treat APD-induced parkinsonism are consistent with this mechanism. L-Dopa reverses the effects of reserpine in animals and humans, presumably by replacement of depleted catecholamines. In APD-induced parkinsonism, oral L-dopa is only partially effective, probably because of persistent blockade of postsynaptic DA receptors caused by continued use of APD. L-Dopa is rarely used in the management of parkinsonism in psychotic patients because of concerns for exacerbating psychosis. Amantadine has been used to treat APD-induced parkinsonism with some success, particularly in patients who fail to respond to anticholinergic drugs (Gelenberg, 1978; Borison, 1983). It does not appear to aggravate psychiatric symptoms when given in moderate doses.

Because of the reciprocal relationship between DA and acetylcholine in basal ganglia, drug-induced parkinsonism, similar to Parkinson's disease, should be characterized by cholinergic sensitivity. Physostigmine can exacerbate phenothiazine-induced parkinsonism (Ambani *et al.*, 1973; Gerlach *et al.*, 1974). Anticholinergic drugs are usually therapeutic, although less so than in the treatment of acute dyskinesias. APD with potent anticholinergic properties, such as thioridizine or clozapine, produce a lower incidence of acute extrapyramidal reactions than APD with little or no anticholinergic effect, such as piperazine phenothiazines, thioxanthenes, or haloperidol. However, because coadministration of an anticholinergic drug with a phenothiazine of low anticholinergic potency still produces more akinesia in laboratory animals than thioridizine or clozapine, other factors, such as relative potency as DA receptor antagonists or regional selectivity of effects in the basal ganglia, are probably also important.

3.3. Treatment

Skillful management of drug-induced parkinsonism begins with awareness of minor clinical manifestations of the syndrome. Subtle forms of akinesia are often mistaken for emotional depression or social withdrawal. If the dose of APD is inappropriately increased or a more potent APD is introduced, symptoms will be exacerbated rather than improved. The use of prophylactic anticholinergic drugs is controversial. Although they may be useful in preventing all of the extrapyramidal syndromes, they appear to be more effective in preventing acute dyskinesia in young individuals than parkinsonism in older individuals (DiMascio and Demirgian, 1970; Keepers *et al.*, 1983). In addition, the incidence of anticholinergic toxicity in the form of confusion, hallucinations, memory disturbance, and urinary retention is greater in older individuals. Routine use of these drugs for prophylactic purposes is therefore not recommended in older patients.

In patients known from previous experience to be unusually sensitive to parkinsonian effects, a relatively low-potency APD such as thoridazine or molindone should be considered. If a high-potency APD is required in such a patient, short-term prophylactic administration of an antiparkinsonism drug is reasonable. Reducing the dose of APD rather than adding an antiparkinsonism drug should be considered in order to avoid the possibility of peripheral or central anticholinergic toxicity. Once patients have been maintained on antiparkinsonism drugs for 3–6 months, the doses should be tapered and discontinued cautiously as many patients no longer require their use (Manos *et al.*, 1981).

In current practice, anticholinergic drugs are the most commonly used agents for treatment of APD-induced parkinsonism. L-Dopa, bromocriptine, and other dopaminergic agonists are of unproved value in this situation and introduce the hazard of exacerbating psychiatric symptoms. Amantadine may be as effective as anticholinergic drugs and may offer the advantages of fewer side effects and greater efficacy in some patients who prove to be resistant to anticholinergic treatment (Gelenberg, 1978; Borison, 1983).

4. NEUROLEPTIC MALIGNANT SYNDROME

Neuroleptic malignant syndrome (NMS) is well described in the French literature of the early 1960s, where it is referred to as "akinetic hypertonic" syndrome (Delay and Deniker, 1968), but until recently it has received scant attention in the American and English literature (Meltzer, 1973). Moreover, NMS is regarded as a rare disorder in this country and is sometimes misdiagnosed as catatonic stupor (Regestein *et al.*, 1977).

The clinical and laboratory features of NMS should allow for prompt diagnosis. Signs usually appear abruptly within several days or weeks after beginning therapy. Occasionally, signs appear after increasing dosage or changing

APD in a patient who has been treated for a longer period of time. High-potency APD's such as piperazine phenothiazines (frequently when given in long-acting parenteral form), haloperidol, and thiothixene are commonly incriminated. However, rare cases also occur following less potent phenothiazines such as thioridazine. Men are affected more frequently than women, and the syndrome appears in psychiatrically normal individuals as well as in patients with Parkinson's disease (Henderson and Wooten, 1981; Toru *et al.*, 1981) and Huntington's chorea (Burke *et al.*, 1981). Clinical features evolve rapidly over the first 24–72 hr to include acute extrapyramidal signs, major disturbances in autonomic function, and stupor. Extrapyramidal signs include widespread muscular rigidity, diffuse and coarse tremor, and dystonic postures of the trunk, face, and extremities. The neck, trunk, arms, and legs become hyperextended, and the hands and feet become fixed in carpopedal spasm. Oropharyngeal hypertonicity results in trismus, sialorrhea, dysarthria, dysphagia, and mutism. Mental status is usually impossible to assess but may be characterized by varying degrees of stupor or coma. Many patients remain awake and alert but unable to communicate because of mutism. Autonomic disturbances include hyperthermia, diaphoresis, tachycardia, hypertension or hypotension, and urinary incontinence. Severe cases may be complicated by dehydration progressing to acute renal failure, with or without myoglobinuria attributable to muscle rhabdomyolysis. Dyspnea, tachypnea, and frank respiratory failure are not uncommon late manifestations that may be associated with pneumonia or pulmonary emboli. The duration of untreated NMS varies between 1 and several weeks and usually requires intensive care and support for much of that time. Mortality is estimated at 15–25% (Caroff, 1980; Cohen *et al.*, 1985) and is usually due to respiratory failure, pulmonary embolism, pneumonia, cardiovascular collapse, or acute renal failure. There is some suggestion that mortality is higher among patients developing NMS from long-acting parenteral fluphenazine than in patients receiving only oral APD (Caroff, 1980). The most characteristic laboratory abnormalities in NMS are leukocytosis and marked elevation of muscle enzymes such as creatine phosphokinase (CPK).

The pathophysiology of NMS is unknown; however, it is probably more attributable to central than peripheral mechanisms. An intriguing observation is that many patients with NMS have been exposed to a similar APD both before and afterward without developing NMS. It is suggested that concomitant factors, such as psychiatric agitation, physical exhaustion, or dehydration, may play a role in predisposition to NMS at a particular time. However, these factors are inconsistent (Smego and Durack, 1982). NMS is frequently compared with malignant hyperthermia (MH) because of the shared features of hyperpyrexia and muscular rigidity. MH is a hereditary muscle disorder in which muscle-relaxing agents, such as succinylcholine and inhalant anesthetics, produce muscle contraction in susceptible individuals by triggering influx of calcium into muscle cytoplasm (Willner and Nagakawa, 1983). Dantrolene effectively reduces MH and recently has also been reported to reverse the hyperpyrexia and muscle necrosis of NMS (Coons *et al.*, 1982; Granato *et al.*, 1983). NMS does not appear to be associated with the same defect in muscle metabolism that occurs in MH (Caroff *et al.*, 1983; Willner and Nagakawa, 1983). Moreover, patients with NMS tolerate succinylcholine and anesthetics without risk (Lotstra *et al.*, 1983), whereas patients with MH have been given APD as preanesthetic agents without precipitating MH.

It has been suggested that central DA blockade is responsible for most of the features of NMS. Clinically, the extrapyramidal signs have the appearance of a severe dystonic reaction occurring simultaneously with severe parkinsonism. Although extreme muscular rigidity may be sufficient to explain the hyperpyrexia of NMS, disturbances in regulation of temperature and other autonomic disturbances may also be due to DA blockade in hypothalamus or hypothalamic connections. The ability of bromocriptine to reverse the extrapyramidal and autonomic signs in NMS and the precipitation of NMS in occasional patients with Parkinson's disease are both consistent with the suggestion that DA receptor blockade plays a major role in its pathophysiology.

Treatment of NMS consists of prompt discontinuation of APD, intensive support of cardiovascular, respiratory, and renal function, and treatment of intercurrent infection. Anticholinergic drugs, diphenhydramine, and diazepam are of no value. Dantrolene may ameliorate rigidity, hyperthermia, and rhabdomyolysis, whereas bromocriptine appears to reverse extrapyramidal and autonomic signs and stupor.

5. AKATHISIA

Akathisia is a state of motor restlessness and subjective distress that occurs following administration of an antipsychotic drug. There is usually a subjective complaint of "anxiety," tension, being driven to move, pulling or drawing sensations in the legs, and an inability to tolerate inactivity. The objective manifestations include a variety of patterns of restless motor activity. When mild, these may be confined to shuffling or tapping movements of the feet while sitting and continuous shifting of weight and rocking of the trunk while standing. In more severe cases, patients appear agitated, are unable to sit, stand, or lie still, and pace or move incessantly. When verbalized, subjective complaints include intense feelings of internal discomfort, tension, and turmoil. The frequent association of akathisia with akinesia and rigidity may result in the paradoxical and striking appearance of incessant walking in patients frozen in fixed postures (Delay and Deniker, 1968). Earlier writers referred to these as "paradoxical behavioral reactions" (Sarwer-Foner, 1960). This syndrome is easy to confuse with psychotic agitation or anxiety, particularly in disturbed or uncommunicative, demented patients (Van Putten, 1975).

Akathisia usually begins within several days to weeks following the start of APD treatment. Unlike acute dyskinesias, the incidence of akathisia continues to increase with time so that about 50% of such reactions occur within 1 month and 90% within 2 or 3 months of the start of sustained neuroleptic treatment (Ayd, 1961). If the offending drug is continued, akathisia may gradually subside spontaneously, but it usually persists for an indefinite period as APD treatment continues. Akathisia typically disappears rapidly when APD are withdrawn. While akathisia has been reported occasionally to persist long after discontinuation of APD, cases of persistent akathisia are more appropriately classified as manifestations of TD in which continuous, restless movements of the lower extremities are particularly prominent. The major difference between most cases of akathisia and akathisialike movements seen in TD is the presence of subjective internal and lower-extremity discomfort and relief by moving, which are characteristic of akathisia.

Akathisia is a common reaction to APD and occurs following administration of the phenothiazines, thioxanthenes, or butyrophenones. As in the other drug-induced syndromes, the reported incidence varies primarily with the potency of the drug and the clinical sensitivity with which the syndrome is detected. Following the potent piperazine phenothiazines incidences between 5 and 50% have been reported, and an average incidence of about 20% has been suggested (Marsden *et al.*, 1975). Nevertheless, clinical experience suggests that since the signs and symptoms of akathisia are typically underreported and often overlooked, they probably occur in the *majority* of patients given an APD and are undoubtedly a major factor in poor compliance with APD treatment.

The pathophysiology of akathisia is obscure. Unlike the other APD-induced syndromes, no specific age predisposition has been noted so that incidence is evidently fairly uniform between the ages of 12 and 65 years (Ayd, 1961; Keepers *et al.*, 1983). Because of its occurrence in postencephalitic parkinsonism and its occasional response to antiparkinsonism drugs, it is assumed to be an extrapyramidal disturbance although there is no other compelling evidence for this assumption. Akathisia is actually relatively resistant to treatment with anticholinergic and antihistaminic drugs and shows no response to physostigmine. Treatment with benzodiazepines is usually of limited benefit. There is some clinical experience to suggest that antiparkinsonism agents and anxiolytics may provide greater benefit in combination as they may have an impact on different components of the syndrome (e.g., movements versus subjective distress, respectively). It has been suggested that akathisia may be analogous to childhood hyperactivity syndromes (Carman, 1972), but adequate studies of the effect of amphetamines in akathisia have not been reported.

It has been hypothesized that akathisia may be due to DA receptor blockade in nonstriatal brain regions (Marsden and Jenner, 1980). In support of this proposal, in contrast to the akinesia produced by DA receptor blockade in striatum, blockade of mesocortical DA receptors in rats reportedly produced locomotor hyperactivity (Carter and Pycock, 1978).

Recently, impressive benefit to akathisia of small doses (30–80 mg/day) of propranolol (a nonspecific β_1 and β_2 antagonist) has been reported (Lipinski *et al.*, 1983, 1984). This treatment led to little change in pulse or blood pressure and was not reproduced by use of a more selective antagonist of β_1 adrenergic receptors, or with an agent with little CNS activity, so that β_2 antagonism may be crucial (Zubenko *et al.*, 1984).

6. TARDIVE DYSKINESIA

6.1. Definition

TD refers to dyskinesia which appears relatively late in the course of treatment with APD (Tarsy and Baldessarini, 1984). The term "tardive" was introduced in the early 1960s (Faurbye *et al.*, 1964) to emphasize a distinction from acute dyskinesia which by that time was well recognized to appear early in APD treatment. Early reports emphasized the orofacial distribution and referred to it as a "buccolinguomasticatory syndrome" (Sigwald *et al.*, 1959). The term "persistent dyskinesia" was introduced to emphasize the sometimes irreversible course of TD (Crane, 1973). However, currently it is recognized that many late-appearing dyskinesias are not permanent. The term TD therefore does not imply irreversibility but instead should be used to designate a dyskinesia which appears relatively late in APD treatment, arbitrarily defined as 3–6 months or more (Tarsy, 1983). Subtypes of TD such as transient TD, withdrawal-emergent TD, and persistent TD have been used to suggest clinical subtypes categorized according to their time course (Schooler and Kane, 1982). Although other criteria for subtyping on the basis of pathophysiology, pharmacological characteristics, topographic distribution, or type of involuntary movements may be forthcoming, current information concerning these aspects is insufficient to permit useful subclassification (Schooler and Kane, 1982).

6.2. Clinical Features

TD appears after some months of exposure to APD and is characterized by a variable mixture of orofacial dyskinesia, athetosis, dystonia, chorea, tics, and facial grimacing. Rhythmical tremor is not part of the syndrome. Orofacial and lingual dyskinesias are characteristic and well-recognized features of TD. Such movements are usually insidious in onset and initially may be detectable only as subtle to-and-fro or lateral movements of the tongue. In some patients, ticlike movements of facial muscles or increased blink frequency are early signs. Later, more obvious protruding, twisting, and curling movements of the tongue; pouting, puckering, sucking, or smacking lip movements; retraction of the corners of the mouth (bridling); bulging of the cheeks; chewing or lateral jaw movements; blepharospasm; and facial grimaces may occur individually or in combination.

In older patients, orofacial dyskinesia is usually the most conspicuous feature of the syndrome. However, involuntary movements of the extremities and trunk are usually also evident. Restless, choreiform, and athetotic limb movements include twisting, spreading, and "piano-playing" finger movements; tapping motions of the feet; and dorsiflexion postures of the great toe. Extremity movements are often more severe in younger individuals and may include dystonic and ballistic postures and movements. Dystonias of the neck and trunk include torticollis, retrocollis, exaggerated truncal lordosis, rocking and swaying, shoulder shrugging, and rotatory or thrusting pelvic movements. Uncommon respiratory dyskinesias such as periodic tachypnea, irregular respiratory rates, and grunting sometimes lead to extensive and unrevealing clinical investigations for pulmonary disorders. Severe and disabling axial dystonias (''tardive dystonia'') may occur, usually in individuals under 50 years of age (Tarsy and Bralower, 1977; Burke *et al.*, 1982). A syndrome resembling TD has also been reported in children, primarily when APD are withdrawn (''withdrawal-emergent symptoms''), characterized by chorea, athetosis, myoclonus, and hemiballism (Polizos *et al.*, 1973). Like other choreoathetotic syndromes, the involuntary movements of TD usually worsen with emotional stress, decrease with drowsiness or sedation, and disappear in sleep. Although patients are often unaware of the more subtle manifestations of TD, failure to complain of severe dyskinesia is more common among institutionalized patients with chronic psychosis or dementia than individuals with less severe psychiatric disorders. On occasion, orofacial dyskinesia may significantly interfere with speech, eating, or respiration, while the more generalized truncal dystonias are severely incapacitating.

6.3. Clinical Course

TD may appear after as little as 3–6 months of APD exposure, although in most published reports dyskinesia was observed after 2 or more years of treatment. Onset is insidious and usually occurs while the patient is still receiving APD. Frequently, however, onset is noted following a reduction in dose or discontinuation of treatment carried out for psychiatric indications or because of a suspicion of early TD. Under these circumstances the dyskinesia is sometimes remarkably severe from the outset. This ''unmasking'' effect of APD withdrawal presumably relates to the hypokinetic or frankly parkinsonian effects of the APD which often result in a delay of recognition of TD.

Sudden withdrawal of APD may be followed by a self-limited dyskinesia that lasts for several days or weeks before subsiding. This is known as ''withdrawal dyskinesia'' (Jacobson *et al.*, 1974; Gardos *et al.*, 1978) and is probably similar to withdrawal-emergent symptoms described in children (Polizos *et al.*, 1973). The relationship of withdrawal dyskinesia to more persistent forms of TD is uncertain. However, since many patients with TD show a slow disappearance of dyskinesia over several months following discontinuation of APD, it is reasonable to consider withdrawal dyskinesia a precursor of more persistent forms of TD (Baldessarini *et al.*, 1980).

Early studies of the natural course of TD following drug withdrawal were carried out among institutionalized patients with prolonged exposure to APD and long duration of TD. Remission rates in such studies varied between 5 and

40% (Baldessarini *et al.*, 1980). Currently, TD is being identified earlier and in younger, outpatient populations. In such patients a remission rate of 50–90% has been observed (Baldessarini *et al.*, 1980), usually occurring within a period of several months but sometimes requiring as long as 1–2 years after drug withdrawal. Although it remains uncertain whether transient or withdrawal dyskinesias constitute a reversible phase in the evolution toward persistent TD, these more optimistic reports indicate the need for careful surveillance, early diagnosis, and, when appropriate, prompt discontinuation of APD treatment.

Since patients with chronic psychosis may decompensate psychiatrically when APD are discontinued, it is often impractical to terminate treatment in patients with TD. The prognosis of TD in patients who continue to receive APD is not clear. However, it was the opinion of the American Psychiatric Association Task Force on TD that in the vast majority of cases, TD either remains fairly static or becomes suppressed by the hypokinetic effects of continued APD treatment (Baldessarini *et al.*, 1980). The chance for remission may be reduced by continued APD treatment (Casey, 1985).

6.4. Differential Diagnosis

Many conditions need to be considered in patients receiving an APD who also manifest signs of dyskinesia (Table 1). Caution must be exercised to avoid an incorrect diagnosis of TD in such patients. In Wilson's and Huntington's diseases, for example, extrapyramidal signs may appear some time after psychiatric symptoms have already been treated with APD.

Initially TD must be distinguished from other APD-induced extrapyramidal

TABLE 1
Differential Diagnosis of Tardive Dyskinesia

1. Stereotyped movements in schizophrenia
2. Spontaneous oral dyskinesias associated with advanced age or dementia
3. Oral dyskinesia associated with edentulism
4. Idiopathic torsion dystonia
5. Focal dystonias (Meige syndrome, spasmodic torticollis)
6. Tourette syndrome
7. Huntington's disease
8. Wilson's disease or other hepatocerebral degenerations
9. Manganese poisoning
10. Fahr's syndrome (idiopathic calcification of the basal ganglia)
11. Postanoxic and postencephalitic states
12. Sydenham's (rheumatic) chorea
13. Endocrine disorders (hyperthyroidism, hypoparthyroidism, oral contraceptive agents)
14. Drug intoxications (L-dopa, amphetamines and other stimulants, anticholinergics, antidepressants, phenytoin, lithium salts)
15. Systemic lupus erythematosus with chorea
16. Brain tumor or other structural abnormality involving thalamus or basal ganglia

syndromes. In patients still receiving APD, this task may be complicated by the coexistence of more than one drug-induced syndrome. Tremor is the only involuntary movement disorder unrelated to TD. This is a reversible sign which is usually associated with bradykinesia or rigidity. Rabbit syndrome sometimes occurs late in treatment and may be unaccompanied by other parkinsonian signs. However, this form of parkinsonian tremor responds to anticholinergic agents and is reversible. Akathisia is usually a reversible early phemenon but also may appear late in treatment with persistence after discontinuation of APD. Acute dyskinesias are readily identified by their dramatic character and onset immediately following APD treatment. However, they sometimes are relatively mild and focal in distribution making a distinction from TD on the basis of appearance alone impossible. Since acute dyskinesia may appear late in treatment when an APD is changed or dose increased or may recur during treatment with long-acting "depot" esters of fluphenazine, confusion with TD is possible.

TD may be difficult to distinguish from stereotyped movements and psychotic mannerisms traditionally associated with chronic schizophrenia (Marsden *et al.*, 1975). Although it has been suggested recently once again that the prevalence, severity, and distribution of abnormal involuntary movements are not greater among APD-treated chronic schizophrenics than untreated patients (Owens *et al.*, 1982), this impression was based on a relatively small sample of untreated patients. The mean prevalence of dyskinesia in 56 studies involving 35,000 APD-treated patients was 20% compared with a mean prevalence of 5% among 19 untreated patient samples totaling 11,000 patients (Kane and Smith, 1982). As a general rule, stereotyped movements of schizophrenia are usually less rhythmic, more variable and complex, and rarely choreoathetotic or dystonic. Spontaneous mouthing movements of the elderly, usually associated with dementia and edentulism, may also be difficult to distinguish from TD. Recent surveys indicate a substantial prevalence of spontaneous orofacial dyskinesias in untreated institutionalized geriatric patients (Kane and Smith, 1982). The common association with dementia, the usual restriction to the mouth and face, and lack of drug exposure should help differentiate these from TD.

Meige syndrome is an idiopathic focal dystonia which usually begins in middle age and is manifested by blepharospasm and oromandibular dystonia indistinguishable in appearance from some forms of TD (Tolosa, 1981). Differentiation from TD requires documenting the lack of APD exposure prior to its appearance. This is sometimes difficult since patients with Meige syndrome have often been treated with APD for their dyskinesia. When TD is manifested by generalized truncal dystonia in a young individual, it may be impossible on the basis of appearance alone to distinguish it from generalized forms of idiopathic torsion dystonia (Burke *et al.*, 1982). In addition to the drug history, a helpful differentiating feature is the slowly progressive course in most idiopathic dystonias contrasted with the rapid onset and static or slowly resolving course seen in TD.

Huntington's disease is usually identifiable by the family history, progressive course, marked gait abnormality, and associated dementia. Unlike

Huntington's disease, TD is rarely a pure chorea, but more commonly includes repetitive and rhythmical athetotic and dystonic movements or postures. For example, abnormal tongue movements in Huntington's disease are less extreme and tend to be sudden and jerky rather than slow, continuous, and twisting as in TD. As a result, patients with Huntington's disease are typically unable to keep their tongues protruded for more than several seconds, while patients with TD, despite more prominent lingual dyskinesia, usually have less difficulty with this maneuver.

In some patients with TD, facial tics and grimacing are particularly prominent and may resemble those of Tourette syndrome. Although irregular respiratory patterns and grunting can occur in TD, the vocal tics and utterances of Tourette syndrome rarely occur. The history of childhood onset, lack of antecedent drug exposure, and a fluctuating course also help to identify Tourette syndrome. Recent reports of symptoms strongly resembling Tourette syndrome appearing after prolonged APD treatment (Klawans *et al.*, 1978) and orolingual dystonia appearing in a patient with Tourette syndrome who was treated with APD (Mizrahi *et al.*, 1980) indicate potential additional complexity for differential diagnosis of the two conditions.

Other differential diagnostic possibilities are listed in Table 1. With increased awareness of TD there is likely to be an increasing tendency to attribute extrapyramidal signs which appear during APD treatment to the drugs themselves. A helpful clue in the diagnosis of TD is that, except for the time of onset and the period during or shortly after APD withdrawal, TD does not appear to be a progressive disorder (Baldessarini *et al.*, 1980) although its severity can vary over time and as a function of behavioral arousal. Motor signs that do *not* accompany TD include tremor, rigidity, akinesia (unless the patient is still receiving APD), cerebellar signs, and pyramidal weakness.

6.5. Risk Factors and Prevalence Rates

Although it seems intuitively reasonable that prolonged treatment with large doses of APD would predispose to TD, this association has not been established in retrospective studies (Kane and Smith, 1982). It remains possible, however, that for those patients who are vulnerable to TD, a dose–response relationship exists at low cumulative doses which cannot be identified in retrospective studies of patients treated for many years with relatively high doses of APD (Baldessarini, 1985). Nearly all studies evaluating age and TD have indicated that patients above age 50 are at a higher risk and may also have a poorer prognosis for eventual remission (Smith and Baldessarini, 1980). The aging brain may be more susceptible to the appearance of oral dyskinesias under a variety of adverse circumstances including prolonged exposure to APD. Female sex also appears to be a risk factor although less consistent than age (Kane and Smith, 1982). Other potential risk factors, such as previous brain damage, psychiatric diagnosis, type of APD, prior acute extrapyramidal signs, and exposure to anticholinergic drugs, have not been established as clinically important (Owens *et al.*, 1982; Smith and Baldessarini, 1980; Baldessarini, 1985).

It has been difficult to establish prevalence rates of TD since criteria and standards for diagnosis vary widely. Although some epidemiological studies have suggested a prevalence rate approaching 50%, these probably reflect increased sensitivity to minor degrees of adventitious movements. It is currently estimated, however, that clinically appreciable cases do occur in at least 10–20% of all patients exposed to APD for more than 1 year, with the rate being even higher in the elderly (Baldessarini *et al.*, 1980, 1985).

6.6. Etiology and Pathophysiology

From a pharmacological point of view, TD is characterized by a state of apparent relative dopaminergic overactivity. TD closely resembles dyskinesias produced by amphetamines, other stimulants, or L-dopa. Administration of L-dopa usually exacerbates TD, while treatment with drugs that deplete or block the action of DA ameliorates the condition. Since in the basal ganglia the effects of DA appear to be counterbalanced by cholinergic effects, the fact that anticholinergic drugs usually exacerbate TD while cholinergic agents produce partial improvement is also compatible with a state of DA overactivity. Such a state of DA overactivity could arise from two potential mechanisms:

In the first, increased sensitivity of postsynaptic DA receptors may play an important role. Since drugs associated with TD produce pharmacological blockade of DA transmission, disuse supersensitivity mediated by changes in DA receptors may result. Prolonged administration of APD to rats and other small animals produces a behavioral state characterized by increased responsiveness to DA agonists (Tarsy and Baldessarini, 1974). This behavioral alteration is paralleled by increased binding of radioactively labeled DA and neuroleptics to receptors in rat striatum (Clow *et al.*, 1979). Administration of APD to *Cebus* monkeys for several months produces orofacial and limb dyskinesias with strong clinical and pharmacological similarities to human TD (Barany *et al.*, 1979). Since the increase in DA behavioral sensitivity and receptor binding observed in small animals and the appearance of dyskinesias in nonhuman primates are transient phenomena which disappear within days to weeks of discontinuing APD treatment, they are appropriate models for transient forms of TD such as withdrawal dyskinesias. The usually prolonged and sometimes irreversible course of TD seems inconsistent with a purely pharmacological state of disuse supersensitivity and suggests instead that long-lasting structural, membrane, or other cellular changes have taken place. In one study, however, although increased DA receptor ligand binding was demonstrated in postmortem brains of APD-treated schizophrenic patients, binding to DA receptors was not increased further in patients who had abnormal involuntary movements in comparison with those without such movements (Crow *et al.*, 1982).

A second possibility is altered availability of other modulating systems that interact with DA mechanisms in the basal ganglia or other motor systems.

This may include losses of functional availability of GABA, acetylcholine, serotonin, or other neurotransmitters or neuromodulators, perhaps by damage to striatal interneurons. The fact that Huntington's disease is associated with loss of cholinergic and GABA-mediated striatal interneurons and may be ameliorated by DA antagonists provides support for this view and has encouraged several therapeutic trials demonstrating weak beneficial effects of acetylcholine or GABA agonists in TD (Growdon *et al.*, 1977; Tamminga *et al.*, 1979).

6.7. Prevention and Treatment

Although there are a large number of reports concerning drug treatments for TD, none has proven to be consistently useful in clinical practice. No treatment to date uniformly benefits all signs of dyskinesia, and most treatments cause only slight to moderate benefit in fewer than half the patients treated (Jeste and Wyatt, 1982). Emphasis, therefore, must rest on prevention, early detection, and management of early and potentially reversible cases.

The use of APD for more than 6 months requires careful evaluation of indications and risks and should be restricted to clinical situations in which other therapies are inadequate. The APD Task Force on TD evaluated indications for short-term and long-term APD treatment (Baldessarini *et al.*, 1980). Long-term APD treatment in psychoneurosis, anxiety states, personality disorders, affective disorders, and chronic pain syndromes has little scientific basis and should be discouraged. The value of maintenance APD treatment for more than 6 months in psychiatric conditions is scientifically supported only for schizophrenia. Even in schizophrenia, however, efforts to maintain patients on lowest effective doses of APD and periodic reevaluation of the need for continued treatment are worthwhile (Baldessarini *et al.*, 1986). After remission of a *first* acute psychotic episode of any type, APD should be gradually decreased and discontinued within a period of several months.

Particular care is taken in the treatment of patients above age 50. Although at this time a clear relationship between drug dose and risk for TD has not been proven, there is growing evidence that many chronically psychotic patients require doses of APD much lower than are commonly used (Baldessarini *et al.*, 1980, 1986), and so it is prudent to seek the least effective dose of APD required for individual treatment. There is no compelling evidence that low potency APD associated with a lower incidence of acute extrapyramidal side effects are less likely to produce TD than high-potency drugs. Recent reports that injectable depot esters of fluphenazine are associated with a higher prevalence of TD are of uncertain significance and may relate to compliance rather than specific pharmacological characteristics (Kane and Smith, 1982).

It is presently uncertain whether drug-induced parkinsonism is a risk factor for subsequent appearance of TD. However, parkinsonism should not be allowed to persist for an extended period. If possible, it is preferable to manage parkinsonism by lowering the dose of APD or changing to a drug with less potent extrapyramidal effects rather than introducing anticholinergic agents.

Anticholinergic drugs do not prevent TD and, in some studies, have been associated with higher incidence of TD (Kane and Smith, 1982), possibly by association with excessive APD dosing. An important practical reason for avoiding parkinsonism is that it masks the signs of underlying dyskinesia.

Since there is increasing evidence that early withdrawal of APD affords a better prognosis for recovery, patients should be carefully examined for dyskinesia at regular intervals with particular attention to the tongue, facial muscles, blink rate, fingers, and toes. Surveillance of the movements of patients who are unaware of being examined is useful in identifying early manifestations of TD. Use of standard dyskinesia rating scales (Baldessarini *et al.*, 1980; Jeste and Wyatt, 1982), such as the abnormal involuntary movement scale or the Rockland scale, is useful to heighten awareness of the signs of TD, to foster careful clinical observation and documentation, and for follow-up comparisons. The value of "drug holidays" in reducing the risk of TD is uncertain (Kane and Smith, 1982). However, they have major potential value in allowing for unmasking of underlying dyskinesia.

6.8. Practical Management

When a patient develops dyskinesia while receiving an APD, ideal management is discontinuation of the medication. The dyskinesia should be described in detail and the patient evaluated to exclude possible alternatives to TD. APD should then be withheld indefinitely in the hope that the dyskinesia will disappear, either slowly or rapidly. In some cases, the dyskinesia becomes even more severe when the drug is stopped. If the dyskinesia fades away within 4–8 weeks, it can be considered a transient dyskinesia and, if it initially appeared on drug discontinuation, a withdrawal dyskinesia. If it continues unchanged for 3 or more months off APD, it can be considered a persistent form of TD, although even persistent cases may eventually remit over months to years of follow-up without APD. Since transient forms of TD may be precursors of persistent TD, patients should not be reexposed to APD unless absolutely necessary. Psychiatric reevaluation to determine whether alternative psychiatric diagnoses or treatments are available is indicated. Patients with affective disorders, neurosis, and chronic anxiety states can usually be treated more appropriately for these conditions without the use of APD.

Before treatment of TD is attempted, the need for suppressing the dyskinesia should be assessed in each patient. Although cosmetically undesirable, the dyskinesia may not be sufficiently disturbing to require specific therapy. If the dyskinesia produces mild symptoms, then low doses of a benzodiazepine or phenobarbital may reduce both the dyskinesia and associated anxiety (Bobruff *et al.*, 1981). A large number of other drugs have been used to treat TD, including cholinergic agents, GABA agonists, and drugs used because of presumed GABA agonist properties, such as sodium valproate and baclofen, but which probably have other mechanisms of action. However, these drugs have usually been effective only in short-term studies, have typically had relatively slight and inconsistent effects (e.g., choline, lecithin), are sometimes associated with unacceptable side effects (e.g., muscimol), or have been useful only in conjunction with continued APD treatment (e.g., baclofen). As a result, their role in clinical practice has remained limited. It is commonly stated that anticholinergic drugs exacerbate TD. Although this is often the case, they commonly have no effect when given orally and may sometimes be effective in severe dystonic forms of TD (Wolf and Koller, 1984). Novel experimental approaches, such as treatment with a low dose of a DA agonist (apomorphine, *N*-propylnorapomorphine, bromocriptine), which presumably reduces DA transmission by activating presynaptic DA autoreceptors, and attempts to desensitize supersensitive postsynaptic DA receptors with gradually elevated doses of L-dopa, are largely of experimental interest at the present time.

In patients with persistent, distressing, and disabling dyskinesia, it may be necessary to resort to cautious use of anti-DA drugs for effective treatment. Before resorting to APD, a trial with a presynaptic DA-depleting agent such as reserpine should be considered. Since reserpine does not bind to postsynaptic DA receptor sites and has rarely been incriminated in TD, it is theoretically preferable to treatment with postsynaptic DA blocker. Although the same has been said for tetrabenazine, there is recent evidence that this drug also acts as a postsynaptic DA receptor blocker (Reches *et al.*, 1982).

It has long been recognized that when reintroduced to patients with TD, APD suppress dyskinesia, either with or without induction of parkinsonism. Such treatment carries with it the theoretical risk of reducing the likelihood of eventual remission of the dyskinesia or further aggravating its severity. At present, however, there is no evidence that once TD appears it will continue to increase in severity with continued APD exposure (Baldessarini *et al.*, 1980, 1985; Casey, 1985). Since patients with severe dyskinesia or psychosis unresponsive to other treatments typically require reintroduction of APD, the apparently nonprogressive character of TD is of some consolation to the patient and family and may be of potential medicolegal significance. In patients for whom reintroduction of APD is necessary for psychiatric indications rather than suppression of dyskinesia, the use of a low-potency APD such as thioridazine or molindone is recommended, although not proven to be safer.

7. SUMMARY

The extrapyramidal movement disorders produced by APD continue to be a significant clinical problem. Acute dyskinesias, parkinsonism, malignant neuroleptic syndrome, akathisia, and TD have been reviewed with emphasis on their clinical features, pathophysiological mechanisms, management, and treatment. Although all are a direct result of treatment with APD, other epidemiological factors may be important in accounting for their variable incidence in various patient populations. Most concepts of their pathophysiology have emphasized the effect of APD on DA neurotransmission, but other mechanisms have been relatively unexplored and, at least in some APD-induced extrapyramidal syndromes, deserve further investigation. The risk of these disorders can be minimized by more conservative clinical use of APD.

8. REFERENCES

Ambani, L. H., Van Woert, M. H., and Bowers, M. B. Jr., 1973, Physostigmine effects on phenothiazine-induced extrapyramidal reactions, *Arch. Neurol.* **29**:444–446.

Ayd, R. F. Jr., 1961, A survey of drug-induced extrapyramidal reactions, *JAMA* **175**:102–108.

Baldessarini, R. J., 1985, Clinical and epidemiological aspects of tardive dyskinesia, *J. Clin. Psychiatry* **46**(4, Sect. 2):8–13.

Baldessarini, R. J., Cole, J. O., Davis, J. M., Gardos, G., Preskorn, S. H., Simpson, G., and Tarsy, D., 1980, Tardive dyskinesia: Summary of a task force report of the American Psychiatric Association, *Am. J. Psychiatry* **137**:1163–1172.

Baldessarini, R. J., Cohen, B. M., and Teicher, M. H., 1986, Pharmacologic treatment of psychoses with receptive agents, in: *Treatment of Acute Psychosis: Current Concepts and Controversies* (S. T. Levy and P. T. Ninan, eds.), Jason Aronson, New York, in press.

Barany, S., Ingvast, A., and Gunne, L. M., 1979, Development of acute dystonia and tardive dyskinesia in *Cebus* monkeys, *Res. Commun. Chem. Pathol. Pharmacol.* **25**:269–279.

Bobruff, A., Gardos, G., Tarsy, D., Rapkin, R. M., Cole, J. O., and Moore, P., 1981, Clonazepam and phenobarbital in tardive dyskinesia, *Am. J., Psychiatry* **138**:189–193.

Borison, R. L., 1983, Amantadine in the management of extrapyramidal side effects, *Clin. Neuropharmacol.* **6**(Suppl. 1):S57–S63.

Bruno, A., and Bruno, S. C., 1966, Effects of L-dopa on pharmacological parkinsonism, *Acta Psychiatr. Scand.* **42**:264.

Burke, R. E., Fahn, S., Mayeux, R., *et al.*, 1981, Neuroleptic malignant syndrome caused by dopamine-depleting drugs, in a patient with Huntington's disease, *Neurology* **31**:1022–1026.

Burke, R. E., Fahn, S., Jankovic, J., Marsden, C. D., Lang, A. E., Gollomp, S., Ilson, J., 1982, Tardive dystonia: Late onset and persistent dystonia caused by antipsychotic drugs, *Neurology* **32**:1335–1346.

Carman, J. S., 1972, Methylphenidate in akathisia, *Lancet* **2**:1093.

Caroff, S. N., 1980, The neuroleptic malignant syndrome, *J. Clin. Psychiatry* **41**:70–83.

Caroff, S., Rosenberg, H., and Gerber, J. C., 1983, Neuroleptic malignant syndrome and malignant hyperthermia, *J. Clin. Psychopharmacol.* **3**:120–121.

Carter, C. J., and Pycock, C. J., 1978, Studies on the role of catecholamines in the frontal cortex, *Br. J. Pharmacol.* **42**:402P.

Casey, D., 1985, Clinical and laboratory studies of tardive dyskinesia, *J. Clin. Psychiatry* **46**(4, Sect. 2):42–47.

Chase, T. N., Schur, J. A., and Gordon, E. K., 1970, Cerebrospinal fluid monoamine catabolites in drug-induced extrapyramidal disorders, *Neuropharmacology* **9**:265–275.

Clow, A., Jenner, P., and Marsden, C. D., 1979, Changes in dopamine mediated behavior during one year's neuroleptic administration, *Eur. J. Pharmacol.* **57**:365–375.

Cohen, B. M., Baldessarini, R. J., Pope, H. J., Jr., and Lipinski, J. F., Jr., 1985, Neuroleptic malignant syndrome, *N. Engl. J. Med.* **313**:1293.

Coons, D. J., Hillman, F. J., and Marshall, R. W., 1982, Treatment of neuroleptic malignant syndrome with dantrolene sodium: A case report, *Am. J. Psychiatry* **139**:944–945.

Crane, G. E., 1973, Persistent dyskinesia, *Br. J. Psychiatr.* **122**:395–405.

Crow, T. J., Cross, A. J., Johnstone, E. C., Owen, F., Owens, D. G. C., and Waddington, J. L., 1982, Abnormal involuntary movements in schizophrenia: Are they related to the disease process or its treatment? Are they associated with changes in dopamine receptors? *J. Clin. Psychopharmacol.* **2**:336–340.

Delay, J., and Deniker, P., 1968, Drug-induced extrapyramidal syndromes, in: *Diseases of the Basal Ganglia* (P. J. Vinken and G. W. Bruyn, eds.), pp. 248–266, North Holland, Amsterdam.

DiMascio, A., and Demirgian, E., 1970, Antiparkinson drug overuse, *Psychosomatics* **11**:596–601.

Faurbye, A., Rasch, P. J., Bender Petersen, P., Brandborg, G., and Pakkenberg, H., 1964, Neurological symptoms in pharmacotherapy of psychoses, *Acta Psychiatr. Scand.* **40**:10–27.

Gagrat, D., Hamilton, J., and Belmaker, R. H., 1978, Intravenous diazepam in the treatment of neuroleptic-induced acute dystonia and akathisia, *Am. J. Psychiatry* **135**:1232–1233.

Gardos, G., Cole, J. O., and Tarsy, D., 1978, Withdrawal syndromes associated with antipsychotic drugs, *Am. J. Psychiatry* **135**:1321–1324.

Garver, D. L., Davis, J. M., and Dekirmenjian, H., 1976, Pharmacokinetics of red blood cell phenothiazine and clinical effects, *Arch. Gen. Psychiatry* **33**:862–866.

Gelenberg, A. J., 1978, Amantadine in the treatment of benztropine-refractory extrapyramidal disorders induced by antipsychotic drugs, *Curr. Ther. Res.* **23**:375–380.

Gelenberg, A., 1982, A proper role for antiparkinson drugs, *Biol. Ther. Psychiatry* **5**:10–11.

Gelenberg, A. J., and Mandel, M. R., 1977, Catatonic reactions to high-potency neuroleptic drugs, *Arch. Gen. Psychiatry* **34**:947–950.

Gerlach, J., Reisby, N., and Randrup, A., 1974, Dopaminergic hypersensitivity and cholinergic hypofunction in the pathophysiology of tardive dyskinesia, *Psychopharmacologia* **34**:21–35.

Gessa, R., Tagliamonte, A., and Gessa, G. K., 1972, Blockade by apomorphine of haloperidol-induced dyskinesia in schizophrenic patients, *Lancet* **2**:981–982.

Granato, J. E., Stern, B. J., Ringel, A., *et al.*, 1983, Neuroleptic malignant syndrome: Successful treatment with dantrolene and bromocriptine, *Ann. Neurol.* **14**:89–90.

Growdon, J. H., Hirsch, M. J., Wurtman, R. J., *et al.*, 1977, Oral choline administration to patients with tardive dyskinesia, *N. Eng. J. Med.* **297**:524–527.

Henderson, V. W., and Wooten, G. F., 1981, Neuroleptic malignant syndrome: A pathogenetic role for dopamine receptor blockade? *Neurology* **31**:132–137.

Hornykiewicz, O., 1973, Parkinson's disease: From brain homogenate to treatment, *Fed. Proc.* **32**:183–190.

Jacobson, G., Baldessarini, R. J., and Manschreck, T., 1974, Tardive and withdrawal dyskinesia associated with haloperidol, *Am. J. Psychiatry* **131**:910–913.

Jeste, D. V., and Wyatt, R. J., 1982, *Understanding and Treating Tardive Dyskinesia*, Guilford Press, New York.

Kane, J. M., and Smith, J. M., 1982, Tardive dyskinesia. Prevalence and risk factors, 1959–1979, *Arch. Gen. Psychiatry* **39**:473–481.

Keepers, G. A., Clappison, V. J., and Casey, D. E., 1983, Initial anticholinergic prophylaxis for neuroleptic-induced extrapyramidal syndromes, *Arch. Gen. Psychiatry* **40**:1113–1117.

Klawans, H. L., Bergen, D., and Bruyn, G. W., 1973, Prolonged drug-induced parkinsonism, *Confin. Neurol.* **35**:368–377.

Klawans, H. K., Falk, D. K., Nausieda, P. A., and Weiner, W. J., 1978, Gilles de la Tourette syndrome after long-term chlorpromazine therapy, *Neurology* **28**:1064–1068.

Kolbe, H., Clow, A., Jenner, P., and Marsden, C. D., 1981, Neuroleptic-induced acute dystonic reactions may be due to enhanced dopamine release on to supersensitive postsynaptic receptors, *Neurology* **31**:434–439.

Korczyn, A. D., and Goldberg, G. J., 1972, Intravenous diazepam in drug-induced dystonic reactions, *Br. J. Psychiatry* **121**:75–77.

Korczyn, A. D., and Goldberg, G. J., 1976, Extrapyramidal effects of neuroleptics, *J. Neurol. Neurosurg. Psychiatry* **39**:866–869.

Liebman, J., and Meale, R., 1980, Neuroleptic-induced acute dyskinesias in squirrel monkeys: Correlation with propensity to cause extrapyramidal side effects, *Psychopharmacology* **68**:25–29.

Lipinski, J. F., Zubenko, G. S., Barreira, P., and Cohen, B. M., 1983, Propranolol in the treatment of neuroleptic-induced akathisia, *Lancet* **2**:685–686.

Lipinski, J. F., Zubenko, G., Cohen, B. M., and Barreira, P., 1984, Propranolol in the treatment of neuroleptic-induced akathisia, *Am. J. Psychiatry* **141**:412–415.

Lotstra, F., Linkowski, P., and Mendlewicz, J., 1983, General anesthesia after neuroleptic malignant syndrome, *Biol. Psychiatry* **18**:243–247.
Manos, N., Gziouzepas, J., and Logothetis, J., 1981, The need for continuous use of antiparkinsonian medication with chronic schizophrenic patients receiving long-term neuroleptic therapy, *Am. J. Psychiatry* **138**:184–188.
Marsden, C.D., and Jenner, P., 1980, The pathophysiology of extrapyramidal side–effects of neuroleptic drugs, *Psychol. Med.* **10**:55–72.
Marsden, C. D., Tarsy, D., and Baldessarini, R. J., 1975, Spontaneous and drug-induced movement disorders in psychotic patients, in: *Psychiatric Aspects of Neurologic Disease* (D. F. Benson and D. Blumer, eds.), pp. 219–266, Grune & Stratton, New York.
Matthysse, S., 1973, Antipsychotic drug actions: A clue to the neuropathology of schizophrenia? *Fed. Proc.* **32**:200–205.
McGeer, P. L., and McGeer, E. G., 1977, Aging and extrapyramidal function, *Arch. Neurol.* **34**:33–35.
Meldrum, B. S., Anlezark, G. M., and Marsden, C. D., 1977, Acute dystonia as an idiosyncratic response to neuroleptics in baboons, *Brain* **100**:313–336.
Meltzer, H. Y., 1973, Rigidity, hyperpyrexia and coma following fluphenazine enanthate, *Psychopharmacologia* **29**:337–346.
Mizrahi, E. M., Holtzman, D., and Tharp, B., 1980, Haloperidol-induced tardive dyskinesia in a child with Gilles de la Tourette's disease, *Arch. Neurol.* **37**:780.
Nybäck, H., Schubert, J., and Sedvall, G., 1970, Effect of apomorphine and pimozide on synthesis and turnover of labelled catecholamines in mouse brain, *J. Pharm. Pharmacol.* **10**:622–624.
Owens, D. G. C., Johnstone, E. C., and Frith, C. D., 1982, Spontaneous involuntary disorders of movement, *Arch. Gen. Psychiatry* **39**:452–461.
Polizos, P., Engelhardt, D. M., Hoffman, S. P., and Waizer, K., 1973, Neurological consequences of psychotropic drug withdrawal in schizophrenic children, *J. Autism Child. Schizophrenia* **3**:247–253.
Reches, A., Burke, R. E., Kuhn, C., Hassan, M., Jackson, V., and Fahn, S., 1982, Tetrabenazine blocks dopaminergic receptors in rat brain, *Ann. Neurol.* **12**:94.
Regestein, Q. R., Alpert, J. S., and Reich, P., 1977, Sudden catatonic stupor with disastrous outcome, *JAMA* **238**:618–620.
Rifkin, A., Quitkin, F., and Klein, D. F., 1975, Akinesia: A poorly recognized drug-induced extrapyramidal behavioral disorder, *Arch. Gen. Psychiatry* **32**:672–674.
Sarwer-Foner, G. J., 1960, Recognition and management of drug-induced extrapyramidal reactions and "paradoxical" behavioral reactions in psychiatry, *Can. Med. Assoc. J.* **83**:312–318.
Schooler, N. R., and Kane, J. M., 1982, Research diagnoses for tardive dyskinesia, *Arch. Gen. Psychiatry* **39**:486–487.
Sigwald, J., Bouttier, D., Raymondeaud, C., and Piot, C., 1959, Quatre cas de dyskinésie facio-bucco-linguo-masticatrice à evolution prolongée secondaire à un traitement par les neuroleptiques, *Rev. Neurol.* **100**:751–755.
Simpson, G. M., and Kunz-Bartholini, E., 1968, Relationship of individual tolerance, behavior, and phenothiazine produced extrapyramidal system disturbance, *Dis. Nerv. Syst.* **29**:269–274.
Skirboll, L. R., Grace, A. A., and Bunney, B. S., 1979, Dopamine auto and postsynaptic receptors: Electro-physiological evidence for differential sensitivity to dopamine agonists, *Science* **206**:80–82.
Smego, R. A., and Durack, D. T., 1982, The neuroleptic malignant syndrome, *Arch. Intern. Med.* **142**:1183–1185.
Smith, J. M., and Baldessarini, R. J., 1980, Changes in prevalence, severity, and recovery in tardive dyskinesia with age, *Arch. Gen. Psychiatry* **37**:1368–1373.
Tamminga, C. A., Crayton, J. W., and Chase, T. N., 1979, Improvement in tardive dyskinesia after muscimol therapy, *Arch. Gen. Psychiatry* **36**:595–598.
Tarsy, D., 1983, History and definition of tardive dyskinesia, *Clin. Neuropharmacol.* **6**:91–100.
Tarsy, D., and Baldessarini, R. J., 1974, Behavioral supersensitivity to apomorphine following

chronic treatment with drugs which interfere with the synaptic function of catecholamines, *Neuropharmacology* **13**:927–940.

Tarsy, D., and Baldessarini, R. J., 1984, Tardive dyskinesia, *Ann. Rev. Med.* **35**:605–623.

Tarsy, D., and Bralower, N., 1977, Tardive dyskinesia in young adults, *Am. J. Psychiatry* **134**:1032–1034.

Tolosa, E. S., 1981, Clinical features of Meige's disease (idiopathic orofacial dystonia), *Arch. Neurol.* **38**:147–151.

Toru, N., Matsuda, O., Makiguch, K., *et al.*, 1981, Neuroleptic malignant syndromelike state following a withdrawal of antiparkinsonian drugs, *J. Nerv. Ment. Dis.* **169**:324–327.

Van Praag, H. M., and Korf, J., 1976, Importance of dopamine metabolism for clinical effects and side effects of neuroleptics, *Am. J. Psychiatry* **133**:1171–1176.

Van Putten, T., 1975, The many faces of akathisia, *Compr. Psychiatry* **16**:43–47.

Willner, H., and Nagakawa, M., 1983, Controversies in malignant hyperthermia, *Semin. Neurol.* **3**:275–282.

Wolf, M. E., and Koller, W. C., 1984, Tardive dystonia: Controlled study of trihexyphenidyl treatment, *Neurology* **34**(Suppl. 1):129.

Zubenko, G. S., Lipinski, J. F., Cohen, B. M., and Barreira, P. J., 1984, Comparison of metoprolol and propranolol in the treatment of akathisia, *Psychiatry Res.* **11**:143–149.

CHAPTER 20

Pharmacotherapy of Movement Disorders in Children and Adolescents

ROBERT C. SCHNACKENBERG

The third edition of the *Diagnostic and Statistical Manual of Mental Disorders* of the American Psychiatric Association (DSM III) (APA, 1980) lists the following groups of movement disorders in children and adolescents: attention deficit disorder with and without hyperactivity, stereotyped movement disorders, and pervasive developmental disorders. The reasons the three disorders are grouped together here is that the children who suffer from these disorders have movements that are not under conscious control, and all become very self-conscious because of their movement disorders. These disorders are more responsive to pharmacotherapy than most other disorders of children and adolescents.

In this chapter emphasis will be given to the pharmacotherapeutic approach to these movement disorders. These disorders will frequently require other interventions besides pharmacotherapy, including individual psychotherapy, family therapy, behavior therapy, special education approaches, group therapy, and residential treatment.

1. ATTENTION DEFICIT DISORDERS

According to DSM III, a child with attention deficit disorder with hyperactivity must exhibit three of the four inattention factors: often fails to finish things he/she starts, often does not listen, is easily distracted, and has difficulty concentrating. The child should have three of the following six: often acts before thinking, shifts excessively from one activity to another, has difficulty organizing work, needs a lot of supervision, frequently calls out in class, and has difficulty waiting his turn. In terms of hyperactivity, the child has at least two of the following five symptoms: runs or climb excessively, has difficulty

ROBERT C. SCHNACKENBERG • Department of Neuropsychiatry and Behavioral Science, University of South Carolina School of Medicine, Columbia, South Carolina 29201.

sitting still, has difficulty staying seated, moves about excessively, and is always on the go. Other criteria include onset before the age of 7, duration of at least 6 months, and the disorder must not be due to schizophrenia, affective disorder, or severe or profound mental retardation.

The etiology of this disorder predisposes a biological underpinning including brain damage and genetic factors. The final common pathway in this disorder is a lack of homeostasis of catecholamine metabolism in the ascending reticular activating system/corpus striatum region (Shayvitz and Bennett, 1982). Methylphenidate appears to have the highest efficacy in this disorder using the time-release form, of 1–2 20-mg tablets a day at breakfast, or the tablet form, 5–10 mg three times a day. The time-release form of methylphenidate will probably replace the tablet form as we gain more experience with it. It is quite inconvenient to give methylphenidate two or three times a day to schoolchildren because of the stigma of taking drugs and the child's dependency on school personnel to administer the noontime dose. Dextroamphetamine as a psychostimulant is not as effective as methylphenidate. It is used in doses of 10–30 mg/day, which may be in the form of 10 mg t.i.d. or a 30-mg spansule given at breakfast time. Another psychostimulant used is pemoline in a dose of 18.75–112.5 mg/day. The administration of the day's dosage at breakfast is an advantage. A disadvantage is that it may take 2 or 3 weeks for the onset of action. Also, children tend to develop tolerance and become refractory to pemoline. The above-mentioned three psychostimulants have been associated with anorexia, weight loss, and insomnia, which are usually reversible.

Some investigators have related the psychostimulant medications to growth and weight loss in the hyperactive child, but these results were inconclusive. Another psychostimulant, caffeine (Schnackenberg, 1973) is useful for selected children, but is more of a narrow-spectrum psychostimulant. Caffeine in a spansule form, from 200 to 300 mg at breakfast, is recommended. Caffeine has the advantage of being less expensive and is not abused by these children. A tricyclic, imipramine, has been helpful with hyperkinesis in doses of 50–150 mg. A combination of carbidopa/levodopa (Langer *et al.*, 1983) has been effective for inattentive schoolchildren who are restless as rated by the teacher. Thioridazine appears to be the best phenothiazine, given in doses of 10–200 mg/day, either divided or given at bedtime. Thioridazine tends to decrease aggressiveness and increase attention and concentration span, comprehension, sociability, and interest. The main problem with thioridazine is oversedation. Among the butyrophenones, haloperidol appears to be most effective in doses of 0.5–16 mg/day at bedtime, or in two daily divided doses. Haloperidol is associated with dystonic reactions. Benzodiazepines such as chlordiazepoxide have been utilized in this syndrome and have been shown to reduce hyperkinesis, but at the cost of an occasional "paradoxical reaction" with loss of control. Lithium has been shown to be ineffective in reducing the symptoms of this syndrome. There has been a recent concern that psychostimulant medications may precipitate Tourette syndrome, which will be described in Section 2 (Comings and Comings, 1984).

Attention deficit disorder and behavioral difficulties may sometimes be early manifestations of an underlying Tourette syndrome. Comings and Comings feel that children who develop tics when treated with psychostimulant medications may already be well along toward developing the symptoms of Tourette syndrome. Psychostimulant medication should be stopped if tics result, but it is advisable to continue the medication if the benefits are great and the tics are mild.

2. STEREOTYPED MOVEMENT DISORDERS (DSM III)

One of the features of these disorders is an abnormality of gross motor movement. These include the following: 307.21, transient tic disorder; 307.22, chronic motor tic disorder; 307.23, Tourette disorder; 307.20, atypical tic disorder; and 307.38, atypical stereotyped movement disorder. For the purposes of this chapter, the cited stereotyped movement disorders are considered to be related along a continuum of severity. As is noted in DSM III, the diagnostic criteria for transient tic disorder include onset during childhood or early adolescence, presence of recurrent involuntary repetitive tics, ability to suppress the movements voluntarily from minutes to hours, variation in intensity of the symptoms over weeks and months, and duration of at least 1 month and not more than 1 year.

For chronic motor tic disorder, DSM III lists the following criteria: presence of recurring, voluntary, rapid tics involving no more than three muscle groups at any one time, unvarying intensity of the tics over weeks or months, ability to suppress the movements voluntarily for minutes to hours, and duration of at least 1 year.

For Tourette disorder, DSM III lists the following criteria: age of onset between 2 and 15 years, presence of recurrent, involuntary, repetitive movements affecting multiple muscle groups, multiple vocal tics, inability to suppress movements from minutes to hours, variations in intensity of the symptoms over weeks and months, and duration of more than 1 year. The etiology studies of tic disorders have focused mainly on Tourette disorder. There is hyperactivity of dopaminergic symptoms in the corpora striata of patients with Tourette disorder. Haloperidol is a potent butyrophenone drug which is effective in blocking dopaminergic pathways in the brain (Lucas, 1979). The recommended dosage is 2–6 mg a day beginning with 0.5 g initially and increasing in increments of 0.5 mg until the desired clinical effect is obtained. Dosage is usually given twice daily and should not generally go beyond 6 mg/day. A minority opinion is that an antiparkinson drug, such as benztropine, 1 mg twice a day should be administered with haloperidol, preventing the side effects of dystonia and parkinsonism. The author considers it most important to prevent the side effects of acute dystonic reactions and parkinsonism and to maintain good rapport with families, although most authors would allow the child to develop these symptoms before beginning the antiparkinson medications.

Other drugs are available that are less effective than haloperidol. They will be mentioned in order for the clinician to have other options. Abuzzahab and Anderson (1973) tabulated the most frequently used treatment modes in the literature, and antidepressants, central nervous system stimulants, anticonvulsants, and antiparkinsons appear to be of help.

However, the phenothiazines appeared to be an improvement in 48% of cases as compared to 89% with haloperidol. It is recommended that chlorpromazine be utilized if haloperidol is not successful. Chlorpromazine should be utilized in a dosage of up to 200 mg/day, usually in divided doses two to four times a day, but occasionally the entire dose is administered at bedtime.

3. PERVASIVE DEVELOPMENTAL DISORDER

DSM III lists the following diagnostic criteria for childhood-onset pervasive developmental disorder: sustained impairment of social relationships: (at least three of the following seven: intensive anxiety, constricted or inappropriate affect, resistance to changing environment, oddities of motor movement such as peculiar posturing, peculiar hand or finger movements, or walking on tip-toe, abnormalities of speech, hyper- or hyposensitivity to sensory stimuli, and self-mutilation); onset after 30 months of age and before 12 years of age; and absence of delusions, hallucinations, incoherence, or marked loosening of associations. No definitive causes have been found for pervasive developmental disorder.

The pharmacotherapy of oddities of motor movement, such as peculiar posturing, peculiar hand or finger movements, or walking on tiptoe, is usually helpful when motor movements are extreme. When indicated, it is recommended that they be used in the following order:

1. Thioridazine, 10–200 mg daily in one to four doses, usually does not require the concomitant administration of antiparkinson medication. Thioridazine may at times cause excess sedation.
2. Chlorpromazine, 10–200 mg/day in one to four doses, may be useful, but it is more likely to have extrapyramidal side effects.
3. Phenothiazines, such as trifluoperazine, 1–20 mg in one to two daily doses, are helpful, but have extrapyramidal side effects.
4. Haloperidol has a potent neuroleptic action, but with a high percentage of extrapyramidal side effects. It is used in doses of 0.5–16 mg one or two times per day.
5. Thioxanthines, such as thiothixene, 1–40 mg, are effective in one or two daily doses. These are less likely to produce hematopoietic and hepatic damage than the phenothiazines. These five neuroleptics (Gualtieri and Quade, 1984) have been associated with tardive kinesia.
6. Molindone, a dihydroindolone, is an activating, stimulating neuroleptic. Extrapyramidal reactions are modest, and it does not cause weight gain in these children, as do all the other neuroleptics mentioned.

Fish recommends the use of diphenhydramine, an antihistamine, 100–300 mg/day in two to four divided doses. The main problem with this antihistamine is its sedative properties.

In summary, the pharmacotherapy of movement disorder in children and adolescents is one of the most advanced areas of pharmacotherapy in children and adolescents. Further research will undoubtedly result in the development of more specific and effective therapeutic agents with less side effects.

4. REFERENCES

Abuzzahab, F. E., and Anderson, F. O., 1973, Gilles de la Tourette's syndrome: International registry, *Minnesota Med.* **56**:492–496.

American Psychiatric Association, 1980, *Diagnostic and Statistical Manual of Mental Disorders*, 3rd edi., Washington, DC.

Commings, D. E., and Commings, B. G., 1984, Tourette's syndrome and attention deficit disorder with hyperactivity: Are they genetically related? *J. Am. Acad. Child Psychiatry* **23**:138.

Fish, B., 1967, Organic therapies, in: *Comprehensive Textbook of Psychiatry* (A. M. Freedman and H. I. Kaplan, eds.), p. 1460, Williams and Wilkins, Baltimore.

Gualtieri, C. T., Quade, D., Hicks, R. E., 1984, *et al.*, Tardive dyskinesia and other clinical consequences of neuroleptic treatment in children and adolescents, *Am. J. Psychiatry* **141**:20–23.

Langer, D. H., Rapaport, T. L., Brown, G. L., *et al.*, 1983, Behavioral effects of carbidopa/levodopa in hyperactive boys, *J. Am. Acad. Child Psychiatry* **21**, 1:10–18.

Lucas, A. R., 1979, Tic: Gilles de la Tourette's syndrome, in: *Basic Handbook of Child Psychiatry*, Vol. III, pp. 677–678, Basic Books, New York.

Schnackenberg, R. C., 1973, Caffeine as a substitute for schedule II psychostimulants, *Am. J. Psychiatry* **130**:796–798.

Shayvitz, B. A., 1982, *A Biochemical Model for Attention Deficit Disorder*, p. 11, Abbott Pharmaceuticals, Chicago.

Index

MIX
Papier aus verantwortungsvollen Quellen
Paper from responsible sources
FSC® C105338

If you have any concerns about our products,
you can contact us on
ProductSafety@springernature.com

In case Publisher is established outside the EU,
the EU authorized representative is:
Springer Nature Customer Service Center GmbH
Europaplatz 3, 69115 Heidelberg, Germany

Printed by Libri Plureos GmbH
in Hamburg, Germany